Neurogenesis in the Adult Brain I

Tatsunori Seki · Kazunobu Sawamoto
Jack M. Parent · Arturo Alvarez-Buylla
Editors

Neurogenesis in the Adult Brain I

Neurobiology

Springer

Editors
Tatsunori Seki, Ph.D.
Professor
Department of Histology
and Neuroanatomy
Tokyo Medical University
Tokyo 160-8402, Japan
sekit@tokyo-med.ac.jp

Jack M. Parent, M.D.
Associate Professor
Department of Neurology
University of Michigan
109 Zina Pitcher Place 5021 BSRB
Ann Arbor, MI 48109, USA
parent@umich.edu

Kazunobu Sawamoto, Ph.D.
Professor
Department of Developmental
and Regenerative Biology
Institute of Molecular Medicine
Nagoya City University Graduate
School of Medical Sciences
1 Kawasumi, Mizuho-cho, Mizuho-ku
Nagoya 467-8601, Japan
sawamoto@med.nagoya-cu.ac.jp

Arturo Alvarez-Buylla, Ph.D.
Professor
Department of Neurosurgery
University of California, San Francisco
San Francisco, CA 94143, USA
abuylla@stemcell.ucsf.edu

ISBN 978-4-431-53932-2 ISBN 978-4-431-53933-9 (eBook)
DOI 10.1007/978-4-431-53933-9
Springer Tokyo Dordrecht Heidelberg London New York

Library of Congress Control Number: 2011928783

Printed on acid-free paper

Springer is part of Springer Science+Business Media (www.springer.com)

Preface

The phenomenon of adult neurogenesis, or persistent generation of neurons in the adult brain, is attracting more and more attention every day. Adult neurogenesis has become increasingly important to studies of brain development and diseases, learning and memory, and aging. A considerable number of papers on the mechanism of adult neurogenesis or linking this process to various physiological and pathological events are published monthly. Most recent popular textbooks in the neurosciences also provide ample coverage of basic principles of adult neurogenesis, a topic which just 20 or 30 years ago was considered tangential, controversial, or unimportant.

The discovery that some populations of neurons continue to be produced postnatally has dramatically changed previous fundamental concepts of neuroscience. For example, it was widely believed that once development is complete and the embryonic and fetal scaffolding for neuronal generation, migration, and integration are dismantled, these processes cannot be reenacted; therefore, once neurons die they never regenerate. This dogma has been overturned, and studies on adult neurogenesis have opened up the possibility of newborn neurons participating in brain tissue repair and processes of learning and memory. Investigations of adult neurogenesis have also led to basic new principles on the identity of neural stem cells, the function of transit amplifying progenitors, and new forms of neuronal migration. Furthermore, some neuropsychiatric disorders are suspected to be associated with defects in adult neurogenesis.

Despite the existence of such a broadly applicable and fundamentally important phenomenon, 20 years ago only a few groups in the world studied adult neurogenesis. It was the pioneering [^{3}H]-thymidine autoradiography studies of Joseph Altman that showed in the 1960s how newborn neurons continue to be formed postnatally in the rodent hippocampus and olfactory bulb. Adult neurogenesis, however, did not become widely accepted and remained a controversial field for more than a decade. In the 1980s a series of rigorous studies, inspired by the neurobiology of song learning in birds, led Fernando Nottebohm and his group to demonstrate the origin, migration, and recruitment of new neurons in song-control nuclei and the rest of the telencephalon of adult canaries. Unfortunately, the history of the field is either ignored or underappreciated by the many neuroscientists who are now interested in or working on adult neurogenesis. To redress that, this book contains historically crucial and memorable articles by Drs. Altman and Nottebohm describing from a very personal perspective the motivations and excitement that triggered these seminal discoveries. They also highlight some of the scientific and funding challenges posed by the strong early opposition to adult neurogenesis. Their two articles should not only be the primary source for neuroscientists interested in the initial discoveries in adult neurogenesis and how they came about, but should also be of value to those interested in science history, funding, and policy.

Novel methods for labeling new neurons – new thymidine analogs such as BrdU, immunohistochemical markers for immature neurons, retrovirus and genetic tagging techniques – resulted in the 1990s in an explosion of studies on the mechanism and function of adult neurogenesis. Stunningly beautiful preparations revealed the entire process of neuronal formation in the adult and revealed the nature and connectivity of individual newly formed neurons. Physiological studies have begun to decipher the unique contribution of new neurons to adult neural circuits. Despite the enormous progress made over the past decade, it is clear that we are still in the early days of understanding the functional meaning and molecular mechanisms of adult neurogenesis. In preparation for this next stage of discovery, we thought it was fitting to compile a collection of thoughtful reviews from leading laboratories working in this area of research.

Contributing researchers describe their current work in 27 chapters that are grouped into two volumes, which cover a wide array of topics concerning adult neurogenesis. The first volume, in addition to the two articles by Drs. Altman and Nottebohm on the history of adult neurogenesis, comprises a comprehensive presentation of the basic biology of adult neurogenesis: basic aspects of neurogenesis, adult neurogenesis in non-mammalian vertebrates, in the mammalian hippocampus and olfactory bulb. In the second volume, clinical implications of adult neurogenesis are considered, including neurogenesis in the adult monkey and human brain, Parkinson's disease, epilepsy, stress, depression, schizophrenia, stroke, brain injury, and neurodegenerative and neuropsychiatric pathology.

A small research group in Japan, the Adult Neurogenesis Kondankai (Conference), working on various aspects of adult neurogenesis initially conceived these volumes. Later, two new editors (A.A.-B. and J.M.P.) joined the project and the original idea was expanded and appropriately shaped. Our goal is not only to provide a comprehensive knowledge base on adult neurogenesis, but also to share our excitement and motivation for an extraordinary field in the neurosciences. We believe that these ingredients will be fundamental to future research in this exciting field towards a better understanding of how adult neurogenesis is maintained and regulated, and how it contributes to plasticity and possibly one day to brain therapy.

During the editing of these two volumes, a massive 9.0-magnitude earthquake struck northeastern Japan on March 11, 2011, and tens of thousands of people died in the resulting tsunami. We dedicate this book to their memory, and offer our deepest condolences to those who lost loved ones.

Tatsunori Seki
Kazunobu Sawamoto
Jack M. Parent
Arturo Alvarez-Buylla

Contents

Contents of Volume II

Contributors

Joseph Altman (*Chapter 1*)
8518 SW 86th Terrace, Ocala, FL, USA

Arturo Alvarez-Buylla (*Chapter 12*)
Department of Neurosurgery, University of California, San Francisco, San Francisco, CA 94143, USA
and
Eli and Edythe Broad Center of Regeneration Medicine and Stem Cell Research, University of California, San Francisco, 513 Parnassus Avenue, Room HSW-1201, Campus Box 0525, San Francisco, CA 94143-0525, USA

Luca Bonfanti (*Chapter 13*)
Department of Veterinary Morphophysiology, University of Turin, Turin, Italy
and
National Institute of Neuroscience-Italy (INN), Turin, Italy
and
Istituto scientifico Fondazione Cavalieri Ottolenghi, Turin, Italy

Angélique Bordey (*Chapter 17*)
Department of Neurosurgery, Yale University School of Medicine, 333 Cedar Street, FMB 422, New Haven, CT 06520-8082, USA
and
Department of Cellular and Molecular Physiology, Yale University School of Medicine, 333 Cedar Street, FMB 422, New Haven, CT 06520-8082, USA

Maureen Saint Georges Chaumet (*Chapter 4*)
Department of Tropical Medicine, Medical Microbiology and Pharmacology, John A, Burns School of Medicine, University of Hawaii, Honolulu, HI 96822, USA

Wei Deng (*Chapter 9*)
Laboratory of Genetics, Salk Institute for Biological Studies, 10010 North Torrey Pines Road, La Jolla, CA 92037, USA

Fred H. Gage (*Chapter 9*)
Laboratory of Genetics, Salk Institute for Biological Studies, 10010 North Torrey Pines Road, La Jolla, CA 92037, USA

José Manuel García-Verdugo (*Chapters 6 and 12*)
Laboratorio de Morfología Celular, Centro de Investigación Príncipe Felipe, Valencia, Spain
and
CIBERNED, Spain

Susana González-Granero (*Chapter 6*)
Unidad de Neurobiología Comparada, Instituto Cavanilles de Biodiversidad y Biología Evolutiva, Universidad de Valencia, c/ Catedrático José Beltrán nº2, CP 46980 Paterna, Valencia, Spain

Young-Goo Han (*Chapter 12*)
Department of Neurosurgery, University of California, San Francisco, San Francisco, CA 94143, USA
and
Eli and Edythe Broad Center of Regeneration Medicine and Stem Cell Research, University of California, San Francisco, 513 Parnassus Avenue, Room HSW-1201, Campus Box 0525, San Francisco, CA 94143-0525, USA

Yuki Hirota (*Chapter 14*)
Department of Developmental and Regenerative Biology, Institute of Molecular Medicine, Nagoya City University Graduate School of Medical Sciences, 1 Kawasumi, Mizuho-cho, Mizuho-ku, Nagoya 467-8601, Japan

Tatsuhiro Hisatsune (*Chapter 8*)
Department of Integrated Biosciences, The University of Tokyo, Kashiwa, Chiba 277-8562, Japan

Shi-hui Huang (*Chapter 14*)
Department of Developmental and Regenerative Biology, Institute of Molecular Medicine, Nagoya City University Graduate School of Medical Sciences, 1 Kawasumi, Mizuho-cho, Mizuho-ku, Nagoya 467-8601, Japan

Yoko Ide (*Chapter 8*)
Department of Integrated Biosciences, The University of Tokyo, Kashiwa, Chiba 277-8562, Japan

Naoko Kaneko (*Chapter 14*)
Department of Developmental and Regenerative Biology, Institute of Molecular Medicine, Nagoya City University Graduate School of Medical Sciences, 1 Kawasumi, Mizuho-cho, Mizuho-ku, Nagoya 467-8601, Japan

Gerd Kempermann (*Chapter 11*)
Genomics of Regeneration, CRTD – Center for Regenerative Therapies Dresden, DFG Forschungszentrum und Excellenzcluster, Tatzberg 47 – 49, 01307 Dresden, Germany

Melissa Lezameta (*Chapter 6*)
Unidad de Neurobiología Comparada, Instituto Cavanilles de Biodiversidad y Biología Evolutiva, Universidad de Valencia, c/ Catedrático José Beltrán nº2, CP 46980 Paterna, Valencia, Spain

Pierre-Marie Lledo (*Chapter 16*)
Laboratory for Perception and Memory, Department of Neuroscience, Institut Pasteur, 75015 Paris, France
and
CNRS, URA2182, 75015 Paris, France

Federico Luzzati (*Chapter 13*)
Department of Animal and Human Biology, University of Turin, Turin, Italy
and
Istituto scientifico Fondazione Cavalieri Ottolenghi, Turin, Italy

Frederic Mercier (*Chapter 4*)
Department of Tropical Medicine, Medical Microbiology and Pharmacology, John A, Burns School of Medicine, University of Hawaii, Honolulu, HI 96822, USA

Zaman Mirzadeh (*Chapter 12*)
Department of Neurosurgery, University of California, San Francisco, San Francisco, CA 94143, USA
and
Eli and Edythe Broad Center of Regeneration Medicine and Stem Cell Research, University of California, San Francisco, 513 Parnassus Avenue, Room HSW-1201, Campus Box 0525, San Francisco, CA 94143-0525, USA

Kinichi Nakashima (*Chapter 3*)
Laboratory of Molecular Neuroscience, Graduate School of Biological Sciences, Nara Institute of Science and Technology, 8916-5 Takayama-cho, Ikoma 630-0101, Japan

Masakazu Namihira (*Chapter 3*)
Laboratory of Molecular Neuroscience, Graduate School of Biological Sciences, Nara Institute of Science and Technology, 8916-5 Takayama-cho, Ikoma 630-0101, Japan

Rokuya Nochi (*Chapter 8*)
Department of Integrated Biosciences, The University of Tokyo, Kashiwa, Chiba 277-8562, Japan

Fernando Nottebohm (*Chapter 2*)
The Rockefeller University, 1230 York Avenue, New York, NY 10065, USA

Paolo Peretto (*Chapter 13*)
Department of Animal and Human Biology, University of Turin, Turin, Italy
and
National Institute of Neuroscience-Italy (INN), Turin, Italy
and
Istituto scientifico Fondazione Cavalieri Ottolenghi, Turin, Italy

Jean-Claude Platel (*Chapter 17*)
Department of Neurosurgery, Yale University School of Medicine, 333 Cedar Street, FMB 422, New Haven, CT 06520-8082, USA
and
Department of Cellular and Molecular Physiology, Yale University School of Medicine, 333 Cedar Street, FMB 422, New Haven, CT 06520-8082, USA

Giovanna Ponti (*Chapter 13*)
Department of Veterinary Morphophysiology, University of Turin, Turin, Italy
and
Istituto scientifico Fondazione Cavalieri Ottolenghi, Turin, Italy

Masato Sawada (*Chapter 14*)
Department of Developmental and Regenerative Biology, Institute of Molecular Medicine, Nagoya City University Graduate School of Medical Sciences, 1 Kawasumi, Mizuho-cho, Mizuho-ku, Nagoya 467-8601, Japan

Kazunobu Sawamoto (*Chapter 14*)
Department of Developmental and Regenerative Biology, Institute of Molecular Medicine, Nagoya City University Graduate School of Medical Sciences, 1 Kawasumi, Mizuho-cho, Mizuho-ku, Nagoya 467-8601, Japan

Jason Schnack (*Chapter 4*)
Department of Tropical Medicine, Medical Microbiology and Pharmacology, John A, Burns School of Medicine, University of Hawaii, Honolulu, HI 96822, USA

Tatsunori Seki (*Chapter 7*)
Department of Histology and Neuroanatomy, Tokyo Medical University, Tokyo 160-8402, Japan

Jan Martin Wojtowicz (*Chapter 10*)
Department of Physiology, University of Toronto, Toronto, ON, Canada

Masahiro Yamaguchi (*Chapter 15*)
Department of Physiology, Graduate School of Medicine, The University of Tokyo, Tokyo 113-0033, Japan

Chunmei Zhao (*Chapter 9*)
Laboratory of Genetics, Salk Institute for Biological Studies, 10010 North Torrey Pines Road, La Jolla, CA 92037, USA

Günther K.H. Zupanc (*Chapter 5*)
Department of Biology, Northeastern University, 134 Mugar Life Science Building, 360 Huntington Avenue, Boston, MA 02115, USA

Part I
History

Chapter 1
The Discovery of Adult Mammalian Neurogenesis*

Joseph Altman

Abstract Visualizing proliferating cells with ^{3}H-thymidine autoradiography, we discovered in the early 1960s that the microneurons (granule cells) of the hippocampal dentate gyrus and the olfactory bulb continue to be produced through adulthood. We later demonstrated that the precursors of hippocampal granule cells proliferate in the dentate subgranular zone, and that this secondary germinal matrix is far more prominent in a carnivore (young cats) than in rodents (rats and guinea pigs). By destroying these proliferating precursor cells with low-level X-rays, we prepared rats that lacked 85% of the normal complement of granule cells, the same proportion that we found to be generated postnatally with quantitative histology and autoradiography. Behavioral tests established that these "degranulated" rats displayed abnormalities comparable to those following extensive hippocampal lesions. We also showed that the granule cells of the olfactory bulb are generated in the persisting subependymal layer of the anterior forebrain and migrate to the olfactory bulb by way of a hitherto unidentified structure, the rostral migratory stream. We discuss why the neuroscience community may have refused to accept these multipronged demonstrations and our laboratory lost its public financing by the mid-1980s.

1.1 Serendipity

Although I decided to become (what is now called) a neuroscientist as a teenager in the early 1940s, and pursued that ambition through difficult times with a pretty clear idea what I hoped to accomplish, the problem of neurogenesis was not the subject I was planning to study. My original idea was to seek a better understanding of human behavior, observing it at its worse in the terrible times

---

* Manuscript submitted in its present format on January 23, 2008.

J. Altman (✉)
8518 SW 86th Terrace, Ocala, FL, USA
e-mail: bayeraltman@earthlink.net

T. Seki et al. (eds.), *Neurogenesis in the Adult Brain I: Neurobiology*,
DOI 10.1007/978-4-431-53933-9_1, © Springer 2011

I lived, and thought that the best approach to that was to study experimentally the neural foundations of the psychological mechanisms that guide behavior. After engaging in an informal and formal "armchair" study of various topics in human and animal behavior, it was in the late 1950s that I succeeded in getting predoctoral training in neurophysiological techniques (working at Mount Sinai Hospital in New York City with the neurosurgeon, Leonard Malis) and postdoctoral training in neuroanatomy (working at the College of Physicians and Surgeons of Columbia University with Malcolm Carpenter). That work led to articles I published on the ascending and descending connections of the feline superior colliculus, and the physiological properties of single collicular neurons. I also published papers on certain facets of the visual behavior (locomotor activity, day–night activity cycle, light aversion) of rats with stereotaxic lesions of the superior colliculus.

While I received my neurophysiological and neuroanatomical training, I also began exploratory work on mapping regional differences in the protein metabolism of the brain following visual stimulation and induced motor activity. The first of these exploratory studies was carried out in collaboration with the neurochemist, Abel Lajtha, who at that time worked at the College of Physicians and Surgeons. We covered one eye of newly hatched chicks, and placed the animals overnight into a box lined with flashing Christmas tree bulbs. The next morning the chicks were injected with a ^{14}C-labeled amino acid and, after a short interval, their brains were removed and the rate of incorporation of the amino acid into a protein fraction was compared in the two dissected optic lobes (the avian homologue of the superior colliculus). The preliminary studies (which we did not think merited publication) suggested a higher rate of protein metabolism in the optic tectum contralateral to the stimulated eye. In the next study, which I carried out in a small laboratory I was establishing at the New York University Medical School, I turned to the use of the novel autoradiographic technique to determine if differences could be obtained in the optic pathways of adult pigeons whose one eye was either blindfolded or removed before the systemic administration of ^{3}H-glycine. While I could not detect laterality differences in the optic pathways in the unilaterally blindfolded pigeons, there were profound differences in the uptake of the radiochemical in the optic tract and the stratum opticum of the optic lobe of the enucleated pigeons contralateral to the removed eye (Altman and Altman 1962). I also observed that the elevated protein metabolism in the affected optic pathway was associated with an increase in the total number of glial cells.

After moving to MIT, I began the autoradiographic approach to the study of brain–behavior relationships by first mapping the regional pattern of CNS protein metabolism in rats injected intraperitoneally with ^{3}H-leucine (Altman 1963a). The quantitative microdensitometric results indicated bilaterally consistent differences in regional label density in various regions throughout the neuraxis. Using the same approach, I also observed elevated uptake of ^{3}H-leucine in motor neurons of the spinal cord in rats that were forced to run at a comfortable speed in a motor-driven activity wheel in comparison with resting rats and rats that were injected with the radiochemical some time after they stopped running (Altman 1963b). In another study with Gopal Das (who joined my laboratory as a graduate student), we observed increased uptake of ^{3}H-leucine in the degenerating optic tract of unilaterally enucleated adult rats in combination with a great increase in the concentration of glial cells (Altman and Das 1964a). However, I was troubled by the inconsistent finding of increased protein metabolism with motor activation but not with visual stimulation, the relative non-specificity of amino acid uptake by single neurons in the exercised rats (such as increases in sensory structures), and by the great variability in individual animals. Accordingly, in an attempt to improve the autoradiographic technique, I also experimented with a more selective radioactive marker than amino acids, i.e., ^{3}H-thymidine. I used the latter to investigate the cellular dynamics of the lesion-induced glial proliferation that we observed histologically with labeled amino acids. Thymidine is the selective and metabolically stable precursor of DNA, and it is incorporated into the nucleus of a cell in large quantities only when the cell is undergoing division. (The early evidence was reviewed in Altman 1969b.) Using ^{3}H-thymidine, I expected to tag the newly forming reactive glial cells. In a pilot study, I made stereotaxic lesions in the lateral geniculate body of rats and concurrently injected a small dose of ^{3}H-thymidine into the lesion site (Altman 1962a). Pairs of animals were then killed 1 day, 1 week, 2 weeks, and 2 months after the operation. The observations revealed that the concentration of intensely labeled proliferating glial cells was highest in those animals that survived for 1 day after ^{3}H-thymidine administration but declined thereafter; however, the number of lightly-labeled cells increased up to 1 month after the injection. This suggested continued division of the already labeled glial cells as a facet of the degenerative and/or regenerative process. Labeled (newly-formed) astrocytes and microglia predominated around the lesion site whereas labeled oligodendrocytes were most numerous in the fiber tracts associated with the damaged lateral geniculate body.

While these results were expected, I was surprised and confused to find a small number of neurons that appeared to be labeled at sites not connected with the lateral geniculate body (Altman 1962b). Is it possible that new neurons are formed in the adult mammalian brain? To clarify this I subsequently injected ^{3}H-thymidine intraperitoneally into normal adult rats and cats (Altman 1963c). In the rats, the autoradiograms showed a variable number of labeled glia cells throughout the brain and spinal cord, and a high concentration of labeled cells in the subependymal layer in the roof of the anterior horn of the lateral ventricle (Fig. 1.1). In addition, radioactive labeling was consistently present in all of the sections over a few granule cells in the dentate gyrus of the hippocampus (Fig. 1.2). In the cats, likewise, there were many labeled glial cells; but, in addition, the nuclei of some small cortical neurons also appeared to be labeled (Fig. 1.3). However, I argued that labeling in the latter (but not in the dentate granule cells) might have been due to the tagging of proliferating perineuronal glia in close proximity of neurons.

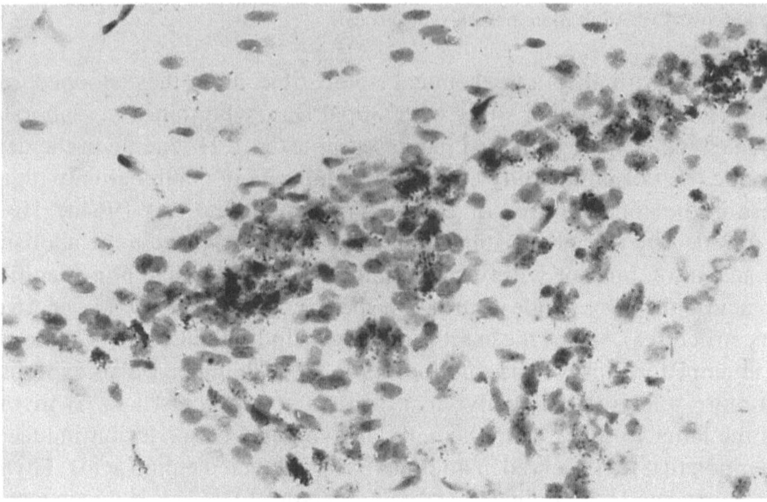

Fig. 1.1 Autoradiogram showing labeled cells in the subependymal layer of the forebrain of an adult rat that was injected intraperitoneally with a single dose of ^{3}H-thymidine (flash labeling) 2 weeks before it was sacrificed. Reproduction of Fig. 5 in Altman (1963c)

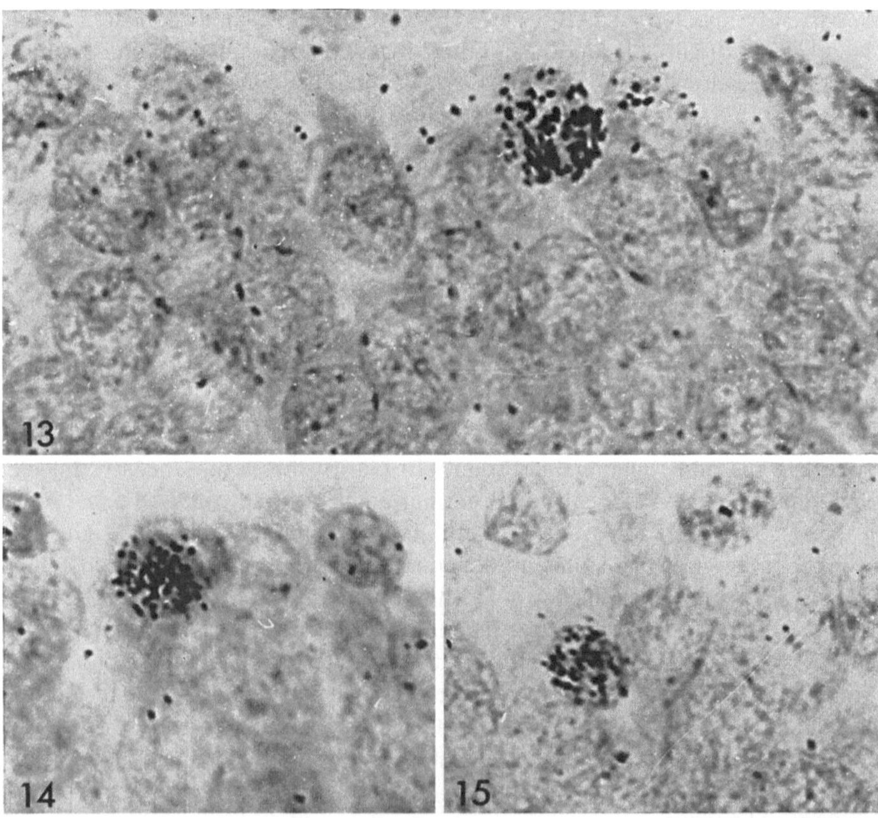

Fig. 1.2 Autoradiogram of labeled granule cells near the hilus of the hippocampal dentate gyrus of an adult rat that was injected intraperitoneally with ^{3}H-thymidine and was killed 2 weeks afterwards. Reproduction of Figs. 13–15 in Altman (1963c)

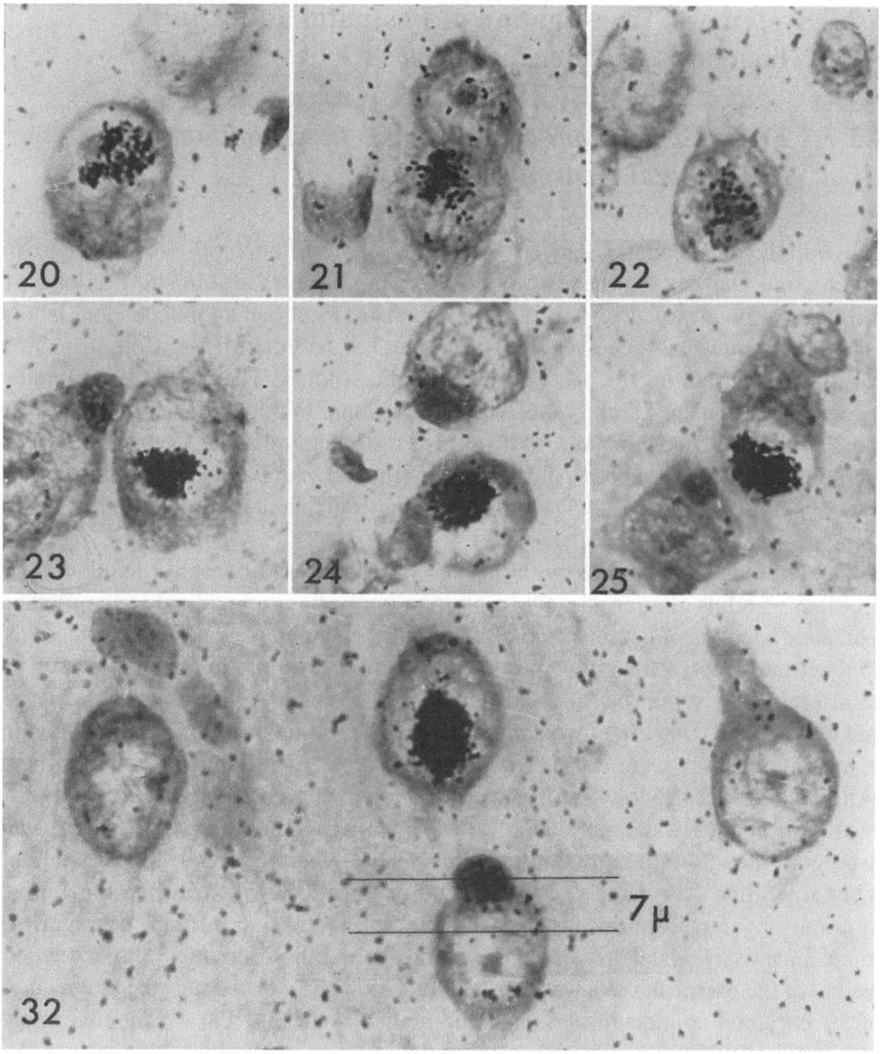

Fig. 1.3 *Upper panel*. Autoradiograms of apparently labeled neocortical (lateral gyrus) neuron nuclei in an adult cat that was injected intraventricularly with ^{3}H-thymidine. *Lower panel*. Autoradiogram to illustrate the possibility that in the 7 µm-thick sections used, the labeling of some of the neurons may have come from closely apposed perineuronal glia. Reproduction of Figs. 20–25 and 32 in Altman (1963c)

1.2 Systematic Investigation of Postnatal and Adult Gliogenesis and Neurogenesis

1.2.1 From Exploratory to Systematic Investigations of Postnatal Neurogenesis

The initial studies that I carried out at MIT were supported by grants to the Psychology Department (headed by Hans-Lukas Teuber). But in short order I was fortunate to be awarded several research grants to establish an independent laboratory. The first award I received was from the U.S. Atomic Energy Commission, as part of its program for the peaceful use of radioisotopes. Thereafter, the laboratory received generous research grants from the National Institute of Mental Health and the National Science Foundation. I was also fortunate in having from the outset the technical help of Elizabeth Altman and William J. Anderson, and in succession the cooperation of Gopal D. Das, several undergraduate and graduate students, and some postdoctoral fellows.

One of our first studies with ^{3}H-thymidine autoradiography dealt with the problem of adult gliogenesis. Since perineuronal glia are involved in the metabolism and functions of neurons, and perifascicular glia may play a role in the electrical conduction of nerve fibers, we posed the question: Might glial proliferation in the adult brain play some role in brain functions associated with behavioral activation and learning? To that effect we compared the rate of glial proliferation in the brains of adult rats that were raised in "impoverished" or "enriched" environments (Altman and Das 1964b). The impoverished rats were reared in isolation in a small cage with opaque walls and devoid of any manipulable objects. The enriched rats lived communally in a large multilevel enclosure where access to food, water, and other amenities was changed every second or third day. This required the animals to climb or jump from one floor to another, move through difficult passageways, run in an activity wheel, and engage in some other skilled activities. While the body weight of the exercising rats was consistently below the isolated rats, the brain weight of the enriched animals turned out to be higher when killed at 4.5 months of age. Moreover, there was a great increase in the number of ^{3}H-thymidine-labeled glia cells in all neocortical areas in the enriched rats when compared with the impoverished rats. This suggested higher rate of glial proliferation in the behaviorally more active animals. Another study dealt with possible differences in the proportion of newly formed dentate granule cells in the hippocampus between enriched and impoverished rats. However, due to serious ethical problems with the student involved in this study, we opted not to publish the results.

We followed these exploratory studies by a large-scale, normative histological and autoradiographic investigation of postnatal neurogenesis. In the first phase of this research (carried out for several years at MIT and continued from 1968 onward at Purdue University) we paid particular attention to postnatal neurogenesis in three brain regions with a cortical organization: the cerebellum, the hippocampus, and the cerebrum including the olfactory bulb (Altman and Das 1965a,b, 1966;

Altman 1966a, 1969a,b). Large groups of rats ranging in age from neonates to adults were injected intraperitoneally with ^{3}H-thymidine and killed at intervals ranging from 2 h (short-survival), several days (intermediate survival), up to 8 months (long survival). The analysis of short-survival autoradiograms showed that the proliferating cells of the external germinal layer of the cerebellum are consistently labeled with ^{3}H-thymidine in up to postnatal day 21 (when this layer disappears). Sequential- and long-survival autoradiograms showed that the labeled cells migrate into the granular and molecular layers where they differentiate as the microneurons of the cerebellar cortex: the granule, basket, and stellate cells.

In contrast to the limitation of postnatal cerebellar neurogenesis to the juvenile period, hippocampal neurogenesis was found to persist through adulthood (Altman and Das 1965a,b). In short-survival autoradiograms, heavily labeled small and dark precursor cells were located at all ages (albeit in decreasing numbers in older animals) in the subgranular zone and hilus of the dentate gyrus (Fig. 1.4, top). After longer survival, the heavily and lightly labeled cells were identified as typical granule cells of the dentate gyrus (Fig. 1.4, bottom). Observations revealed that the differentiating granule cells settle in a regular order in the granular layer, with the early-differentiating (heavily labeled) neurons positioned superficially, and the later-generated neurons (lightly labeled due to the dilution of the radiochemical after repeated multiplication) inside near the hilus (Fig. 1.5). Quantitative studies established that the number of undifferentiated precursor cells increased up to postnatal day 15, and then declined, whereas the number of differentiated granule cells began to increase by day 8 and remained very high as late 300 days (Fig. 1.6). In adult rats injected with ^{3}H-thymidine the percentage of labeled precursor cells was highest 4 days after injection and declined by 2 weeks; in contrast, the percentage of labeled granule cells remained relatively constant up to 8 months after injection (Fig. 1.7). With reference to the reduction in the addition of granule cells to the dentate gyrus as a function of age, a sharp decline was noted in the number of and percentage of labeled granule cells between days 10 and 30, and a steady decline between days 30 and 240 (Fig. 1.8). Finally, in an attempt to determine whether postnatal hippocampal neurogenesis is something unique to altricial rats (rodents with a relatively short period of intrauterine development), we also studied postnatal neurogenesis in the precocial guinea pig, a rodent with a prolonged fetal life (Altman and Das 1967), and in a carnivore, the cat. Importantly, we found that the subgranular zone containing the proliferating precursors of dentate granule cells is more prominent in the guinea pig than in the rat and, even more so, in the cat (Figs. 1.9 and 1.10). The postnatal production of hippocampal granule cells appeared to be a general phenomenon in mammals.

Shirley Bayer joined our laboratory in 1970. As the first step in preparation for her scientific career, she undertook a quantitative assessment of the proportion of dentate granule cells acquired during the postnatal period (Bayer and Altman 1974, 1975). Using a procedure that we called the "progressively delayed cumulative labeling technique," we administered ^{3}H-thymidine daily to large groups of rats on postnatal days 0–3, 4–7, 8–11, 12–15 and 16–19, and sacrificed the animals at 60 days. In this procedure, which labels virtually all the proliferating cells of the developing brain,

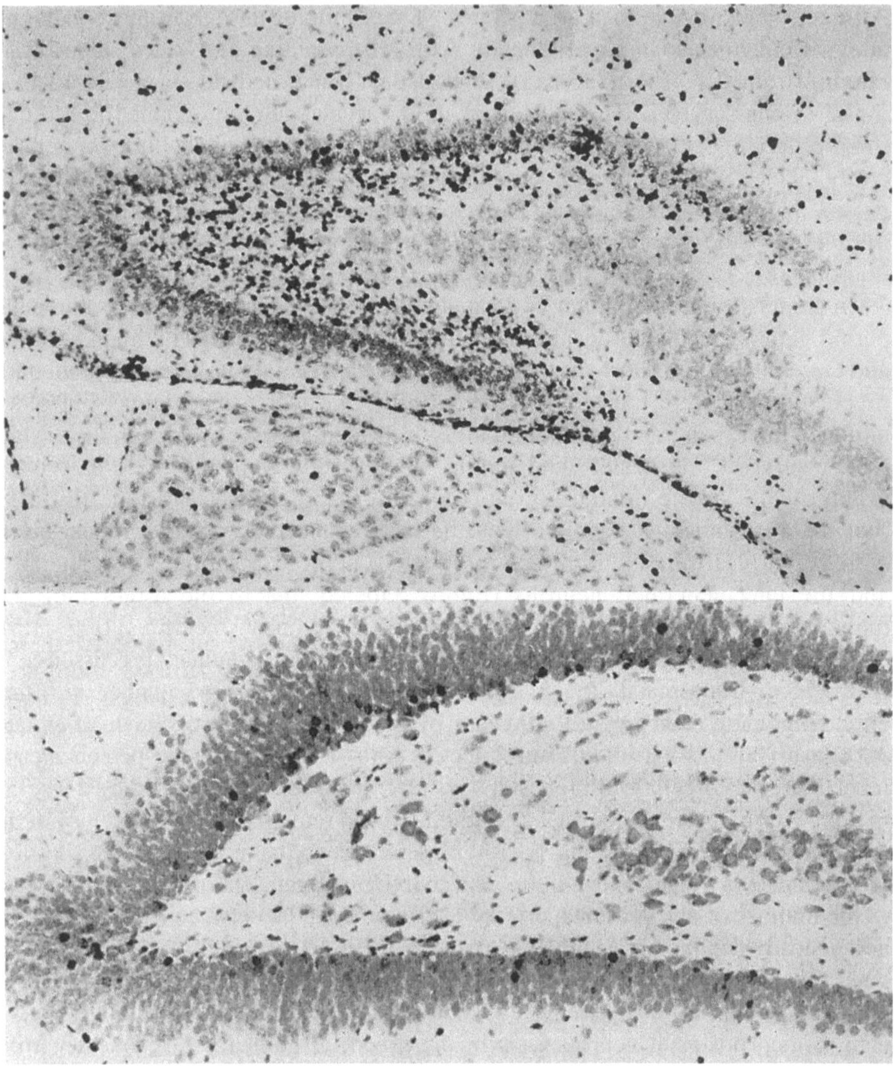

Fig. 1.4 *Top*. Autoradiogram of labeled cells in the hippocampal subgranular zone and hilus of a rat pup repeatedly injected with ^{3}H-thymidine on the fifths, sixth, seventh and eighth days (cumulative labeling), and killed on the ninth day. Note the relative thinness of the dentate gyrus at this age with many unlabeled precursor cells and a few labeled granule cells. *Bottom*. Autoradiogram of labeled "deep" granule cells (near the hilus) in a rat injected with ^{3}H-thymidine on postnatal day 10 (flash labeling) and killed 2 months later. Note the high proportion of unlabeled "superficial" granule cells. These are assumed to have been generated before postnatal day 10. Reproduction of Fig. 21 in Altman (1966a), and Fig. 1A (flipped) in Altman and Das (1965a)

instead of counting the labeled dentate granule cells, she counted the percentages of unlabeled cells and labeled cells as a function of advancing postnatal age, on the assumption that the unlabeled cells are those that differentiated (became postmitotic) before the injection. The results indicated that, as judged by the labeling pattern at

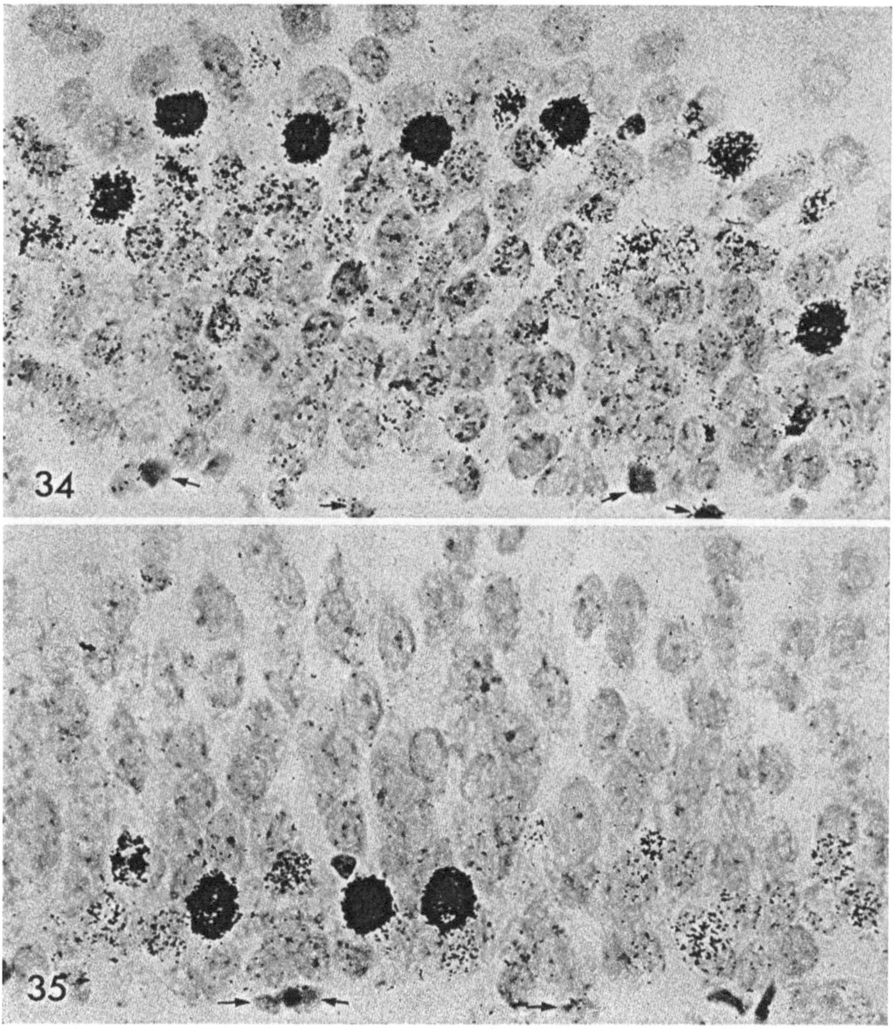

Fig. 1.5 Autoradiograms showing the outside-in pattern of settling of the postnatally generated granule cells. *Top*. Labeling pattern of dentate granule cells in a young-adult rat that was injected with a single dose of ^{3}H-thymidine on postnatal day 2 and killed 2 months later. Note that the outermost granule cells are unlabeled. These are the granule cells that were generated before the injection. The next row of heavily labeled cells are the progeny of precursor cells that differentiated (became postmitotic) soon after the injection. The lightly labeled granule cells below them were generated later, as indicated by the dilution of the administered radioactive thymidine, as the daughter cells underwent further divisions before differentiation. *Bottom*. Pattern of granule cell labeling in a rat that was injected on postnatal day 13 and killed 2 months later. Note that the earlier generated outer granule cells are unlabeled, and heavily and lightly labeled granule cells are confined to the deep row close to the hilus. *Arrows* point the small, dark precursor cells of the subgranular zone. Reproduction of Figs. 34 and 35 in Altman (1966a)

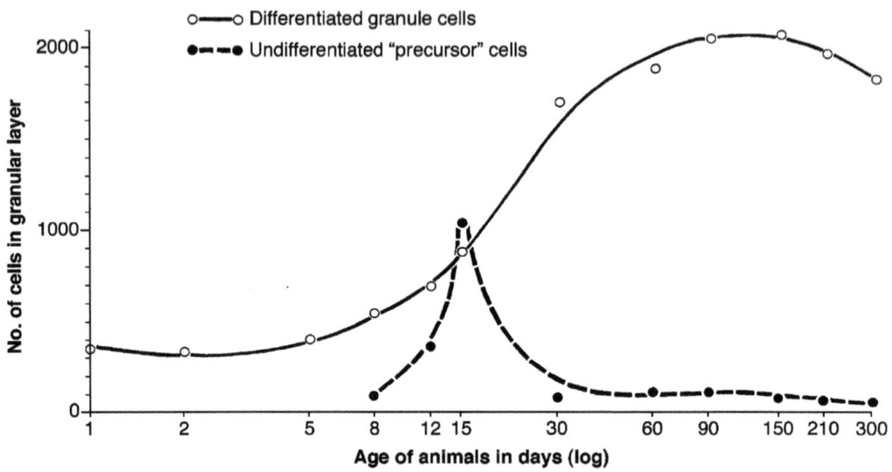

Fig. 1.6 Number of undifferentiated precursor cells and of differentiated granule cells in 25 μm^2 sample areas in the hippocampal dentate gyrus, as a function of age. Note the great increase in the number of granule cells from the second week onward. Reproduction of Fig. 7B in Altman and Das (1965a)

Fig. 1.7 Percentage of labeled precursor cells and granule cells in the hippocampal dentate gyrus of rats injected with ^{3}H-thymidine as adults (4 months) and survived for periods ranging from 4 days to 8 months. Reproduction of Fig. 5 in Altman and Das (1965a)

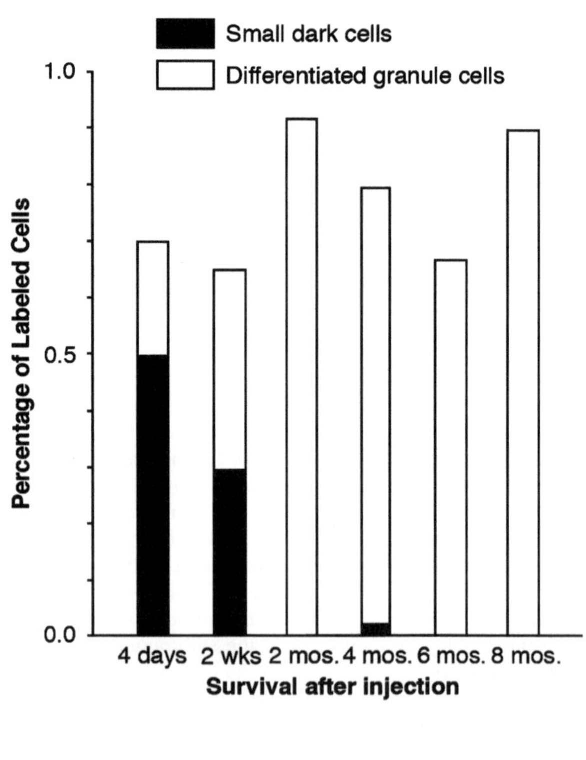

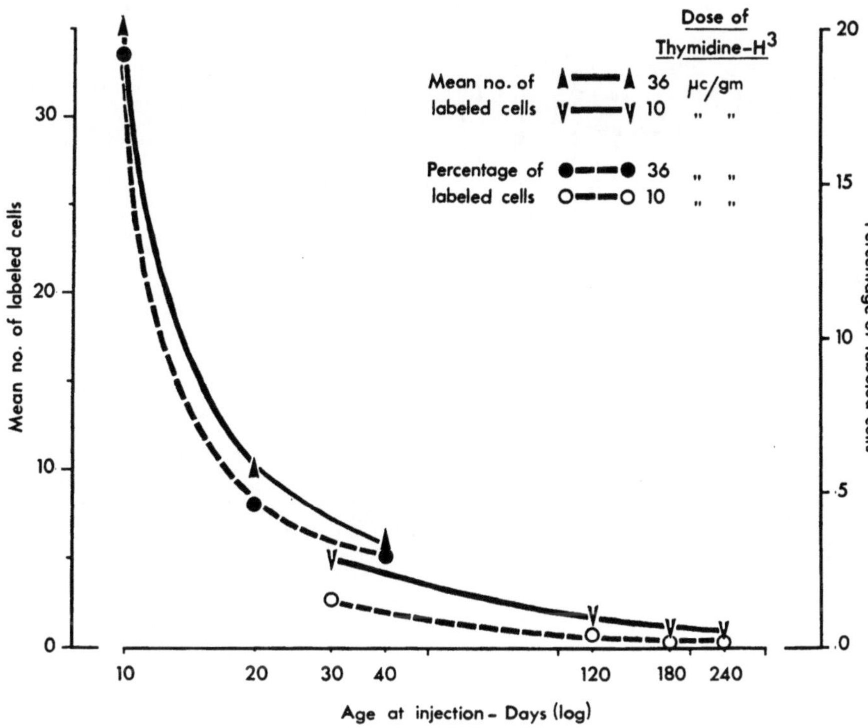

Fig. 1.8 Number and percentage of labeled granule cells in the hippocampal dentate gyrus as a function of postnatal age at the time of injection. Reproduction of Fig. 2 in Altman and Das (1965a)

60 days, 15% of the dentate granule cells are generated prenatally, while 85% are generated postnatally (Fig. 1.11, top). The postnatal acquisition of dentate granule cells is highest during the first week, declines somewhat by the second week, and 16% of the granule cells are generated after postnatal day 16. Counting all mature granule cells in matched sections of the dorsal hippocampus in a large number of rats established that the population increased tenfold from 3 to 70 days. Do the dentate granule cells acquired during adulthood replace dying neurons or do they expand that neuronal population? Shirley Bayer examined this question several years later in a computerized volumetric study (Bayer et al. 1982; Bayer 1982). The results indicated a linear increase in the total number of dentate granule cells in the right hippocampus of rats aged 30, 120, 200 and 365 days of age (Fig. 1.12). The estimated increase between 1 month and 1 year of age was in the range of 35–43%.

1.2.2 Hippocampal Granule Cell Hypoplasia Produced by X-Irradiation

Our next step was to substantiate our autoradiographic results with other experimental procedures. In a series of earlier studies dealing with the postnatal

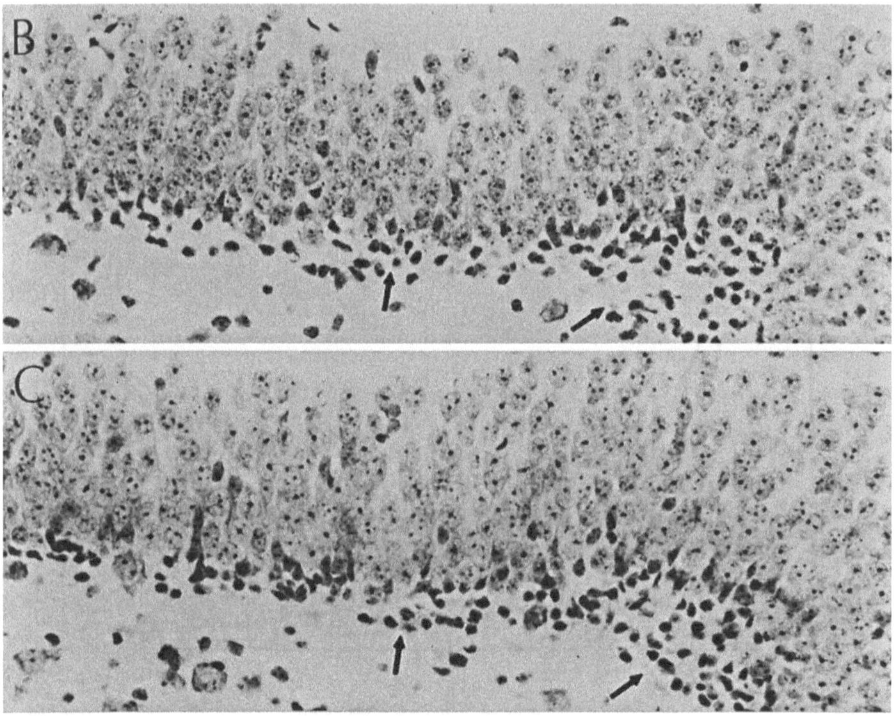

Fig. 1.9 The prominent subgranular zone of the hippocampal dentate gyrus (arrows) in 6 day-old (*top*) and 18 day-old (*bottom*) guinea pigs. Reproduction of Fig. 7B, C in Altman and Bayer (1975)

development of the cerebellar cortex, we found that exposure of the cerebellum to a single dose of 150–200 R X-ray, decimated its subpial proliferative matrix, the external germinal layer (EGL). Evidently, the EGL composed of the precursors of cerebellar microneurons is extremely radiosensitive. However, we also found that within a few days after irradiation, the EGL regenerated and produced a near-normal complement of granule, basket, and stellate cells (Altman et al. 1969a,b; Altman and Anderson 1971, 1972). Because we wanted to study the behavioral role of microneurons, we prevented this regeneration by exposing the cerebellum to repeated daily doses of X-ray from birth until about days 10–15. This procedure produced a hypoplastic cerebellum that contained the full complement of prenatally generated deep neurons and Purkinje cells but was virtually devoid of granule, basket, and stellate cells. In a series of behavioral studies we then showed that by preventing the acquisition of cerebellar microneurons, the rats displayed many of the symptoms previously described following ablation of the cerebellum (e.g., Altman et al. 1971; Brunner and Altman 1973; Pellegrino and Altman 1979). Performing the same procedure by exposing the hippocampus to successive daily doses of low-level X-ray we obtained similar results (Fig. 1.13). Repeated counts showed that with the right number of daily exposures of the hippocampus we could consistently produce rats in which the dentate gyrus contained only the 15%

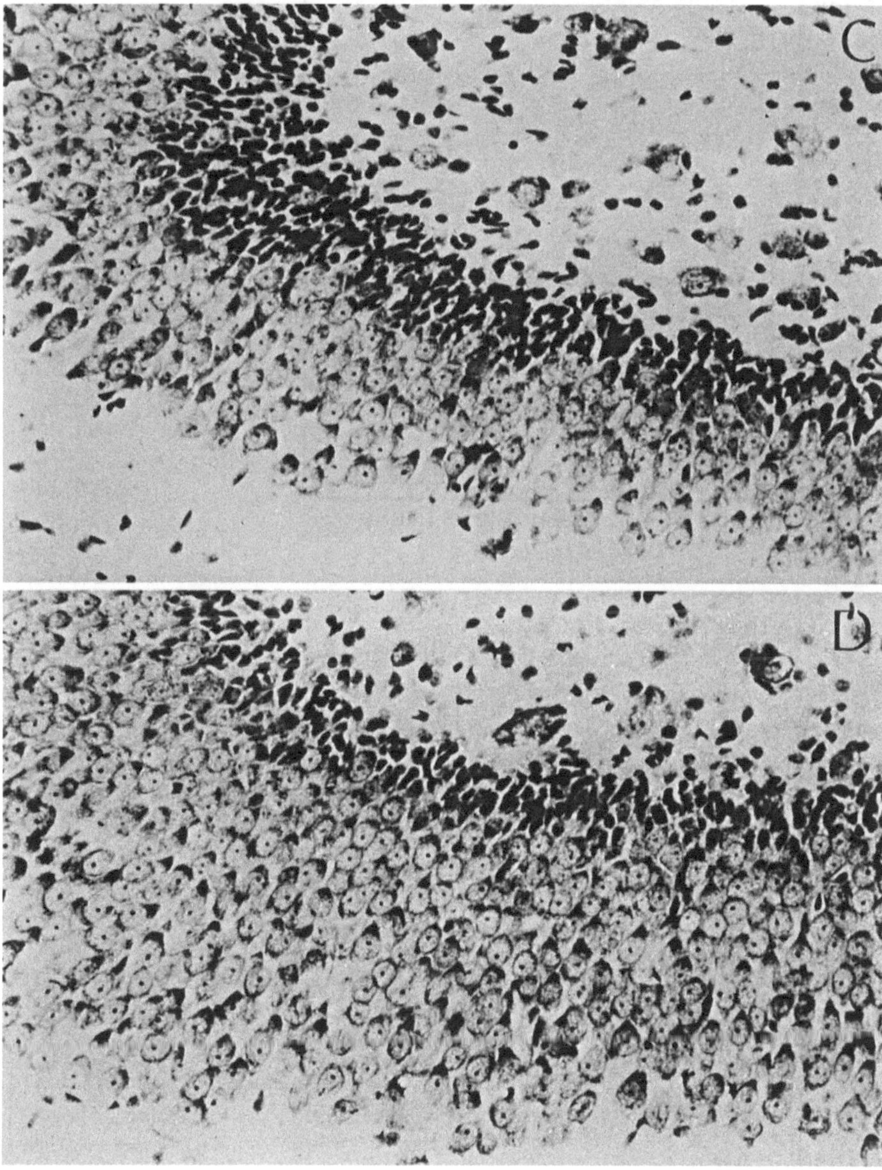

Fig. 1.10 The prominent subgranular zone of the hippocampal dentate gyrus in cats aged 30 day-old (*top*) and 60 day-old (*bottom*) cats. Postnatal hippocampal neurogenesis may be more prominent in carnivores than in rodents. Reproduction of Fig. 8C, D in Altman and Bayer (1975)

of the normal complement of granule cells (e.g., Bayer et al. 1973; Bayer and Altman 1974, 1975; Altman and Bayer 1975). That is, we obtained a perfect match between the histological and autoradiographic estimates of the proportion of granule cells generated postnatally and that obtained with X-irradiation.

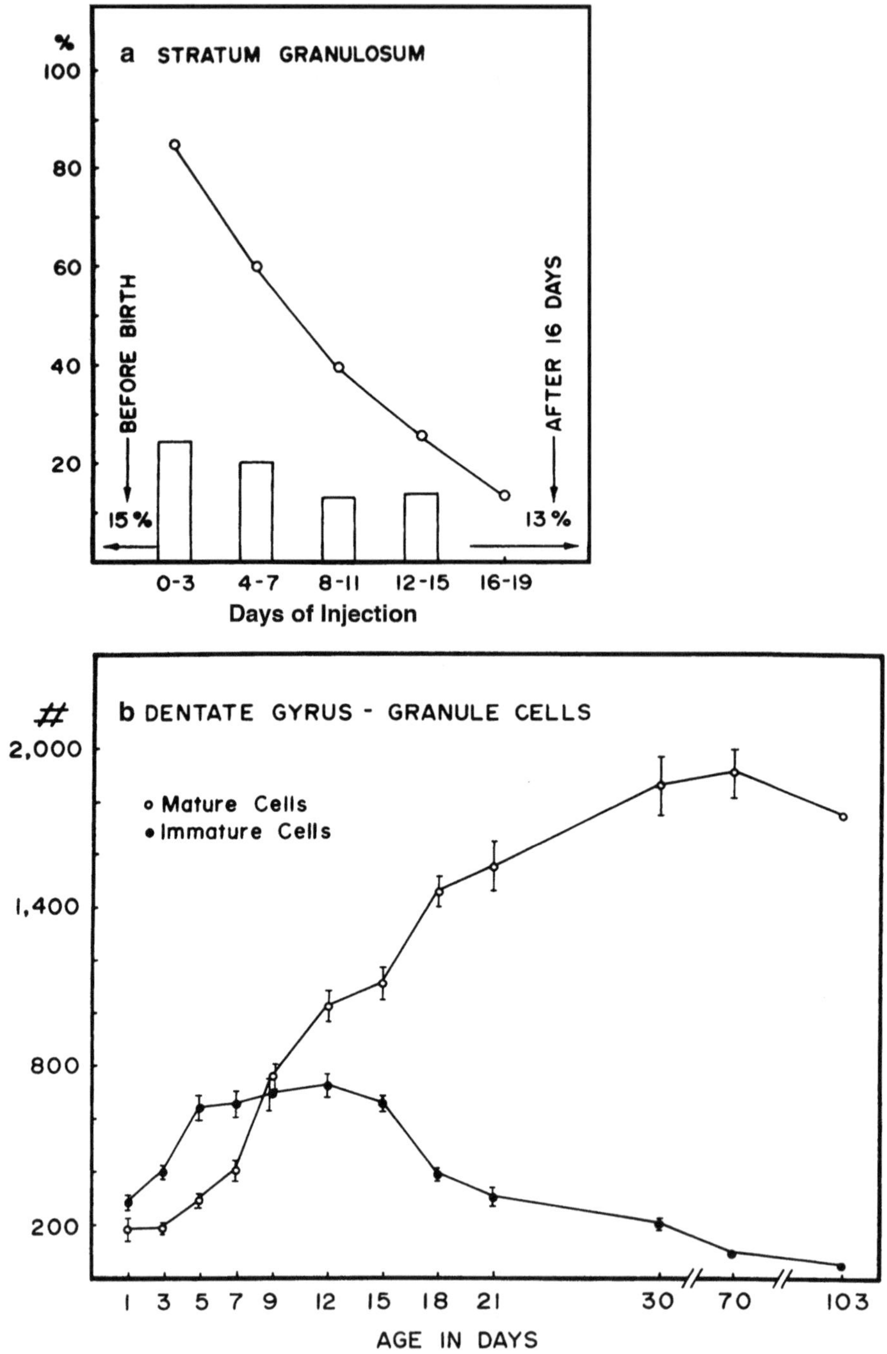

Fig. 1.11 *Top*. Estimate of the percentage of dentate granule cells formed in 4 day-blocks, using the "progressively delayed cumulative labeling technique" (see text). The results indicate that, as judged at 2 months of age, 85% of the granule cells in the rat hippocampus are generated postnatally, 15% prenatally. *Bottom*. Estimate of the number of immature and mature dentate granule cells in matched sections of the dorsal hippocampus. Each point, up to day 70, represents the mean from six rats. There is close to a tenfold increase in the number of mature granule cells between postnatal days 3 and 70. Reproduction of Figs. 9A and 13A in Bayer and Altman (1974)

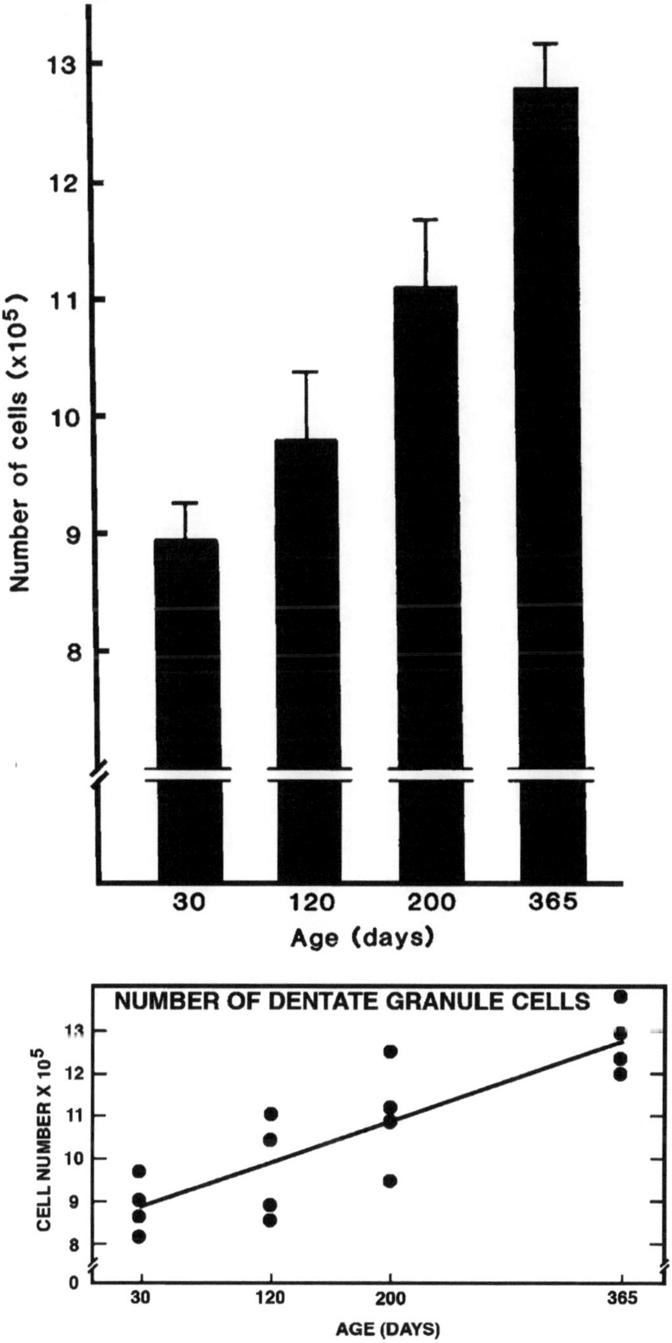

Fig. 1.12 *Top*. Estimate of the total number of dentate granule cells in the right hippocampus of rats aged 30, 120, 200 and 365 days. Each *bar* represents the mean from four rats. Reproduction of Fig. 2A in Bayer et al. (1982). *Bottom*. Range in the number of granule cells in individual rats as a function of age. Reproduction of Fig. 2B in Bayer (1982)

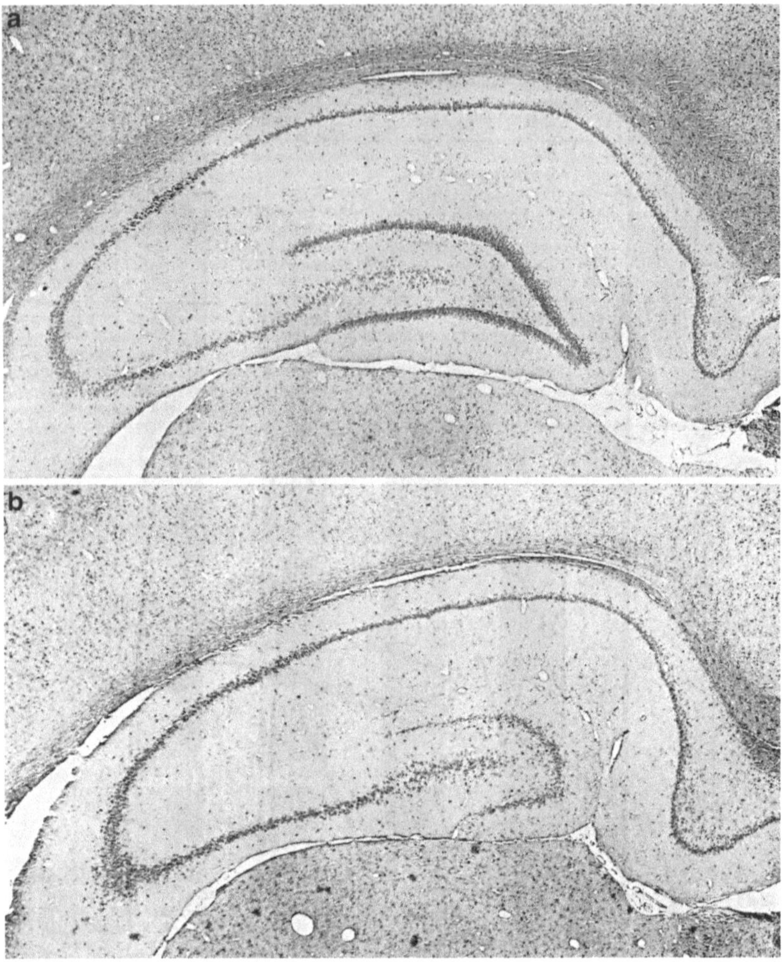

Fig. 1.13 The dentate gyrus in a normal rat (*top*) and a rat irradiated with eight doses of low-level X-ray between postnatal days 2 and 15 (*bottom*). Note the selective reduction in the population of granule cells in the irradiated rat. Reproduction of Fig. 2 in Bayer et al. (1973)

1.2.3 Behavioral Effects of Hippocampal Granule Cell Hypoplasia

In an early exploratory study, we found differences in the number of ^{3}H-thymidine labeled cells in the dentate granular layer (we made no distinction between undifferentiated and differentiated cells) in rats that were "handled" daily from 2 to 11 days after birth and "unhandled" rats (Altman et al. 1968). These rats were flash-labeled with the radiochemical on day 11 and were killed 3 h, or 3 or 30 days thereafter. At all ages, the concentration of labeled granular layer cells was somewhat higher in the handled rats than in the unhandled rats, suggesting influence by that treatment on hippocampal cell proliferation (Fig. 1.14). Several years later we

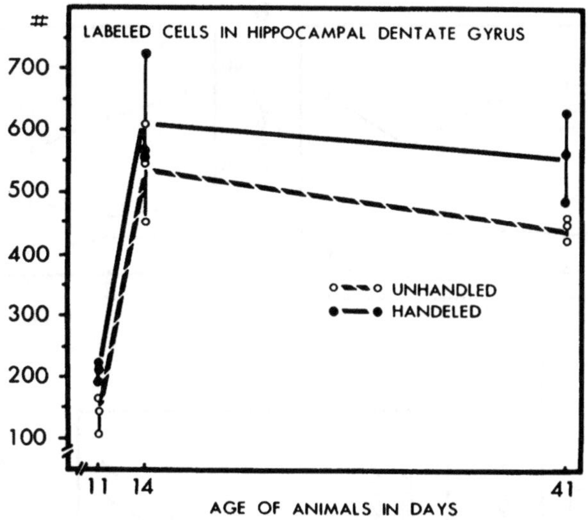

Fig. 1.14 Results of an exploratory study of the number of labeled cells in the dentate gyrus of the dorsal hippocampus in unhandled rats and rats that were handled daily from day 2 to day 11 and survived for 3 h, 3 days and 30 days thereafter. No distinction was made in this study between small precursors cells and differentiated granule cells. Reproduced from Altman et al. (1968)

carried out a series of behavioral studies in adult rats, comparing irradiated rats with hippocampal hypoplasia (which involved handling them daily in preparation for the irradiation) with normal rats (Bayer et al. 1973). Using a variety of tests, these studies consistently showed that granule cell hypoplasia had the same effect on the behavior of adult rats as was previously reported after extensive hippocampal lesions. In the first of these studies we used two control groups: one that was wrapped in the same way ("handled," "stressed") as the experimental group but was not irradiated, and another that was left undisturbed ("unhandled"). Among the behavioral changes we found the following. Rats with granule cell hypoplasia traversed more squares and reared more often in open field tests than did the two control groups (Fig. 1.15a,b). This suggested hyperactivity. To a much lesser extent, the wrapped rats were also hyperactive. In the alternation test in a T-maze, normal rats tend to randomly explore one or the other arm on their first trial but will explore on their second trial, with a high degree of probability, the opposite arm. This behavior is not seen after hippocampal lesions. Correspondingly, we found in rats with granule cell hypoplasia (unlike the two control groups) they randomly chose either arm of the maze on their second trial (Fig. 1.12c). This suggested short-term memory deficit for what arm the animal has chosen on its first trial. The same deficit was suggested by the failure of the irradiated rats in a passive avoidance task, which involves the withholding of a natural response (feeding from a cup) that leads to punishment (Fig. 1.12d), and their performance in a two-way active avoidance apparatus (shuttle box), in which the irradiated rats (unlike normal rats) readily return to the compartment where they were previously shocked (Fig. 1.12e). Similar abnormalities were obtained with other avoidance learning and extinction

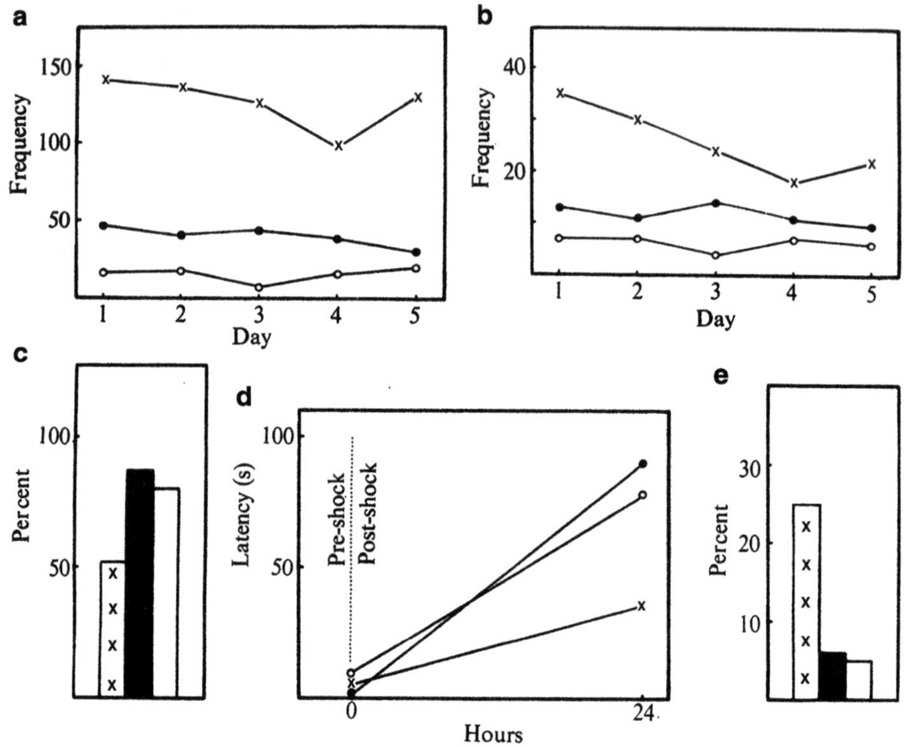

Fig. 1.15 Behavioral effects of hippocampal granule cell hypoplasia in the irradiated rats (*x* in graphs or bars) on several behavioral tests in comparison with two groups of controls (*opaque* versus *open circle* or *bar*). See text for details. (**a**) Increase in the number of squares traversed in an open field. (**b**) Increase in the number of rearing responses in the open field. (**c**) Random alternation in a spontaneous alternation test in a T-maze. (**d**) Reduced latency in approaching a food cup after being shocked. (**e**) Reduced avoidance response in a two-way avoidance task. Reproduction of Fig. 3 in Bayer et al. (1973)

tasks (Brunner et al. 1974; Haggbloom et al. 1974). In another study, we examined the effects of hippocampal granule cell hypoplasia on the initial learning and the subsequent reversal of a series of tactile and visual discrimination tasks graded in term of their difficulty (Gazzara and Altman 1981). The results indicated that the irradiated rats were not handicapped on easy sensory discrimination tasks, irrespective of the modality used, but were handicapped in difficult tasks in which the intensity or detectability of the stimuli was low. We suggested that the rats with hippocampal hypoplasia might have been inattentive because of their hyperactivity. These behavioral studies, in combination with the results obtained after cerebellar irradiation alluded to above, led to our formulation of the concept of a type of attentional deficit disorder due to microneuronal hypoplasia (Altman 1986, 1987). While there have been some supportive follow up studies (e.g., Diaz-Granados et al. 1994; Highfield et al. 1998), this hypothesis deserves further investigation.

1.2.4 Postnatal and Adult Neurogenesis in the Neocortex and the Olfactory Bulb

In an early application of the autoradiographic technique, Smart and Leblond discovered that cells of the subependymal layer of the mouse cerebrum are labeled with ^{3}H-thymidine (Smart 1961; Smart and Leblond 1961). Confirming that in the rat (Altman 1963c), I also found that after prolonged survival there was a rapid decline in the percentage of intensely labeled subependymal cells and an increase in the percentage of lightly labeled cells (Fig. 1.16, top). This was in sharp contrast to the locally multiplying glia in the white matter and gray matter of the cortex where, after an initial spurt, the concentration of labeled cell remained relatively constant (Fig. 1.16, bottom). Evidently, the cortical subependymal layer is a proliferative germinal matrix. What types of cells does this proliferative matrix generate? At least a partial answer to this question came when I discovered that an extension of the anterior cortical subependymal layer is a source of a large stream of spindle shaped, darkly staining cells that migrate to the olfactory bulb. I called this system, the rostral migratory stream (RMS, Altman 1969a). Analysis of sequential autoradiograms in rats that were flash-labeled with ^{3}H-thymidine on day 30 indicated that the RMS contains cells that migrate into the olfactory bulb and settle there as the granule cells of the granular and periglomerular layers (Fig. 1.17). Quantification of the labeling pattern in an adult rat indicated the following pattern (Fig. 1.18). One hour after injection, the percentage of labeled cells is highest in the subependymal layer near the fountain of the RMS, is lower in the vertical limb of the RMS, and is negligible in the horizontal limb and the olfactory bulb. Three days after the injection, the percentage of labeled cells becomes high in the vertical limb of the RMS but is still low in the olfactory bulb. Six days after the injection, the percentage of labeled RMS cells becomes low near the lateral ventricle, is higher in the vertical limb, and is highest in the olfactory bulb. Twenty days after injection there were no longer any labeled cells in the RMS but the olfactory bulb had a high concentration of labeled cells (as seen in Fig. 1.17). I concluded that the RMS cells migrated into the olfactory bulb and settled there some time between 3 and 20 days after administration of the radioactive thymidine. This study also established that, although diminishing in size, the rostral migratory stream is still prominent in adult rats (Fig. 1.19). A later investigation (Rosselli-Austin and Altman 1979) showed that there is a steep increase in the ratio between the postnatally generated granule cells and the prenatally generated mitral cells in the rat olfactory bulb between the second week of life and 90 days of age (Fig. 1.20, top). Linda Rosselli-Austin's study indicated that 89% of the olfactory bulb granule cells are acquired postnatally, and that the labeled granule cells survive up to 1 year of age (Fig. 1.20, bottom).

In summary, our studies carried out in the 1960s and 1970s established definitively that a high proportion of the cerebellar, hippocampal, and olfactory bulb microneurons are generated postnatally. We also showed that whereas the generation

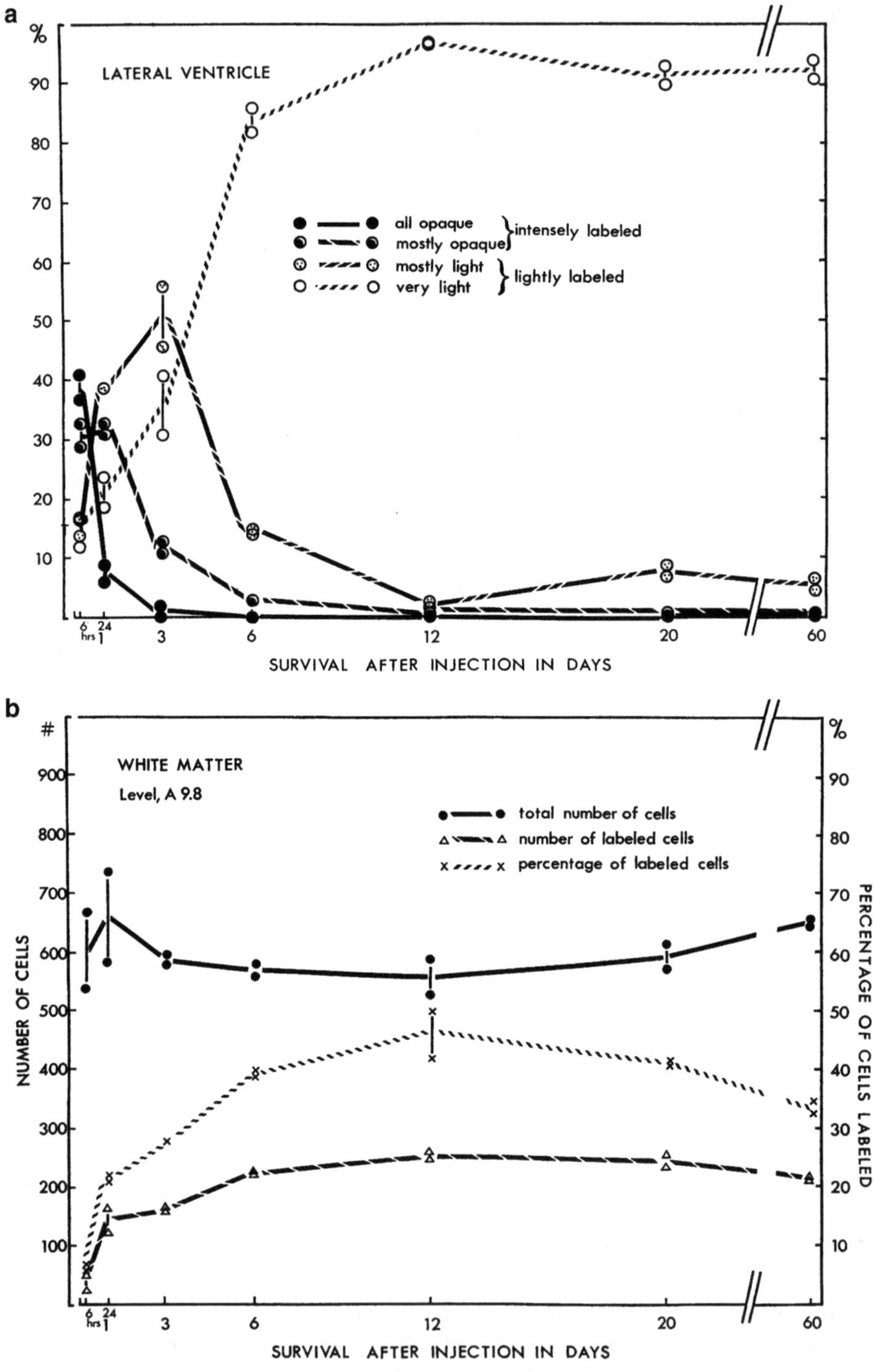
a
%
LATERAL VENTRICLE
90
80
70
60
50
40
30
20
10
all opaque
mostly opaque
intensely labeled
mostly light
very light
lightly labeled
6 hrs
24
1
3
6
12
20
60
SURVIVAL AFTER INJECTION IN DAYS
b
#
%
WHITE MATTER
Level, A 9.8
total number of cells
number of labeled cells
percentage of labeled cells
NUMBER OF CELLS
PERCENTAGE OF CELLS LABELED
900
800
700
600
500
400
300
200
100
90
80
70
60
50
40
30
20
10
6 hrs
24
1
3
6
12
20
60
SURVIVAL AFTER INJECTION IN DAYS

of cerebellar microneurons is limited to the infantile period (postnatal day 21 in the rat), the production of hippocampal and olfactory bulb microneurons continues, albeit at a reduced rate, through adulthood.

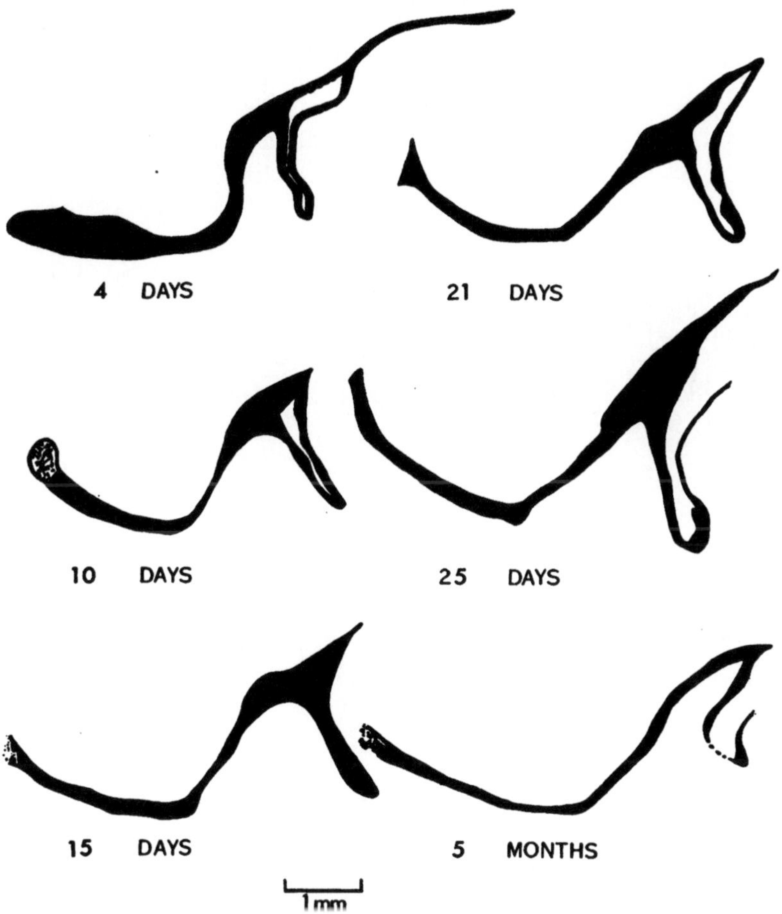

Fig. 1.17 Tracings of the rostral migratory stream (RMS) in sagittal sections of the forebrain in rats ranging in age from postnatal day 4 to 5 months. The olfactory bulb portion of the RMS is not shown in the older animals. Reproduction of Fig. 2 in Altman (1969a)

Fig. 1.16 *Top*. Percentage of labeled cells in the subependymal layer of the lateral ventricle of rats that flash-labeled with ^{3}H-thymidine on postnatal day 13 and killed at different hourly and daily intervals thereafter. Note the rapid decline in precursor cells with high levels of the radiochemical (all opaque, mostly opaque) and increase of cells in which the radiochemical became diluted (mostly light, very light) due to continuous cell divisions. *Bottom*. Percentage of labeled glia (most of them heavily labeled) after survival ranging from 6 h to 60 days, and the total number of glia in unit areas in matched sections of the cortical white matter. The initial increase in the number and percentage of labeled glia cells was followed by relatively little change over time. This indicates a much lower rate of cell proliferation at this site than in the subependymal layer. Reproduction of Figs. 3 and 7 in Altman (1966c)

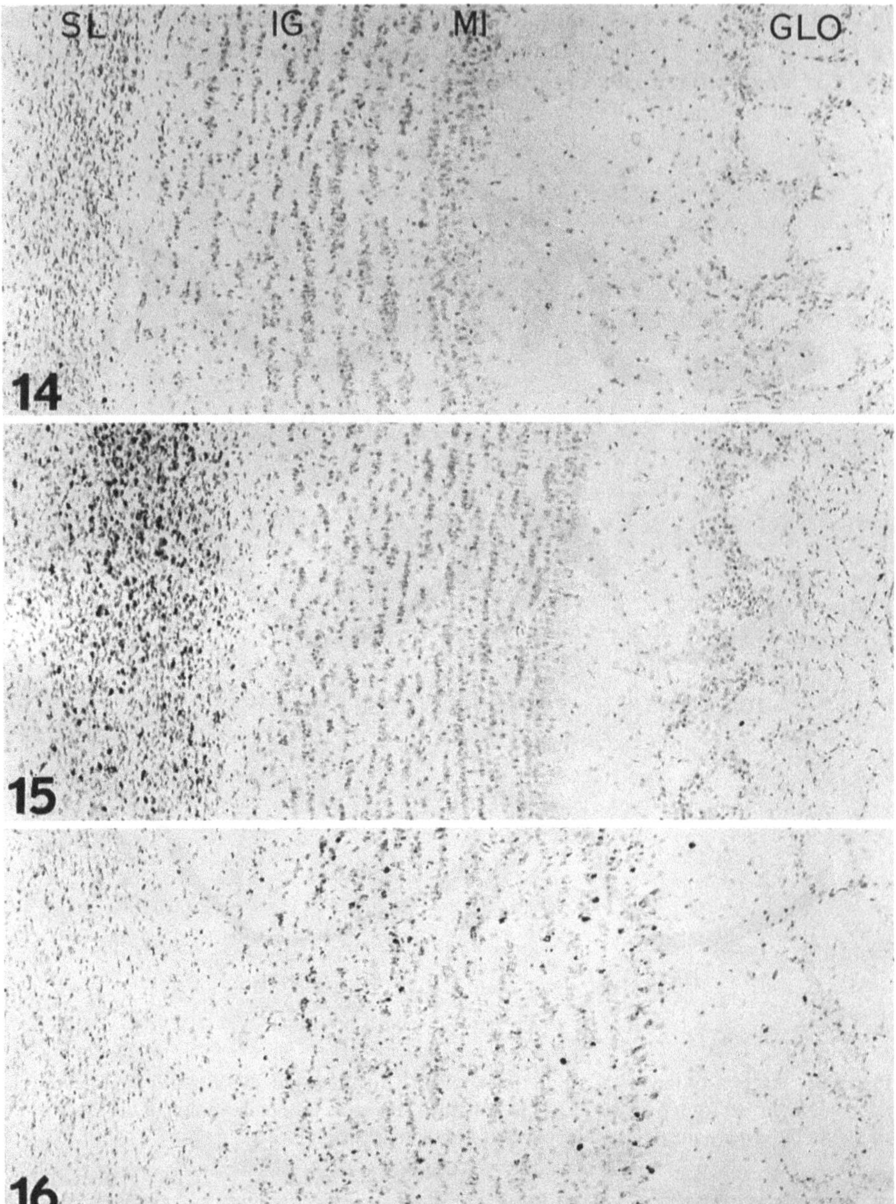

Fig. 1.18 Pattern of cell labeling in the olfactory bulb of rats that were flash-labeled with ^{3}H-thymidine on postnatal 30 and killed 1 h (*top*), 3 days (*middle*), and 20 days (*bottom*) after the injections. There are few labeled cells in any of the layers of the olfactory bulb after 1-h survival, suggesting minimal local multiplication. Labeled cells abound in the RMS after 3-day survival, and they have apparently migrated into the olfactory bulb after 20-day survival. Abbreviations: *GLO* glomerular layer, *IG* internal granular layer, *MI* mitral cell layer, *SL* RMS of the subependymal layer. Reproduction of Figs. 14–16 in Altman (1969)

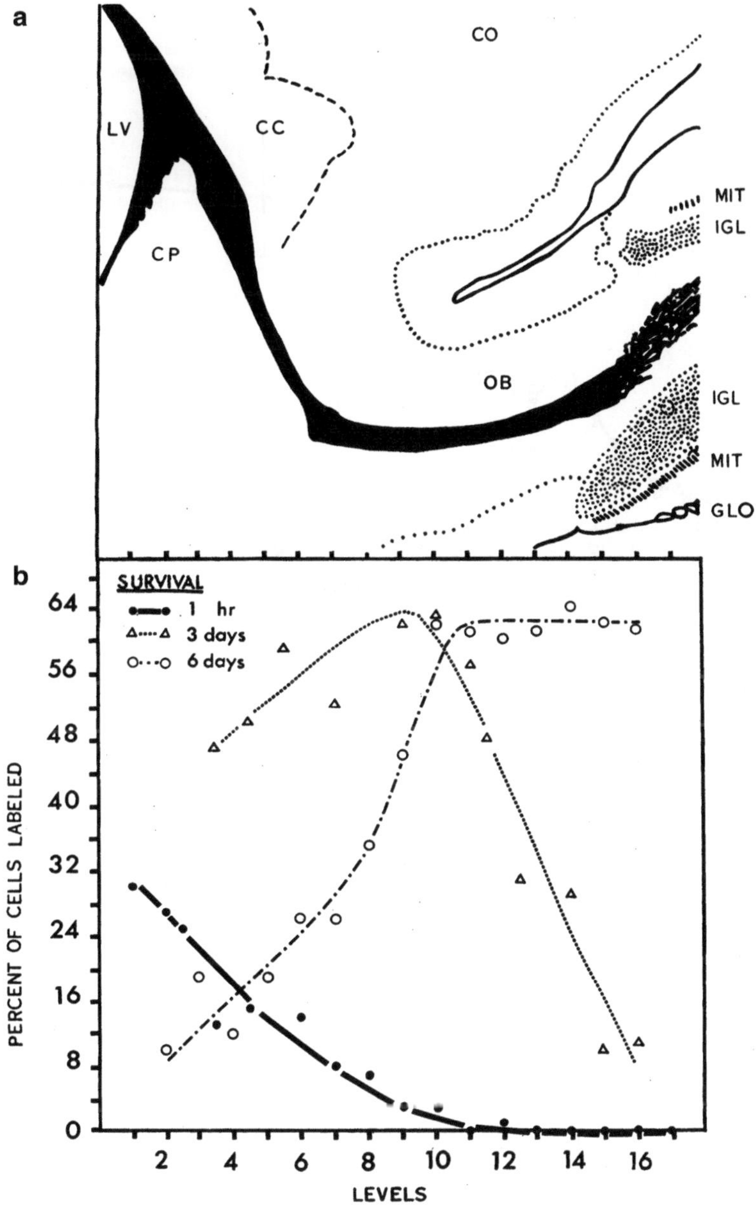

Fig. 1.19 *Top*. Sagittal view of the RMS (*black*) in a young–adult rat. Abbreviations: *CC* corpus callosum, *CO* cerebral cortex, *CP* caudate-putamen, *GLO* glomerular layer, *IGL* internal granular layer, *LV* lateral ventricle, *MIT* mitral cell layer, *OB* olfactory bulb. *Bottom*. Percentage of labeled cells at different levels of the RMS, from caudal (near the lateral ventricle) to rostral (olfactory bulb) in young–adult rats that survived for 1 h, 3 days and 6 days after flash labeling with ^{3}H-thymidine. One hour after injection the concentration of labeled cells was highest near the subependymal layer and the vertical limb of the RMS but was negligible in the olfactory bulb. Six days after injection the concentration of labeled cells decreased caudally and was highest rostrally in the horizontal limb of the RMS and in the olfactory bulb. These findings demonstrated that the RMS contains a high proportion of cells that migrate into the olfactory bulb. Reproduction of Fig. 5 in Altman (1969a)

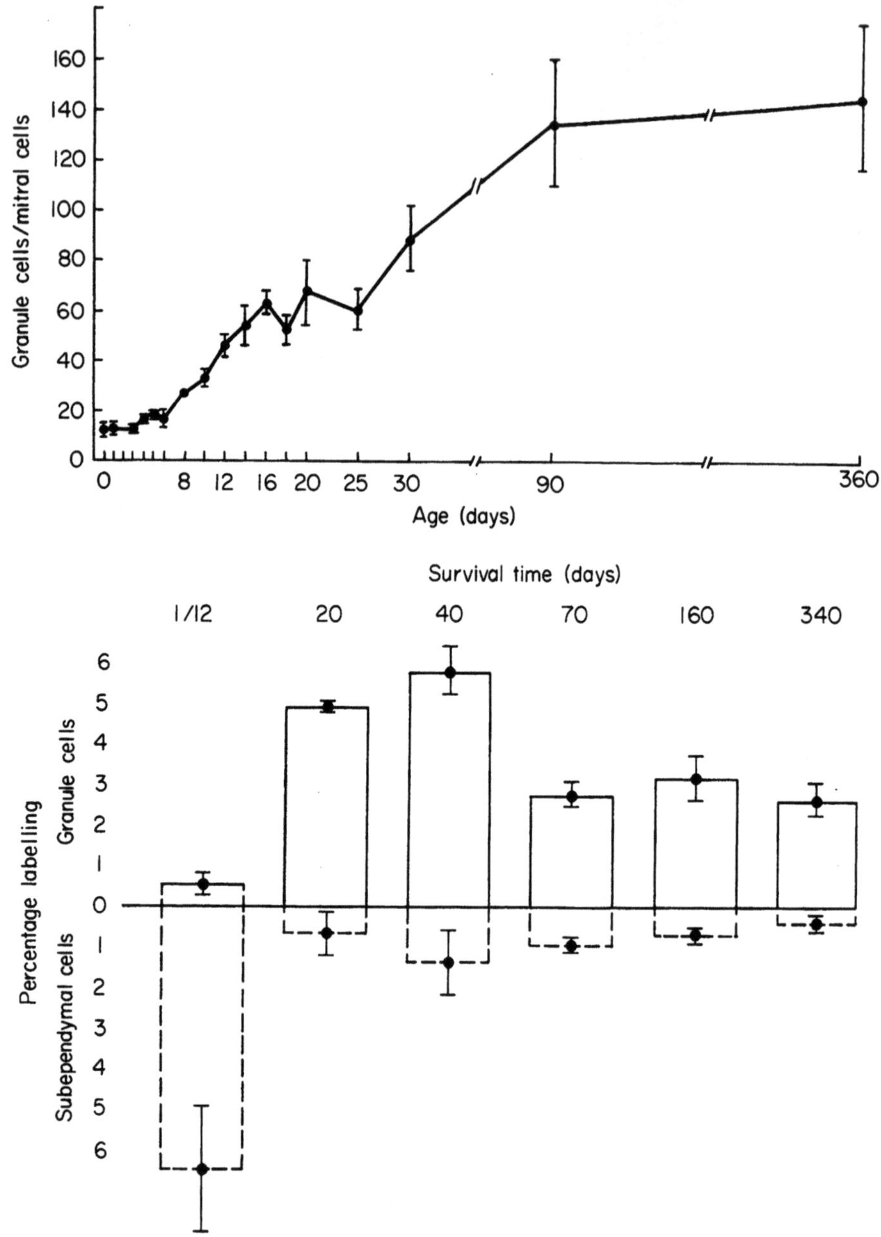

Fig. 1.20 *Top*. Age changes in the ratio of the postnatally generated granule cells of the olfactory bulb and the prenatally generated mitral cells from birth to 1-year of age. *Bottom*. Percentage of labeled subependymal (RMS) cells and granule cells as a function of survival in rats that were flash-labeled with ^{3}H-thymidine at 30-days of age and survived up to 340 days thereafter. Reproduction of Figs. 8 and 9 in Rosselli-Austin and Altman (1979)

1.2.5 The Relationship Between Prenatal and Postnatal Neurogenesis

From the outset, our approach to adult neurogenesis has been a developmental one. We conceptualized central nervous system (CNS) neurogenesis as a prolonged process, one that begins during early embryonic development and proceeds through late embryonic, early fetal, perinatal, infantile, juvenile, and adult periods with distinctive age-related features and properties. It was with that in mind that we have switched by the late 1970s from the confined analysis of neurogenesis in selected brain regions to a comprehensive histological and autoradiographic analysis of the prenatal and postnatal development of the entire CNS. We started this longitudinal project with rats as subjects (e.g., Bayer and Altman 1991; Altman and Bayer 1997) and then, in the early 1990s, we turned to humans. The latter resulted in a comprehensive account of the prenatal development of the human central nervous system (Bayer and Altman 1991; Altman and Bayer 2002; Bayer and Altman 2002–2007). This investigation remains to be extended to the postnatal period.

An example of how this longitudinal approach has shed some light on adult hippocampal neurogenesis is the study in which we tracked the development of the rat hippocampus from the earliest stages of CNS development, i.e., when the future forebrain consists only of a proliferative "stockbuilding" neuroepithelium (Altman and Bayer 1990a,b,c). As we conceive of it, the neuroepithelium (NEP) is the primary germinal matrix of the CNS that is composed of a population of pluripotent neural stem cells, as well as precursors of neurons and neuroglia with progressively reduced fate potentials. Initially, the future hippocampal NEP can be delineated in histological and thymidine-autoradiographic sections from the rest of the cortical NEP only by its medial position. But after a few days, three morphologically different components become discernible in this medial portion of the cortical NEP: the ammonic NEP, the dentate NEP, and the fimbrial GEP (glioepithelium) (Fig. 1.21). Analysis of sequential autoradiograms indicates that the cell-dense and intensely-labeled ammonic NEP is the source of the early-generated large pyramidal cells (hippocampal macroneurons) that will form Ammon's horn. The pyramidal cells sojourn for several days in the intermediate zone (a simpler field than the "stratified transitional field" found in the neocortex; Altman and Bayer 2002) before settling in the stratum pyramidale. The dentate NEP around a ventricular indentation, the dentate notch, initially contains fewer cells and has a different labeling pattern than the ammonic NEP. But soon a stream of proliferating and migrating spindle-shaped cells leave the dentate NEP to form the dentate gyrus. By the time of birth, early-settling granule cells form the thin external leaf of the dentate stratum granulosum but the bulk of these precursors cells sojourn in the hylar subgranular zone. The subgranular zone is most prominent and mitotically active during the juvenile period but persists as a secondary neurogenic matrix into adulthood. Finally, the fimbrial GEP is the primary germinal source of the oligodendrocytes of the progressively expanding hippocampal fiber system, the alveus and the fornix.

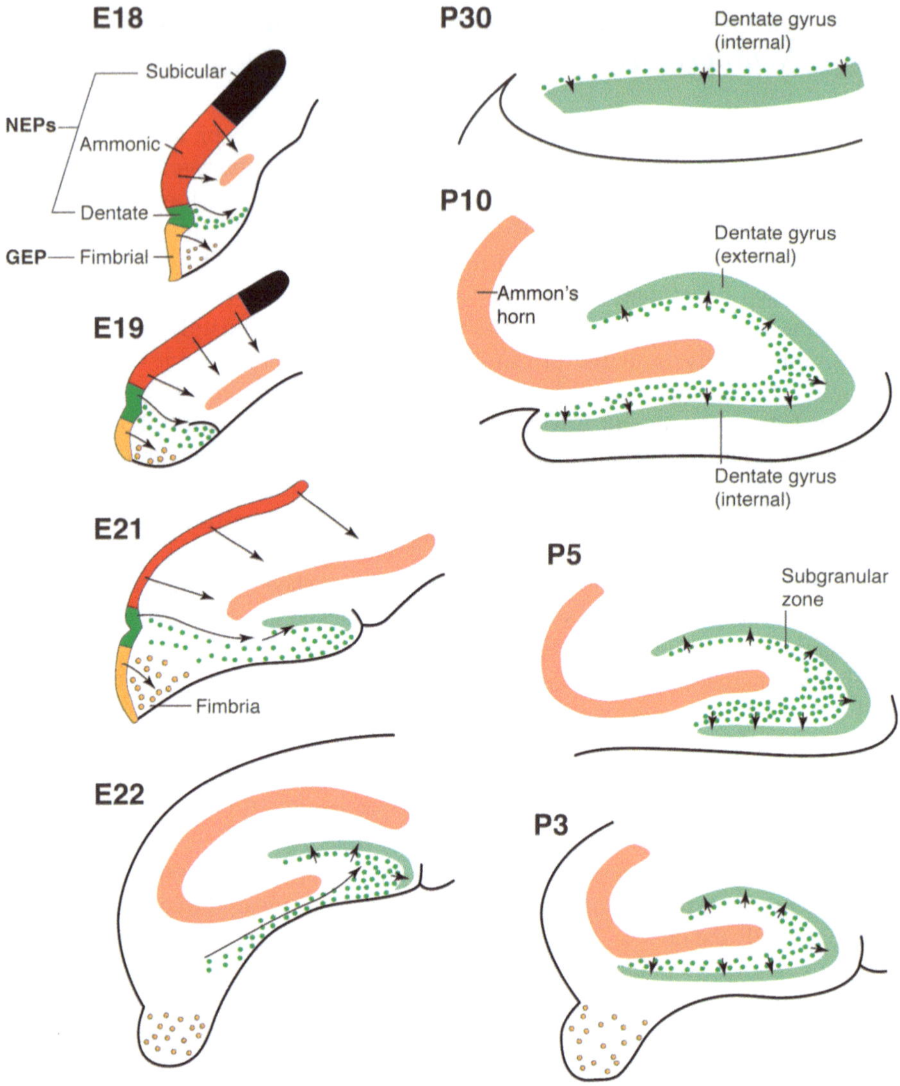

Fig. 1.21 Schematic summary diagram of the primary and secondary germinal matrices of the hippocampus. The primary germinal matrix consists of the subicular, Ammonic and dentate NEPs, and the fimbrial GEP (glioepithelium). These persist until the perinatal period. The dentate NEP is the source of migrating precursor cells that form the subgranular zone, a secondary germinal matrix that is the source of the (mostly) postnatally differentiating granule cells, one that persists as a neurogenic zone through adulthood. Modified after Fig. 16 in Altman and Bayer (1990b), Fig. 13 in Altman and Bayer (1990c), and Fig. 11 in Altman and Bayer (1993)

1.3 Peer Reactions: Initial Excitement, Followed by Marginalization and Subsequent Disqualification

1.3.1 A Short Period of Excitement

Charles Gross deserves credit to be among the first neuroscientists who recently tried to reconstruct the history of the discovery of postnatal neurogenesis in the mammalian brain (Gross 2000). As a colleague of mine at MIT in the 1960s, he had personal knowledge of some of the events that transpired when I began to publish papers on this subject. Regrettably, his interpretation of the nature of my original demonstration is not quite accurate, and his reconstruction of peer reaction to it differs substantially from what I can recollect and can at least partially document. For instance, he states in reference to our work, "the available techniques were not really adequate for an unambiguous demonstration that the adult-generated cells were neurons rather than glia" (Gross 2000, p. 68). As I see it, no neuroanatomist looking at our early illustrations, unless blinded by prejudice, should have failed to see that the reduced silver grains in our autoradiograms were unambiguously over dentate granule cells (see, Figs. 1.2 and 1.5). Neuroanatomists familiar with the cytology of the hippocampus know that the stratum granulosum is a distinctive neuronal layer that is densely packed with granule cells distinguished by pale nuclei, a layer that is virtually devoid of smaller and more darkly staining neuroglia cells. And what more direct and unambiguous way is there to demonstrate nuclear DNA duplication and cell proliferation than using ^{3}H-thymidine, a selective marker that can be tracked over several generations as it is diluted with successive cell divisions (Fig. 1.5)? In his review of the discovery of neurogenesis in adult birds, Fernando Nottebohm graciously states a few years later that "Altman's observations and suggestions on postnatal neurogenesis were original and his wording was cautious … Yet at the time, his claims met stiff resistance." Nottebohm justifies this peer reaction by saying, "It has been often said that the standard of proof is proportional to the importance and novelty of a claim" (Nottebohm 2002, pp. 737–738), and he lists, again, among the several possible objections that the critics may have entertained that the labeled cells were not neurons.

More recently, Gerd Kempermann (2006) reproduced my earliest illustration of ^{3}H-thymidine-labeled dentate granule cells (Altman 1963c), and refers to them as "the first known depiction of adult neurogenesis" (Kempermann 2006, p. 37). Kempermann continued, with reference to a subsequent paper I published with the late Gopal Das (Altman and Das 1965a,b): "One strength of this study was that it sought evidence of neuronal development and did not deliver a mere snapshot in time." Another strength was, Kempermann wrote, the demonstration with a second method "that postnatal and adult hippocampal neurogenesis caused a sixfold increase in the number of granule cells between 6 days after birth and the age of 3 months" (Kempermann 2006, p. 38). What Kempermann might have added is that, in fact, we used extensively an additional methodology to support postnatal hippocampal neurogenesis, as well as a series of studies exploring its possible functional importance. As I mentioned earlier, we used focal irradiation of the hippocampus with low-level x-ray to destroy the radiosensitive precursors of dentate neurons, a procedure that consistently produces a hippocampus with only about 15% of the normal complement of granule cells, the same percentage that we were estimating to form prenatally with quantitative histological and autoradiographic techniques. With regard to the behavioral significance of the 85% of granule cells that are acquired postnatally, our research showed that the prevention of this development produces behavioral abnormalities that mimic partial or total destruction of the hippocampus.

My recollection of my standing in the scientific community during the 1960s also differs from Gross's account. Gross says, referring to me: "he was a self-taught postdoctoral fellow working on his own in a Psychology Department ..." (Gross 2000, p. 68), and, referring to the dogma of "no-new-neurons in the adult brain," speculates about "the difficulty that unknown and junior scientists have in challenging such traditions" (Gross 2000, p. 72). As a matter of fact, I did have the proper academic credentials; I was known in the neuroscientific community; and most of the scientists who tried to stop my work were younger and many of them were less well established than I was. To begin with, on the strength of my preceding training in neurophysiology and neuroanatomy (which I alluded to above), and possibly in recognition of my accomplishments, I was appointed to the faculty of MIT in 1962, with the rank of Associate Professor. The position involved the usual academic obligations as well teaching several courses in behavioral neurobiology. To support the latter task, I used a mimeographed textbook (entitled, *The Biology of Behavior*), which was later published and became widely distributed nationally and internationally (Altman 1966b).

Charles Gross leaves the impression in his historic review that because I was an unknown scientist my early publications could be ignored by the neurobiological community. However, my recollection is that they were not ignored at all but created considerable publicity. The discovery of postnatal neurogenesis was widely reported in the press at the time and it created considerable interest, if not excitement, among many researchers for years thereafter. For instance, I still have a copy of an article in the *Washington Post* about the possible importance of the discovery of adult neurogenesis in mammals, written by Joshua Lederberg (pioneer molecular biologist, later president of the Rockefeller University). Referring to one of my papers (Altman 1967), Lederberg wrote: "In a nutshell, Dr. Altman has obtained good evidence for the continued multiplication of neurons in young rats or kittens, especially in a region of the brain called the hippocampus" (Lederberg 1968, p. A15). [To be precise, I actually argued for the multiplication of the "precursors" of neurons.] And in reference to another paper that appeared in *Developmental Psychobiology* (Altman et al. 1968), Lederberg wrote approvingly about the possibility of "stretching out the period of the brain's development" by "intentionally handling or playing with young rats for just 15 min daily during their first 11 days of life." "The main point we have to ponder," he wrote, "is the complicated interplay between the environmental experience of the rat (or human infant?) and the development of the actual structure of the brain" (Lederberg 1968, p. A15).

As I remember it, I got too much rather than too little attention from the scientific community during the 1960s. There were endless invitations to give lectures at different universities throughout the United States. Perhaps after repeating the same lectures over and over again my performance became lackluster. After all, what I wanted was not publicity but to get back to the laboratory and continue the exciting but time-consuming research we were engaged in. Nor were we isolated once back in the laboratory. I recall visits, some lengthy ones, from many scientists from the USA as well as abroad. Among the visitors I recall John Eccles from Australia, Otto Creutzfeldt from Germany, Jean Piaget from Switzerland, Jerzy Konorski from Poland, Michel Jouvet from France, and several others. I also recall William Windle (then the editor of *Experimental Neurology*) coming by, asking questions and examining our histological slides, and encouraging me to send our forthcoming papers on neuronal regeneration to his journal. I recall Maxwell Cowan (who was to become the editor of the *Journal of Comparative Neurology*) visiting the laboratory and assuring me that, unlike the retiring editor, he will see to it that our papers will be promptly put through the review process. I also recall an invitation from Dominic Purpura (then editor of *Brain Research*) to submit papers to his journal. Moreover, I was invited by the outgoing editor of the *Journal of Comparative and Physiological Psychology* to take over the editorship of the journal (which I politely declined) and from the editors of *Experimental Brain Research* to become the editor of the developmental neurobiology section of that journal (which I accepted). I was grateful for this initial expression of support from the scientific community, and with the generous financial support that we were receiving from three government agencies we were able to pursue our research goals and disseminate the data we gathered.

But then things started to change in the late 1960s, although it took me several more years to realize that something was amiss. The first wake-up call came when I was supposed to be granted tenure at MIT and my promotion was denied. According to the letter I received, some outside referees did not consider my promotion justified. That was no problem. By that time I intended to leave the Boston area for personal reasons and had received expressions of interest as well as several firm and attractive offers from a number of universities to join their faculty. I accepted the offer from the Department of Biological Sciences of Purdue University and joined it as a tenured professor in 1968. I selected that Department not only because of the generous support I was offered but also because I thought that moving to a rural milieu will give me and members of my laboratory the peace and quiet that we needed to pursue our research. Indeed, we succeeded in establishing a productive laboratory, one that remained in operation at Purdue University (West Lafayette) and later at Indiana/Purdue University (Indianapolis) for over three decades. We continued to receive generous public funding for many years and had little difficulty in publishing the papers that we were submitting to various peer-reviewed journals. Indeed, the papers must have found an audience. According to a note in *Current Contents* in 1981, analysis of the *Science Citation Index* database for 1965–1978 indicated that I was among the 1,000 most cited authors in all the sciences (Garfield 1981). The list contained only 24 individuals in the neurosciences under the categories: "neurology," "neurobiology," and "neuropharmacology." I would not argue that the number of citations could in any sense be construed to reflect scientific merit or significance. I merely want to point out that our work was widely read and quoted by our peers during that period.

1.3.2 Marginalization

I now know that there was a concerted attempt by some influential members of the neuroscientific community to marginalize us, but at the time I did not pay much attention to it. Reaching for some textbooks on my bookshelf published during the 1970s on developmental neurobiology, I find the flat statement by Marcus Jacobson: "there is no convincing evidence of neuron production in the brain of adult mammals ..." (Jacobson 1970, p. 33). Jacobson modifies this dogmatic statement somewhat in the second edition of his book: "There is no evidence showing that neurons are formed in the subventricular zone in adult mammals, although granule cells continue to be formed in various other germinal zones for a relatively short period after birth" (Jacobson 1978, pp. 72–73). While he refers to several of our papers on adult neurogenesis, he misquotes them to the effect that "granule cells of the fascia dentate originate from day 10 of gestation to postnatal day 20 ..." (Jacobson 1978, p. 73). That position was advocated by Angevine (1965) not by us. Another introductory text on brain development (Lund 1978), makes no reference at all to adult neurogenesis and quotes none of our papers on that subject.

This neglect of our work continued during the 1980s. As I continue checking my bookshelf, I find no reference to adult neurogenesis in the textbook on developmental neurobiology by Purves and Lichtman (1985) or in the chapter dealing with the development of the CNS in Eric Kandel's first edition of the *Principles of Neural Science* (Schacher 1981). In the third edition of that textbook, published in 1991, Jessell states: "Neurogenesis ceases early in the development of the mammalian brain, but persists into adulthood in some vertebrates, such as fish and birds. For example, Fernando Nottebohm and his colleagues found that the number of neurons in certain nuclei in the brains of adult songbirds changes cyclically on a seasonal basis." Denying postnatal mammalian neurogenesis, Jesssell qualifies the preceding statement by saying that "...neurons can be generated from undifferentiated progenitor cells in the brain of some adult nonmammalian vertebrates. The underlying mechanisms of such differentiation are not understood well enough to determine whether cells in the adult mammalian central nervous system have a similar potential" (Jessell 1991, p. 268). Reference to our published evidence of postnatal neurogenesis was omitted not only by these widely circulated introductory textbooks (of course, I find textbooks that did refer to our work) but also in some more advanced publications. For instance, a book of reprints in developmental neurobiology, with 47 journal articles from the 1960s and 1970s, contains none of our papers (Patterson and Purves 1982). The book contains three reprints authored or co-authored by Pasko Rakic but there is no reference in any of them to our work (although he was fully familiar with our demonstrations). There was obviously a movement afoot to marginalize us.

1.3.3 Disqualification

Open criticism of a scientific claim that runs counter to an established dogma is fully justified; indeed, such a claim must be subjected to a thorough scrutiny before the established paradigm is prematurely abandoned. However, I am not aware of any public criticism or rebuttal of the data we have presented. Moreover, the dismissal of our claim for postnatal neurogenesis has increased, rather than decreased, as we gathered more and more evidence in its favor, and linked the phenomenon of postnatal neurogenesis to its prenatal antecedents. Instead of open criticism, there appears to have been a clandestine effort by a group of influential neuroscientists to suppress the evidence we have presented and, later on, to silence us altogether by closing down our laboratory. I make this accusation for the following reasons: (a) by the early 1980s we were starting to have difficulties in getting our grant applications approved; (b) by the mid-1980s we lost *all* our grant support; and (c) by the early 1990s we had several of our submitted papers outright rejected. [The latter included a series of three papers in which we reported the discovery of the "stratified transitional field" in the developing rat cerebral cortex. The papers were cited as "Altman J, Bayer SA, submitted, 1991a,b,c" in Bayer and Altman (1991).] In fact, we have never since been able to get our grant applications approved, not even when the topic of our research shifted from adult neurogenesis to the comprehensive embryological study of CNS development in rats (Bayer and Altman 1991; Altman and Bayer 1995, 1997) and, later, in humans (Altman and Bayer 2001, 2002; Bayer and Altman 2002–2007).

What might have possessed a group of influential scientists to seek to destroy a dedicated and productive laboratory? I do not have a factual answer to these questions because I have never been informed of the reasons why we were disqualified. What I know unfortunately that the laboratory lost all its public support by the mid-1980s, at about the time when Pasko Rakic published his widely quoted paper in which he denied neurogenesis in the brains of adult primates (Rakic 1985). We were not the only targets of Rakic's public disapproval of adult neurogenesis. I recall a meeting organized in 1984 in New York City by Fernando Nottebohm, called *Hope for a New Neurology*, with a focus on adult neurogenesis. (I was not invited to speak at that meeting but my wife, Shirley Bayer, was and I have accompanied her). At that occasion Rakic berated Michael Kaplan who reported his work on postnatal neurogenesis in rats. Kaplan, an enthusiastic young scientist at the time, later described that experience as the "death of a dogma and a research career" (Kaplan 2001). In the late 1970s Kaplan presented his combined electron microscopic and autoradiographic evidence that the radioactively labeled, newly formed hippocampal and olfactory bulb granule cells of adult rats had axons, dendrites and synapses and, hence, met all the ultrastructural criteria of neurons

(Kaplan and Hinds 1977). Instead of being congratulated for presenting this new evidence, Kaplan was severely castigated. Having rejected the evidence for postnatal neurogenesis in rodents in the 1970s, Rakic later admitted the possibility of that phenomenon in lower mammals but, as noted, denied that in primates. He stated categorically: "Systematic analysis of autoradiograms prepared from postpubertal rhesus monkeys given single and multiple injections of tritium-labeled thymidine and killed 3 days to 6 years later displayed a slow turnover of glial cells but failed to reveal any radiolabeled neurons. Therefore, unlike neurons of some nonprimate species, all neurons of the rhesus monkey brain are generated during prenatal and early postnatal life" (Rakic 1985, p. 1054). But Rakic was wrong again. Several reports were published by the late 1990s of postnatal neurogenesis in the monkey hippocampus (e.g., Gould et al. 1999) and also in humans (Eriksson et al. 1998). As the evidence mounted, Rakic joined these authors (now using BrdU rather than ^{3}H-thymidine) and reported that there is "continuous generation of neurons, oligodendrocytes, and astrocytes in the hippocampal dentate gyrus of adult macaque monkeys" (Kornack and Rakic 1999, p. 5768). Why did Rakic fail earlier to obtain autoradiographic evidence for postnatal neurogenesis in primates? He owed the scientific community an explanation but offered none. Instead, he gave the historical record a new twist by stating that "Unlike cells in most tissues … most neurons of the mammalian brain are entirely generated during early development … One exception, which was first suspected in rodents 30 years ago [quoting Altman and Das 1965b] and later substantiated [quoting Kaplan and Hinds 1977] is the granule neurons of the dentate gyrus of the hippocampus" (Kornack and Rakic, 1999, p. 5768). Note the disingenuous term "suspected" to characterize our multipronged demonstrations. Kaplan is now "in" (after all, he is no longer a competitor) but Altman is still "out."

How could a coterie of scientists succeed in making a productive laboratory terminally lose its public funding? Were we blacklisted? I consider that unlikely. We have never been accused of doing anything wrong and have remained productive to this day. The criteria for granting government research awards, as I heard it over and over again when I served on grant-awarding panels, are "scientific merit" and "past performance." And what might have justified professional administrators, serving as trustees of public funds, to try to close down a productive laboratory without an open public inquiry? I suspect that both sociological and psychological factors contributed to our disqualification. The sociological factors are easy to reconstruct; the psychological factors are more difficult and speculative.

To begin with the sociology. Neurobiology was still a Small Science when I entered the field in the early 1960s. There were relatively few neuroscientists around, the expense of doing neurobiological research was far more modest than it is today, and any capable and hard working researcher could accomplish a lot with the assistance of a few collaborators and technicians. However, within a few decades neurobiology turned into a Big Science, with many more applicants seeking public funding than the granting agencies could possibly support. Universities admitted ever more graduate students to be trained in the neurosciences, research laboratories increased in number and size, and consequently there arose a fierce competition for research and training grants by the 1980s. As this transformation took place, many neuroscientists who started out as bona fide researchers turned into administrators and public relation managers. Instead of sitting at the workbench, the job of the "principal investigator" became to coordinate what went on in his or her laboratory, edit papers written by associates who did the research that he or she no longer had the time (and sometimes the training) to perform, and spend endless hours politicking and paper work to keep the laboratory financially solvent. The leader of a research group had to engage in public relations work at home to drum up interest in the work the laboratory was pursuing; write progress reports and write and rewrite grant applications; spend days traveling from one place to another to attend meetings and conferences to find outside supporters and confederates; and, above all, establish good relationship with the administrators of public funds to obtain preferential treatment. While initially it did not much matter whom you knew but what you knew, increasingly it became more and important whom you knew, and how many of them you knew, to get funded.

Perhaps the idea of training so many neuroscientists, all of whom could not possibly be supported by available funds, was that those with the greatest ability and perseverance will prevail and those less well qualified will leave the field and pursue some other career. Unfortunately, instead of the scientifically best qualified, faculty positions were increasingly occupied by those with an administrative bent and great political skills; people who could attracted large sums of money and support their institution. As a consequence of this selection process, the panels of granting agencies and the editorial boards of journals – which became flooded with grant applications and manuscripts – increasingly became filled with a new breed of scientists, individuals skilled in forming alliances to support one another's projects and getting ahead of their competitors. I have personally witnessed that just a few derogatory remarks made by one or two members of a panel judging a grant application meant that the unfortunate applicant received a "priority" rating that was officially "approved" but was not funded. The same may happen when a biased editor sends a submitted manuscript to a reviewer known to be hostile to the author or his group.

Let me now turn to two possible psychological factors contributing to our disqualification. First, we bucked the trend by practicing Small Science in an environment that increasingly favored Big Science. By spending endless hours in the laboratory and doing very little public relations work inevitably led to our isolation. Having failed to spend the necessary time and effort in the market place, we failed to recruit a cadre of confederates and supporters. Students and postdoctoral fellows quickly learned while they listened to popular speakers making their rounds, and dominating endless symposia and conferences, as to who was "in" and who was "out"; whom to quote or not quote in your bibliography to make it more likely that it will be reviewed by a peer sympathetic to your approach or findings; and what line of research to pursue in light of what is favored or not favored by the granting agencies at any given time. Again, in my personal experience, I watched how most of my former students and even associates, realizing that we were out of favor, stopped working on problems related to postnatal neurogenesis (which they received their training for) and found other projects to commit themselves to or abandoned their research career altogether. But there may have been another psychological factor that has actually contributed to our becoming outcasts. Big Science needs administrators. In addition to those who turn to administration with the selfish motive of exercising power and influence, there are undoubtedly those who sacrifice their research career for the public good. But that sacrifice may have unfortunate consequences. What scientist would not rather make a great discovery himself or herself than be the cheerleader of the group that makes such a discovery? The researcher turned administrator may console himself or herself that in an age of Big Science that is no longer possible. But what if that is contradicted by an individual or a small group of researchers who come up with a new discovery? They become the envy of the advocates of Big Science. They are liable to be distrusted and their claim discounted as unlikely to be true.

1.4 Prospering Under Adverse Conditions and Switching to the Comprehensive Study of CNS Development

How did our laboratory survive, indeed prosper, without research grants and how could we remain productive while facing peer indifference and outright hostility? And why did we turn by the early 1980s from focusing on postnatal and adult neurogenesis to the study of the embryonic development of the CNS? To answer the first question I have to divulge some personal matters that I have never before aired in public. To the second question I have a more objective answer.

Our laboratory survived and prospered, notwithstanding peer hostility and the withdrawal of all public funding, because several circumstances worked in our favor. The most important of these was that by the time we could no longer had public funds we have amassed a large collection of fully-processed embryos and brains. Financing the work ourselves, we were in a position to analyze the available materials and, instead of submitting papers to mainstream journals (a costly and painful enterprise) we started to summarize our results in books that also yielded some royalty. But there was also a psychological factor; namely, that I was well prepared by my earlier life experiences not to be discouraged by public indifference and hostility. When I committed myself to the study of brain–behavior relationships as a teenager (and wrote my first programmatic essay on the subject) I was barred from finishing my formal education in quasi-fascist Hungary because I belonged to a disenfranchised religious/ethnic minority. When the Wehrmacht occupied Hungary, I was incarcerated in a forced-labor camp and worked on a railroad gang. After some time there, I escaped and lived clandestinely in Budapest, which very soon came under siege. My survival was aided, I tend to believe, by a syndrome that I developed, what I call "aparanoia." Blissfully neglecting the fact that they were after my kind, I did not hide but walked with head erect through the streets, past bayoneted soldiers and gendarmes looking for Jews and deserters. I convinced myself that I was going to survive no matter what the Nazis' intentions were. When the Red Army liberated us in 1945, I resumed my formal education. However,

I could not tolerate the rising Communist dictatorship and fled the country in 1946. I became a stateless, displaced person in West Germany, waiting there for many years before I could secure the requisite documents to migrate to Australia. While that sounds like a misfortune, those were very good years. I was able to attend lectures and seminars at a prominent German university and spent endless hours in its library. I continued informally with my education in Australia, where I worked through the first half of the 1950s as a college librarian. In my spare hours there, I wrote the bulk of the manuscript that was later published as a textbook in behavioral neurobiology (Altman 1966a,b,c).

When I migrated to the United States in 1955, I embarked on a research career in neurobiology by getting formal predoctoral and postdoctoral training at different facilities in New York City (described above). In 1959 I organized a small laboratory at the New York University Medical School and worked there with Elizabeth Altman and some student assistants. In the early 1960s we moved to the Massachusetts Institute of Technology (also described above) and after working there for many years, the greatly expanded laboratory and staff was transferred to Purdue University in 1968. At Purdue I had the great fortune that Shirley Bayer joined the laboratory as a graduate student in 1970. Although we came from a very different background, we soon established a close personal and professional partnership. For many years we both received generous public grants, and the laboratory became very productive. Then abruptly we lost all our funding and had to face all the consequences that lack of financial support produces in the academic world. How did we react to that misfortune? We thought of fighting our disqualification and did some protesting. But, not being able to accomplish anything, we decided that it made better sense to finance our research ourselves rather than waste our time and energy in writing and rewriting grant applications and battling our adversaries. This is probably not the course of action that many scientists might have taken in a similar situation but I was adequately prepared to work in isolation.

To turn to second question: Why did Shirley Bayer and I turn by the mid-1980s from the study of postnatal and adult neurogenesis to the embryonic development of the CNS in rats and humans? Perhaps peer indifference contributed to our reduced enthusiasm for that work. But a more important consideration was our conviction that adult neurogenesis must be studied as a facet of CNS development. We maintained all along that adult neurogenesis is the end phase in the continuum of CNS development. Accordingly, we embarked in the mid-1970s on a long-term project to re-examine the whole course of CNS development by combining traditional histological techniques with ^{3}H-thymidine autoradiography. By the end of the 1980s we had prepared close to 2,000 histologically and autoradiographically processed prenatal and postnatal rat embryos and dissected brains. This enabled us to carry out a detailed investigation of the prenatal and postnatal development of the rodent cerebellum, spinal cord, brainstem, diencephalon, hippocampus and basal ganglia, as well as some components of the cerebral allocortex and neocortex (references to these studies are listed in Altman and Bayer 1982; Bayer and Altman 1991, 2004; Altman and Bayer 1995, 1997). We then turned in the mid-1990s to the study of human CNS development. To accomplish that, we worked for months at the National Museum of Health and Medicine in Washington, DC, which houses the Carnegie, Minot, and Yakovlev Collections of prenatal and postnatal human brains. (We did seek grant support for this effort but got none; except that the high daily fee to use the facility was graciously waived.) We took over 10,000 low- and higher-power photomicrographs there in preparation for a 5-volume *Atlas of Human Central Nervous System Development*. That task was completed recently (Bayer and Altman 2002–2007).

Our developmental approach, we believe, has shed some light on the nature and significance of postnatal and adult neurogenesis (Altman and Bayer 2007). It brought support for our concept of two neurogenic production sites in the developing CNS: the primary neuroepithelium (NEP) and the secondary germinal matrices (SGM). The NEP is composed of stem cells and progenitor cells of the macroneurons that form the core structures and gross circuitry of the CNS. It originates early during embryonic development as the open neural plate. After the neural plate folds and fuses dorsally, the lumen of the cephalic vesicles become filled with cerebrospinal fluid to form the narrow protoventricles. The protoventricles then expand to form the superventricles of the telencephalon, diencephalon, mesencephalon, and rhombencephalon. Because the stem cells and progenitor cells of the NEP have to shuttle to the ventricular lumen to undergo mitosis, the size and configuration (eminences, protuberances, invaginations) of the variegated shorelines of the superventricles are major determinants of the number of neurons generated for different brain structures at different NEP compartments (mosaics). We provided some evidence that the NEP is initially composed of stockbuilding precursor cells that expand and form the region-specific compartments, and postulated that their proliferation and progressive fate-restriction are dependent on two important regulatory mechanisms: hypothetical trophic factors contained in the cerebrospinal fluid system, and reciprocal signaling between the NEP and the concurrently developing peripheral sense organs and motor systems. In the next stage of CNS development, the stockbuilding NEP cells start to generate exiting daughter cells of two types, postmitotic cells that differentiate as neurons and neuroglia, and distinctive precursor cells that retain their proliferative capacities. The postmitotic young projection and relay neurons migrate radially, tangentially and, at some sites, follow tortuous routes and sojourn in intermediate zones before they settle down. These early-generated neurons form the brain parenchyma by filling the expansion spaces of a unique, hitherto unrecognized tissue, the initially expanding then shrinking superarachnoid reticulum. In mammals, the periventricular NEP is principally active during the prenatal period.

The NEP, as noted, is also the source of progenitor cells that form the SGMs. Unlike the NEP, the SGMs persist for varying periods postnatally and generate the locally-arborizing, short-axoned microneurons that become important constituents of the fine circuitry of certain brain structures. Among the SGMs of the developing mammalian brain are the following. (a) The bulk of the subependymal layers of the forebrain that generate the microneurons of the cerebral cortex and the neostriatum during the late embryonic and perinatal periods. (b) The subpial external germinal layer of the cerebellum that generates the microneurons of the cerebellar cortex during the infantile and juvenile periods. (c) The interstitial subgranular zone of the hippocampus that generates the microneurons of the hippocampal dentate gyrus throughout life. (d) The circumscribed subependymal layer of the anterior cerebrum that, likewise, generates the microneurons of the olfactory bulb throughout the entire life span of all mammals. While the macroneurons that are generated in mammals during gestation are minimally affected by extrauterine events, the postnatally generated microneurons of the cerebellum, olfactory bulb, and hippocampus are subject to considerable environmental influences. That led to our earlier suggestion that they may play a major role in responding to environmental factors, perhaps by mediating certain types of learning processes (Altman 1967, 1970).

1.5 The Changed Paradigm in the Study of Adult Neurogenesis

Scientific theories – unlike political attitudes, religious faiths or aesthetic judgments – are not matters of personal or group preferences. A person's or a group's political views are understandably guided by economic and social interests, and idiosyncratic or conventional preferences. Religious beliefs are based on social tradition and indoctrination, and a faith in a supernatural world inaccessible to the senses. Aesthetic choices may be a matter of personal taste or of changing fads, fashions or artistic trends. But scientific theories cannot be sustained by appeal to such subjective values. Scientific theories must comply with the objective facts ascertained by empirical observations and research findings; they stand or fall as new data gathered confirm them or refute them. Of course, science is pursued by human beings with different interests, backgrounds and tastes, and therefore personal and social factors necessarily play a role what a scientist studies and what theories he or she will favor. But sooner or later the facts will prevail and no matter how powerful or prestigious the supporters of a refuted theory, the theory will eventually be abandoned. In the past, with few scientists around, false scientific theories, such as the Ptolemaic system of the structure of the universe, could be upheld for some time after it has been shown to be inadequate. Nicolas Copernicus' *De revolutionibus orbium coelestium* was published in 1543, providing good astronomical evidence that the earth rotates around the sun, but his theory was not widely accepted until 1687, when Isaac Newton, using Johann Kepler's additional observations and calculations, developed his theory of gravitation in his *Philosophiae naturalis principia mathematica.* In the meantime the ecclesiastical authorities censured and put under house arrest Galileo Galilei, and burned Giordano Bruno at stake, because they dared to argue in favor of the heliocentric theory. As science has become such a large enterprise with so many practitioners, untenable theories fall by the wayside much faster nowadays than they did in the past. The theory of adult neurogenesis, of course, does not have either the same global significance as the theory of gravitation nor does it challenge the biblical account of the Earth's place in the solar system. But the theory of adult neurogenesis does have important theoretical and practical implications. It supports an emerging new view of the CNS, one with greater dynamic properties and regenerative potentials than was hitherto contemplated. Because of the rapid increase in the pace of scientific progress, it took only about 20–30 years that the old dogma of "no postnatal neurogenesis" was abandoned by all but a few diehard scientists.

The study of adult vertebrate neurogenesis became a respectable research endeavor in the 1980s when Fernando Nottebohm and his collaborators reported their findings about the seasonal turnover of neurons in the forebrain of songbirds (Goldman and Nottebohm 1983; Alvarez-Buylla and Nottebohm 1988; see Chap. 2 for details). And the study of mammalian neurogenesis became a truly popular subject soon thereafter when two new techniques were introduced. One was the increased use of bromodeoxyuridine (BrdU), instead of ^{3}H-thymidine, to label the proliferative precursors of neurons. The other was the introduction of various molecular markers for the identification of the proliferating, differentiating and maturing neurons. These two techniques were very important innovations. The BrdU procedure is much simpler and faster than ^{3}H-thymidine autoradiography. Indeed, I suspect that one of the reasons why more investigators have not studied adult neurogenesis in the 1970s and 1980s was that they were discouraged by the difficulty of effectively using the liquid emulsion technique for high-resolution ^{3}H-thymidine autoradiography. We spent several years to develop a reliable and consistent methodology. This included the use of ^{3}H-thymidine of the right specific activity; injection of the optimal dose of the radiochemical in relation to the animal's body weight; proper and consistent dilution of the nuclear emulsion; slow drying of the emulsion in the dark before packaging and refrigeration to avoid mechanical artifacts; very long exposure period (we consistently used 3 months) to get optimal results; and, last but not least, use of counterstaining techniques that do not remove part or all of the emulsion. Significantly, I am not aware of reports of false positive results with ^{3}H-thymidine but only of false negatives (e.g., Rakic 1985). And I must also note that with all its advantages, the BrdU technique has its pitfalls. Most importantly, BrdU is a toxic substance that produces morphological and behavioral abnormalities, triggers cell death, and leads to the formation of teratomas (Kolb et al. 1999; Sekerkova et al. 2004; Taupin 2006). In contrast, with survival almost as long as 1 year after the administration of our standardized dose of ^{3}H-thymidine, we have observed no deleterious somatic effect in our experimental animals and little or no reduction in the number or proportion radioactively tagged granule cells (see, for instance, Fig. 1.20). Hence the comparative validity and reliability of these two techniques warrants further investigation.

The molecular markers that have been added to the armamentarium of the traditional techniques for the identification of neurons and neuroglia have also been very useful in the study of adult neurogenesis. Validated molecular markers allow investigators to distinguish small neurons from different types of neuroglia, and identify stages in the differentiation of young neurons. However, there are again pitfalls because molecular markers are only *indirect* indicators whether or not a cell is a true neuron. *Direct* evidence requires the application of traditional anatomical, physiological and behavioral techniques. By definition, a cell is not a bona fide neuron unless it has dendrites and an axon; is capable of producing generator potentials and action potentials; has synapses that allow information transmission from sense organs to neurons, from one neuron to the next, and from motor neurons to muscles or other effectors; and serves as a link in the coordination of behavior. Since differentiating, migrating and maturing young neurons may lack some of these features and properties, the cell identified with a neuronal marker may be a potential neuron, a neuron in the process of differentiation, or a cell that has some neuronal features but may never become a true functional neuron. Let me illustrate the pitfalls of reliance on markers in the categorization of developing CNS cells with reference to one of the earliest and most commonly used molecular marker in neuroscience, i.e., glial fibrillary acidic protein (Bignami and Dahl 1974). This protein (GFAP) is widely claimed to be a definitive marker for astrocytes. However, it has been known for some time that GFAP is present in germinal matrices where precursor cells generate both neurons and neuroglia (Levitt et al. 1981), and the possibility has been raised that a marker that specifically reacts in the mature brain with glial filaments may also react with transient filamentous elements in progenitor cells destined to become neurons (Bennett 1987). For example, the precursors of granule cells in the hippocampal subgranular zone are GFAP-positive (Seri et al. 2001; Namba et al. 2005). The assumption that GFAP is a definitive marker of astrocytes led some investigators to conclude that the hippocampal granule cells are progeny of astrocytes (e.g., Seri et al. 2001). However, a more reasonable conclusion would be that CNS progenitor cells, irrespective whether they will produce neurons or neuroglia, share the property of expressing GFAP (e.g., Steiner et al. 2004). That would invalidate the currently popular theory that "radial glia" are progenitors of neurons (e.g., Anthony et al. 2004; Englund et al. 2005). The currently available evidence supports the view that both neurons and neuroglia derive from NEP and SGM precursor cells that share many molecular properties.

It will require the work of a professional science historian to reconstruct how attitudes have changed over the period when the study of postnatal and adult neurogenesis was not considered a topic deserving public support and the time when it became a popular subject matter studied by hundreds of investigators. Typing the term "hippocampal neurogenesis" into the search engine of Google Scholar (the date is 01-15-08) the query retrieves 5,150 references; "subgranular zone" retrieves 2,750 references; and "rostral migratory stream" retrieves 2,330 references. Having identified these phenomena (and coined the terms) in the decade between 1965 and 1975, we feel gratified that by 2008 they have captured the attention of so many investigators. (Never mind, that so many of the articles consulted do not credit us for these discoveries but attribute them to others. Why get involved in old controversy and risk unfavorable treatment by a reviewer in the opposing camp? Quoting an article recently published in mainstream journals is a safer bet. After all that passed the Imprimatur of a semi-official Establishment journal.) We also feel gratified that, judging by the increasing number of references to them, some of our later empirical and conceptual contributions are also becoming accepted. For instance, we have advocated for decades the replacement of the narrowly conceived term "ventricular zone" with the broader term "neuroepithelium." We did that to stress the fact that the NEP is not just a component of the maturing brain but the source of all its neural elements, one that forms and expands greatly before the brain has any differentiated neurons. Likewise, we have advocated the replacement of the term "external granular layer" of the cerebellum with "external germinal layer" to stress two facts. First, that this subpial tissue is not composed of "granule" cells, which is a universally accepted term for such differentiated neurons as the granule cells of the cerebellum, hippocampus and olfactory bulb. Second, the term "germinal" reminds us that this transient proliferative matrix is an important component of a late-generated neurogenic system of the developing cerebellum.

Much of the foregoing is about the past. As we are entering a new period in neuroscience, what Nottebohm called the New Neurology, it is more rewarding to look forward and contemplate what the future may bring. The substantive chapters that follow this personal memoir, will acquaint the reader with some of the recent advances in the study of adult neurogenesis. Adult neurogenesis has been identified by now in all vertebrates, fishes, reptiles, birds, and mammals. Important investigations are being carried out in various laboratories about the physiological properties of the postnatally acquired hippocampal and olfactory bulb granule cells, their migration and integration into the circuitry of these structures, and their role in how animals adapt to their environment and learn. Also of great promise are the exploratory studies concerned with the possible role of adult neurogenesis in stress disorders and depression, and such brain abnormalities and pathologies as schizophrenia, Parkinsonism, stroke and ischemia. There is no way to predict what these ongoing studies will bring to light but the full examination of the dynamic properties and remedial potentials of the adult brain is certainly a well-justified scientific enterprise.

Acknowledgment I thank Dr. Tatsunori Seki for the invitation to present this personal memoir, and dedicate this contribution to Dr. Shirley Bayer who shares credit for all that we have accomplished together in the last 3–4 decades.

References

Altman J (1962a) Autoradiographic study of degenerative and regenerative proliferation of neuroglia cells with tritiated thymidine. Exp Neurol 5:302–318

Altman J (1962b) Are new neurons formed in the brains of adult mammals? Science 135:1127–1128

Altman J (1963a) Regional utilization of leucine-H^3 by normal rat brain: Microdensitometric evaluation of autoradiograms. J Histochem Cytochem 11:741–750

Altman J (1963b) Differences in the utilization of tritiated leucine by single neurones in normal and exercised rats: An autoradiographic investigation with microdensitometry. Nature 199:777–780

Altman J (1963c) Autoradiographic investigation of cell proliferation in the brains of rats and cats. Anat Rec 145:573–591

Altman J (1966a) Autoradiographic and histological studies of postnatal neurogenesis. II. A longitudinal investigation of the kinetics, migration and transformation of cells incorporating

tritiated thymidine in infant rats, with special reference to postnatal neurogenesis in some brain regions. J Comp Neurol 128:431–474
Altman J (1966b) Organic Foundations of Animal Behavior. New York: Holt, Rinehart and Winston
Altman J (1966c) Proliferation and migration of undifferentiated precursor cell in the rat during postnatal gliogenesis. Exp Neurol 16:263–278
Altman J (1967) Postnatal growth and differentiation of the mammalian brain, with implications for a morphological theory of memory. In, Quarton G, Melnechuk T, Schmitt FO (eds) The Neurosciences: A Study Program, pp. 723–743. New York: Rockefeller Univ Press
Altman J (1969a) Autoradiographic and histological studies of postnatal neurogenesis. IV. Cell proliferation in the anterior forebrain, with special reference to persisting neurogenesis in the olfactory bulb. J Comp Neurol 137:433–458
Altman J (1969b) DNA metabolism and cell proliferation. In, Lajtha A (ed) Handbook of Neurochemistry, vol. 2, pp. 137–182. New York: Plenum Press
Altman J (1970) Postnatal neurogenesis and the problem of neural plasticity. In, Himwich WA (ed) Developmental Neurobiology, pp. 197–237. Springfield IL: Thomas
Altman J (1986) An animal model of minimal brain dysfunction. In, Lewis M (ed) Learning Disabilities and Prenatal Risk, pp. 241–304. Urbana: Univ Illinois Press
Altman J (1987) Morphological and behavioral markers of environmentally induced retardation of brain development: An animal model. Environ Health Perspect 74:153–168
Altman J, Altman, E (1962) Increased utilization of an amino acid and cellular proliferation demonstrated autoradiographically in the optic pathways of pigeons. Exp Neurol 6:142–151
Altman J, Anderson WJ (1971) Irradiation of the cerebellum in infant rats with low-level X-ray: Histological and cytological effects during infancy and adulthood. Exp Neurol 30:492–509
Altman J, Anderson WJ (1972) Experimental reorganization of the cerebellar cortex. I. Morphological effects of elimination of all microneurons with prolonged X-irradiation started at birth. J Comp Neurol 146:355–406
Altman J, Bayer SA (1975) Postnatal development of the hippocampal dentate gyrus under normal and experimental conditions. In, Isaacson RL, Pribram KH (eds) The Hippocampus, vol. 1, pp. 95–122. New York: Plenum Press
Altman J, Bayer SA (1982) Morphological development of the rat cerebellum and some of its mechanisms. Exp Brain Res Suppl 6:8–46
Altman J, Bayer SA (1990a) Mosaic organization of the hippocampal neuroepithelium and the multiple germinal sources of dentate granule cells. J Comp Neurol 301:325–342
Altman J, Bayer SA (1990b) Prolonged sojourn of developing pyramidal cells in the intermediate zone of the hippocampus and their settling in the stratum pyramidale. J Comp Neurol 301:343–364
Altman J, Bayer SA (1990c) Migration and distribution of two populations of hippocampal granule cell precursors during the perinatal and postnatal periods. J Comp Neurol 301:365–381
Altman J, Bayer SA (1993) Are new neurons formed in the brains of adult mammals? A progress report, 1962–1992. In, Cuello AC (ed) Neuronal Cell Death and Repair, pp. 203–225. Amsterdam: Elsevier (Restorative Neurology, vol. 6)
Altman J, Bayer SA (1995) Atlas of Prenatal Rat Brain Development. Boca Raton, FL: CRC Press
Altman J, Bayer SA (1997) Development of the Cerebellar System in Relation to its Evolution, Structure and Functions. Boca Raton, FL: CRC Press
Altman J, Bayer SA (2001) Development of the Human Spinal Cord: An Interpretation Based on Experimental Studies in Animals. New York: Oxford Univ Press
Altman J, Bayer SA (2002) Regional differences in the stratified transitional field and the honeycomb matrix of the developing human cerebral cortex. J Neurocytol 31:613–632
Altman J, Bayer SA (2007) Concluding essay. In, Bayer SA, Altman J (eds) The Human Brain During the Early First Trimester, pp. 426–489. Boca Raton, FL: CRC Press (Atlas of Human Central Nervous System Development, vol. 5)
Altman J, Das GD (1964a) Autoradiographic and histological investigation of changes in the visual system of rats after unilateral enucleation. Anat Rec 148:535–545

Altman J, Das GD (1964b) Autoradiographic examination of the effects of enriched environment on the rate of glial multiplication in the adult rat brain. Nature 204:1161–1163
Altman J, Das GD (1965a) Autoradiographic and histological evidence of postnatal hippocampal neurogenesis in rats. J Comp Neurol 124:319–336
Altman J, Das GD (1965b) Postnatal origin of microneurones in the rat brain. Nature 207:953–956
Altman J, Das GD (1966) Autoradiographic and histological studies of postnatal neurogenesis. J Comp Neurol 126:337–390
Altman J, Das GD (1967) Postnatal neurogenesis in the guinea pig. Nature 214:1098–1101
Altman J, Das GD, Anderson WJ (1968) Effects of infantile handling on morphological development of the rat brain: An exploratory study. Dev Psychobiol 1:10–20
Altman J, Anderson WJ, Wright KA (1969a) Early effects of X-irradiation of the cerebellum in infant rats: Decimation and reconstitution of the external granular layer. Exp Neurol 24:196–216
Altman J, Anderson WJ, Wright KA (1969b) Reconstitution of the external granular layer of the cerebellar cortex in infant rats after low-level X-irradiation. Anat Rec 163:453–472
Altman J, Anderson WJ, M Strop (1971) Retardation of cerebellar and motor development by focal X-irradiation during infancy. Physiol Behav 7:143–150
Alvarez-Buylla A, Nottebohm F (1988) Migration of young neurons in adult avian brain. Nature 335:353–354
Angevine JB (1965) Time of neuron origin in the hippocampal region: An autoradiographic study in the mouse. Exp Neurol Suppl 2:1–70
Anthony TE, Klein C, Fishell G, Heintz N (2004) Radial glia serve as neuronal progenitors in all regions of the central nervous system. Neuron 41:881–890
Bayer SA (1982) Changes in the total number of dentate granule cells in juvenile and adult rats: A correlated volumetric and ^{3}H-thymidine autoradiographic study. Exp Brain Res 46:315–323
Bayer SA, Altman J (1974) Hippocampal development in the rat: Cytogenesis and morphogenesis examined with autoradiography and low-level X-irradiation. J Comp Neurol 158: 55–80
Bayer SA, Altman J (1975) Radiation-induced interference with postnatal hippocampal cytogenesis in rats and its long-term effects on the acquisition of neurons and glia. J Comp Neurol 163:1–20
Bayer SA, Altman J (1991) Neocortical Development. New York: Raven Press
Bayer SA, Altman J (2002–2007) Atlas of Human Central Nervous System Development (5 volumes). Boca Raton, FL: CRC Press (Taylor & Francis Group)
Bayer SA, Altman J (2004) Development of the telencephalon: Neural stem cells, neurogenesis, and neuronal migration. In, Paxinos G (ed) The Rat Nervous System (Third ed), pp. 27–73. New York: Elsevier
Bayer SA, Brunner RL, Hine R, Altman J (1973) Behavioural effects of interference with the postnatal acquisition of hippocampal granule cells. Nature New Biol 242:222–224
Bayer SA, Yackel JW, Puri PS (1982) Neurons in the rat dentate gyrus granular layer substantially increase during juvenile and adult life. Science 216:890–892
Bennett G (1987) Changes in intermediate filament composition during neurogenesis. Curr Topics Dev Biol 21:151–182
Bignami A, Dahl D (1974) Astrocyte-specific protein and neuroglial differentiation: An immunofluorescence study with antibodies to the glial fibrillary acidic protein. J Comp Neurol 153:27–38
Brunner RL, Altman J (1973) Locomotor deficits in adult rats with moderate to massive retardation of cerebellar development during infancy. Behav Biol 9:169–188
Brunner RL, Hagbloom SJ, Gazarra RA (1974) Effects of hippocampal X-irradiation-produced granule-cell agenesis on instrumental runway performance in rats. Physiol Behav 13:485–494

Diaz-Granados JL, Greene PL, Amsel A (1994) Selective activity enhancement and persistence in weanling rats after hippocampal x-irradiation in infancy. Behav Neural Biol 61:251–259
Englund C, Fink A, Lau C et al (2005) Pax6, Tbr 2, and TBR1 are expressed sequentially by radial glia, intermediate progenitor cells, and postmitotic neurons in developing neocortex. J Neurosci 25:247–251
Eriksson PS, Perfilieva E, Bjork-Eriksson T et al (1998) Neurogenesis in the adult human hippocampus. Nat Med 4:1313–1317
Garfield E (1981) The 1,000 contemporary scientists most-cited 1965–1978. Part I. The basic list and introduction. Curr Contents 41:5–14
Gazzara RA, Altman J (1981) Early postnatal X-irradiation of the hippocampus and discrimination learning in adult rats. J Comp Physiol Psychol 95:484–495
Goldman SA, Nottebohm F (1983) Neuronal production, migration and differentiation in a vocal control nucleus of the adult female canary brain. Proc Natl Acad Sci USA 80:2390–2394
Gould E, Reeves AJ, Graziano MSA et al (1999) Neurogenesis in the neocortex of adult primates. Science 286:548–552
Gross CG (2000) Neurogenesis in the adult brain: Death of a dogma. Nat Rev Neurosci 1:67–73
Haggbloom, SJ, Brunner RL, Bayer SA (1974) Effects of hippocampal granule-cell agenesis on acquisition of escape from fear and one-way active-avoidance responses. J Comp Physiol Psychol 95:86:447–457
Highfield DA, Hu D, Amsel A (1998) Alleviation of x-irradiation-based deficit in memory-based learning by D-amphetamine: Suggestions for attention deficit-hyperactivity disorder. Proc Natl Acad Sci USA 95:5785–5788
Jacobson M (1970) Developmental Neurobiology. New York: Holt, Rinehart and Winston
Jacobson M (1978) Developmental Neurobiology. Second ed. New York: Plenum Press
Jessell TM (1991) Reactions of neurons to injury. In, Kandel E et al (eds) Principles of Neural Science (Third ed), pp. 258–269. New York: Elsevier
Kaplan MS (2001) Environment complexity stimulates visual cortex neurogenesis: Death of a dogma and a research career. Trends Neurosci 24:617–620
Kaplan MS, Hinds JW (1977) Neurogenesis in the adult rat: Electron microscopic analysis of light radioautographs. Science 197:1092–1094
Kempermann G (2006) Adult Neurogenesis: Stem Cells and Neuronal Development in the Adult Brain. New York: Oxford Univ Press
Kolb B, Pedersen B, Ballermann M et al (1999) Embryonic and postnatal injections of bromodeoxyuridine produce age-dependent morphological and behavioral abnormalities. J Neurosci 19:2337–2346
Kornack DR, Rakic P (1999) Continuation of neurogenesis in the hippocampus of the adult macaque monkey. Proc Natl Acad Sci USA 96:5768–5773
Lederberg J (1968) Stretching out the period of the brain's development. Washington Post, October 26, p. A15
Levitt P, Cooper ML, Rakic P (1981) Coexistence of neuronal and glial precursor cells in the cerebral ventricular zone of the fetal monkey: An ultrastructural immunoperoxidase analysis. J Neurosci 1:27–39
Lund RD (1978) Development and Plasticity of the Brain: An Introduction. New York: Oxford Univ Press
Namba T, Mochizuki H, Onodera M et al (2005) The fate of neural progenitor cells expressing astrocytic and radial glial markers in the postnatal rat dentate gyrus. Eur J Neurosci 22:1928–1941
Nottebohm F (2002) Neuronal replacement in adult brain. Brain Res Bull 57:737–749
Patterson PH, Purves D (1982) Readings in Developmental Neurobiology. Cold Spring Harbor, NY: Cold Spring Harbor Laboratory
Pellegrino LJ, Altman J (1979) Effects of differential interference with postnatal cerebellar neurogenesis on motor performance, activity level, and maze learning of rats. J Comp Physiol Psychol 93:1–33

Purves D, Lichtman JW (1985) Principles of Neural Development. Sunderland, MA: Sinauer Associates
Rakic P (1985) Limits of neurogenesis in primates. Science 227:1054–1056
Rosselli-Austin L, Altman J (1979) The postnatal development of the main olfactory bulb of the rat. J Dev Physiol 1:295–313
Schacher S (1981) Determination and differentiation in the development of the nervous system. In, Kandel ER, Schwartz JH (eds) Principles of Neural Science, pp. 503–520. New York: Elsevier
Sekerkova G, Ilijic E, Mugnaini E (2004) Bromodeoxyuridine administered during neurogenesis of the projection neurons causes cerebellar defects in the rats. J Comp Neurol 470:221–239
Seri B, García-Verdugo JM, McEwen BS et al (2001) Astrocytes give rise to new neurons in the adult mammalian hippocampus. J Neurosci 21:7153–7160
Smart I (1961) The subependymal layer of the mouse brain and its cell production as shown by radioautography after thymidine-H^3 injection. J Comp Neurol 116:325–347
Smart I, Leblond CP (1961) Evidence for division and transformations of neuroglia cells in the mouse brain, as derived from radioautography after injection of thymidine-H^3. J Comp Neurol 116:349–367
Steiner B, Kronenberg G, Jessberger S et al (2004) Differential regulation of gliogenesis in the context of adult hippocampal neurogenesis in mice. Glia 46:41–52
Taupin P (2006) BrdU immunohistochemistry of studying adult neurogenesis: Paradigms, pitfalls, limitations, and validation. Brain Res Rev 53:198–214

Chapter 2
Song Learning in Birds Offers a Model for Neuronal Replacement in Adult Brain

Fernando Nottebohm

I dedicate this chapter to a true pioneer, Joseph Altman. He was first to publish evidence that some kinds of neurons of the mammalian brain are produced post-natally and even in adulthood.

The discovery of neurogenesis in adult canaries came as a surprise because it was found in a context where it had not been contemplated, the study of vocal learning. To everybody's disbelief, the new, spontaneously produced neurons replaced numerically others that had died, a process of spontaneous brain self-repair or rejuvenation. I will describe how these discoveries came about and how they have helped us understand the natural history of neurogenesis and neuronal replacement in adult brain. Adult neurogenesis may also shed light on a basic issue of brain function: what limits learning?

2.1 The Study of Vocal Learning

The study of vocal learning in birds was revolutionized in the early 1950s by the use of an instrument, the sound spectrograph, that converted sounds recorded on tape into a two dimensional display, the sound spectrogram (Fig. 2.1), that read like a musical notation with time on the horizontal axis and frequency on the vertical one (Thorpe, 1954). This technique enabled the collection and comparison of sounds gathered during the entire vocal development of individual birds and the side-by-side comparison of the sounds produced by different individuals. The sound-spectrograph enabled the classical work of Thorpe (1958), Marler and Tamura (1964), Marler (1970a) and Konishi (1965), which showed that vocal learning in oscine songbirds: (1) occurred during a sensitive period, usually before sexual maturity; (2) required intact hearing; (3) was preceded by the memorization of an external model, that then guided changes in vocal output until that model was matched. Both Thorpe (1955; Thorpe and Pilcher, 1958) and Marler (1970b)

F. Nottebohm (✉)
The Rockefeller University, 1230 York Avenue, New York, NY 10065, USA
e-mail: nottebo@rockefeller.edu

T. Seki et al. (eds.), *Neurogenesis in the Adult Brain I: Neurobiology*,
DOI 10.1007/978-4-431-53933-9_2, © Springer 2011

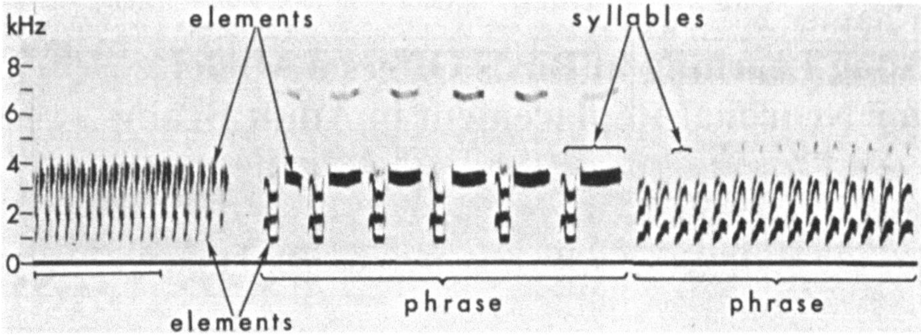

Fig. 2.1 (Figure 1 from Nottebohm and Nottebohm, 1978). A sound spectrogram of canary song. Canaries sing by repeating many times a same "syllable", then switching to the next syllable, that is also repeated many times, and in this manner eventually producing the full repertoire of syllables that characterizes the song of that individual. Each syllable is composed of 1–3 elements, each followed by a brief silent gap. Each sound is produced at is "fundamental" frequency as well as at higher "harmonics". In this visual display of sound, frequency in Kilohertz is represented in the vertical axis and time, in seconds, in the horizontal axis. The *horizontal bar* on the *lower left* represents a 0.5 s duration

emphasized that song imitation was preceded by a stage of vocal experimentation, called "subsong", that was reminiscent of infant babbling. These new tools and observations suggested that songbirds would provide a good animal model for studying vocal learning in general, and this drew a lot of interest.

There are two general issues about vocal learning: how it occurs in the individual, including its ontogeny, and how it evolved. Both are related, but we do not know how. Because of this uncertainty, the origins of vocal learning have always been difficult to imagine. It is widely accepted that in the process of evolution new traits emerge from variability in pre-existing ones. Unlearned vocal repertoires, as those found in domestic fowl, *Gallus domesticus* (Konishi, 1963) and ring-doves, *Streptopelia risoria* (Nottebohm and Nottebohm, 1971), develop normally even when individuals are deafened a few days after hatching. How does one go from this situation, thought to be the primitive one, to one in which vocal ontogeny becomes more dependent on hearing – and eventually on imitation – and how does one do this transition so that all intervening evolutionary stages serve animal communication? That was a paradox that, early on, caught my attention (Nottebohm, 1972a).

2.2 Two Sound Sources and Left Hemispheric Dominance

The appeal of vocal learning in birds as a model system became even stronger when work with chaffinches, *Fringilla coelebs*, (the subjects of Thorpe's 1958 article) showed functional asymmetry in their vocal organ, the syrinx. The chaffinch syrinx consists of right and left anatomically symmetrical and functionally independent halves, each with its own supply of airflow, muscular control and innervation (Fig. 2.2).

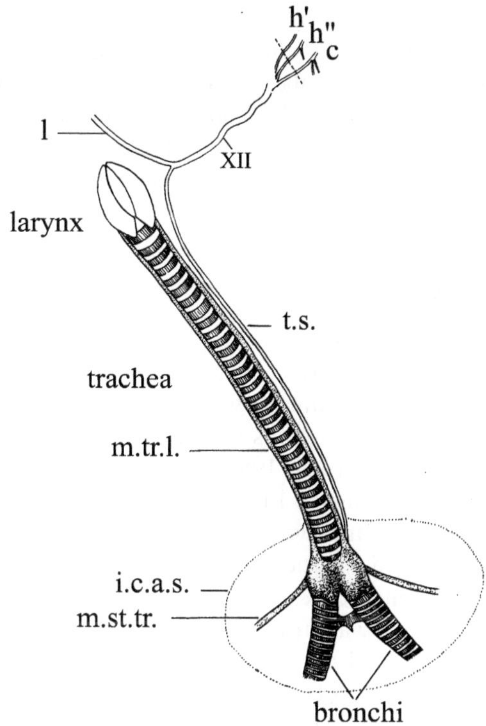

Fig. 2.2 (Figure 3 from Nottebohm, 1971b). Schematic drawing of the vocal tract of songbirds. The syrinx, vocal organ of birds, is at the confluence of the two bronchi and the trachea. It is surrounded by an air space, the interclavicular air sac (icas). The syrinx has several pairs of intrinsic muscles, represented here just as a stippled muscle mass; the sternotrachealis muscle (m.st.tr.) anchors the syrinx to the sternum (not shown). The tracheolateralis muscle (m.tr.l) runs along the full length of the tracheal and its contraction or relaxation allow the trachea to conform to the neck's posture; the rostral pull of m.tr.l is countered by the caudal pull of the m.st.tr. The larynx is open during phonation, but otherwise does not generate sound. The muscles of trachea and syrinx are innervated by the tracheosyringeal nerve (t.s.), a branch of the hypoglossus nerve (XII). Section of the ts. nerve on one side, or section of its roots (h', h", c), blocks vocal control on the ipsilateral side of the syrinx

Yet I found that most of the sounds of chaffinch song were produced by the left syringeal half, which therefore was dominant for song. This situation was uncannily reminiscent of handedness in humans. Moreover, this dominance could be reversed if the left syringeal half was denervated early on. This suggested that peripherally both sides were equipotential but that a central bias – like left cerebral dominance for speech and handedness in humans – determined which syringeal half normally did most of the singing (Nottebohm, 1971b, 1972b; DeVoogd et al., 1991). We now had three human traits – vocal learning, left hemispheric dominance and handedness – for which the basic biology could be investigated in songbirds.

The next step was to extend the findings of left syringeal dominance in chaffinches to another songbird, such as the canary, *Serinus canaria*, that could be bred in captivity. I was eager then to go into the brain to see how the traits that interested me – vocal learning, hemispheric dominance – were represented there and how they developed. A stereotaxic atlas of the canary brain was produced (Stokes et al., 1974), modeled after that of pigeons, *Columba livia*, and using the same terminology (Karten and Hodos, 1967). Once this work was done we set out to uncover the brain pathways that control the performance of the syrinx.

2.3 The Song System

The results were gratifying. Over the next 10 or so years a system of anatomically discreet nuclei and connecting pathways was identified and shown to control vocal output and vocal learning in songbirds (Nottebohm et al., 1976, 1982; Okuhata and Saito, 1987; Bottjer et al., 1984, 1989). I called this array of nuclei and pathways "the song system". It was the first time that we had, for any vertebrate, an array of nuclei and pathways that was associated, apparently exclusively, with the acquisition and production of a particular learned behavior. This came as a surprise because, under the influence of Karl Lashley (1950), the neural substrate for learned skills in vertebrates was still thought to be widely distributed throughout the brain. Our findings placed the song system conceptually closer to the functional centers conceived by von Holst (1935) and von Holst and St. Paul (1960) and to the simple networks of invertebrates that were proving so helpful in studying simple patterns of behavior (e.g. Huber, 1960; Bullock, 1961; Roeder, 1962; Wiersma, 1962; Wilson, 1961) and that eventually, in the sea slug *Aplysia*, became so important for understanding simple learning paradigms (Kandel, 2006).

The song system can be thought of as consisting of four modules, shown schematically in Fig. 2.3 and in greater detail in Fig. 2.4. The full name of the forebrain nuclei that are part of the song system appears in the legend for Fig. 2.4. ***Module #1*** is composed of brainstem nuclei that control the muscles for respiration and phonation. ***Module #2*** is a descending motor pathway in caudal forebrain that tells module #1 what to do (Nottebohm et al., 1976; Yu and Margoliash, 1996; Hahnloser et al., 2002). ***Module #3*** is a forebrain module that loops through basal ganglia (Area X) and thalamus (DLM) and then returns to anterior forebrain (LMAN), from where it projects to nucleus RA of module #2; module #3 is necessary for song learning but not for the production of learned song (Bottjer et al., 1984; Scharff and Nottebohm, 1991; Brainard and Doupe, 2000; Kao et al., 2005; Oelveczky et al., 2005; Andalman and Fee, 2009; but see Aronov et al., 2008). ***Module #4*** consists of the ascending auditory pathway and its relays to modules #2 and #3 (Vates et al., 1996). Modules 2 and 3 have their origin in a same nucleus, the High Vocal Center (HVC), which also receives auditory input (Cardin and Schmidt, 2004; Dave et al., 1998; Janata and Margoliash, 1999; Katz and Gurney, 1981; Margoliash, 1986; Vates et al., 1996; Williams and Nottebohm, 1985). The blend of auditory and motor capabilities, the absence of muscle representation and its high position in the descending vocal pathways, make HVC reminiscent of Broca's area in humans and

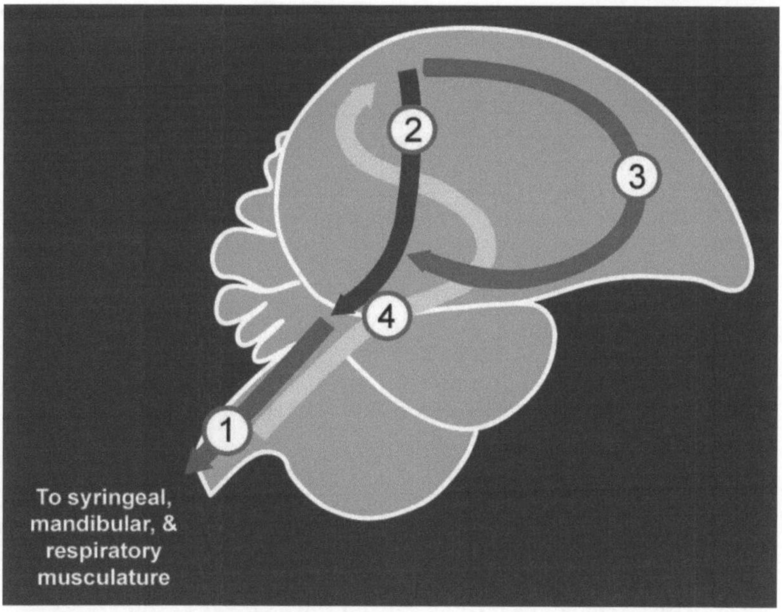

Fig. 2.3 (Figure 4 from Nottebohm, 1993). Schematic diagram of a songbird's brain (rostral to right, caudal to left). The four numbered *arrows* stand for the four song system "modules" described in the text

justify its name as the high vocal center. This name is justified, too, by elegant neurophysiological work that strongly suggests that HVC, more than any other nucleus of the song system, plays a key role in the patterning of learned song (Yu and Margoliash, 1996; Hahnloser et al., 2002; Long and Fee, 2008). Moreover, recent evidence shows synaptic changes occur in this nucleus as new song is learned (Roberts et al., 2010), supporting an earlier correlation between the complexity of learned song and dendritic spine density in HVC (Airey et al., 2000). In addition to the four modules shown, there are feedback loops between different parts of the song system (Vates and Nottebohm, 1995; Vates et al., 1997).

2.4 Sexual Dimorphism

Song is a seasonal behavior and, perhaps not surprisingly, the song system of songbirds has androgen and estrogen receptors that make it very sensitive to gonadal hormones (Arnold et al., 1976). This might have prepared us for the next observation, which, however, came as a surprise. Once careful anatomical details were on hand from sufficient individuals, we realized that the song system of canaries was sexually dimorphic: Area X (homologous to the striatum of mammalian basal ganglia, see Fig. 2.3) was 3.8 times larger in male than in female canaries, while

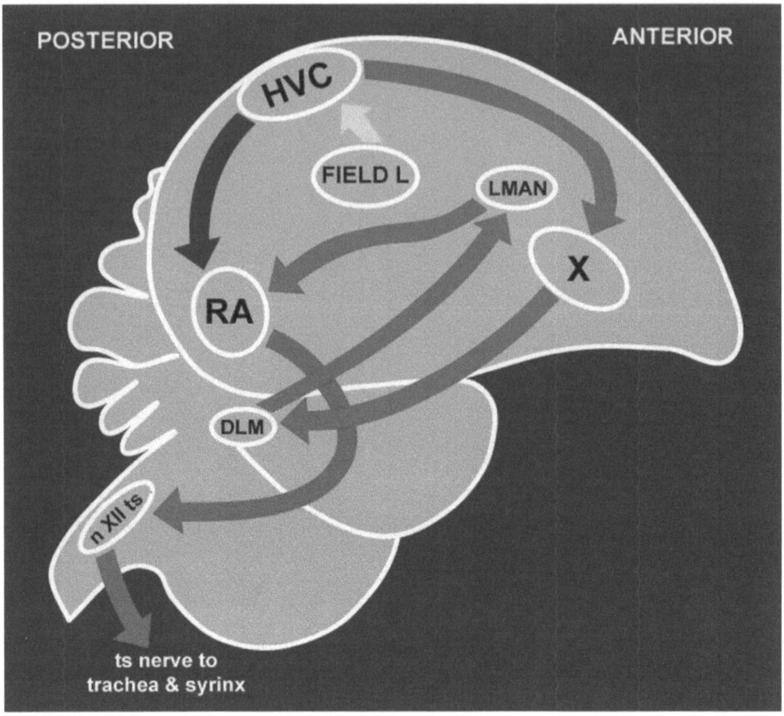

Fig. 2.4 More detailed, but still schematic, representation of the song system's pathways. The *shading* of the various *arrows* conforms to that used in Fig. 2.3 and shows which pathways are part of the different modules. The ascending auditory pathway (#4) is here represented as a *short arrow* that touches HVC. All pathways shown here are ipsilateral

the male / female ratios for HVC and RA were, respectively 3.2 and 2.7. We thought this dimorphism might reflect the fact that whereas male canaries sing relatively large and complex learned song repertoires and the sounds produced are very stereotyped (Waser and Marler, 1977; Marler and Waser, 1977; Nottebohm et al., 1986), female canaries sing much less and when they sing, their song is simpler and the sounds more variable (Nottebohm, 1980b). In zebra finches, *Taeniopygia guttata*, in which females never sing, the male/female differences in size of song nuclei were twice as large as in canaries (Nottebohm and Arnold, 1976).

These observations raised the obvious question. Did the male and female song nuclei have different developmental histories or did the differences observed in adulthood reflect current levels of circulating hormones. There was a precedent for the latter explanation in the rodent preoptic area (Raisman and Field, 1971). Our work and that of others indicates that both mechanisms play a role in the sexual dimorphism of the song system. The size of song nuclei, including HVC, could be made to double by systemic administration of testosterone to adult female canaries over a period of 3–4 weeks. During this time the females started to sing more and more and the sounds produced became more stereotyped and like those of adult males. However the number of syllable types in their song remained small and their song nuclei were still significantly smaller than those of adult males (Nottebohm, 1980a). Closer inspection revealed that in nucleus RA (Fig. 2.3) the hormone-driven growth in female volume was associated with the growth of dendrites and formation of new synapses (DeVoogd and Nottebohm, 1981; DeVoogd et al., 1985; Canady et al., 1988).

In addition to the role of hormones in adulthood, the sexual dimorphism observed in the song system of temperate area songbirds owes much to differences that emerge during development. The song nuclei of male and female zebra finches – the species now most commonly used in studies of vocal learning in birds – are very similar until day 30 after hatching. After then, many of the cells in the female nuclei atrophy and die and the female song control nuclei shrivel; this effect is prevented by treating the female zebra finches with estrogen before day 30; when these females are given testosterone as adults they sing like males (Gurney and Konishi, 1980; Gurney, 1981; Simpson and Vicario, 1991). In addition to the role of gonadal hormones during ontogeny and in adulthood, other genetic differences between males and females contribute also to the sexual dimorphism of the song system (Wade and Arnold, 1996; Agate et al., 2003, 2004), but that is a story beyond the scope of the present review. What matters, for the present story, is that by 1980 we knew that the anatomy of the adult song system of canaries was very sensitive to testosterone.

2.5 The Late Ontogeny of the Song System

Even as studies of the anatomy and hormonal sensitivity of the song system raced ahead, we also focused on the behavior of song learning – its timing and manner of occurrence. Male canaries in our breeding colonies followed the natural photoperiod of New York State. They started breeding in late March and stopped in mid July. Starting at approximately 40 days after hatching, juveniles produced variable, low amplitude patterns of frequency modulation called "subsong", reminiscent of the babbling of infants. These sounds occurred in a non-communicatory context, often as the young bird with feathers fluffed and eyes closed seemed to doze. By 3 months of age these sounds were louder and by 4 most of the adult sounds were already present, though still delivered in a variable manner as "plastic song". Stereotyped adult song was first used in early spring, when the birds were in full breeding condition. In all these respects, canaries were not different from many other songbirds, such as the chaffinch (Thorpe, 1958), that acquire their learned song during their first year of life. Our canaries differed, though, in one important respect, they continued to change their song in successive years and for this reason we called them "open-ended learners" (Nottebohm and Nottebohm, 1978; Nottebohm et al., 1986, 1987).

It was of great interest to me that telencephalic nuclei such as HVC and RA grew dramatically during the juvenile period of song acquisition. By day 15 after hatching the brain of a young male canary had already achieved full adult weight, yet the volume of HVC 30 days after hatching was 20% that of an adult 1-year-old male. RA at that time had 30% of its adult volume. Both RA and HVC showed a very marked spurt of growth from the thirtieth to the 60th day after hatching. At 60 days the volume of HVC was close to 50% that of an adult 1-year-old male (Nottebohm, 1980b, 1989). If these changes in volume accompanied the learning of a new song, what happened in successive years? Would new episode's of song learning be accompanied by further anatomical changes in HVC and RA?

2.6 Brain Seasonal Changes

The blood testosterone levels of our adult male canaries dropped in mid-summer and soon thereafter the birds started molting their feathers. During the molt, male canaries sang very little if at all and the song of those that sang tended to be variable and reminiscent of that of juveniles (Nottebohm et al., 1987). The size of HVC and RA of males in breeding condition killed in April, when 12 months old, was 1.99 and 1.77 times larger, respectively, than that of other males of the same cohort killed in mid-September, after the end of the molt. Since another study had shown that the size of these two nuclei did not differ significantly between 1-, 2- and 3-year-old males (Nottebohm et al., 1981), I inferred that the reduction in volume observed in late summer was temporary. The results of this seasonal comparison were published in *Science* under the title of "A brain for all seasons" (Nottebohm, 1981). The late summer reduction in volume seen in the song nuclei of the males seemed like the inverse of the doubling in volume seen in the testosterone-treated females. In addition, I indicated that there were also whole brain changes, such that the weight of brains collected in September was 15% lighter than that of brains collected in April. What to make of this?

Towards the end of the 1981 *Science* article I wrote: "I hypothesize that the acquisition of a new motor coordination or of a new auditory-motor integration is made possible or facilitated by the growth of new dendritic segments and the consequent opportunity to form new synapses. The plasticity offered by such a scheme is potentially twofold: to allow for the formation of new interneuronal relations and to bring into existence synapses that have not yet been altered by previous patterns of use. Seasonal changes in the volume of HVC and RA may reflect the amount of plastic substrate that can be exploited for such learning purposes. According to this hypothesis the plastic substrate for vocal learning is renewed once yearly, a growing, then a shedding of synapses, much the way trees grow leaves in the spring and shed them in the fall." This interpretation was in line with the Zeitgeist in the neurosciences at that time because it assumed that the plasticity required for learning was vested *solely* in the synapses. It was in line, too, with our observation that in females testosterone induced dendritic growth and the addition of new synapses.

2.7 Space for Learning and Memory

The seasonal ebb and flow of the brain's space for song fitted well with another observation. Extensive song recordings from 25 adult male canaries of the Belgian Wasserschlager strain were scored for the number of different "song syllables" produced by each bird – "syllables" being the units of repetition produced in canary song (Fig. 2.1). The birds were killed after their song was recorded and the volume of their HVC estimated from serial histological sections. An interesting correlation emerged (Fig. 2.5): a large HVC was a poor predictor of how many different syllables a canary might sing, but birds with a large syllable count tended to have a large HVC. Conversely, if HVC was relatively small, the associated syllable count tended to be relatively small (Nottebohm et al., 1981). This relation between the complexity of learned song and the amount of brain space for learned song was subsequently corroborated in another songbird (Canady et al., 1984) and in studies comparing many species (DeVoogd et al., 1993).

These space/learning relations suggested that brain space for a particular learned skill might be in short supply. However, if birds such as the canary were able to learn new song repertoires every year, then perhaps there was always more space available. The seasonal changes in volume observed in HVC and RA offered a

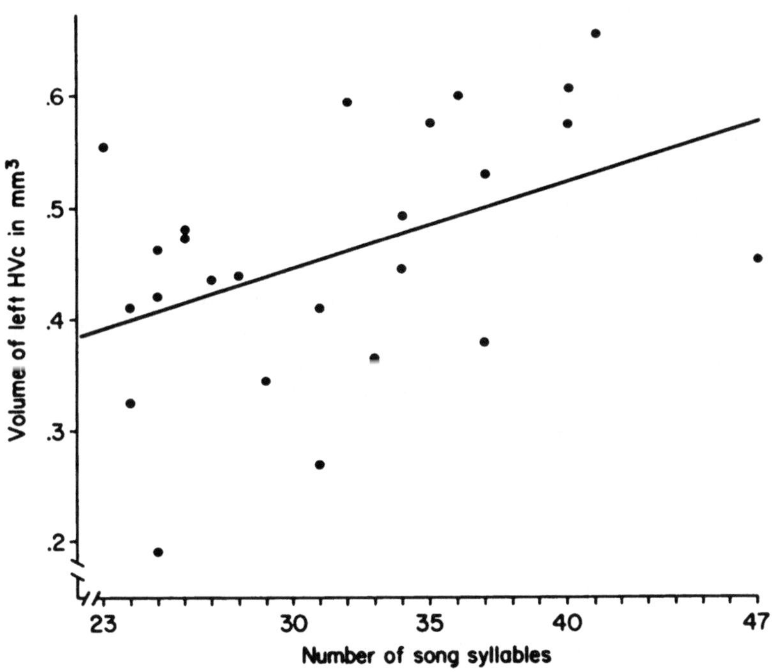

Fig. 2.5 (Figure 1 from Nottebohm et al., 1981). Regression of size of syllable repertoire on the volume of left HVC for 25 adult male canaries

solution to this conundrum: perhaps as songs were discarded, so was the substrate that held them; then, as a new substrate was generated, new songs could be learned once again. This interpretation suggested that a better understanding of the "space" eliminated would provide a clue to how and where the memories of learned song were stored.

2.8 Always the Same Neurons?

Our thinking so far had gone along with the commonly accepted view that learning could be explained by changes in synaptic number and synaptic efficacy. This idea had been advanced initially by two Italians, Lugaro and Tanzi, whose thoughts were quoted and further developed by Santiago Ramon y Cajal in his Croonian Lectures to the Royal Society of London . Cajal said: "Mental training cannot better the organization of the brain by adding to the number of cells; we know that nervous elements have lost the property of multiplication past the embryonic life; but it is possible to imagine that mental exercise facilitates a greater development of the protoplasmic apparatus and of the nervous collaterals in the part of the brain in use. In this way, pre-existing connections between groups of cells could be reinforced by multiplication of the terminal branches of protoplasmic appendix and nervous collaterals. But the preexisting connections could also be reinforced by the formation of new collaterals and protoplasmic expansions." (Cajal, 1894, p. 466). These "connections" are now called "synapses". Two researchers Jerzy Konorski (1948) and Donald Hebb (1949) formalized the role of synapses in learning and subsequent workers spelled out the chemistry of how these synaptic changes come about and how long they last (review in Kandel, 2006).

Then, one evening, as I was enjoying a hot, end of day shower I found myself thinking once again about the seasonal changes in HVC and RA volume, and how these changes were more pronounced in HVC than in RA. What were the underlying mechanisms and need they be the same in both cases? We really had no evidence that the dendrites of males grew and withdrew seasonally, this was just an inference based on the observation of dendritic growth in females treated with testosterone. We did not know if in our seasonal comparisons we were always dealing with the same collection of neurons. Could neurons come and go? Might there be conditions under which neurons were born and conditions under which neurons were eliminated and replaced? It was fun to think in this manner, perhaps because it was so different from all I had been taught and because if this budding idea were true, the implications would be big.

There was in my laboratory at the time a very bright and resourceful doctoral student, Steve Goldman, who had not yet chosen a topic for his thesis work. I told him of my fantasy about neuronal comings and goings and to this day remember the amused smile that crossed his face. He knew of a quick way to test it, but, he asked me, "do not talk about this to anyone. People will wonder about the lab if word gets out." I agreed to keep mum and he told me about tritiated thymidine. 3H-thymidine could be used as a marker of DNA synthesis and therefore as a presumptive marker of birth date as DNA doubled during the S-phase preceding mitosis. He added that this method was widely used by embryologists staging the order in which different populations of cells were produced (Messier et al., 1958; review in Korr, 1980). 3H-thymidine administered systemically remained in the blood for a short time before it was metabolized and it was only during that brief period that it was available for uptake by cells. This tool would tell us whether new neurons were added when the size of song nuclei doubled.

2.9 3H-Labeled Cells in Adult HVC

We tested for adult neurogenesis in adult female canaries in which, we knew, testosterone treatment would roughly double the volume of HVC and RA. Each bird received six intramuscular injections of 50 uCi of 3H-thymidine at 8 h intervals over a 2 day period, starting at various intervals after the hormone implant. Most birds were killed 37 days after onset of hormone treatment. Brain levels showing HVC and RA were sectioned at 6 um intervals and the sections incubated for autoradiography. After incubation they were stained with cresyl violet. Six kinds of cells were recognized: (1) large (10–18 um soma diameter) cells with clear nuclei and dark-staining, central nucleoli and scant cytoplasm, which were tentatively identified as neurons; (2) smaller cells (6–10 um) containing light nuclei with stippled heterochromatin, small eccentric nucleoli and scant cytoplasm, which we considered astrocytes; (3) small cells (5–8 um) with deeply basophilic cytoplasm, small nuclei and dark karyoplasm, considered to be oligodendrocytes; (4) thin, fusiform perivascular cells, clearly endothelial; (5) very small (4–6 um) cells of variable morphology lining the lateral ventricle over HVC, which were presumed to be ependymal and subependymal cells jointly referred to as ventricular zone cells; (6) cells of uncertain identify. These rather simple criteria, we felt, might under-estimate neuron numbers, but we preferred to err on the conservative side. Using this material, a cell was considered labeled if it had five or more exposed silver grains per nucleus – background was exceedingly low in our material. To validate our light-microscopic criteria for neuronal identification, we cut the brains of some of the 3H-thymidine-treated birds at alternating 1 um and 100 nm intervals. We reasoned that a light microscopic neuronal profile that was one micron thick was very unlikely to include over a labeled cell contaminating material from another cell type, e.g. a glia, that contributed the label, and that an adjacent 100 nm section could be used for electro-microscopy to see if the ultrastructural details of the labeled cell conformed with those of a neuron. Electron microscopy revealed details in labeled cells identified as "neurons" – extensive rough endoplasmic reticulum, homogeneous nucleoplasm, long processes filled with microtubules and in one case an apparent axonal hillock – that were compatible with this identification, but we failed to see in these cells any unambiguous synaptic profiles. The other cell types looked very different under the electron microscope. Our counts indicated that 0.9–2.0% of the cells that we identified under the light microscope as "neurons" were labeled per day of 3H-thymidine treatment when the birds were killed 18 or more days after the last 3H-thymidine injection, with no labeled "neurons" in the HVC of two birds that received their 3H-thymidine 1–2 days before they were killed. In the latter two cases, however, the ventricular zone above HVC was blanketed with small cells whose nuclei were tritium labeled, while within HVC only endothelial and glia cells were labeled. Under none of the survival times we used were there any 3H-labeled "neurons" in RA.

We inferred from these results that if new neurons were recruited into adult HVC, they were not produced by the division of existing HVC neurons, because if that had been the case, labeled neurons would have occurred in HVC even at short survival times. The short-term heavy labeling of cells in the ventricular zone above HVC suggested that this might be the place of birth of the new neurons. Unexpectedly, the numbers of labeled neurons were similar in the females that received testosterone as in those that received cholesterol or empty silastic implants. Since male and female carduelines were known to continue to modify their calls in adulthood (Mundinger, 1970), we speculated that the new neurons of females might be involved with this kind of vocal learning. Finally, since the numbers of cells we called neurons seemed to be recruited at a rather high rate, yet the size of the female HVC did not change significantly between successive years (Nottebohm et al., 1981), we thought that perhaps the new cells replaced numerically older ones that had died, a phenomenon for which there was a precedent in the olfactory epithelium (Graziadei and Monti Graziadei, 1978).

I invited Pasko Rakic, the Yale developmental neurobiologist and a great skeptic of adult neurogenesis, to be a member of Steve Godman's thesis committee and he graciously accepted. He came to the laboratory and looked over the histological material. He accepted the results, but we all acknowledged that there were still loose ends and that much more work needed to be done. Those results were scrutinized, too, by the great embryologist Viktor Hamburger, who communicated them for publication in the Proceedings of the National Academy of Sciences and that is how the first report of adult neurogenesis in songbirds came to be (Goldman and Nottebohm, 1983).

Our results and interpretations were tentative because: (1) There could be in the brain cells that by some criteria looked neuronal, yet played some other role. (2) We were not sure that the labeling we saw resulted from DNA synthesis that accompanied mitotic events. Successive 3H-thymidine injections might create labeling cycles whereby tritium incorporated into DNA during spontaneous, on-going DNA repair induced damage that was followed by more DNA repair and 3H-labeling, followed by more damage and repair and so on. (3) Even if the cells were neurons, we had no evidence that they were part of functioning circuits and that they played a specific role in behavior. (4) The whole phenomenon could be a laboratory artifact that resulted from the highly restricted and monotonous conditions under which our captive birds were kept. Because of these uncertainties, our claims were no less vulnerable to criticism than those of Joseph Altman, of whose work we had learned while preparing our results for publication. Altman and others had claimed to have evidence of post-natal and adult neurogenesis in mammals (Altman and Das, 1965; Altman, 1969; Bayer et al., 1982; Kaplan and Hinds, 1977), but their work had met strong resistance. To overcome this resistance, our evidence would have to be more compelling.

2.10 Hope for a New Neurology

I organized in 1984 a conference in New York on the topic of "Hope for a New Neurology", underwritten by Jane and Peter Pattison's Institute for Child Development Research. The intention was to review what was known about a number of brain neurodegenerative disorders, evidence for recovery following lesion, the possible therapeutic value of transplants and the emerging field of adult neurogenesis. Among those invited was Shirley Bayer, Joseph Altman's wife, who continued his work. Altman, who had retired, was in the audience. I have often regretted that I did not invite Joseph Altman himself to be a speaker, for he would have had so much to say about a subject he knew so well. Also there was Pasko Rakic, whose work on brain development I admired. I knew Rakic was not convinced that adult neurogenesis was real and I wanted him present as a critic. It was in that context that he said: "Extraordinary claims require extraordinary evidence". Of course, the claims he found extraordinary were extraordinary only when placed next to the 100-year old dogma that adult neurogenesis couldn't, shouldn't and didn't happen. Rakic felt that evidence for adult neurogenesis in mammals would be believable only when the facts in its favor could not be explained in any other way. Rakic (1985a,b) felt that he knew what he was talking about. He had looked for evidence of post-natal or adult neurogenesis in close to 10,000 sections from the brains of 10 Rhesus macaques of both sexes that had received injections of 3H-thymidine at ages ranging from 6 months to 5 years. He did not find a single "heavily labeled" (15–20 exposed silver grains) cell that met his criteria of neuronal identity. His definition of "heavily labeled" was that of a neuron with at least half the maximum number of exposed silver grains found over the nucleus of other cell types known to continue to divide in adulthood. I admired Rakic's integrity and high standards of proof, but could not understand how the same man that was willing to accept the possibility that adult neurogenesis occurred in birds still argued so vehemently against its occurrence in mammals. Possibly he thought about these matters as a medical man. If adult neurogenesis did not occur in primates – and he had found zero evidence of it – and only rarely if at all in other mammals, then the data from other species were irrelevant to "hopes for a new neurology". His was a defensible position.

2.11 The Role of Joseph Altman

The negative results reported by Rakic pertained to macaques, and he did not rule out the possibility that earlier reports of post-natal neurogenesis in rodents and cats might be correct. Among those earlier reports were those of Joseph Altman (1962), to whom I alluded earlier. It is of interest that although Altman gained attention in the minds of many as first to challenge Cajal's assertion – and that of many neuroanatomists that came after him – that no new neurons were produced after the end of embryonic development, he (Altman) thought it had been done before. To quote from lines 6–10 of Altman and Das (1965), "It was recognized in the nineteenth century (Lahousse, 1988; Schaper, 1894; Cajal, 1911) that a class of small nerve cells, the granule cells, shows a high rate of proliferative activity in the cerebellar cortex in young animals. Likewise, good quantitative evidence was obtained some time ago (Sugita, 1918) of an increase in the total number of nerve cells in the cerebral cortex of rats up to the 20th day after birth." All the same, Altman's initial and subsequent findings on adult neurogenesis (e.g. Altman, 1963) were resisted and this lingering resistance was in evidence at the time of the Hope for a New Neurology conference on April 16–18, 1984. Reasons for resistance were technical and not trivial. For example, it was possible that in some cases "label" over the nucleus of a neuron came from an overlying fragment of nuclear glia. Thin (1 um), contiguous labeled sections of "neurons" and adjacent thinner ones inspected under the electron microscope were necessary to establish the source of label and to bolster the neuronal identification. In 1984 the evidence of this kind for cells born in adult mammalian brain was sparse (Kaplan and Hinds, 1977) or inadequate. The electron-microscopic evidence of post-natal neurogenesis presented by Kaplan (1985) at the Hope for a New Neurology conference came from a 9-day-old rat, hardly an adult. Moreover, it was possible that DNA duplication occurred in differentiated cells leading to polyploidy, without this necessitating the formation of new neurons. It was even possible, as Altman acknowledged, that the 3H-label seen in the nucleus of cells resulted not from the synthesis of new DNA during the S-phase of mitosis, but from a slow turnover of DNA. It is of interest that Altman (1962) countered this concern by pointing out that some neurons showed intense labeling as soon as 1 day after 3H-thymidine injection, which is now considered also too soon for a cell to be born, migrate and differentiate into its adult neuronal phenotype. Finally, skeptics adduced that the neuronal identity of labeled cells had not been sufficiently established. Perhaps, under some conditions, synaptic contacts were made on glia? The point is, a skeptic can always find some reason to doubt what it does not want to believe and it is up to the scientist making the claim to meet all these potential difficulties so that, at the end, the interpretation offered is the only one, or most likely one to account for all known facts.

As suggested above, some of the original skepticism may have been justified. In addition, Altman reported in his 1962 paper in Science that bilateral electrolytic lesions in the lateral geniculate body of "young adult" rats, followed by intraperitoneal injections of 3H-thymidine, resulted in the labeling of neurons in cortex, including "a few labeled pyramidal cells". This claim, without intervening electrolytic lesions, appears again in his 1963, Anatomical Record publication, where labeled cortical neurons (including a photograph of a labeled pyramidal cell) appeared after administration of 3H-thymidine to young adult rats and adult cats of unknown age. Some 48 years later, the topic of endogenous addition of new neurons to the adult mammalian cortex remains controversial, with only a small scatter of positive entries (Gould et al., 1999; Magavi et al., 2000; Dayer et al., 2005) and a persistence of negative reports (Breunig et al., 2007; Koketsu et al., 2003; Kornack and Rakic, 2001; Rakic, 2002).

There was also an interesting semantic issue. Cajal (1894) had stated that the central nervous system ceased to add new neurons "past the embryonic life", but presumably this need not mean that there was a sudden stop at the moment of hatching or when a mammal was born. Some species are born at earlier developmental stages than others – with extremes, for example in marsupials. Even among rodents, mice and rats are born at an earlier developmental stage than guinea pigs. In short, in some species processes associated with embryogeny continue in parts of the brain that are still developing after birth. Two examples of this were known at the time Altman started his work – one was the cerebellum (Uzman, 1960; Miale and Sidman, 1961); the other was the hippocampus (Angevine, 1965). A third area, the subependymal layer in the anterior reaches of the lateral ventricle was known to continue to show mitotic activity into adolescence and adulthood (Bryans, 1959; Messier et al., 1958). Altman's focus on these three areas showed, after 1965, that new cells that he identified as granular neurons ("microneurons") continued to be added post-natally to cerebellum, olfactory bulb and dentate gyrus, and that this addition continued into adulthood in the case of the latter two structures. His inferences from these observations were extensively presented in his reviews (Altman, 1967, 1969 and 1970). Altman's overarching theory was that macroneurons formed early in the development of the brain and established "hard-wired" circuits. Microneurons – in some cases at least – developed late because the fine details of their connectivity were influenced by events that happened late, when the young animals first encountered their post-natal environment. He was not dogmatic about this view, but rather, echoing the style of his 1962 title ("Are new neurons formed in the brains of adult mammals?"), his thoughts this time also came with a question mark, and I quote from his 1967 review (p.741): "Is the proliferation of the precursors of microneurons delayed in altricial mammals until birth because their function is a post-natal one – that of adjusting the genetically specified programs of behavior in accordance with information about specific environmental conditions as it is acquired by the growing animal?" It was a good question that placed order on the facts known to him. Moreover, Altman used his insights about the late development of some brain circuits to explain why some kinds of learning might be difficult, or absent in young altricial mammals (Altman and Bulut, 1976). We now remember Joseph Altman for the things that, in the 1960s, he got right: (1) Olfactory bulb granule cells continue to be produced after birth and into adulthood and are born in the subventricular zone of the lateral ventricles, from where they migrate rostrally to the bulb; (2) Granule cells of the dentate gyrus continue to be produced post-natally and into adulthood, as was confirmed by the beautiful work of his wife, Shirley Bayer, that presented her results at the 1984 Hope for a New Neurology conference; (3) post-natal neurogenesis in cerebellum; (4) the late neurogenesis that occurs in some brain systems of altricial mammals can explain the late occurrence of some kinds of learning. He was the first, in modern times, to challenge the 100 year old dogma that neurogenesis stops at birth and, despite strong opposition, persevered with this claim and was right.

Moreover, though the information accumulated by Altman during the 1960s was in the public domain, he was one of few that not only believed in it, but also used it to formulate bold new hypotheses. His titles reflected caution, preferring the use of "post-natal" to that of "adult" neurogenesis (Altman, 1967, 1970).

In some ways, the great battle fought in 1984 between Pasko Rakic and those that wanted to discard the "no new neurons in adulthood" dogma was a matter of emphasis. It was a matter of the greatest interest to neurosurgeons and neurologists to know what kinds of repair potential were spontaneously available in the brain of their patients, and what kinds might be "turned on". That is why observations on primates or other mammals were so important, because they were the closest to potential clinical applications. But there was also a deeper issue. Was neurogenesis in the adult vertebrate brain possible at all? Where, if anywhere in the adult brain, were new neurons produced? From what cell types? Could neurons born in adult brain migrate from birth site to work site? Would they be able to join existing circuits? If these feats were possible in the brains of some vertebrates, then the "no neurons in adulthood" dogma was overly severe and the study of adult neurogenesis became not only more interesting but also a source of hope.

By the time of the 1984 "Hope for a New Neurology" conference postnatal neurogenesis was known to occur in a number of cold-blooded vertebrates whose body, eyes and brain continued to grow after sexual maturity (Anderson and Waxman, 1985; Birse et al., 1980; Easter, 1983; Graziadei and DeHan, 1973; Johns, 1982; Leonard et al., 1978; Raymond and Easter, 1983); in addition, adult neurogenesis had been shown to occur in the olfactory epithelium of mammals (Graziadei and Monti Graziadei, 1978, 1985). These reports had not met the level of skepticism generated by Joseph Altman's claims about new neuron formation in the brain of adult mammals. Perhaps the evidence from cold-blooded vertebrates and from the olfactory mucosa was not seen as relevant to the "no neurons in adulthood" dogma and so no big battle was fought over those observations.

Ironically, both sides could claim a measure of victory in the disputes over neurogenesis in the brain of warm-blooded vertebrates. No embryologist doubted in 1984 or doubts now that the vast majority of the neurons present in the central nervous system of eutherian mammals are produced during embryogeny and not later. By this measure, the "no new neurons in adulthood dogma" was 99.9% right. Yet, if the exceptions discovered by Altman were confirmed, the consequences for the brain sciences could be great. We found the field polarized in this manner when we started our work on neurogenesis in songbirds and we went after the basic data that would test the reality of this phenomenon and, if real, its manner of occurrence.

2.12 Are They Really Adult-Born Neurons?

The procedure we used to test for adult neurogenesis in female canaries also yielded positive results in males. Moreover, in both cases new neurons occurred through much of the telencephalon, so the phenomenon was not restricted to the song system. However, there were no neurons that met stringent criteria for 3H-labeling outside of telencephalon: none in the diencephalons, mesencephalon, cerebellum or brain stem. Since the telencephalon had been traditionally associated with learning and higher brain functions, maybe we had to think of the new cells in that context (Nottebohm, 1984). But we first had to resolve two questions: (1) was the labeling observed evidence of birth? (2) Were the new cells really neurons? The answer to the first of these questions came from a low-tech observation. We got the answer to the first of these questions by comparing the number of exposed silver grains over the nuclei of endothelial, glia and neuronal nuclei in HVC. The numbers were very similar for all three cell types. True, peaks of label (45 and 40 exposed silver grains, respectively) occurred over the nuclei of glial and endothelial cells, compared to 25 exposed silver grains in neurons, but the rest of the histogram distributions were very similar in the three cell types. Perhaps glia and endothelia collected more label because of their greater proximity to capillary blood? What mattered was that there was not an order of magnitude difference in amount of labeling when comparing glia and endothelia vs. neurons. The distributions of label argued, instead, for a common process, the formation of new DNA during the S-phase preceding mitosis (Nottebohm, 1985).

But were the labeled "neurons", neurons? Our initial evidence from electron microscopy had been a bit disappointing in that we found no synaptic profiles on the soma of HVC labeled "neurons" from adult female canaries (Goldman and Nottebohm, 1983). This issue was revisited by Gail Burd, an electron-microscopist in my laboratory. The protocol she followed was similar to that used by Steve Goldman, but she used HVC samples from adult male canaries, rather than females as in the previous study. Not only did she show in this material that contiguous 1-μm thick sections of a same "neuron" showed nuclear labeling – further reducing the likelihood of glial "contamination" – but she found that these cells received in their somata three different kinds of synapses, characterized by synaptic vesicles of different size and density. Electron-microscopic examination of these cells revealed also other standard features of neuronal ultra-structure (Burd and Nottebohm, 1985). We felt this was encouraging, but still not conclusive.

The next test was more stringent and, as so often in my career, I was lucky that the right person, John Paton, a neurophysiologist of impeccable standards, was at that time in my laboratory. The approach, again, was simple: we knew that if we injected adult canaries twice daily with 3H-labeled thymidine for a period of 2 weeks and killed the birds 1 month after the last injection, approximately 10% of the neurons in HVC reached our labeling criterion. This meant that if we used in such birds an electrode to penetrate cells in HVC, one in ten of neurons randomly entered would be labeled. We used beveled micropipettes containing a solution of horseradish peroxidase (HRP) in KCl. The pipette was slowly advanced into HVC. Penetrations of cells were accompanied by a negative d-c shift in electrical potential and cells were identified as neurons by the presence of spontaneous or depolarization-induced action potentials. Many of the neurons penetrated responded to noise bursts and the latencies of any time-locked synaptic potentials were noted. Then, over a period of 20 min, the HRP was iontophoretically injected into the cell and the stereotaxic coordinates of the cell injected were recorded. Then the bird was killed and the brain prepared for histology. The HRP reached into all corners of the cells injected, but did not spill out, giving a beautiful, Golgi-like, rendering of soma and processes, down to the level of synaptic spines. The full anatomy of these cells was photographed and drawn, as seen in 100 um-thick sections. When these cells were later cut into thinner sections and the sections were covered with a photographic emulsion, the nuclei of seven of the cells had over them a rich collection of exposed silver grains. We inferred that, as expected, approximately one in ten of the HVC neurons that had been randomly penetrated had been born during (or after) the 2 weeks of 3H-thymidine treatment that had ended 30 days prior to electrode penetration; four of these seven cells had responded to sound. We now had direct evidence that at least some of the labeled HVC cells that we had identified as neurons were, indeed, neurons and that some of these cells had been recruited into existing, functioning circuits. I believe this experiment turned the tide on how other scientists felt about adult neurogenesis (Paton and Nottebohm, 1984).

Four years later this result was confirmed using a totally different and much simpler approach. This time we used a combination of 3H-thymidine as a birth-date marker and fluorogold as a retrograde tracer. By making injections of flurogold in Area X or RA in canaries pretreated with 3H-thymidine, we were able to show that whereas the great majority of HVC neurons that project to Area X (Fig. 2.4) are born before hatching, those that project to RA (HVC→RA projection neurons) are born after hatching and during the very period when song is learned as an auditory-motor skill (Alvarez-Buylla et al., 1988a) as well as later in adulthood.

Yet, it seemed wise to remain skeptical. For adult neurogenesis to be an indisputable fact, we had to show where the new cells came from and how they reached their destination. It is only when all parts of a story fit together, that one is confident that this is the story that explains all observed facts.

2.13 Birth and Migration of New Neurons

Again, good fortune interceded, this time in the person of Arturo Alvarez-Buylla (pre-viewed in the above ref.), an extraordinarily able experimentalist whose thesis work focused on the origin and migration of new neurons in adult avian brain. We already suspected that, as during embryogeny, the new neurons were born in ventricular walls (Goldman and Nottebohm, 1983), but we needed stronger evidence that this was so.

Soon after the initial, 1983, publication, Daniel Buskirk joined my laboratory and started a search for monoclonal antibodies that might identify cell types or phenomena involved in the production / recruitment of new neurons into adult brain. It was the kind of approach that some belittle as a “fishing expedition” and that, with a bit of luck can generate key clues. Mice were injected intraperitoneally with a suspension of homogenized adult canary brain. Spleen cells from these mice were fused with mouse myeloma cells and plated into tissue culture dishes, followed by screening for positive hybridomas. We then screened for antibodies that recognized markers restricted to the telencephalon and associated with the ventricular zone. Our "expedition" came up with the nicest "fish" we could have hoped for: a monoclonal antibody, 40E-C, that stained cells with small bodies in the walls of the lateral ventricle and long, undividing processes that dove into the forebrain's parenchyma. Follow up work by Alvarez-Buylla showed that this antibody recognized vimentin. We had stumbled upon radial glia, a cell type familiar to embryologists but not known to occur in adult birds (Alvarez-Buylla et al., 1987) or mammals.

But this observation got still better. When the antibody that stained for vimentin was combined with cresyl violet, we were able to see an uncanny association between the vimentin-positive radial fibers and small, darkly staining elongated cells that were closely apposed to these fibers and showed their same orientation (Fig. 2.6). This kind of cell had not been described before in the parenchyma of adult telencephalon. We thought these might be young, migrating neurons (Alvarez-Buylla et al., 1988b). To test this interpretation, adult male canaries received two intramuscular injections of 3H-thymidine 12 h apart and were killed at various intervals thereafter. One day after the last injection there were no labeled neurons, but lots of labeled cells in the ventricular wall of the lateral ventricle. By 3 days, still no labeled neurons, but small, elongated, darkly staining labeled cells were seen a short distance from the ventricular zone. This collection of labeled cells continued to grow and move further away from the lateral

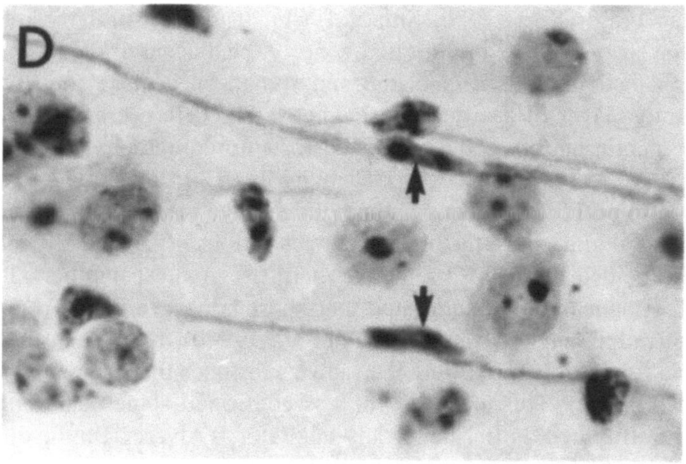

Fig. 2.6 (Figure 2D from Alvarez-Buylla et al. 1988b). A cresyl violet stain that reveals cell nuclei and an antibody stain that recognizes vimentin allow us to see the relation between young migrating neurons (small, elongated nucleus) and the vimentin positive fibers of radial glia in adult canary forebrain

ventricle in successive days. By day 20 it had reached the furthest corners of the telencephalon and on that day, too, the first labeled, differentiated neurons were seen. Thereafter, the cohort of small, elongated, darkly staining labeled cells dwindled and, at the same time the number of labeled neurons steadily increased. We felt that we now had further evidence that the neurons were born in the ventricular zone, and from there migrated, as spindle-shaped, young neurons, guided by radial glia fibers (Fig. 2.6), until they reached, presumably, a signal that triggered the change from the migratory to the sedentary phenotype. When we compared the maximum count (at 20 days) of migratory young neurons to the number of adult, labeled neurons, we inferred that only one third or less of the cells that migrated away from the ventricular zone were still present, as neurons, when our counts stopped, 40 days after 3H-thymidine injection. As during embryogeny, the system overproduced neurons and only a fraction was incorporated for longer durations of time. We felt that the story of adult neurogenesis had, with these details, been firmed up a bit more (Alvarez-Buylla and Nottebohm, 1988).

2.14 Neuronal Stem Cells

We had identified the presumptive birth-site. But who were the parents? To answer this question, adult canaries received a single injection of 3H-thymidine and were killed 1 h later. Virtually all of the labeled ventricular zone cells were in the walls of the forebrain's lateral ventricle. Apparently, cells on the walls of other ventricles seldom divided in adulthood, as already reported for the adult mammalian brain (Korr, 1980; Smart, 1961). In other, similarly treated adult canaries the strip of tissue (ventricular zone, VZ) adjacent to the lateral ventricle was excised and its cells dissociated and spread on a glass slide, where they formed a monolayer of easily identified cells. The cells were then reacted with the anti-vimentin antibody and incubated for autoradiography. At 1 h survivals, 80% of the VZ cells labeled with 3H-thymidine (20 times above background) were vimentin positive and a good many of these had the long, undividing process typical of radial glia. Since the highest concentrations of labeled VZ cells of adult canaries coincided with those regions where the greatest number of migrating neurons first appeared, we suggested that neurons were probably born there by division of radial cells (Alvarez-Buylla et al., 1990b). If so, then the very cells responsible for neurogenesis were the same as the ones that seemed to guide their migration (Fig 2.6). This arrangement was confirmed some years later in work that Alvarez-Buylla did after he left my laboratory (Alvarez-Buylla et al., 1998). Subsequently, Alvarez-Buylla and co-workers showed that the neuronal stem cells of adult mammalian brain are also glia – radial glia and astrocytes (Doetsch et al., 1999; Seri et al., 2001; Alvarez-Buylla et al., 2002; Merkle et al., 2004;

Alvarez-Buylla and Lim, 2004). Furthermore, work in developing mammalian brain has also shown that radial glia are progenitors of neurons (Kriegstein and Alvarez-Buylla, 2009). Time and again I have been gratified by how well our findings in birds generalized to mammals.

2.15 The Relation Between Behavior, Adult Neurogenesis and Neuronal Replacement

My initial interest had been the biology of vocal learning. This had led to studies of adult neurogenesis. But what was the relation between neurogenesis and behavior? The fact that in canaries HVC neuron numbers did not increase with age suggested that when new HVC neurons were recruited they replaced – numerically at least – older ones that died (Nottebohm, 1985). A similar conclusion had been reached by Bayer (1985) in her study of the dynamics of neuronal recruitment and total neuronal numbers in the olfactory bulb of adult rodents. But what were the behavioral correlates of replacement, if any? For example, did the incidence of neurogenesis in HVC change seasonally, accompanying changes in song?

To test this idea, adult canaries received intramuscular injections of 3H-thymidine every month of the year and were killed 1 month later and the proportion of labeled neurons in HVC was counted. This proportion showed two yearly peaks, in October and March. These peaks were preceded 2 months earlier by peaks in the ratio of pycnotic cells to total number of neurons in HVC, suggesting that we were dealing with a process of replacement (Kirn et al., 1994) (Fig. 2.7). Pycnotic staining is thought to identify dying cells, but of course, by the time a cell is that far gone it is not possible to know if it had been a neuron.

Two earlier studies had shown that the song of adult male canaries became particularly unstable during September and February, when many new song syllables were added (Nottebohm et al., 1986, 1987). These changes in behavior occurred between the peaks in HVC cell death and the peaks in new neuron recruitment. The relative magnitude of the changes in behavior (higher in summer than in winter) bore a close relation to the relative magnitude of the neuronal recruitment peaks, also higher in late summer than in late winter (Kirn et al., 1994) (Fig. 2.6).

These results of our 1994 study suggested that song stereotypy was weakened by the death of neurons in HVC and that the resulting vocal variability allowed the birds to master new sounds. The subsequent increase in new neuron and total neuronal numbers occurred as stereotypy was, once more, reinstated. Other observations, described in the next section, suggest how the cellular and behavioral changes come about.

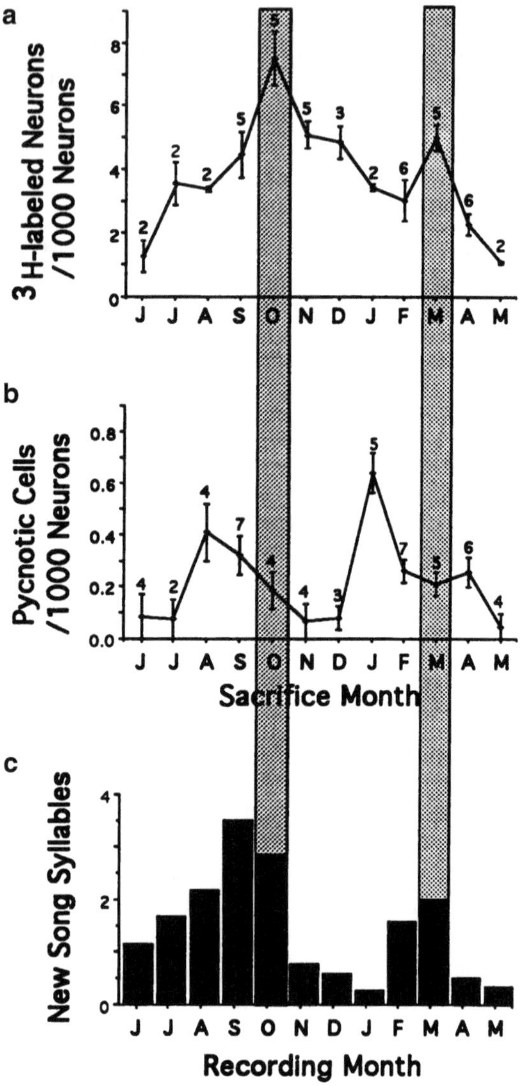

Fig. 2.7 (Figure 3 from Kirn et al., 1994). Monthly relation between neuronal recruitment and cell death in nucleus HVC and appearance of new song syllables in 1–2 year old male canaries. (**a**) Mean (plus or minus standard error) of 3H-labeled neurons per 1,000 HVC neurons in birds killed at different times of year 1 month after 3H-thymidine injection. (**b**) Number of pycnotic cells per 1,000 HVC neurons. For (**a**) and (**b**) the number of birds in each monthly sample is indicated above the *error bar* and the first letter of each month indicates when the birds were killed. (**c**) Mean number of new syllable types that appeared in the song of six adult male canaries each month during their second year of life. The *shaded bars* linking (**a–c**) emphasize timing of the two peaks in neuronal recruitment and their relation to peaks in cell death and syllable acquisition

2.16 Use and Disuse as Regulators of New Neuron Survival

Work on neurogenesis emphasizes counts of neurons labeled with a birth-date marker. For practical reasons, those counts must often wait until the new cells have reached their destination and acquired an adult neuronal phenotype. By that time it is not possible to say how many of the new neurons were born and what fraction survived. The number present at the time of the count is the number "recruited" into that part of the brain and this number, as we shall see, continues to change.

The blood testosterone levels of adult canaries fall after midsummer and at that time the song of these birds becomes more hesitant and variable (Nottebohm et al., 1987). How do these two changes relate to the new neurons? Testosterone does not regulate the production of new HVC neurons, but does promote the survival of those born in adulthood (Rasika et al., 1994). In this manner daylength, gonadal function, reproduction, song and neuronal replacement are part of a seasonal choreography that recurs every year. Though there is no direct proof for this, it seems plausible that the song syllables no longer in use and the cells that encode them are discarded to make room for new cells that can then engage in the mastery of new song syllables (Kirn et al., 1994). This view is supported by the observation that a significant fraction of the HVC→RA neurons present in the HVC of adult male canaries in early spring disappears by early fall, while those present in early fall are still present in the following spring (Kirn et al., 1991; Kirn and Nottebohm, 1993; Nottebohm et al., 1994).

We wondered to what extent the effect of testosterone on new neuron survival was mediated directly by its action on androgen sensitive neurons (Arnold et al., 1976), and how much was mediated indirectly by promoting higher levels of singing. The two variables were teased apart by looking at the performance of adult male canaries in early fall, a time when, unlike in the spring, amount of singing is not driven by testosterone. Two groups were set up: birds in one group had intact testis, those in the other had been castrated in early September. The rate of singing between these two groups, tested for 1 month starting 15 days after the date of castration, did not differ. All birds received 3H-thymidine systemically during post-castration days 10–14. When castrates and intacts that sang comparable amounts were compared, the number of 3H-labeled HVC neurons was 2.6 times higher in intact controls than in castrates. However, in castrates with plasma testosterone levels that were undetectable, the mean amount of singing was positively related to the number of new neurons. This outcome suggests that singing and gonadal factors promote, separately, the recruitment of new neurons into HVC circuits and that when they exert this effect together they do so in an additive manner (Alvarez-Borda and Nottebohm, 2002).

The next experiment tested directly how singing contributes to new neuron survival. New HVC neurons of adult canaries are already in place 8 days after these cells are born (Kirn et al., 1999); this relatively brief period probably results from the proximity of their birth site – the floor (ventricular zone) of the lateral ventricle overlying HVC (Scott and Lois, 2007) – to their final destination. Yet, even after the cells have made the transition from their migratory to their sedentary stage they are not yet "safe". Their survival must be confirmed by "use". We compared the number of new HVC neurons in two groups of adult male canaries matched for age. In one group the birds were able to sing unhindered for the full 38 days that intervened between injection of the birth date marker bromodeoxyuridine (BrdU) and the day they were killed. In the other group, singing was discouraged by a person that sat in the room where the birds were housed; this person waved a hand at any bird that started to sing. This protocol was followed during daylight hours from day 31 to day 38 after BrdU treatment. When the birds were killed, the number of new

HVC neurons was 63% higher in the birds that sang freely during all 38 days before they were killed. This effect presumably reflects how pathway use affects new neuron survival. In that same study we showed that the expression of brain-derived neurotrophic factor (BDNF) in the HVC of adult male canaries allowed to sing was directly proportional to the number of songs produced per hour. This increased expression is accompanied by higher concentrations of BDNF in HVC (Li et al., 2000). An earlier study had shown that BDNF infused into the HVC of adult canaries promotes the survival of new HVC neurons (Rasika et al., 1999). Under normal conditions, both singing and rising testosterone levels up-regulate BDNF expression and thereby promote the survival of new HVC neurons (Rasika et al., 1999; Alvarez-Borda and Nottebohm, 2002; Alvarez-Borda et al., 2004).

2.17 New Neurons as a Vehicle for Learning

I pointed out earlier that song nuclei HVC and RA (Fig. 2.4) develop late during ontogeny. In the case of HVC, but not RA, many new neurons continue to be added during the very weeks and months when juvenile canaries first acquire their song. As discussed earlier, many of these new neurons are HVC→RA *projection* neurons (Alvarez-Buylla et al., 1988a, 1990a, 1992). In zebra finches, too, the neurons that project from HVC to RA are added at the time song is first learned (Nordeen and Nordeen, 1988). This is important. It will be remembered that Joseph Altman had promoted the hypothesis that only inter-neurons were added after birth (the granule cells of cerebellum, hippocampus and olfactory bulb), but clearly that is not the case in the song system. Moreover, the very elegant work of Hahnloser et al. (2002) with zebra finches strongly suggests that the sparse firing of HVC→RA projection neurons encodes the program for learned song (Long and Fee, 2008). If so, the memory of learned song may be in the connections formed by the new cells at the time song is acquired. Taken together, these observations suggest that in zebra finches and canaries availability of uncommitted cells is important for learning song as a sensory-motor skill, whether this be in juveniles or, in the case of adult canaries, seasonally.

Given these correlations, does the recruitment of HVC→RA neurons in adulthood differ between species that, like the zebra finch learn their song only once, before sexual maturity, and canaries that modify their song every year? The answer is yes. Many more new neurons are added to the HVC of adult canaries than to that of zebra finches (Alvarez-Buylla et al., 1990a), which brings us to the next issue. Zebra finches normally master their learned song before sexual maturity (Immelmann, 1969), which occurs around post-hatching day 80; most of the pupil's modification of vocal output takes place between post-hatching days 45 and 65. After then and until day 80 or 90 the changes that occur have to do with increased song stereotypy. New songs presented after day 65 are not imitated (Boehner, 1990). However, new songs presented after day 65 are imitated if the bird had, until then, been denied access to a live model it could imitate (Eales, 1985).

2.18 Manipulations of Learning Load Alter New Neuron Recruitment

The recruitment of HVC→RA neurons in socially reared juvenile zebra finches allowed to imitate a model drops markedly after post-hatching day 65 and by day 150 is only one-fifth of that on day 65. However, the recruitment rate remains significantly higher in isolates unable to imitate an external model because none is available (Fig. 2.8). A possible interpretation is that when an expected peak in memory load, which normally occurs during the sensitive period for song learning, is delayed, juvenile rates of new neuron recruitment remain high, as if waiting for this event (Wilbrecht et al., 2006).

We now know that the vocal variability required for song learning depends, probably, not just on new neurons but also on the "noisy" output from LMAN (Fig. 2.4) to RA (Kao et al., 2005; Oelveczky et al., 2005) and on rising BDNF levels in nucleus RA (Kittelberger and Mooney, 2004). When that output from LMAN to RA is blocked, immediate song stereotypy follows (Scharff and Nottebohm, 1991; Brainard and Doupe, 2001) and no further vocal learning is possible (Bottjer et al., 1984). However, we do not know if and how the events in HVC affect the role of LMAN in driving output variability. The neurons in HVC that project to Area X (Fig. 2.4) and that are born before hatching (Alvarez-Buylla et al., 1988a,b) may hold the clue for integrating events in HVC and LMAN, but I will not enter into those details here (Mooney, 2009).

The HVC→RA→nXIIts descending pathway (Fig. 2.4) of the male zebra finch song system is much the same and predominantly uncrossed on each side of the brain and both sides play an important role in the production of learned song (Williams et al., 1992). The number of 30-day-old HVC→RA neurons counted in

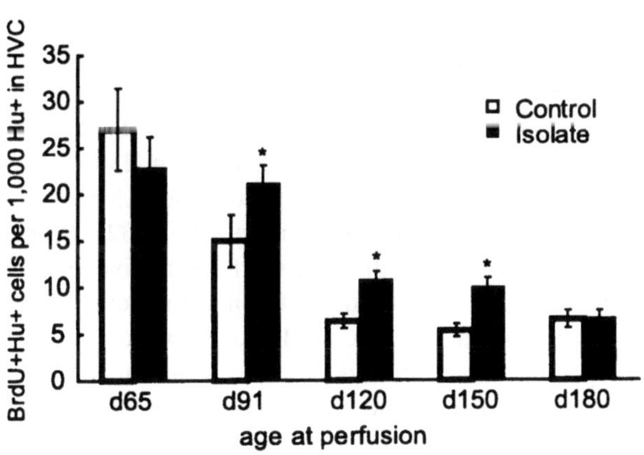

Fig. 2.8 (Figure 1b from Wilbrecht et al., 2006). An excess of new neurons is added to nucleus HVC after post-hatching day 65 and until post-hatching day 150, a time when isolates, but not socially reared controls, can still imitate new songs

90-day-old zebra finches was the same in birds that had learned their song by imitation as in birds that had been prevented from learning by early deafening or bilateral lesion of nucleus LMAN (Wilbrecht et al., 2002a,b). One might be tempted to conclude, from this, that the recruitment of new HVC→RA neurons at that age is not affected by learning, but two other sets of data suggest this would be the wrong conclusion. One set we have already discussed: recruitment of these cells remains higher in isolates that were not able to imitate an external model. The other set has to do with a nerve-cutting experiment. As discussed earlier, the syrinx of songbirds has right and left independent sound sources. If the right or left syringeal half of male zebra finches is denervated before the onset of song learning, then that side of the syrinx and the higher pathways that innervate it are not able play a major role in song imitation. Nonetheless, these birds are able to produce close copies of a model song. In them, the HVC on the intact side must assume a disproportionately important role in the acquisition and production of learned song. The number of 30-day-old neurons counted in the HVC of the intact side at 90 days of age was in these birds 60% higher that on the denervated side, whose numbers were the same as in intact controls. Presumably during song learning the intact side acquired most of the skill normally shared by the two sides (Wilbrecht et al., 2002a). This difference was eliminated if unilateral nerve section was accompanied by deafening or by bilateral lesions of LMAN. Apparently the act of song learning by reference to an external model only affected the recruitment of neurons involved with song learning. Thus, while there is a developmental program that arrives at a "normal count" of neurons in adult HVC regardless of experience (see also Burek et al., 1991), an extra learning load (Wilbrecht et al., 2002a) or the delay of an expected learning load (Wilbrecht et al., 2006) can alter this count during the time song would be normally mastered.

If the above results, taken together, seem a bit confusing, this is probably because we are not yet very advanced in how we think about what HVC does. This point is emphasized by the next experiment. Male zebra finches reared and housed socially until 4–5 months of age were given systemic injections of 3H-thymidine. Two hours after the last injection individual birds were moved into a larger aviary that they shared either with just a female or with 20 other pairs of adult zebra finches. The injected birds were killed 40 days after this change in social setting and the number of 3H-labeled neurons in HVC was counted. To our surprise, the birds housed with the larger group had 2.5 more labeled neurons than those housed just with a mate (Lipkind et al., 2002). Since the change in social setting occurred well after the end of the sensitive period for new song learning we are left to wonder why the difference in social setting affected new neuron survival. Not only is this a reminder of how poorly we understand the behavior we study, but also a reminder of the importance of including in our studies natural variables, such as the complexity of the social group in which an animal lives. Moreover, it is important to remember that much as HVC plays a key role in the production of learned song (encoding), it may also be involved in song perception (decoding), as suggested by the "motor theory for song perception in birds" (Williams and Nottebohm, 1985).

2.19 Evidence from Other Songbird Systems

Most of our work on adult neurogenesis and neuronal replacement focused on song learning because an understanding of how vocal learning came about continued to be a central concern for my laboratory. However, I was worried that laboratory work on just one system might be misleading and therefore I wanted to extend it to other systems, other species, and particularly to animals living in the wild. The latter goal was met by a study of neuronal recruitment and turnover in the hippocampus of free-ranging chickadees, *Parus atricapillus.* This study showed that large neurons continued to be added in adulthood to the hippocampus of these forest songbirds, particularly at times of year (later summer, early fall) when chickadees cached many food items that they later had to retrieve. Recruitment of these cells was twice as high in free-ranging individuals as in others caught in the wild and then housed in a large outdoors aviary (Fig. 2.9). Unlike reports in the hippocampus of laboratory bred and laboratory housed rodents, the new cells recruited into the chickadee hippocampus lived only for a few weeks or months (Fig. 2.10). Perhaps after then, the spatial information held by those cells was no longer relevant to the animal's needs (Barnea and Nottebohm, 1994, 1996).

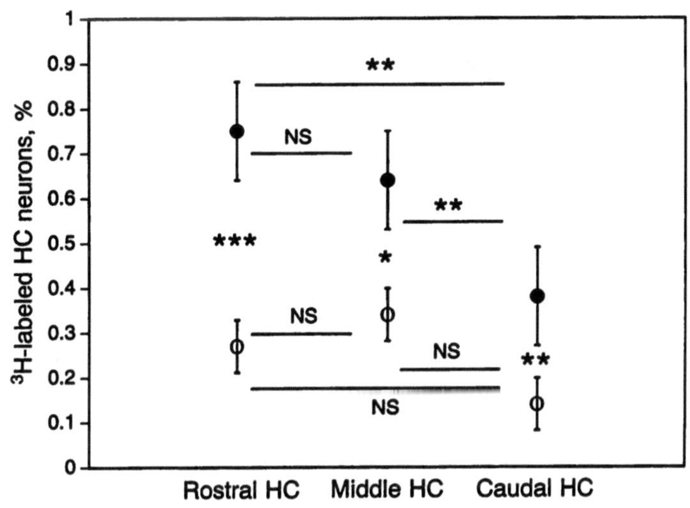

Fig. 2.9 (Figure 6 from Barnea and Nottebohm, 1994). Mean (plus or minus standard error of the mean) percentage of 3H-labeled hippocampal complex (HC) neurons in adult wild-caught chickadees. Six of these birds had been living in the wild until killed in November (*filled circles*); four had been housed in a large outdoor aviary during the 3 months before they were killed, also in November (*open circles*). Each of the 10 birds received a single dose of 50 uCi of 3H-thymidine and was killed 6 weeks later. Two trends are shown in this figure: (1) a higher percentage of 3H-labeled HC neurons was present rostrally than caudally in the free-ranging birds, a twofold difference; (2) the percentage of new HC neurons was twofold higher in the free-ranging than in the captive birds. Differences between groups were not significant (NS) or reached significance levels of $P<0.01$ (*), $P<0.001$ (**), or $P<0.0001$ (***)

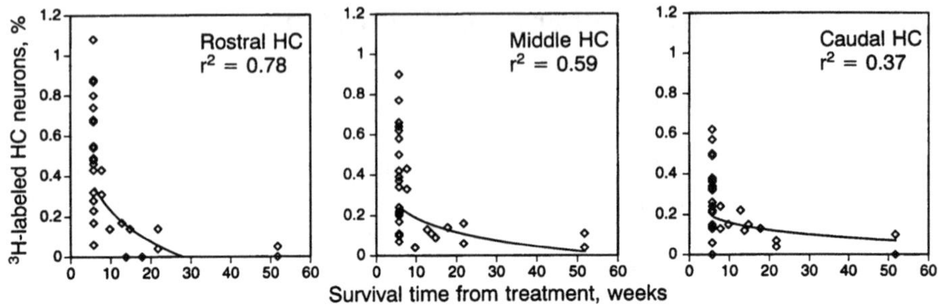

Fig. 2.10 (Figure 5 from Barnea and Nottebohm, 1994). Survival curves of 3H-labeled neurons in the rostral, middle and caudal HV of free-ranging chickadees treated with a single dose of 3H-thymidine at different times of year and then killed 6–52 weeks later. Many fewer new neurons are present after the 6-week survival

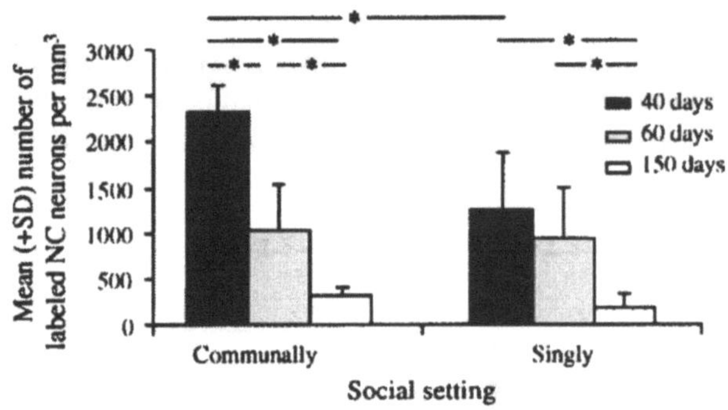

Fig. 2.11 (Figure 3 from Barnea et al., 2006). Mean (plus or minus SD) number of 3H-labeled neurons per mm³ in the caudal nidopallium (NC) of adult zebra finches that received systemic injection of 3H-thymidine and were then housed singly or communally in large outdoors aviaries (* indicates P = or < than 0.05). Birds were killed 40, 60 or 150 days after the change in social setting. Differences present at 40 days disappear at longer survivals

Another study focused on how changes in the social conditions under which adult zebra finches were housed affected the survival of neurons born immediately before the social change. This approach had already yielded evidence that social setting affected the survival of new neurons in HVC, which I mentioned at the end of the previous section. In a follow up, we looked at a part of the caudal forebrain, the nidopallium caudale, NC, thought to integrate auditory and non-auditory sensory information. A change to a more complex social environment – sharing an outdoor aviary with many zebra finches vs. sharing it with no other birds had a marked effect on up-regulating NC neuronal recruitment. As in the case of the chickadees, the increased recruitment was followed by the eventual demise of most cells so that by 150 days only a small fraction of the originally recruited cohort still remained (Fig. 2.11). It was of interest, too, that survival of the new neurons depended on part of the brain sampled, e.g. more rostral or more caudal, as if in some parts of the brain memories were up-dated more frequently than in others (assuming memory storage in these cells!) (Barnea et al., 2006).

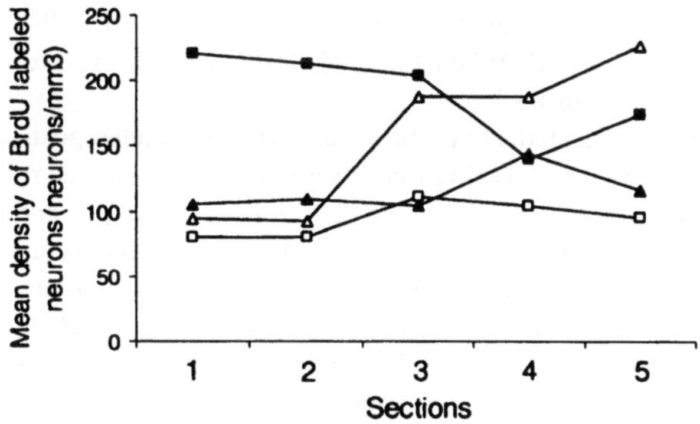

Fig. 2.12 (Figure 5 from Adar et al., 2008). Mean density of BrdU-labeled neurons per mm^3 in five sections of nidopallium of adult male zebra finches. BrdU was administered systemically at 4 months of age and birds were moved to a novel social setting either 30 or 90 days after the last injection of BrdU. The new social setting consisted of being housed with a female the male had not encountered before or 40 other zebra finches of both sexes the bird had not encountered before. Birds in either of these two settings occupied outdoors aviaries of the same size. All birds were killed 40 days after entering the new social setting. In the four groups shown here *open symbols* correspond to birds that encountered social change 1 month after the last BrdU injection; *filled symbols*, birds that encountered social change 90 days after BrdU injection. *Triangles*, complex social setting; *squares*, housed as pairs. Group sizes were six or seven birds. The most rostral section is 1, the most caudal, 5

A further study of the NC of adult zebra finches (Adar et al., 2008) showed that position and age of "new" neurons at the time of a social change as well as the nature of the social change influenced the survival of the new cells. A complex social setting favored the survival of caudal 1-month old neurons; a simple social setting favored the survival of rostral, 3-months old neurons. Perhaps the rostrally placed, older cells were too set in their ways to cope with a surge in information load, a situation that, however, encouraged the survival of younger, caudal neurons (Fig. 2.12). Apparently a same stimulus can prolong the survival of some cells and doom others. Studies that focus on a single cohort of same age cells, obtained from a single anatomical location may miss the rather complex strategies a brain might use to manage its replaceable neurons. A replaceable neuron vested with information may be a humbug or a precious asset, depending on past and present memory loads.

2.20 Overview

The song system is likely to yield the first description of how the vertebrate brain acquires a complex learned behavior and the role of late born neurons in this process.

I reviewed how my laboratory discovered evidence of widespread neuronal recruitment in adult avian brain. That evidence was scrutinized and challenged and held suspect until reliability of the birth dating process, reliability of neuronal identification, evidence of function, and origin of the new cells and manner of migration were established. When these matters were resolved, one big question remained: Why new neurons?

Joseph Altman suggested a hypothesis: experience gleaned post-natally sculpted the fine wiring of circuits handling that information. He saw this role vested in the late born granule cells of mammalian cerebellum, hippocampus and olfactory bulb. We have found that in songbirds the new neurons added to the forebrain of juveniles and adults have a much wider distribution than in mammals and are not confined to granule cells or inter-neurons. Our best-studied example, the HVC→RA neurons, are projection neurons. As in Altman's hypothesis, however, they are involved with a late-developing function, song learning.

If late-born neurons are required for the mastery of late-developing skills, then we might predict that some of the neurons required for vocal learning in humans would also develop late, as language is acquired, and that new waves of these neurons would be recruited when new languages are learned and perhaps this would also apply to manual skills, to the mastery of music and to some sports. Clearly, there is need for much more comparative data. But even if there is a correlation between late developing skills and post-natally born neurons, there was no prediction that neuronal replacement would be encountered in the juvenile and adult brain or that there would be specific types of cells that belonged to the replaceable category. As we became aware of replaceable neurons, many new questions came to mind: what was special about these cells, about their functions, about the information they held and the way they held it? Did these cells age in a particularly fast way, were they susceptible to death by over-use? Moreover, why was adult neurogenesis so much more common in birds that in mammals?

During my formative years I had been taught that neural plasticity had to do with the nature of connections between existing neurons. Information was stored in the synapses and could be altered by changing the number and efficacy of synapses between cells that were already in place. Since these changes could, in principle, happen again and again, the adult brain had virtually unlimited space for all its learning needs, which could be satisfied by neurons already in place. Neuroscientists had not, by the mid 1980s, predicted the existence of ephemeral neurons, of cells born in adulthood, in captive or free ranging animals, that would be recruited for a few weeks or months or a year and would then disappear and be replaced by others.

Why, then does the healthy brain replace some of its neurons? Might it be that in some systems the whole neuron, rather than each synapse, is the unit of learning? Might it be that the very susceptibility of synapses to change makes them unreliable repositories for memory? Might long-lasting changes in gene expression be required to lock up irreversibly a cell's functional modifications associated with learning thus securing for long periods a particular memory? If so, might this come at the expense of that cell participating in new learning events?

The need to replace neurons could result, too, from age restrictions on what a neuron can do. Young neurons may be able to sculpt their connections in ways that older neurons cannot, regardless of participation in earlier learning events.

An overproduction of new neurons with modifiable connections may provide many alternatives from which to choose by use and disuse, with only a fraction of the new cells persisting for longer periods of time. Restriction by age and restriction by learning could be complementary processes and the consequence would still be that new neurons have a learning flexibility that older neurons lack. Learned song, the circuits that support it and the cells within these circuits that continue to be added offer good material to test the role of new neurons and the manner in which this role is achieved.

If neurons, rather than synapses are the units of long-term memory, new principles of brain economics emerge. Synapses are too numerous to count, but neuron numbers are finite. Assigning memory space to skills mastered may deny "space" to memories still to be acquired. May be neuronal replacement clears the brain of idle cells or of unwanted memories and of the cells that hold them, and makes room for new cells and the new memories they may store. The 10-year life span of a songbird is five times longer than that of a mouse yet the brain of a canary and that of a mouse weigh approximately the same, one gram. Neuronal replacement may be much more common in birds than in mammals because the brain of the former must be able to learn many more things over a longer span of time.

But even as I offer the above speculations, I should warn that the link between song learning and the recruitment of new HVC→RA neurons remains to be tested and until then should remain suspect. Recent, unpublished observations by Clare Walton, a doctoral student in my laboratory, show that the number of HVC→RA neurons in adult male zebra finches doubles during the first 2 years after song is mastered, though the song itself changes relatively little after sexual maturity. Do the new neurons partake in song production? Do they learn this role? We do not know. It is not inconceivable that neurogenesis in juvenile and adult brain has little or nothing to do with learning, but is related to other, still unimagined needs.

The account I have offered is almost painfully rustic in the extent to which it leaves out the myriads of questions that come to mind about the molecular biology of adult neurogenesis and neuronal replacement. Some of these questions occurred to me too and I would have liked to address them, if funding had been available. However, it is probably just as well that I did not, for I now realize that this will be the work of another life time. Lucky those that will carry on with these studies.

The discovery of adult neurogenesis, of its mechanisms and of its use in neuronal replacement offer, potentially, a whole new approach to brain rejuvenation and brain repair. Knowing that young neurons can be added to old brains and that in some systems can replace, numerically, other neurons that have died should allow a new approach to the study of brain aging. As molecular ways to control the production, migration, differentiation, connections and survival of new neurons become available, it is unthinkable that the clinical potential of replaceable neurons will go unexplored. A quarter century after the 1984 Hope for a New Neurology conference in New York, I am confident that the adult brain's potential to produce and replace neurons will eventually be harnessed for clinical purposes. But that is not why I did this work. I did it because I fell in love with the song of birds and wanted to understand how vocal learning came to be.

References

Adar, E., Nottebohm, F., Barnea, A. 2008. The relationship between nature of social change, age and position of new neurons and their survival in adult zebra finch brain. J. Neurosci. 28:5394–5400.

Agate, R.J., Grisham, W., Wade, J., Mann, S., Wingfield, J., Schanen, C., Palotie, A., Arnold, A.P. 2003. A neural, not gonadal, origin of brain sex differences in a gynandromorphic finch. Proc. Natl. Acad. Sci. USA 100:4873–4878.

Agate, R.J., Choe, M., Arnold, A.P. 2004. Sex differences in structure and expression of the sex chromosome genes CHD1Z and CHD1W in zebra finches. Mol. Biol. Evol. 21:384–396.

Airey, D.C., Kroodsma, D.E., DeVoogd, T.J. 2000. Differences in the complexity of song tutoring cause differences in the amount learned and in dendritic spine density in a songbird telencephalic song control nucleus. Neurobiol. Learn. Mem. 73:274–281.

Altman, J. 1962. Are new neurons formed in the brains of adult mammals? Science 135:1127–1128.

Altman, J. 1963. Autoradiographic investigation of cell proliferation in the brains of rats and cats. Anat. Rec. 145:573–591.

Altman, J. 1967. Postnatal growth and differentiation of the mammalian brain, with implications for a morphological theory of memory. In "The Neurosciences, a Study Program", G.C. Quarton, T. Melnechuck, F.O. Schmidt, eds., pp. 723–743. The Rockefeller University Press, New York.

Altman, J. 1969. DNA metabolism and cell proliferation. In "Structural Neurochemistry", A. Lajtha, ed. Plenum Press, New York/London, Handbook of Neurochemistry, vol II, pp. 137–182.

Altman, J. 1970. Postnatal neurogenesis and the problem of neural plasticity. In "Developmental Neurobiology", pp.197–237.

Altman, J., Bulut, F.G. 1976. Organic maturation and the development of learning capacity. In "Neural Mechanisms of Learning and Memory", M.R. Rosenzweig, E.L. Bennett, eds., pp. 236–240. MIT Press.

Altman, J., Das, G.D. 1965. Autoradiographic and histological evidence of postnatal hippocampal neurogenesis in rats. J Comp Neurol 124:319–335.

Alvarez-Borda, B., Nottebohm, F. 2002 Gonads and singing play separate, additive roles in new neuron recruitment in adult canary brain. J. Neurosci. 22:8684–8690.

Alvarez-Borda, B., Haripal, B., Nottebohm, F. 2004. Timing of brain-derived neurotrophic factor exposure affects life expectancy of new neurons. Proc. Natl. Acad. Sci. USA 101:3957–3961.

Alvarez-Buylla, A., Lim, D.A. 2004. For the long run: maintaining germinal niches in the adult brain. Neuron 41:683–686.

Alvarez-Buylla, A., Nottebohm, F. 1988. Migration of young neurons in adult avian brain. Nature 335:353–354.

Alvarez-Buylla, A., Buskirk, D.R., Nottebohm, F. 1987. Monoclonal antibody reveals radial glia in adult avian brain. J. Comp. Neurol. 264:159–170.

Alvarez-Buylla, A., Theelen, M., Nottebohm, F. 1988a. Birth of projection neurons in the higher vocal center of the canary forebrain before, during and after song learning. Proc. Natl. Acad. Sci. USA. 85:8722–8726.

Alvarez-Buylla, A., Theelen, M., Nottebohm, F. 1988b. Mapping of radial glia and of a new cell type in adult canary brain. J. Neurosci. 8:2707–2712.

Alvarez-Buylla, A., Kirn, J.R., Nottebohm, F. 1990a. Birth of projection neurons in adult avian brain may be related to perceptual or motor learning. Science 249:1444–1446.

Alvarez-Buylla, A., Theelen, M., Nottebohm, F. 1990b. Proliferation "hot spots" in adult avian ventricular zone reveal radial cell division. Neuron 5:101–109.

Alvarez-Buylla, A., Ling, C-Y., Nottebohm, F. 1992. High vocal center growth and its relation to neurogenesis, neuronal replacement and song acquisition in juvenile canaries. J. Neurobiol. 23:396–406.

Alvarez-Buylla, A., Garcia-Verdugo, J.M., Mateo, A.S., Merchant-Larios, H. 1998. Primary neural precursors and intermitotic nuclear migration in the ventricular zone of adult canaries. J. Neurosci. 18:1020–1037.

Alvarez-Buylla, A., Seri, B., Doetsch, F. 2002. Identification of neural stem cells in the adult vertebrate brain. Brain Res. Bull. 57:751–758.
Andalman, A.S., Fee, M.S. 2009. A basal ganglia-forebrain circuit in the songbird biases motor output to avoid vocal errors. Proc. Natl. Acad. Sci. USA 106:12518–12523.
Anderson, M.J., Waxman, S.G. 1985. Neurogenesis in adult vertebrate spinal cord in situ and in vitro: a new model system. In "Hope for a New Neurology", F. Nottebohm, ed., Ann. N.Y. Acad. Sci., vol 457, pp. 213–233.
Angevine, J.B. 1965. Time of neuron origin in the hippocampal region. An autoradiographic study in the mouse. Exp. Neurol. Suppl. 2:1–70.
Arnold, A.P., Nottebohm, F., Pfaff, D.W. 1976. Hormone concentrating cells in vocal control and other areas of the brain of the zebra finch. J. Comp. Neurol. 165:487–512.
Aronov, D., Andalman, A.S., Fee, M.S. 2008. A specialized forebrain circuit for vocal babbling in the juvenile songbird. Science 320:630–634.
Barnea, A., Nottebohm, F. 1994. Seasonal recruitment of new neurons in the hippocampus of adult, free-ranging black-capped chickadees. Proc. Natl. Acad. Sci. USA 91:11217–11221.
Barnea, A., Nottebohm, F. 1996. Recruitment and replacement of hippocampal neurons in young and adult chickadees: an addition to the theory of hippocampal learning. Proc. Natl. Acad. Sci. USA 93:714–718.
Barnea, A., Mishal, A., Nottebohm, F. 2006. Social and spatial changes induce multiple survival regimes for new neurons in two regions of the adult brain: an anatomical representation of time? Behav. Brain Res. 167:63–74.
Bayer, S.A. 1985. Neuron production in the hippocampus and olfactory bulb of the adult rat brain: addition or replacement? In "Hope for a New Neurology", F. Nottebohm, ed., Ann. N.Y. Acad. Sci., vol 457, pp. 163–172.
Bayer, S.A., Yackel, J.W., Puri, P.S. 1982. Neurons in the rat dentate gyrus granular layer substantially increase during juvenile and adult life. Science 216:890–892.
Birse, S.C., Leonard, R.B., Coggeshall, R.E. 1980. Neuronal increase in various areas of the nervous system of the guppy, *Lebistes*. J. Comp. Neurol. 194:291–301.
Boehner, J. 1990. Early acquisition of song in the zebra finch, Taeniopygia guttata. Anim. Behav. 39:369–374.
Bottjer, S.W., Miesner, E.A., Arnold, A.P. 1984. Forebrain lesions disrupt development but not maintenance of song in passerine birds. Science 224:901–903.
Bottjer, S.W., Halsema, K.A., Brown, S.A., Miesner, E.A. 1989. Axonal connections of a forebrain nucleus involved with vocal learning in zebra finches. J. Comp. Neurol. 279:312–326.
Brainard, M.S., Doupe, A.J. 2000. Interruption of a basal ganglia-forebrain circuit prevents plasticity of learned vocalizations. Nature 404:762–766.
Brainard, M.S., Doupe, A.J. 2001. Postlearning consolidation of birdsong: stabilizing effects of age and anterior forebrain lesions. J. Neurosci. 21:2501–2517.
Breunig, J.J et al. 2007. Everything that glitters isn't gold: a critical review of postnatal neural precursor analyses. Stem Cell Res. 1:612–627.
Bryans, W.A. 1959. Mitotic activity in the brain of the adult rat. Anat. Rec. 133:65–71.
Bullock, T.H. 1961. The origins of patterned nervous discharge. Behaviour 17:48–59.
Burd, G.D., Nottebohm, F. 1985. Ultrastructural characterization of synaptic terminals formed on newly generated neurons in a song control nucleus of the adult canary forebrain. J. Comp. Neurol. 240:143–152.
Burek, M.J., Nordeen, K.W., Nordeen, E.J. 1991. Neuron loss and addition in developing zebra finch song nuclei are independent of auditory experience during song learning. J. Neurobiol. 22:215–223.
Cajal, S.R. 1894. The Croonian Lecture. La fine structure des centres nerveux. Proc. R. soc. Lond. B 55:444–467.
Cajal, S.R. 1911. Histologie du Systeme Nerveux, vol. 2, pp. 80–119. Maloire, Paris.
Canady, R.A., Kroodsma, D.E., Nottebohm, F. 1984. Population differences in complexity of a learned skill are correlated with brain space involved. Proc. Natl. Acad. Sci. USA 81:6232–6234.

Canady R.A., Burd, G.B., DeVoogd, T.J., Nottebohm, F. 1988. Effect of testosterone on input received by an identified neuron type of the canary song system: a Golgi/EM/Degeneration study. J. Neurosci. 8:3770–3784.
Cardin, J.A., Schmidt, M.F. 2004. Auditory responses in multiple sensorimotor song system nuclei are co-modulated by behavioral state. J. Neurophysiol. 91:2148–2163.
Dave, A.S., Yu, A.C., Margoliash, D. 1998. Behavioral state modulation of auditory activity in a vocal motor system. Science 282:2250–2254.
Dayer, A.G., Cleaver K.M., Abouantoun, T., Cameron, H.A. 2005. New GABAergic interneurons in the adult neocortex and striatum are generated from different precursors. J. Cell Biol. 168:415–427.
DeVoogd, T.J., Nottebohm, F. 1981. Gonadal hormones induce dendritic growth in the adult brain. Science 214:202–204.
DeVoogd, T.J., Nixdorf, B., Nottebohm, F. 1985. Synaptogenesis and changes in synaptic morphology related to acquisition of a new behavior. Brain Res. 329:304–308.
DeVoogd, T.J., Pyskaty, D.J., Nottebohm, F. 1991. Lateral asymmetries and testosterone-induced changes in the gross morphology of the hypoglossal nucleus in adult canaries. J. Comp. Neurol. 307:65–76.
DeVoogd, T.J., Krebs, J.R., Healy, S.D., Purvis, A. 1993. Relations between song repertoire size and the volume of brain nuclei related to song: comparative evolutionary analyses amongst oscine birds. Proc. R. Soc. Lond. B 254:75–82.
Doetsch, F., Caille, I., Lim, D.A., Garcia-Verdugo, J.M., Alvarez-Buylla, A. 1999. Subventricular zone astrocytes are neural stem cells in the adult mammalian brain. Cell 97:703–716.
Eales, L.A. 1985. Song learning in zebra finches: some effects of song model availability on what is learnt and when. Anim. Behav. 33:1293–1300.
Easter, S.S., Jr. 1983. Postnatal neurogenesis and changing connections. Trends Neurosci. 6:53–56.
Goldman, S.A., Nottebohm, F. 1983. Neuronal production, migration and differentiation in a vocal control nucleus of the adult female canary brain. Proc. Natl. Acad. Sci. USA 80: 2390–2394.
Gould, E., Reeves, A.J., Graziano M.S.A., Gross, C.G. 1999. Neurogenesis in the neocortex of adult primates. Science 286:548–552.
Graziadei, P.P.C., DeHan, R.S. 1973. Neuronal regeneration in frog olfactory system. J. Cell Biol. 59:525–530.
Graziadei, P.P.C., Monti Graziadei, G.A. 1978. Continuous cell renewal in the olfactory system. In "Handbook of Sensory Physiology", M. Jacobson, ed., vol IX, pp. 55–83. Springer Verlag, Berlin.
Graziadei, P.P.C., Monti Graziadei, G.A. 1985. Neurogenesis and plasticity of the olfactory sensory neurons. In "Hope for a New Neurology", F. Nottebohm, ed., Ann. N.Y. Acad. Sci., vol 457, pp. 143–161.
Gurney, M.E. 1981. Hormonal control of cell form and number in the zebra finch song system. J. Neurosci. 1:658–673.
Gurney, M.E., Konishi, M. 1980. Hormone induced sexual differentiation of brain and behavior in zebra finches. Science 208:1380–1383.
Hahnloser, R.H., Kozhevnikov, A.A., Fee, M.S. 2002. An ultra-sparse code underlies the generation of neural sequences in a songbird. Nature 419:65–70.
Hebb, D.O. 1949. The organization of behavior. John Wiley & Sons, New York.
Huber, F. 1960. Untersuchungen ueber die Funktion des Zentralnervensystems und insbesondere des Gehirns bei der Fortbewegung und der Lauterzeugung der Grillen. Z. vergl. Physiol. 44:60–132.
Immelmann, K. 1969. Song development in the zebra finch and other estrildid finches. In "Bird Vocalizations", R.A. Hinde, ed., pp. 61–74. Cambridge Univ. Press, London & New York.
Janata, P., Margoliash, D. 1999. Gradual emergence of song selectivity in sensorimotor structures of the male zebra finch song system. J. Neurosci. 19:5108–5118.

Johns, P.R. 1982. Formation of photoreceptors in larval and adult goldfish. J. Neurosci. 2:178–198.
Kandel, E.R. 2006. In search of memory. W.W. Norton & Co., New York and London.
Kao, M.H., Doupe, A.J., Brainard, M.S. 2005. Contributions of an avian basal ganglia-forebrain circuit to real-time modulation of song. Nature 433:638–643.
Kaplan, M.S. 1985. Formation and turnover of neurons in young and senescent animals: an electron microscopic and morphometric analysis. In "Hope for a New Neurology", F. Nottebohm, ed., Ann. N.Y. Acad. Sci., vol 457, pp. 173–192.
Kaplan, M.S., Hinds, J.W. 1977. Neurogenesis in the adult rat: electron microscopic analysis of light autoradiographs. Science 197:1092–1094.
Karten, H.J., Hodos, W. 1967. A stereotaxic atlas of the brain of the pigeon (*Columba livia*). The Johns Hopkins Press, Baltimore.
Katz, L.C., Gurney, M.E. 1981. Auditory responses in the zebra finch's motor system for song. Brain Res. 221:192–197.
Kirn, J.R., Nottebohm, F. 1993. Direct evidence for loss and replacement of projection neurons in adult canary brain. J. Neurosci. 13:1654–1663.
Kirn, J.R., Alvarez-Buylla, A., Nottebohm, F. 1991. Production and survival of projection neurons in the forebrain vocal center of adult male canaries. J. Neurosci. 11:1756–1762.
Kirn, J., O'Loughlin, B., Kasparian, S., Nottebohm, F. 1994. Cell death and neuronal recruitment in the high vocal center of adult male canaries are temporally related to changes in song. Proc. Natl. Acad. Sci. USA 91:7844–7848.
Kirn, J.R., Fishman, K., Sasportas, A., Alvarez-Buylla, F., Nottebohm, F. 1999. Fate of new neurons in adult canary high vocal center during the first 30 days after their formation. J. Comp. Neurol. 411:487–494.
Kittelberger, J.M., Mooney, R. 2004. Acute injections of brain-derived neurotrophic factor in a vocal premotor nucleus reversibly disrupt adult birdsong stability and trigger syllable deletion. J. Neurobiol. 62:406–424.
Koketsu, D., Mikami, A., Miyamoto, Y., Hisatsune, T. 2003. Nonrenewal of neurons in the cerebral neocortex of adult macaque monkeys. J. Neurosci. 23:937–942.
Konishi, M. 1963. The role of auditory feedback in the vocal behavior of the domestic fowl. Z. Tierpsychol. 20:349–367.
Konishi, M. 1965. The role of auditory feedback in the control of vocalization in the white-crowned sparrow. Z. Tierpsychol. 22:770–783.
Konorski, J. 1948. Conditioned reflexes and neuron organization. Cambridge Univ. Press, London.
Kornack, D.R., Rakic, P. 2001. Cell proliferation without neurogenesis in adult primate neocortex. Science 294:2127–2130.
Korr, H. 1980. Proliferation of different cell types in the brain. Adv. Anat. Embryol. Cell Biol. 61:1–69.
Kriegstein, A., Alvarez-Buylla, A. 2009. The glial nature of embryonic and adult neural stem cells. Ann. Rev. Neurosci. 32:149–184.
Lahousse, E. 1888. Recherches sur l'ontogenese du cervelet. Arch. de Biol. 8:43–110.
Lashley, K.S. 1950. In search of the engram. In "Physiological Mechanisms in Animal Behavior". Symposia Soc. Exp. Biol., vol IV, Cambridge Univ. Press, pp. 454–482.
Leonard, R.B., Goggeshall, R.E., Willis, W.D. 1978. A documentation of an age related increase in neuronal and axonal numbers in the stingray. J. Comp. Neurol. 179:13–22.
Li, X-C., Jarvis, E.D., Alvarez-Borda, B., Lim, D.A., Nottebohm, F. 2000. A relationship between behavior, neurotrophin expression and new neuron survival. Proc. Natl. Acad. Sci. USA 97:8584–8589.
Lipkind, D., Nottebohm, F., Rado, R., Barnea, A. 2002. Social change affects the survival of new neurons in the forebrain of adult songbirds. Behav. Brain Res. 133:31–43.
Long, M.A., Fee, M.S. 2008. Using temperature to analyse temporal dynamics in the songbird motor pathway. Nature 456:189–194.

Magavi, S.S., Leavitt, B.R., Macklis, J.D. 2000. Induction of neurogenesis in the neocortex of adult mice. Nature 405:951–955.
Margoliash, D. 1986. Preference for autogenous song by auditory neurons in a song system nucleus of the white-crowned sparrow. J. Neurosci. 6:1643–1661.
Marler, P. 1970a. A comparative approach to vocal learning: song learning in white-crowned sparrows. J. Comp. Physiol. Psychol. 71 (monogr.):1–25.
Marler, P. 1970b. Birdsong and speech development: could there be parallels? Am. Sci. 58:669–673.
Marler, P., Tamura, M. 1964. Culturally transmitted patterns of vocal behavior in sparrows. Science 146:1483–1486.
Marler, P., Waser, M.S. 1977. The role of auditory feedback in canary song development. J. Comp. Physiol. Psychol. 91:8–16.
Merkle, F.T., Tramontin, A.D., Garcia-Verdugo, J.M., Alvarez-Buylla, A. 2004. Proc. Natl. Acad. Sci. USA 101:17528–17532.
Messier, B., Leblond, C.P., Smart, I. 1958. Presence of DNA synthesis and mitosis in the brain of young adult mice. Exptl. Cell. Res. 14:224–226.
Miale, I.L., Sidman, R.L. 1961. An autoradiograpic analysis of histogenesis in the mouse cerebellum. Exp. Neurol. 4:277–296.
Mooney, R. 2009. Neural mechanisms for learned birdsong. Learn. Mem. 16:655–669.
Mundinger, P. 1970. Vocal imitation and individual recognition of finch calls. Science 168:480–482.
Nordeen, E.J., Nordeen, K.W. 1988. Sex and regional differences in the incorporation of neurons born during song learning in zebra finches. J. Neurosci. 8:2869–2874.
Nottebohm, F. 1971. Neural lateralization of vocal control in a passerine bird. I. Song. J. Exp. Zool. 177:229–261.
Nottebohm, F. 1972a. The origins of vocal learning. Am. Nat. 106:116–140.
Nottebohm, F. 1972b. Neural lateralization of vocal control in a passerine bird. II. Subsong, calls and a theory of vocal learning. J. Exp. Zool. 179:35–49.
Nottebohm, F. 1980a. Testosterone triggers growth of brain vocal control nuclei in adult female canaries. Brain Res. 189:429–436.
Nottebohm, F. 1980b. Brain pathways for vocal learning in birds: a review of the first 10 years. In "Progress in Psychobiology and Physiological Psychology", J.M.S. Sprage, A.N.E. Epstein, eds., vol. 9, pp. 85–124. Academic Press, New York.
Nottebohm, F. 1981. A brain for all seasons: cyclical anatomical changes in song control nuclei of the canary brain. Science 214:1368–1370.
Nottebohm, F. 1984. Birdsong as a model in which to study brain processes related to learning. Condor 86:227–236.
Nottebohm, F. 1985. Neuronal replacement in adulthood. In "Hope for a New Neurology", F. Nottebohm, ed., Ann. N.Y. Acad. Sci., vol 457, pp. 143–161.
Nottebohm, F. 1989. From birdsong to neurogenesis. Sci. Am. 260:74–79.
Nottebohm, F. 1993. The search for neural mechanisms that define the sensitive period for song learning in birds. Netherlands J. Zoology.
Nottebohm, F., Arnold, A.P. 1976. Sexual dimorphism in vocal control areas of the songbird brain, Science 194:211–213.
Nottebohm, F., Nottebohm, M.E. 1971. Vocalizations of breeding behavior of surgically deafened ring doves, *Streptopelia risoria*. Anim. Behav. 19:313–327.
Nottebohm, F., Nottebohm, M.E. 1978. Relationship between song repertoire and age in the canary, *Serinus canarius*. Z. Tierpsychol. 46:298–305.
Nottebohm, F., Stokes, T.M., Leonard, C.M. 1976. Central control of song in the canary, *Serinus canarius*. J. Comp. Neurol. 165:457–486.
Nottebohm, F., Kasparian, S., Pandazis, C. 1981. Brain space for a learned task. Brain Res. 213:99–109.
Nottebohm, F., Kelley, D.B., Paton, J.A. 1982. Connections of vocal control nuclei in the canary telencephalon. J. Comp. Neurol. 207:344–357.

Nottebohm, F., Nottebohm, M.E., Crane, L.A. 1986. Developmental and seasonal changes in canary song and their relation to changes in the anatomy of song-control nuclei. Behav. Neural. Biol. 46:445–471.

Nottebohm, F., Nottebohm, M.E., Crane, L.A., Wingfield, J.C. 1987. Seasonal changes in gonadal hormone levels of adult male canaries and their relation to song. Behav. Neural. Biol. 47:197–211.

Nottebohm, F. 1993. The search for neural mechanisms that define the sensitive period for song learning in birds. Netherlands J. Zoology.

Nottebohm, F., O'Loughlin, B., Gould, K., Yohay, C., Alvarez-Buylla, A. 1994. The life span of new neurons in a song control nucleus of the canary brain depends on time of year when these cells are born. Proc. Natl. Acad. Sci. USA 91:7849–7853.

Oelveczky, B.P., Andalman, A.S., Fee, M.S. 2005. Vocal experimentation in the juvenile songbird requires a basal ganglia circuit. PLoS Biol. 3:902–908, e153.

Okuhata, S., Saito, N. 1987. Synaptic connections of thalamo-cerebral vocal nuclei of the canary. Brain Res. Bull. 18:35–44.

Paton, J.A., Nottebohm, F. 1984. Neurons generated in the adult brain are recruited into functional circuits. Science 225:1046–1048.

Raisman, G., Field, P.M. 1971. Sexual dimorphism in the preoptic area of the brain. Science 173:731–733.

Rakic, P. 1985a. Limits of neurogenesis in primates. Science 227:154–156.

Rakic, P. 1985b. DNA synthesis and cell division in the adult primate brain. In "Hope for a New Neurology", F. Nottebohm, ed., Ann. N.Y. Acad. Sci., vol 457, pp. 193–211.

Rakic, P. 2002. Neurogenesis in adult primate neocortex: an evaluation of the evidence. Nat. Rev. Neurosci. 3:65–71.

Rasika, S., Nottebohm, F., Alvarez-Buylla, A. 1994. Testosterone increases the recruitment and/or survival of new high vocal center neurons in adult female canaries. Proc. Natl. Acad. Sci. USA 91:7854–7858.

Rasika, S., Alvarez-Buylla, A., Nottebohm, F. 1999. BDNF mediates the effects of testosterone on the survival of new neurons in an adult brain. Neuron 22:53–62.

Raymond, P.A., Easter, S.S. 1983. Postembryonic growth of the optic tectum in goldfish. I. Location of germinal cells and numbers of neurons produced. J. Neurosci. 3:1077–1091.

Roberts, T.F., Tschida, K.A., Klein, M.E., Mooney, R. 2010. Rapid spine stabilization and synaptic enhancement at the onset of behavioural learning. Nature 463:948–952.

Roeder, K.D. 1962. Neural mechanisms of animal behavior. Am. Zool. 2:105–115.

Schaper, A. 1894. Die morphologische und histologische Entwicklung des Kleinhirns. Morphol. Jahrb, 21:625–670.

Scharff, C., Nottebohm, F. 1991. A comparative study of the behavioral deficits following lesions of various parts of the zebra finch song system: implications for vocal learning. J. Neurosci. 11:2896–2913.

Scott, B.B., Lois, C. 2007. Developmental origin and identity of song system neurons born during vocal learning in songbirds. J. Comp. Neurol. 502:202–214.

Seri, B., Garcia-Verdugo, J.M., McEwen, B.S., Alvarez-Buylla, A. 2001. Astrocytes give rise to new neurons in the adult mammalian hippocampus. J. Neurosci. 21:7154–7160.

Simpson, H.B., Vicario, D.S. 1991. Early estrogen treatment of female zebra finches masculinizes the brain pathway for learned vocalizations. J. Neurobiol. 22:777–793.

Smart, I. 1961. The subependymal layer of mouse brain and its cell production as shown by radioautography after thymidine-H3 injection. J. Comp. Neurol. 116:325–348.

Stokes, T.C., Leonard, C.M., Nottebohm, F. 1974. The telencephalon, diencephalon and mesencephalon of the canary, *Serinus canaria*, in stereotaxic coordinates. J. Comp. Neurol. 156:337–374.

Sugita, N. 1918. Comparative studies on the growth of the cerebral cortex. V., pts. I and 11. J. Comp. Neur. 30:61–117.

Thorpe, W.H. 1954. The process of song learning in the chaffinch as studied by means of the sound spectrograph. Nature 173:465–469.
Thorpe, W.H. 1955. Comments on the "bird fancyer's delight" together with notes on imitation in the subsong of the chaffinch. Ibis 97:247–251.
Thorpe, W.H. 1958. The learning of song patterns by birds, with especial reference to the song of the chaffinch, *Fringilla coelebs*. Ibis 100:535–570.
Thorpe, W.H., Pilcher, P.M. 1958. The nature and characteristics of sub-song. Br. Birds 51:509–514.
Uzman, L.L. 1960. The histogenesis of the mouse cerebellum as studied by its tritiated thymidine uptake. J. Comp. Neurol. 114:137–159.
Vates, G.E., Nottebohm, F. 1995. Feedback circuitry within a song learning pathway. Proc. Natl. Acad. Sci. USA 92:5139–5143.
Vates, G.E., Broome, B.M., Mello, C.V., Nottebohm, F. 1996. Auditory pathways of caudal telencephalon and their relation to the song system of adult male zebra finches (*Taeniopygia guttata*). J. Comp. Neurol. 366:613–642.
Vates, G.E., Vicario, D.S., Nottebohm, F. 1997. Reafferent thalamo-"cortical" loops in the song system of oscine songbirds. J. Comp. Neurol. 380:275–290.
von Holst, E. 1935. Ueber den Prozess der zentralnervoesen Koordination. Pflueg. Arch. ges. Physiol. 236:149–158.
von Holst, E., von Saint Paul, U. 1960. Vom Wirkungsgefuege der Triebe. Naturwissenschaften 47:409–422.
Wade, J., Arnold, A.P. 1996. Functional testicular tissue does not masculinize development of the zebra finch song system. Proc. Natl. Acad. Sci. USA 93:5264–5268.
Waser, M.S., Marler, P. 1977. Song learning in canaries. J. Comp. Physiol. Psychol. 91:1–7.
Wiersma, C.A.G. 1962. The organization of the arthropod central nervous system. Am. Zool. 2:67–78.
Wilbrecht, L., Crionas, A., Nottebohm, F. 2002a. Experience affects recruitment of new neurons but not adult neuron number. J. Neurosci. 22:825–831.
Wilbrecht, L., Petersen, T., Nottebohm, 2002b. Bilateral LMAN lesions cancel differences in HVC neuronal recruitment induced by unilateral syringeal denervation. J. Comp. Physiol. A 188:909–915.
Wilbrecht, L., Williams, H., Gangadhar, N., Nottebohm, F. 2006. High levels of new neuron addition persist when the sensitive period for song learning is experimentally prolonged. J. Neurosci. 26:9135–9141.
Williams, H., Nottebohm, F. 1985. Auditory responses in avian vocal motor neurons: a motor theory for song perception in birds. Science 229:279–282.
Williams, H., Crane, L.A., Hale, T.K., Esposito, M.A., Nottebohm, F. 1992. Right side dominance for song control in the zebra finch. J. Neurobiol. 23:1006–1020.
Wilson, D.M. 1961. The central nervous control of flight in the locust. J. Exp. Biol. 38:471–490.
Yu, A.C., Margoliash, D. 1996. Temporal hierarchical control of singing in birds. Science 273:1871–1875.

Part II
Basic Aspects of Neurogenesis

Chapter 3
Fate Specification of Neural Stem Cells

Masakazu Namihira and Kinichi Nakashima

Abstract Neural stem cells (NSCs) possess the ability to self-renew and to differentiate along neuronal and glial lineages. At the molecular level, these processes are currently defined in terms of a dynamic interplay between extracellular cues, including cytokine signaling, and intracellular programs such as epigenetic modification, including histone methylation and acetylation and DNA methylation. This review discusses recent advances in our understanding of the molecular mechanisms that control the specification of neuronal, astroglial, and oligodendroglial fates in NSCs of the developing and adult central nervous system.

3.1 Introduction

Neural stem cells (NSCs) are defined as cells that possess the ability to self-renew and to generate the three major cell types in the central nervous system (CNS): neurons, astrocytes, and oligodendrocytes (reviewed in Gage, 2000; Okano, 2002; Temple, 2001). The molecular mechanisms controlling the specification of neural cell fates have been the focus of intense research in recent years. Mechanisms that specify cell fate in adult progenitors are receiving particular attention because of the lessons they may teach us about how to treat neurological diseases by controlling the behavior of transplanted or endogenous progenitors.

During brain development, telencephalic neuroepithelial cells, including NSCs, divide symmetrically in early gestation to increase their own numbers (Temple, 2001). These cells then undergo neurogenesis through mostly asymmetrical division, each giving rise to two distinct daughter cells: another NSC with the same potential as its mother cell, and a neuron. Toward the end of the neurogenic phase, NSCs eventually acquire multipotentiality, becoming able to generate astrocytes

M. Namihira and K. Nakashima (✉)
Laboratory of Molecular Neuroscience, Graduate School of Biological Sciences, Nara Institute of Science and Technology,
8916-5 Takayama-cho, Ikoma 630-0101, Japan
e-mail: kin@bs.naist.jp

T. Seki et al. (eds.), *Neurogenesis in the Adult Brain I: Neurobiology*,
DOI 10.1007/978-4-431-53933-9_3, © Springer 2011

and oligodendrocytes in addition to neurons (Qian et al., 2000). Adult NSCs have been found in the two principal neurogenic regions, the subgranular zone (SGZ) of the hippocampus and the subventricular zone (SVZ), both in vivo and in vitro (Cameron et al., 1993; Gage et al., 1998; Garcia-Verdugo et al., 1998). Even in certain nonneurogenic regions, including the spinal cord, there exist cells harboring the traits of NSCs when cultured in vitro (Gage, 2000).

Although the mechanisms of NSC fate determination are not yet fully understood, it is gradually becoming apparent that both extracellular cues including cytokine signaling and intracellular programs such as epigenetic gene regulation are deeply involved in their fate specification. The term "epigenetic mechanism" refers to any heritable influence (in the progeny of cells or of individuals) on chromosome or gene function that is not accompanied by a change in DNA sequence (Yoder et al., 1997), and includes processes such as DNA methylation, histone modification, and noncoding RNA expression. These epigenetic modifications ensure appropriate gene activation at each step of NSC differentiation.

In this review, we focus on molecular mechanisms that regulate the specification of neuronal, astroglial, and oligodendroglial fates in NSCs. We discuss fate specification not only in adult NSCs but also in those of the developing brain and spinal cord, because it is conceivable that our knowledge of mechanisms operating in embryonic NSC may also pertain to those in adult NSCs. The regulation of progenitor proliferation, self-renewal, and specification of neuronal subtype identities have been excellently reviewed elsewhere (Campbell, 2005; Molofsky et al., 2004) and are not discussed here.

3.2 Molecular Mechanisms Regulating Neuronal Fate

We focus first on the major transcription factors and epigenetic regulation governing neuronal fate determination. Accumulating evidence suggests that different steps of adult neurogenesis are also regulated by physiological and pathological stimuli, including an enriched environment, stress, learning, and seizures (Ming and Song, 2005); these topics are covered by other authors in this book.

3.2.1 Neurogenic bHLH

Recent experiments have shown that proliferation, specification, and differentiation of NSCs are controlled to a large extent by transcription factors with basic helix-loop-helix (bHLH) motifs. The bHLH transcription factors (named after the structural motif that mediates their DNA-binding and dimerization functions) (Murre et al., 1989) are a large family, with 125 members encoded in the human genome (Ledent et al., 2002). In an NSC that is destined to become a neuron, the transition from proliferation to neurogenesis is accompanied by an increase in proneural bHLH factor expression. Neurogenesis is mediated by two broad categories of bHLH factors: proneural bHLH factors [e.g., NEUROGENIN1, NEUROGENIN 2 (NGNs), and MASH], which are involved in initiating neurogenesis, and neuronal differentiation bHLH factors (e.g., NEUROD), which facilitate terminal differentiation. These bHLH proteins bind DNA as heterodimers complexed with E proteins (E12, E47, E2-2, HEB) (Fig. 3.1). The bHLH domain mediates interaction with DNA sequences that contain the core hexanucleotide motif CANNTG, known as the E box (reviewed in Bertrand et al., 2002). Transactivation is mediated by the interaction of bHLH heterodimers with coactivators, such as p300/CBP and PCAF (Fig. 3.1), which then recruit a large complex that includes the basal transcriptional machinery. These coactivators also facilitate transcription by acetylating histones, leading to the unraveling of DNA from a tightly packed structure to one that is accessible to the transcriptional machinery. Conversely, the bHLH proteins are negatively regulated by another set of bHLH factors, such as HES-1 and HES-5 (where Hes is a homologue of hairy and enhancer of split) (Fig. 3.1). These

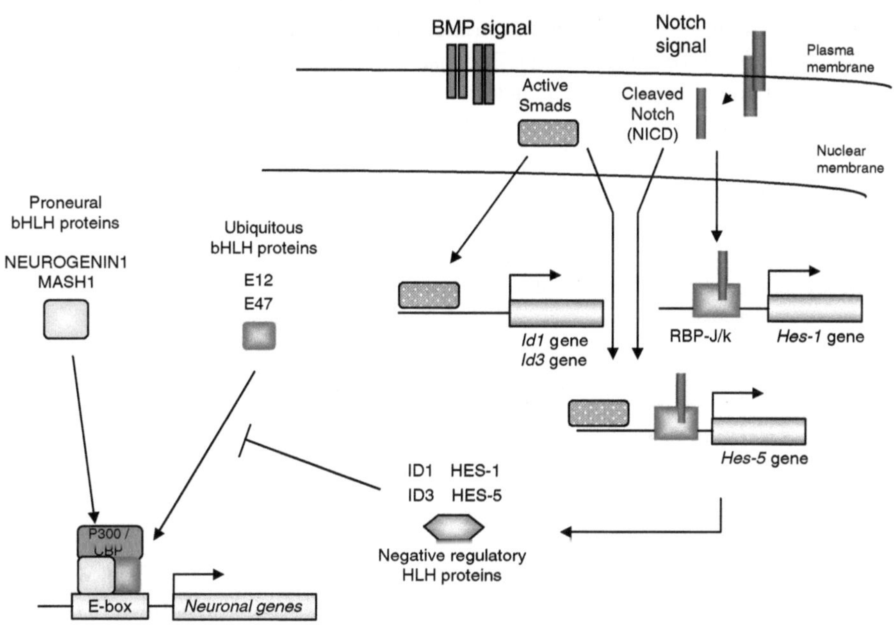

Fig. 3.1 Regulation of neuronal differentiation by bHLH factors. Several tissue-specific basic helix-loop-helix (bHLH) factors, such as NEUROGENIN1, dimerize with E proteins, such as E12, and bind to DNA elements termed E-boxes (CANNTG) to promote the expression of target neuronal genes. Activation of BMP and Notch signaling upregulates the expression of negative HLH factors. The ID and HES proteins, which inhibit the function of proneural bHLH transcription factors by competitively binding to their heterodimerization partners E12 and E47, leading to suppression of neurogenesis of neural stem cells (NSCs). When Notch is activated by its ligands, such as Delta, proteolysis occurs and the intracellular domain of Notch (NICD) is translocated into the nucleus and complexes with RBP-J/k. This complex activates transcription of *Hes-1* and *Hes-5*. Bone morphogenetic protein (BMP)-activated Smads also induces *Hes-5* expression but it requires the NICD/RBP-Jk complex, implying cross-talk between Notch and BMP signaling

mammalian bHLH proteins are distant relatives of the product of the *Drosophila* pair-rule gene hairy (Akazawa et al., 1992; Ohtsuka et al., 1999). The expression of *Hes-1* and *Hes-5* is known to be induced by Notch signaling and has been suggested to inhibit transcriptional activity of neurogenic bHLH proteins by competitive binding to their heterodimer partners E12 and E47 (Fig. 3.1) (Akazawa et al., 1992; Ohtsuka et al., 1999). Another type of negative regulator of HLH proteins is the ID (for inhibitor of differentiation) family of proteins, which have an HLH domain but lack a basic region (Norton et al., 1998). ID proteins inhibit the function of tissue-specific bHLH factors by a mechanism analogous to that used by HES proteins (see Fig. 3.1) (Norton et al., 1998).

An important subject for future studies of neuronal differentiation is the nature of the regulatory mechanisms that initiate the expression of proneural bHLH genes. Although members of the HES family are known to repress the expression of *Mash1* (Chen et al., 1997), the signaling pathways that upregulate proneural factor activity are largely unknown. NGN1, NGN2, and MASH1 all have consensus sequences that correspond to glycogen synthase kinase 3 (GSK3) phosphorylation sites, which may represent a common mechanism to regulate the activity of these bHLH factors (Moore et al., 2002). Because two families of morphogens, Wnts and FGFs, are known to regulate the activity of GSK3, it is tempting to speculate that bHLH factors couple extrinsic signals and intrinsic pathways through GSK3-mediated phosphorylation. Moreover, the promoters of *Ngn1* and *Ngn2* can be directly activated by a β-catenin/T-cell factor (TCF) complex in transfected cells (Hirabayashi et al., 2004; Israsena et al., 2004), and *Ngn2* expression is lost in mice with a cortical-specific deletion of β-catenin (Machon et al., 2003), suggesting that Wnt signaling promotes neurogenesis by directly inducing expression of *Ngn1/2*.

3.2.2 Epigenetic Modification in Neuronal Differentiation of NSCs

3.2.2.1 Chromatin Modification

Chromatin modification plays a critical role in the regulation of cell type-specific gene expression. The cell has two basic tools at its disposal to dynamically regulate chromatin state: covalent modification of chromatin, which includes histone modifications such as acetylation and methylation at specific positions, and ATP-dependent chromatin remodeling.

One of the best characterized histone modifications is lysine acetylation, which is mediated by two groups of enzymes: histone acetyltransferases (HATs) and histone deacetylases (HDACs). HATs induce acetylation of N-terminal histone tails, which is considered to decrease the interaction of the positively charged histone tails with the negatively charged phosphate backbone of DNA and hence result in relaxation of the nucleosomes. HDACs catalyze the reverse reaction; in the deacetylated state, histones package DNA into condensed chromatin, which in turn prevents access of transcriptional activators to their target sequences, thus resulting in transcriptional repression (Hsieh and Gage, 2004). The well-known antiepileptic valproic acid (VPA) functions as an inhibitor of HDACs. VPA has been shown to induce neuronal differentiation of adult NSCs (Fig. 3.2a) (Hsieh et al., 2004a). In addition, VPA inhibits glial cell differentiation of NSCs, even under conditions that favor lineage-specific differentiation. Among the VPA-upregulated neuron-specific genes is *NeuroD*, whose overexpression in NSCs results in the induction and suppression of neuronal and glial differentiation,

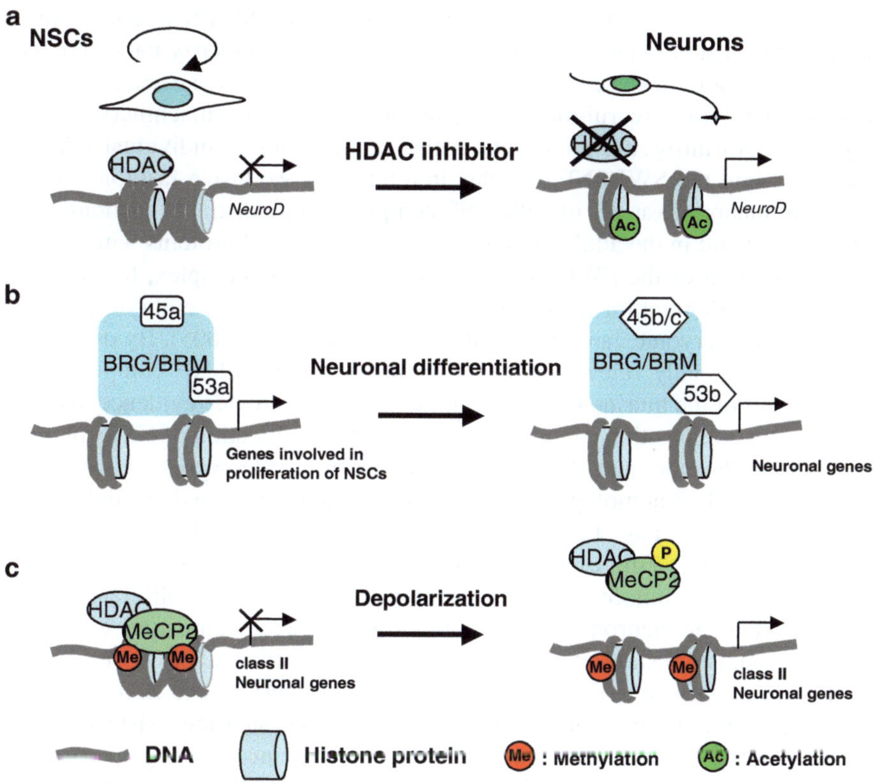

Fig. 3.2 Epigenetic modification regulating neuronal differentiation. (**a**) Histone deacetylase (HDAC) inhibitors, such as valproic acid (VPA), promote neuronal fate through the induction of neurogenic transcription factors including NeuroD. (**b**) Brahma (BRM)-associated factor (BAF) chromatin remodeling complexes, also known as SWI/SNF complexes, comprise a catalytic subunit (BRG/BRM) and additional core subunits. Neural progenitor-specific BAF (npBAF) complexes contain two additional subunits, BAF45a and BAF53a. During neuronal differentiation, these are exchanged for neuron-specific homologous proteins, BAF45b/c and BAF53b, thus forming postmitotic neuron-specific BAF (nBAF) complexes. Although npBAF complexes specify gene expression patterns necessary for NSC proliferation, the switch in cofactor composition is critical for neuronal differentiation. (**c**) Class II neuronal genes are bound by CoREST (not indicated) and MeCP2 complexes on adjacent methylated CpGs in their promoters throughout neuronal differentiation. Upon membrane depolarization, MeCP2 becomes phosphorylated and dissociates from the chromatin of target genes, resulting in maximal levels of the gene expression

respectively. Taken together, these results suggest that VPA simultaneously promotes neuronal fate and inhibits glial fate through the induction of neurogenic transcription factors, including NEUROD (Hsieh et al., 2004a).

ATP-dependent chromatin remodeling is carried out by enzyme complexes that use ATP hydrolysis to slide histone octamers around which nucleosomal DNA is wrapped, thus exposing the DNA and facilitating the binding of sequence-specific transcription factors (reviewed in de la Serna et al., 2006). Among these, the best understood chromatin remodelers belong to the family of Brahma-related gene (BRG)/Brahma (BRM)-associated factor (BAF) complexes, also known as SWI/SNF complexes. Similar to most large enzyme machines, SWI/SNF complexes are composed of a set of central subunits that provides the core enzymatic activity of the complex, along with several associated factors which serve specialized roles in substrate targeting, recruitment of additional regulatory machinery, or both (Fig. 3.2b). An intriguing possibility is that homologues of individual BAFs are assembled into the SWI/SNF complex in a mutually exclusive fashion, allowing for a combinatorial variety of SWI/SNF complexes with distinct functions during development and in the adult. Association of neural bHLH proteins with BRG1, a catalytic subunit of the SWI/SNF chromatin remodeling complex, has also been shown to be important for their neurogenic activity in *Xenopus* embryos and in the mammalian embryonic carcinoma cell line P19 (Seo et al., 2005). By demonstrating two functionally relevant cofactor swaps in the SWI/SNF complex during development of the murine brain, Lessard et al. (2007) provide evidence that SWI/SNF subunit composition can indeed regulate fate specification of NSCs in mouse brain during development (Fig. 3.2b). Examination of the expression patterns of BAF45 and BAF53 homologues during brain development revealed a tight switch: around embryonic day 12–13 (E12–13), BAF45a and BAF53a give way to BAF45b/c and BAF53b, respectively. This event occurs around the same time that neural stem/progenitor cells in the developing brain cease to proliferate and start to differentiate into neurons. Indeed, immunofluorescence staining for these BAFs in E12.5 spinal cord showed mutually exclusive expression patterns, with BAF45a and BAF53a confined to proliferative regions and BAF45b/c and BAF53b mainly expressed in the differentiated zone. These data suggested the existence of two distinct SWI/SNF complexes during development: a neural progenitor-specific complex, termed npBAF (BAF45a and BAF53a), and a postmitotic neuron-specific complex, nBAF (BAF45b/c and BAF53b) (Lessard et al., 2007) (see Fig. 3.2b). Lessard et al. also showed that neural progenitor-specific BAF45a gain-of-function led to an increase in the mitotic index of cells both in vivo and in cell culture, whereas BAF45a loss-of-function, BAF53a loss-of-function, or both reduced the cell proliferation rate in cultured cells without affecting the survival or extent of neuronal differentiation. Cortical cultures and neurospheres prepared from these *Brg*-deficient embryos showed reduced mitotic index and self-renewal capacity, respectively (Lessard et al., 2007). These findings clearly show that BAF complexes exert their effect on cell proliferation and differentiation by producing substantial changes in gene expression programs in a chromatin-dependent manner (Fig. 3.2b).

The RE1-silencing transcription factor REST (Chong et al., 1995), also known as neuron-restrictive silencer factor NRSF (REST/NRSF) (Schoenherr and Anderson, 1995), plays a fundamental role in the progression of pluripotent cells to lineage-restricted neural progenitors (Ballas et al., 2005; Ballas and Mandel, 2005). REST/NRSF is a zinc finger protein and blocks transcription of its target genes by binding to a specific consensus 21- to 23-base-pair (bp) RE1-binding site/neuron-restrictive silencer element (RE1/NRSE) that is present in a large number of genes encoding proteins relevant to fundamental neuronal traits (Schoenherr et al., 1996), such as ion channels, synaptic vesicle proteins, and neurotransmitter receptors. Accordingly, REST/NRSF prevents premature expression of terminal differentiation genes in NSCs (for a review, see Schoenherr et al., 1996). REST/NRSF bound to the RE1/NRSE site of neuronal genes in embryonic stem (ES) cells, and neuronal gene chromatin in ES cells was maintained in a poised state via a specialized REST/NRSF repressor complex, including CoREST, HDAC1, and MeCP2, that permitted basal expression of neuronal genes at a very low level. When ES cells differentiate into neural progenitors, REST/NRSF protein was downregulated by proteasomal degradation and the REST/NRSF repressor complex disappeared from the RE-1/NRSE site in the promoter regions of neuronal genes, with the result that the genes became derepressed and activated in neurons (Ballas et al., 2005). It is suggested that the loss of REST/NRSF in neurons leads to the expression of two classes of RE-1/NRSE-containing neuronal genes (Ballas et al., 2005). Class I genes were expressed by default, relying solely on dissociation of the REST/NRSF repressor complex from the RE-1/NRSE site for maximal expression. In contrast, class II genes, typified by *brain-derived neurotrophic factor (bdnf)* and *calbindin* genes, contain, in addition to the REST/NRSF complex at the RE-1/NRSE site, CoREST and MeCP2 complexes on adjacent methylated CpGs in their promoters that remained bound throughout differentiation (Fig. 3.2c). Upon depolarization of neural cells, while CoREST remained bound to the chromatin, MeCP2 became phosphorylated and dissociated, together with HDAC and mSin3, from the chromatin of class II genes, leading to a maximal level of expression (Fig. 3.2c). The *calbindin* and *bdnf* genes are representative of a large class of genes whose expression is upregulated in neurons by disruption of HDAC activity and, in some cases, also by depolarization (see Fig. 3.2c). Many genes in this class are implicated in the functional plasticity of neurons in vivo (Ballas and Mandel, 2005).

3.2.2.2 ncRNA

Noncoding (nc) RNAs, including small interfering (si) RNA and micro (mi) RNA, appear to play important roles in various biological processes such as X-inactivation in mammals, transcriptional and posttranscriptional gene silencing in plants, and lineage-specific differentiation in hematopoietic stem cells (Hsieh and Gage, 2004). A novel small double-stranded (ds) ncRNA has been reported to play a critical role in mediating neuronal differentiation (Kuwabara et al., 2004). The sequence of this ncRNA matches RE1/NRSE, the binding site for REST/NRSF. These dsRNA molecules are specifically expressed in adult NSCs early in their differentiation. At the onset of neuronal differentiation, NRSE dsRNAs are proposed to interact with the REST/NRSF complex and trigger a conformational change, which prevents its association with corepressor proteins such as HDACs and methyl DNA-binding proteins, thus converting REST/NRSF from a repressor to an activator. This type of dsRNA has been named small modulatory (sm) RNA, and the discovery of smRNA extends the important contribution of ncRNAs as key regulators in cell behavior at both transcriptional and posttranscriptional levels.

miRNAs regulate several aspects of development. Among miRNAs expressed predominantly in neural tissues, miR-124a has been shown to participate in the differentiation of NSCs into neurons by mediating degradation of nonneuronal gene transcripts (Conaco et al., 2006). This study indicated that miR-124a expression is regulated by REST/NRSF. As described earlier, REST/NRSF is known to recruit HDACs via corepressors associated with its repressor domains to suppress target gene expression (Andres et al., 1999; Battaglioli et al., 2002; Roopra et al., 2000). Therefore, in neural progenitors expressing REST/NRSF, suppression of miR-124a expression by REST/NRSF permits the existence of nonneuronal gene transcripts in the cells. In differentiated neurons, in which REST is no longer present on the miR-124a gene chromatin, miR-124a is expressed, causing rapid degradation of nonneuronal transcripts. REST/NRSF thus links transcriptional and posttranscriptional events to fine-tune the balance of phenotype between neuronal and nonneuronal cells.

3.3 Molecular Mechanisms Regulating Glial Fates

3.3.1 Astrocytes

Astrocytes have traditionally been viewed as a neural cell population with structural and supportive roles such as supplying essential substrates to neurons and participating in the blood–brain barrier (Kettenmann and Ransom, 1988). It is now recognized not only that certain astrocytic cell subpopulations have stem cell properties (Doetsch, 2003), but also that others even modulate directly the development and activity of the synapse (Christopherson et al., 2005). In many cases, astrocyte differentiation is characterized both in vivo and in vitro by the expression of the two most typical astrocyte-specific markers, the intermediate filament glial fibrillary acidic protein (GFAP) and the calcium-binding protein S100β. Regulation of the *gfap* and *S100β* promoters has been used as the main model to examine the molecular mechanisms controlling astrocyte fate specification and differentiation. These genes are not thought to have a role in selection of the astroglial lineage, but it is assumed that the mechanisms involved in regulating their expression also operate on yet-unidentified genes involved in specification of astroglial fate. A number of signaling pathways have been shown to promote the commitment of progenitors to the astroglial fate. These different pathways synergize extensively, either at the level of their signaling cascades or directly on the *gfap* promoter (Fig. 3.3a).

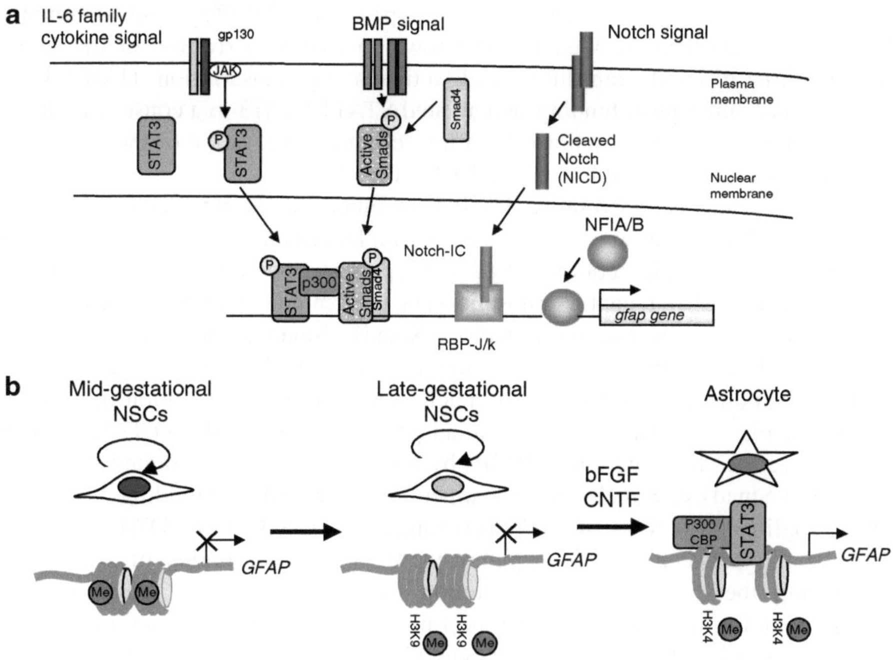

Fig. 3.3 Molecular mechanisms of astrocyte differentiation. (**a**) JAK-STAT, BMP-Smad, and Notch signaling pathways and transcriptional factor NFIA/B regulate the astrocyte-specific gene *gfap*. Synergistic integration of leukemia inhibitory factor (LIF) and BMP2 signaling is achieved by complex formation of their respective downstream transcription factors, STAT3 and Smads, together with p300. (See text for details.) (**b**) Epigenetic modification regulating astrocyte differentiation in the developing brain. At midgestation, when neuroepithelial cells differentiate only into neurons, the STAT3-binding site in the astrocyte-specific *gfap* promoter is highly methylated. Therefore, even if the cells are surrounded by STAT3-activating cytokine, STAT3 cannot activate target gene transcription. As gestation proceeds, the STAT3-binding site becomes demethylated, enabling STAT3 to bind its recognition sequence and to induce gene expression. Thus, DNA methylation in a cell type-specific gene promoter is a critical determinant for NSCs to acquire multipotentiality for differentiation during development. In late-gestational NSCs, cotreatment with fibroblast growth factor 2 (FGF2) and ciliary neurotrophic factor (CNTF) leads to a marked increase in the fraction of cells that express *gfap*. FGF2 facilitates access of the STAT/CBP complex to the *gfap* promoter by inducing H3K4 methylation and suppressing H3K9 methylation around the STAT3-binding site

3.3.1.1 Signaling Pathways Promoting Astroglial Fate

Cytokines in the interleukin-6 (IL-6) family have been suggested to have the potential to induce astrocytic differentiation of NSCs (Bonni et al., 1997). IL-6 members [i.e., IL-6, IL-11, leukemia inhibitory factor (LIF), oncostatin M (OSM), ciliary neurotrophic factor (CNTF), cardiotrophin-1, and cardiotrophin-like cytokine] share the membrane glycoprotein gp130 in their functional receptor complexes as a critical component for signal transduction (Elson et al., 2000; Lelievre et al., 2001; Taga and Kishimoto, 1997). Upon binding of an IL-6 family member to its specific receptor, gp130 dimerizes either with itself or with another partner, such as the LIF or OSM receptor, depending on the type of cytokine (Taga and Kishimoto, 1997). This dimerization of gp130 is known to lead to activation of the Janus tyrosine kinase/signal transducer and activator of transcription (JAK-STAT) pathway. The subsequent binding of activated STAT1/STAT3 to a consensus site in the *gfap* promoter plays a major role in transcriptional activation of the *gfap* gene (Bonni et al., 1997; Nakashima et al., 1999) (Fig. 3.3a).

Bone morphogenetic proteins (BMPs), members of the transforming growth factor-β (TGF-β) superfamily, signal through a heterotetrameric serine/threonine kinase receptor complex composed of two type I and two type II receptor molecules. Activated BMP receptors phosphorylate C-terminally located serine residues in the downstream transcription factors Smad1, Smad5, and Smad8 (Wrana, 2000), which then bind with the common mediator Smad protein Smad4, and the resultant complex accumulates in the nucleus to activate transcription of specific genes (see Fig. 3.2). The inhibitory Smad proteins Smad6 and Smad7 have been shown to repress the function of BMPs by inhibiting receptor-mediated phosphorylation of Smad1 or Smad5, or by competing with Smad4 for binding to Smad1. The astrogliogenic activity of BMPs is strongly enhanced by JAK-STAT signaling. The synergy between these two pathways reflects the formation of a complex comprising the downstream transcription factors STAT1/STAT3 and Smad1/Smad4 bound to nearby sites in the *gfap* promoter, as well as the recruitment of the coactivator P300, resulting in the synergistic activation of *gfap* (see Fig. 3.3a) (Nakashima et al., 1999).

BMPs have been shown to promote neuronal differentiation in early cortical explant cultures at E12–E13 in the mouse (Li et al., 1998). In older progenitor cultures (E14 and later), however, BMP2 and BMP4 instead promote astrocyte differentiation and repress the alternate neuronal and oligodendrocytic fates (Gross et al., 1996; Nakashima et al., 2001). Inhibition of neurogenesis by BMP2 involves induction of the antineurogenic transcription factors *Id1*, *Id3*, and *Hes-5* (Nakashima et al., 2001). Although *Id1* and *Id3* are direct targets of the BMP-signaling pathway, BMP collaborates with the Notch-signaling pathway to induce *Hes-5* expression; BMP-activated Smad1 forms a complex with Notch-ICD (Notch-intracellular domain) and the coactivators P300 and P/CAF on the *Hes-5* promoter, resulting in enhanced expression of *Hes-5* (Takizawa et al., 2003) (see Fig. 3.1). HES and ID proteins sequester E proteins, the dimerization partners of the proneural bHLH proteins Mash1 and NGNs, thereby blocking their activity and promoting the degradation of Mash1 (Vinals et al., 2004). Inhibition of oligodendrogenesis by BMPs also involves the induction of ID proteins that dimerize with oligodendrocyte-inducing bHLH transcription factors (OLIGs; see below) and their E protein partners, and thus block their oligodendroglial-promoting activity (Samanta and Kessler, 2004).

BMPs and BMP receptors are expressed in the adult SVZ, and BMP signaling also promotes astroglial differentiation and inhibits neurogenesis in this system. However, the BMP inhibitor Noggin is expressed by ependymal cells adjacent to the SVZ, suggesting that Noggin creates a neurogenic niche in the SVZ that promotes neurogenesis through inhibition of BMP signaling (Lim et al., 2000).

The activation of Notch signaling not only inhibits neurogenesis but also promotes gliogenesis, as first demonstrated in the peripheral nervous system and retina (Morrison et al., 2000; Furukawa et al., 2000). Similar findings were then reported for embryonic telencephalic NSCs (Chambers et al., 2001; Grandbarbe et al., 2003) and adult hippocampal NSCs (Tanigaki et al., 2001), suggesting that this is a general function of Notch. Analysis of Notch receptor activity in the mouse with a fluorescent reporter (Kohyama et al., 2005) or with an antibody specific for an activated form of Notch1 (Tokunaga et al., 2004) showed that the pathway is indeed active in radial glial cells as well as transiently in immature *gfap*-positive astrocytes. Notch has been proposed to induce astrogliogenesis directly through activation of the *gfap* promoter by a transcriptional complex involving the Notch-ICD and the DNA-binding protein RBP-J/k (also termed CSL, CBF1, and suppressor of hairless) (see Fig. 3.3a) (Ge et al., 2002). Interestingly, RBP-J/k represses *gfap* transcription in self-renewing progenitors when bound with the corepressor N-CoR (Hermanson et al., 2002), suggesting that activation of *gfap* transcription at the onset of astrogliogenesis involves the replacement of N-CoR by Notch-ICD, resulting in a functional switch for RBP-J/k from transcriptional repressor to activator. Importantly, although Notch activation has already started at midgestation, when active neurogenesis is occurring, it maintains progenitors in an undifferentiated state rather than promoting astrogenesis, providing evidence that Notch function is influenced by the cellular context, a recurrent theme with signaling pathways specifying neural cell fates.

3.3.1.2 Transcription Factors Inducing Astrogenesis

Recently, Deneen et al. reported that nuclear factor I/A (NFIA) and B (NFIB), members of a family of CCAAT box element-binding transcription factors (Gronostajski, 2000), are both necessary and sufficient to promote glial fate specification in embryonic spinal cord progenitors in vivo (Deneen et al., 2006). Their previous studies had shown that adult mice deficient in NFIA or NFIB exhibited a reduction in cortical GFAP expression (das Neves et al., 1999; Steele-Perkins et al., 2005), as well as a reduction in the number of midline glia (Shu et al., 2003). They then reported that in E18.5 embryos lacking either NFIA or NFIB, a reduction in spinal cord GFAP expression was also observed (Deneen et al., 2006). Furthermore, misexpression of NFIA or NFIB was sufficient to accelerate GFAP expression in astrocytic precursors by several days, in vivo and in vitro. These data argue that NFIA/B promote the terminal differentiation of astrocytes. This role of NFIA/B is likely to be direct, as functional NFI-binding sites have been identified in the *gfap* promoter (see Fig. 3.3a) (Cebolla and Vallejo, 2006). It will be important to determine the relationship between NFIA/B and other regulators of *gfap* expression, such as in the JAK/STAT and BMP/SMAD pathways (Bonni et al., 1997; Nakashima et al., 1999).

In addition to NFIA/B, stem cell leukemia (*Scl*, also known as Tal1), which encodes a bHLH transcription factor, is important for the regulation of both astrocyte versus oligodendrocyte cell fate acquisition and V2b versus V2a interneuron cell fate acquisition in the p2 domain of the developing vertebrate spinal cord (Muroyama et al., 2005). In the developing spinal cord, after V2 interneuron production, p2 progenitors give rise to a wave of astrocyte precursors (Zhou and Anderson, 2002; Stolt et al., 2003). A significant decrease in p2 domain-associated S100β-expressing cells in conditionally knocked-out mice, in which the *Scl* gene is specifically ablated in NSCs, was observed at E14 compared with controls. In contrast, these mice showed an increased number of oligodendrocyte progenitor cells (OLPs) that were normally located in the dorsal part of the spinal cord at E14. Furthermore, by forced expression of mouse *Scl* in the chick neural tube, ectopic development of S100β-expressing cells and robust induction of other astrocyte-specific marker genes were shown at the pMN domain, where motor neurons develop under normal conditions. These findings suggest that SCL function is necessary and sufficient for p2-associated astrocyte development, and that it represses the production of OLPs in the developing spinal cord (Muroyama et al., 2005).

3.3.1.3 Epigenetic Modification Regulating Astrocyte Differentiation

Methylation of genomic DNA at CpG dinucleotides is one of the major epigenetic modifications of mammalian genomes, implicated in processes such as genomic imprinting, X chromosome inactivation, aging, tumorigenesis, and the regulation of tissue-specific gene expression (Robertson and Jones, 2000). In these biological processes, CpG methylation contributes in a general way to transcriptional suppression by favoring the formation of inactive chromatin or by preventing transcriptional regulators from binding to their target DNA elements. It has become increasingly evident that DNA methylation also participates in the acquisition of multipotentiality of NSCs during development.

As already described, NSCs lack multipotentiality in early gestation and only differentiate into neurons during midgestation. NSCs then gradually acquire multipotentiality and differentiate into astrocytes and oligodendrocytes in late gestation (Temple, 2001). DNA methylation in astrocyte-specific gene promoters may be a critical cell-intrinsic determinant of NSC differentiation into astrocytes in the fetal brain (see Fig. 3.3b) (Takizawa et al., 2001). The IL-6 family of cytokines, including LIF and CNTF, efficiently induce astrocytic differentiation of NSCs through activation of the JAK/STAT pathway, as indicated earlier (Bonni et al., 1997; Nakashima et al., 1999). However, in contrast to cultures of neuroepithelial cells prepared from mouse telencephalon at E14 (relatively late gestation), GFAP was not induced in a culture of telencephalic neuroepithelial cells at E11 (midgestation) even when stimulated with the JAK/STAT pathway-activating cytokine LIF (Takizawa et al., 2001). Furthermore, the methylation target sequence CpG occurs in the STAT-binding element (TTCGGAGAA) itself within the *gfap* promoter, and this element is conserved in the mouse, rat, and human (Bonni et al., 1997; Nakashima et al., 1999). Given these findings, the methylation status of this particular STAT-binding site was compared between E11 and E14 neuroepithelial cells. These experiments revealed that the STAT-binding site was hypermethylated in E11 neuroepithelial cells, which do not respond to LIF to induce *gfap* expression, but was barely methylated in E14 neuroepithelial cells, which have the potential to express *gfap* in response to LIF stimulation (see Fig. 3.3b) (Takizawa et al., 2001). Moreover, STAT3 does not bind to the methylated form of its target sequence, explaining the lack of response of the *gfap* promoter to cytokine stimulation (Takizawa et al., 2001). S100β is also an astrocytic marker and is expressed during earlier stages of astrocytic differentiation than is GFAP. A particular cytosine residue in the *S100β* promoter becomes demethylated, correlating with the onset of *S100β* expression in the brain at E14 (Namihira et al., 2004). In addition, another astrocyte-inducing cytokine, BMP2, increases histone acetylation around the CpG site in neuroepithelial cells at E14, but not at E11 when *S100β* expression is not yet observed in vivo. Thus, it is conceivable that DNA methylation plays an important role in defining the timing of NSC fate specification (see Fig. 3.3b).

The importance of this methylation in vivo was subsequently demonstrated by analysis of a conditional knockout of the DNA methyltransferase 1 (*dnmt1*) gene in NSCs (Fan et al., 2005). The brains of these mice displayed decreased numbers of neurons, precocious astrogenesis, and aberrant upregulation of the *gfap* and *S100β* genes. Interestingly, genes in the gp130-JAK-STAT pathway were also derepressed in *dnmt1*-deficient NSCs, and STAT3 inhibition abolished the observed increase in gliogenesis (Fan et al., 2005). A subsequent study provided further support for the idea that methylation of genes in the gp130-JAK-STAT pathway regulated the gliogenic potential of cortical NSCs, and also demonstrated that the cytokines themselves enhanced the expression of their cognate pathway molecules, thereby defining a positive feed-forward loop (He et al., 2005). These data suggest that DNA methylation regulates the timing and magnitude of astrocytic differentiation through both modulation of JAK-STAT activity and direct inhibition of glial marker genes via inactive chromatin modifications (Fan et al., 2005).

Recently, fibroblast growth factor-2 (FGF2), which by itself is not capable of inducing astrocyte-specific gene expression, has been reported to positively affect the ability of CNTF, an IL-6 cytokine family member, to induce expression of GFAP (see Fig. 3.3b) (Song and Ghosh, 2004). FGF2 promotes access of a complex involving STAT3 and a HAT, CREB-binding protein (CBP), to the *gfap* promoter by inducing H3K4 methylation while suppressing H3K9 methylation at the STAT-binding site. These findings suggest that extracellular signals can modulate the accessibility of transcription factors to gene promoters by changing in local histone modification (Fig. 3.3b).

3.3.2 Oligodendrocytes

Once most cortical neurons have migrated to their correct positions, they are surrounded by astrocytes to form functional synapses. At approximately this time, oligodendrocytic differentiation begins. Oligodendrocytes tend to arise from specific locations within the CNS, in particular the ventral region of the neural tube, and migrate into the white matter of the CNS to myelinate axons (reviewed in Noctor et al., 2004).

3.3.2.1 Signaling Pathways Regulating Oligodendroglial Fate

Injection of neutralizing antibodies against Sonic hedgehog (Shh) into wild-type and Shh-knockout mice suggested that Shh is important for the generation of oligodendrocytes and their precursors. Furthermore, ectopic administration of Shh, or transplantation of notochord, where Shh is produced, to dorsal spinal cord leads to ectopic oligodendrocyte generation, suggesting that Shh plays a critical role in oligodendrocytic differentiation (Pringle et al., 1998). Shh promotes oligodendrogenesis by inducing the expression of OLIG1 and OLIG2, two essential determinants of oligodendrocytic fate, in ventral telencephalic progenitors (Lu et al., 2000; Yung et al., 2002). Gain- and loss-of-function experiments in mice have shown that Shh is both necessary and sufficient to induce OLIG1 and OLIG2 expression in the telencephalon (Lu et al., 2000; Nery et al., 2001; Zhou et al., 2000).

Platelet-derived growth factor (PDGF) has also been shown to be an important factor, maintaining proliferation of oligodendrocyte precursors as well as stimulating their differentiation into mature oligodendrocytes (Lee et al., 2000). However, it remains unclear whether PDGF also has an earlier role in specification of OLPs. PDGF receptor is expressed by a subset of stem cells in the adult SVZ, and PDGF signaling in the SVZ promotes the generation of oligodendrocytes, possibly at the expense of neuronal production (Jackson et al., 2006). Whether PDGF is indeed involved in a fate choice between oligodendroglial and neuronal lineages in stem cells of the adult SVZ, and whether it has a similar role in the embryonic CNS, remains to be determined. Recently, insulin-like growth factor (IGF)-I has been reported to promote the differentiation of adult NSCs to an oligodendroglial lineage (Hsieh et al., 2004b). The function of IGF-I in oligodendrocyte differentiation is, at least in part, attributable to inhibition of BMP signaling, which prevents NSCs from differentiating into oligodendrocytes.

3.3.2.2 Transcription Factors Regulating Oligodendrocyte Differentiation

The bHLH factors OLIG1 and OLIG2 are essential and sufficient for oligodendrocyte lineage specification (Ross et al., 2003; Sauvageot and Stiles, 2002). Compound disruption of *Olig1* and *Olig2* resulted in a complete loss of the oligodendrocyte lineage throughout the CNS (Zhou and Anderson, 2002). Disruption of *Olig2* alone is sufficient to completely inhibit oligodendrocyte development in the spinal cord; however, localized clusters of early oligodendrocyte progenitors form in the ventral forebrain, midbrain, diencephalon, cerebellum, and medulla of *Olig2* null mice, suggesting that *Olig1* can partially compensate for loss of *Olig2* in the developing brain (Lu et al., 2002; Takebayashi et al., 2002).

Loss of *Olig* gene function in *Olig1/2* double mutant mice also resulted in ectopic generation of astrocytes, and forced expression of *Olig2* in progenitor cells was sufficient to repress GFAP expression and inhibit astrocytic differentiation (Gabay et al., 2003; Zhou and Anderson, 2002). OLIG2 has been shown to block astrocytic differentiation in progenitor cells treated with cytokines, by sequestering

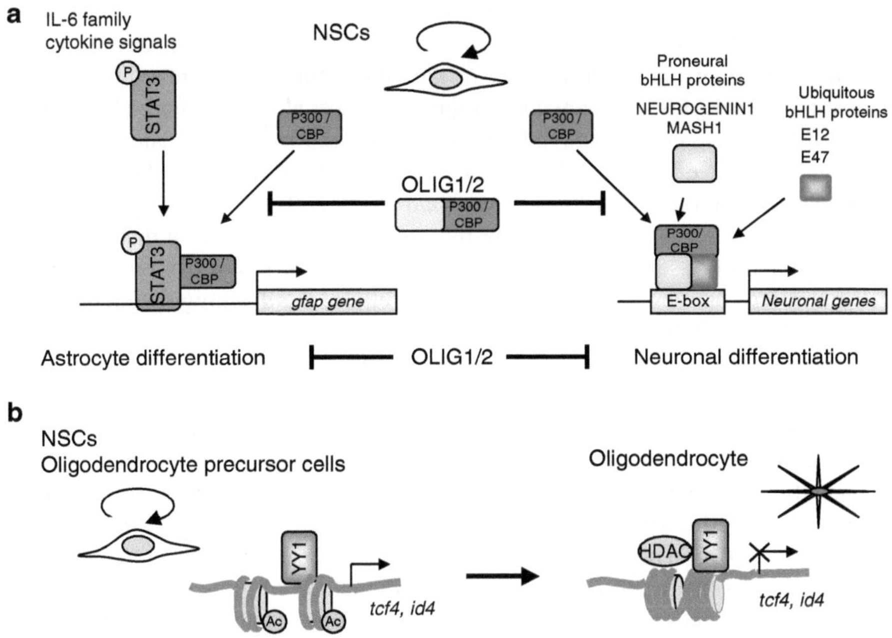

Fig. 3.4 Mechanisms regulating oligodendrocyte differentiation. (**a**) bHLH proteins OLIG1 and OLIG2 are major determinants of oligodendroglial fate. (See text for details.) (**b**) In proliferating progenitors, transcriptional inhibitors (*Id4*, *Tcf4*) are expressed and inhibit myelin gene expression. As the progenitors begin differentiation, HDAC1 is recruited to the promoters of these inhibitors to form a repressive complex with YY1. The decreased levels of these inhibitory molecules allow myelin gene expression to begin

the coactivator P300 from the astrogliogenic transcription factor STAT3 (Fig. 3.4a) (Fukuda et al., 2004), through a mechanism reminiscent of the inhibition of astrogliogenesis by NGN1 (Sun et al., 2001). Astrocytic differentiation in the postnatal telencephalon must thus require downregulation of *Olig* genes. This step may involve BMPs, which are produced in the dorsal telencephalon and have been shown to inhibit oligodendrogenesis in progenitor cultures, through induction of ID proteins that inactivate OLIG proteins by dimerization (Samanta and Kessler, 2004; Yung et al., 2002). Another mechanism that may inactivate OLIG2 during astrocytic differentiation is its nuclear export, promoted by the serine/threonine protein kinase Akt downstream of cytokine signaling (Setoguchi and Kondo, 2004).

The inhibition of neuronal differentiation by OLIG2 involves a competition with NGN proteins for dimerization with E proteins and for binding to shared DNA-binding sites in the promoters of postmitotic motor neuron genes (Fig. 3.4) (Lee et al., 2005). In the embryonic telencephalon, *Olig2* is similarly expressed in ventral progenitors that generate both oligodendrocytes and neurons (Furusho et al., 2006; Takebayashi et al., 2000; Yung et al., 2002). However, the role of OLIG2 in neuronal specification in the developmental telencephalon is less well characterized and is probably less important than its role in oligodendrogenesis.

Interestingly, OLIG2 is expressed in the early postnatal SVZ in both astrocytes and oligodendrocytes, but not in neurons (Marshall et al., 2005). OLIG2 is then downregulated in mature astrocytes in the adult, but may be re-expressed by a fraction of astrocytes following brain injury (Buffo et al., 2005). Forced expression of OLIG2 in SVZ progenitors induces both oligodendrocytes and astrocytes, while dominant negative versions of the protein block the production of both cell types and instead promote neurogenesis (Marshall et al., 2005). This astrogliogenic role of OLIG2 in the postnatal brain may result in the generation of a distinct type of astrocyte and represent a mechanism of specification different from that operating for embryonic astrocytes, which, in contrast, require the downregulation of OLIG2 (Marshall et al., 2005).

Pharmacological inhibitors of HDACs prevent the expression of myelin genes in vitro (Marin-Husstege et al., 2002) and developmental myelination in vivo (Shen et al., 2005), suggesting that histone deacetylation is necessary for the initiation of the transcriptional program leading to myelin gene expression. Recently, He et al. screened the promoters of genes regulated by TSA, an inhibitor of HDAC activity during oligodendrocyte differentiation, for the presence of common binding motifs (He et al., 2007). They found that the consensus sequence NNCCATNN (Yant et al., 1995) for the transcription factor Yin Yang 1 (YY1) was present in the promoter region motifs of more than 30% of the genes downregulated by histone deacetylation during oligodendrocyte differentiation (He et al., 2007). YY1 is a zinc finger protein with the ability to bind DNA (Seto et al., 1991) and functions as a chromatin modifier (Thomas and Seto, 1999). The results of He et al. also indicate that, at the molecular level, YY1 acts as a lineage-specific repressor of transcriptional inhibitors for myelin gene expression (*Tcf4* and *Id4*) by recruiting histone deacetylase-1 to their promoters during oligodendrocytic differentiation (Fig. 3.4b) (He et al., 2007). Furthermore, conditional ablation of *Yy1* in the oligodendrocyte lineage in vivo induces a phenotype characterized by defective myelination, ataxia, and tremor. At the cellular level, lack of *Yy1* arrests differentiation of oligodendrocyte progenitors after they exit from the cell cycle. This YY1-mediated repressive event occurs only when progenitors differentiate into oligodendrocytes and not when they become astrocytes. Collectively, their data showed that YY1 and HDACs of its partner are the essential components of oligodendrocytic differentiation and maturation in the developing mouse CNS (Fig. 3.4b) (He et al., 2007).

3.4 Closing Remarks

Fate specification in NSCs requires a multitude of tightly coordinated steps. In the telencephalon of the developing mammalian brain, this is well illustrated by the intricate coordination of the JAK-STAT pathway, which involves temporally regulated extracellular signals, autoregulatory loops controlling various components of the signaling machinery, cross-talk with other gliogenic pathways at various steps, and regulation of STAT1/3 transcription factor binding by modifications of both DNA and chromatin. By introducing the concept of epigenetics, we have become able to provide mechanistic explanations for many biological phenomena that formerly could only be described as cell context dependent. However, the importance of histone modifications, DNA methylation, and ncRNAs in the selection of neuronal and glial fates, and the mechanisms involved, remain largely unexplored. Methods to identify and manipulate NSCs in vivo and in vitro are improving rapidly, and these developments should ensure that our understanding of how neural cell fates are specified will progress rapidly as well.

In a recent series of studies, human fibroblasts were reprogrammed in vitro into pluripotent stem cell-like cells [termed induced pluripotent stem (iPS) cells] through retroviral transduction of combinations of transcription factors (Takahashi et al., 2007; Park et al., 2008). One theme of future therapeutic applications stem cell technology will therefore be the development of robust and reliable differentiation protocols for human iPS cells. Because of their ability to self-renew and their remarkable potential for differentiation into neural derivatives, NSCs hold great promise for therapeutic applications in neurological dysfunctions such as Parkinson's disease, amyotrophic lateral sclerosis, and spinal cord injury (Lindvall and Kokaia, 2006; Okano, 2002). Establishing methods for the generation of NSCs from human iPS cells and understanding the basic mechanisms regulating fate specification of NSCs will provide the foundation for cell replacement therapy using transplanted cells.

Acknowledgements We thank Dr. Ian Smith for helpful comments and critical reading of the manuscript. We are very grateful to M. Ueda for her excellent secretarial assistance.

References

Akazawa C, Sasai Y, Nakanishi S et al (1992) Molecular characterization of a rat negative regulator with a basic helix-loop-helix structure predominantly expressed in the developing nervous system. J Biol Chem 267:21879–21885.

Andres ME, Burger C, Peral-Rubio MJ et al (1999) CoREST: a functional corepressor required for regulation of neural-specific gene expression. Proc Natl Acad Sci USA 96:9873–9878.

Ballas N, Mandel G (2005) The many faces of REST oversee epigenetic programming of neuronal genes. Curr Opin Neurobiol 15:500–506.

Ballas N, Grunseich C, Lu DD et al (2005) REST and its corepressors mediate plasticity of neuronal gene chromatin throughout neurogenesis. Cell 121:645–657.

Battaglioli E, Andres ME, Rose DW et al (2002) REST repression of neuronal genes requires components of the hSWI.SNF complex. J Biol Chem 277:41038–41045.

Bertrand N, Castro DS, Guillemot F (2002) Proneural genes and the specification of neural cell types. Nat Rev Neurosci 3:517–530.

Bonni A, Sun Y, Nadal-Vicens M et al (1997) Regulation of gliogenesis in the central nervous system by the JAK-STAT signaling pathway. Science 278:477–483.

Buffo A, Vosko MR, Erturk D et al (2005) Expression pattern of the transcription factor Olig2 in response to brain injuries: implications for neuronal repair. Proc Natl Acad Sci USA 102:18183–18188.

Cameron HA, Woolley CS, McEwen BS et al (1993) Differentiation of newly born neurons and glia in the dentate gyrus of the adult rat. Neuroscience 56:337–344.

Campbell K (2005) Cortical neuron specification: it has its time and place. Neuron 46:373–376.

Cebolla B, Vallejo M (2006) Nuclear factor-I regulates glial fibrillary acidic protein gene expression in astrocytes differentiated from cortical precursor cells. J Neurochem 97:1057–1070.
Chambers CB, Peng Y, Nguyen H et al (2001) Spatiotemporal selectivity of response to Notch1 signals in mammalian forebrain precursors. Development (Camb) 128:689–702.
Chen H, Thiagalingam A, Chopra H et al (1997) Conservation of the Drosophila lateral inhibition pathway in human lung cancer: a hairy-related protein (HES-1) directly represses achaete-scute homolog-1 expression. Proc Natl Acad Sci USA 94:5355–5360.
Chong JA, Tapia-Ramirez J, Kim S et al (1995) REST: a mammalian silencer protein that restricts sodium channel gene expression to neurons. Cell 80:949–957.
Christopherson KS, Ullian EM, Stokes CC et al (2005) Thrombospondins are astrocyte-secreted proteins that promote CNS synaptogenesis. Cell 120:421–433.
Conaco C, Otto S, Han JJ et al (2006) Reciprocal actions of REST and a microRNA promote neuronal identity. Proc Natl Acad Sci USA 103:2422–2427.
das Neves L, Duchala CS, Tolentino-Silva F et al (1999) Disruption of the murine nuclear factor I-A gene (Nfia) results in perinatal lethality, hydrocephalus, and agenesis of the corpus callosum. Proc Natl Acad Sci USA 96:11946–11951.
de la Serna IL, Ohkawa Y, Imbalzano AN (2006) Chromatin remodelling in mammalian differentiation: lessons from ATP-dependent remodellers. Nat Rev Genet 7:461–473.
Deneen B, Ho R, Lukaszewicz A et al (2006) The transcription factor NFIA controls the onset of gliogenesis in the developing spinal cord. Neuron 52:953–968.
Doetsch F (2003) The glial identity of neural stem cells. Nat Neurosci 6:1127–1134.
Elson GC, Lelievre E, Guillet C et al (2000) CLF associates with CLC to form a functional heteromeric ligand for the CNTF receptor complex. Nat Neurosci 3:867–872.
Fan G, Martinowich K, Chin MH et al (2005) DNA methylation controls the timing of astrogliogenesis through regulation of JAK-STAT signaling. Development (Camb) 132:3345–3356.
Fukuda S, Kondo T, Takebayashi H et al (2004) Negative regulatory effect of an oligodendrocytic bHLH factor OLIG2 on the astrocytic differentiation pathway. Cell Death Differ 11:196–202.
Furukawa T, Mukherjee S, Bao ZZ et al (2000) rax, Hes1, and notch1 promote the formation of Muller glia by postnatal retinal progenitor cells. Neuron 26:383–394.
Furusho M, Ono K, Takebayashi H et al (2006) Involvement of the Olig2 transcription factor in cholinergic neuron development of the basal forebrain. Dev Biol 293:348–357.
Gabay L, Lowell S, Rubin LL et al (2003) Deregulation of dorsoventral patterning by FGF confers trilineage differentiation capacity on CNS stem cells in vitro. Neuron 40:485–499.
Gage FH (2000) Mammalian neural stem cells. Science 287:1433–1438.
Gage FH, Kempermann G, Palmer TD et al (1998) Multipotent progenitor cells in the adult dentate gyrus. J Neurobiol 36:249–266.
Garcia-Verdugo JM, Doetsch F, Wichterle H et al (1998) Architecture and cell types of the adult subventricular zone: in search of the stem cells. J Neurobiol 36:234–248.
Ge W, Martinowich K, Wu X et al (2002) Notch signaling promotes astrogliogenesis via direct CSL-mediated glial gene activation. J Neurosci Res 69:848–860.
Grandbarbe L, Bouissac J, Rand M et al (2003) Delta-Notch signaling controls the generation of neurons/glia from neural stem cells in a stepwise process. Development (Camb) 130:1391–1402.
Gronostajski RM (2000) Roles of the NFI/CTF gene family in transcription and development. Gene (Amst) 249:31–45.
Gross RE, Mehler MF, Mabie PC et al (1996) Bone morphogenetic proteins promote astroglial lineage commitment by mammalian subventricular zone progenitor cells. Neuron 17:595–606.
He F, Ge W, Martinowich K et al (2005) A positive autoregulatory loop of Jak-STAT signaling controls the onset of astrogliogenesis. Nat Neurosci 8:616–625.
He Y, Dupree J, Wang J et al (2007) The transcription factor Yin Yang 1 is essential for oligodendrocyte progenitor differentiation. Neuron 55:217–230.
Hermanson O, Jepsen K, Rosenfeld MG (2002) N-CoR controls differentiation of neural stem cells into astrocytes. Nature (Lond) 419:934–939.
Hirabayashi Y, Itoh Y, Tabata H et al (2004) The Wnt/beta-catenin pathway directs neuronal differentiation of cortical neural precursor cells. Development (Camb) 131:2791–2801.

Hsieh J, Gage FH (2004) Epigenetic control of neural stem cell fate. Curr Opin Genet Dev 14:461–469.

Hsieh J, Nakashima K, Kuwabara T et al (2004a) Histone deacetylase inhibition-mediated neuronal differentiation of multipotent adult neural progenitor cells. Proc Natl Acad Sci USA 101:16659–16664.

Hsieh J, Aimone JB, Kaspar BK et al (2004b) IGF-I instructs multipotent adult neural progenitor cells to become oligodendrocytes. J Cell Biol 164:111–122.

Israsena N, Hu M, Fu W et al (2004) The presence of FGF2 signaling determines whether beta-catenin exerts effects on proliferation or neuronal differentiation of neural stem cells. Dev Biol 268:220–231.

Jackson EL, Garcia-Verdugo JM, Gil-Perotin S et al (2006) PDGFR alpha-positive B cells are neural stem cells in the adult SVZ that form glioma-like growths in response to increased PDGF signaling. Neuron 51:187–199.

Kettenmann H, Ransom BR (1988) Electrical coupling between astrocytes and between oligodendrocytes studied in mammalian cell cultures. Glia 1:64–73.

Kohyama J, Tokunaga A, Fujita Y et al (2005) Visualization of spatiotemporal activation of Notch signaling: live monitoring and significance in neural development. Dev Biol 286:311–325.

Kuwabara T, Hsieh J, Nakashima K et al (2004) A small modulatory dsRNA specifies the fate of adult neural stem cells. Cell 116:779–793.

Ledent V, Paquet O, Vervoort M (2002) Phylogenetic analysis of the human basic helix-loop-helix proteins. Genome Biol 3:RESEARCH0030.

Lee JC, Mayer-Proschel M, Rao MS (2000) Gliogenesis in the central nervous system. Glia 30:105–121.

Lee SK, Lee B, Ruiz EC et al (2005) Olig2 and Ngn2 function in opposition to modulate gene expression in motor neuron progenitor cells. Genes Dev 19:282–294.

Lelievre E, Plun-Favreau H, Chevalier S et al (2001) Signaling pathways recruited by the cardiotrophin-like cytokine/cytokine-like factor-1 composite cytokine: specific requirement of the membrane-bound form of ciliary neurotrophic factor receptor alpha component. J Biol Chem 276:22476–22484.

Lessard J, Wu JI, Ranish JA et al (2007) An essential switch in subunit composition of a chromatin remodeling complex during neural development. Neuron 55:201–215.

Li W, Cogswell CA, LoTurco JJ (1998) Neuronal differentiation of precursors in the neocortical ventricular zone is triggered by BMP. J Neurosci 18:8853–8862.

Lim DA, Tramontin AD, Trevejo JM et al (2000) Noggin antagonizes BMP signaling to create a niche for adult neurogenesis. Neuron 28:713–726.

Lindvall O, Kokaia Z (2006) Stem cells for the treatment of neurological disorders. Nature 441:1094–1096.

Lu QR, Yuk D, Alberta JA et al (2000) Sonic hedgehog-regulated oligodendrocyte lineage genes encoding bHLH proteins in the mammalian central nervous system. Neuron 25:317–329.

Lu QR, Sun T, Zhu Z et al (2002) Common developmental requirement for Olig function indicates a motor neuron/oligodendrocyte connection. Cell 109:75–86.

Machon O, van den Bout CJ, Backman M et al (2003) Role of beta-catenin in the developing cortical and hippocampal neuroepithelium. Neuroscience 122:129–143.

Marin-Husstege M, Muggironi M, Liu A et al (2002) Histone deacetylase activity is necessary for oligodendrocyte lineage progression. J Neurosci 22:10333–10345.

Marshall CA, Novitch BG, Goldman JE (2005) Olig2 directs astrocyte and oligodendrocyte formation in postnatal subventricular zone cells. J Neurosci 25:7289–7298.

Ming GL, Song H (2005) Adult neurogenesis in the mammalian central nervous system. Annu Rev Neurosci 28:223–250.

Molofsky AV, Pardal R, Morrison SJ (2004) Diverse mechanisms regulate stem cell self-renewal. Curr Opin Cell Biol 16:700–707.

Moore KB, Schneider ML, Vetter ML (2002) Posttranslational mechanisms control the timing of bHLH function and regulate retinal cell fate. Neuron 34:183–195.

Morrison SJ, Perez SE, Qiao Z et al (2000) Transient Notch activation initiates an irreversible switch from neurogenesis to gliogenesis by neural crest stem cells. Cell 101:499–510.

Muroyama Y, Fujiwara Y, Orkin SH et al (2005) Specification of astrocytes by bHLH protein SCL in a restricted region of the neural tube. Nature 438:360–363.

Murre C, McCaw PS, Vaessin H et al (1989) Interactions between heterologous helix-loop-helix proteins generate complexes that bind specifically to a common DNA sequence. Cell 58:537–544.

Nakashima K, Yanagisawa M, Arakawa et al (1999) Synergistic signaling in fetal brain by STAT3-Smad1 complex bridged by p300. Science 284:479–482.

Nakashima K, Takizawa T, Ochiai W et al (2001) BMP2-mediated alteration in the developmental pathway of fetal mouse brain cells from neurogenesis to astrocytogenesis. Proc Natl Acad Sci USA 98:5868–5873.

Namihira M, Nakashima K, Taga T (2004) Developmental stage dependent regulation of DNA methylation and chromatin modification in a immature astrocyte specific gene promoter. FEBS Lett 572:184–188.

Nery S, Wichterle H, Fishell G (2001) Sonic hedgehog contributes to oligodendrocyte specification in the mammalian forebrain. Development 128:527–540.

Noctor SC, Martinez-Cerdeno V, Ivic L et al (2004) Cortical neurons arise in symmetric and asymmetric division zones and migrate through specific phases. Nat Neurosci 7:136–144.

Norton JD, Deed RW, Craggs G et al (1998) Id helix-loop-helix proteins in cell growth and differentiation. Trends Cell Biol 8:58–65.

Ohtsuka T, Ishibashi M, Gradwohl G et al (1999) Hes1 and Hes5 as notch effectors in mammalian neuronal differentiation. EMBO J 18:2196–2207.

Okano H (2002) Stem cell biology of the central nervous system. J Neurosci Res 69:698–707.

Park IH, Zhao R, West JA et al (2008) Reprogramming of human somatic cells to pluripotency with defined factors. Nature 451:141–146.

Pringle NP, Guthrie S, Lumsden A et al (1998) Dorsal spinal cord neuroepithelium generates astrocytes but not oligodendrocytes. Neuron 20:883–893.

Qian X, Shen Q, Goderie SK et al (2000) Timing of CNS cell generation: a programmed sequence of neuron and glial cell production from isolated murine cortical stem cells. Neuron 28:69–80.

Robertson KD, Jones PA (2000) DNA methylation: past, present and future directions. Carcinogenesis 21:461–467.

Roopra A, Sharling L, Wood IC et al (2000) Transcriptional repression by neuron-restrictive silencer factor is mediated via the Sin3-histone deacetylase complex. Mol Cell Biol 20:2147–2157.

Ross SE, Greenberg ME, Stiles CD (2003) Basic helix-loop-helix factors in cortical development. Neuron 39:13–25.

Samanta J, Kessler JA (2004) Interactions between ID and OLIG proteins mediate the inhibitory effects of BMP4 on oligodendroglial differentiation. Development 131:4131–4142.

Sauvageot CM, Stiles CD (2002) Molecular mechanisms controlling cortical gliogenesis. Curr Opin Neurobiol 12:244–249.

Schoenherr CJ, Anderson DJ (1995) The neuron-restrictive silencer factor (NRSF): a coordinate repressor of multiple neuron-specific genes. Science 267:1360–1363.

Schoenherr CJ, Paquette AJ, Anderson DJ (1996) Identification of potential target genes for the neuron-restrictive silencer factor. Proc Natl Acad Sci USA 93:9881–9886.

Seo S, Herr A, Lim JW et al (2005) Geminin regulates neuronal differentiation by antagonizing Brg1 activity. Genes Dev 19:1723–1734.

Seto E, Shi Y, Shenk T (1991) YY1 is an initiator sequence-binding protein that directs and activates transcription in vitro. Nature 354:241–245.

Setoguchi T, Kondo T (2004) Nuclear export of OLIG2 in neural stem cells is essential for ciliary neurotrophic factor-induced astrocyte differentiation. J Cell Biol 166:963–968.

Shen S, Li J, Casaccia-Bonnefil P (2005) Histone modifications affect timing of oligodendrocyte progenitor differentiation in the developing rat brain. J Cell Biol 169:577–589.

Shu T, Butz KG, Plachez C et al (2003) Abnormal development of forebrain midline glia and commissural projections in Nfia knock-out mice. J Neurosci 23:203–212.

Song MR, Ghosh A (2004) FGF2-induced chromatin remodeling regulates CNTF-mediated gene expression and astrocyte differentiation. Nat Neurosci 7:229–235.
Steele-Perkins G, Plachez C, Butz KG et al (2005) The transcription factor gene Nfib is essential for both lung maturation and brain development. Mol Cell Biol 25:685–698.
Stolt CC, Lommes P, Sock E et al (2003) The Sox9 transcription factor determines glial fate choice in the developing spinal cord. Genes Dev 17:1677–1689.
Sun Y, Nadal-Vicens M, Misono S et al (2001) Neurogenin promotes neurogenesis and inhibits glial differentiation by independent mechanisms. Cell 104:365–376.
Taga T, Kishimoto T (1997) Gp130 and the interleukin-6 family of cytokines. Annu Rev Immunol 15:797–819.
Takahashi K, Tanabe K, Ohnuki M et al (2007) Induction of pluripotent stem cells from adult human fibroblasts by defined factors. Cell 131:861–872.
Takebayashi H, Yoshida S, Sugimori M et al (2000) Dynamic expression of basic helix-loop-helix Olig family members: implication of Olig2 in neuron and oligodendrocyte differentiation and identification of a new member, Olig3. Mech Dev 99:143–148.
Takebayashi H, Nabeshima Y, Yoshida S et al (2002) The basic helix-loop-helix factor olig2 is essential for the development of motoneuron and oligodendrocyte lineages. Curr Biol 12:1157–1163.
Takizawa T, Nakashima K, Namihira M et al (2001) DNA methylation is a critical cell-intrinsic determinant of astrocyte differentiation in the fetal brain. Dev Cell 1:749–758.
Takizawa T, Ochiai W, Nakashima K et al (2003) Enhanced gene activation by Notch and BMP signaling cross-talk. Nucleic Acids Res 31:5723–5731.
Tanigaki K, Nogaki F, Takahashi J et al (2001) Notch1 and Notch3 instructively restrict bFGF-responsive multipotent neural progenitor cells to an astroglial fate. Neuron 29:45–55.
Temple S (2001) The development of neural stem cells. Nature 414:112–117.
Thomas MJ, Seto E (1999) Unlocking the mechanisms of transcription factor YY1: are chromatin modifying enzymes the key? Gene 236:197–208.
Tokunaga A, Kohyama J, Yoshida T et al (2004) Mapping spatio-temporal activation of Notch signaling during neurogenesis and gliogenesis in the developing mouse brain. J Neurochem 90:142–154.
Vinals F, Reiriz J, Ambrosio S et al (2004) BMP-2 decreases Mash1 stability by increasing Id1 expression. EMBO J 23:3527–3537.
Wrana JL (2000) Regulation of Smad activity. Cell 100:189–192.
Yant SR, Zhu W, Millinoff D et al (1995) High affinity YY1 binding motifs: identification of two core types (ACAT and CCAT) and distribution of potential binding sites within the human beta globin cluster. Nucleic Acids Res 23:4353–4362.
Yoder JA, Walsh CP, Bestor TH (1997) Cytosine methylation and the ecology of intragenomic parasites. Trends Genet 13:335–340.
Yung SY, Gokhan S, Jurcsak J et al (2002) Differential modulation of BMP signaling promotes the elaboration of cerebral cortical GABAergic neurons or oligodendrocytes from a common sonic hedgehog-responsive ventral forebrain progenitor species. Proc Natl Acad Sci USA 99:16273–16278.
Zhou Q, Anderson DJ (2002) The bHLH transcription factors OLIG2 and OLIG1 couple neuronal and glial subtype specification. Cell 109:61–73.
Zhou Q, Wang S, Anderson DJ (2000) Identification of a novel family of oligodendrocyte lineage-specific basic helix-loop-helix transcription factors. Neuron 25:331–343.

Chapter 4
Fractones: Home and Conductors of the Neural Stem Cell Niche

Frederic Mercier, Jason Schnack, and Maureen Saint Georges Chaumet

Abstract Throughout life, neural stem cells proliferate and differentiate in restricted zones of the brain, termed niches, to produce new neurons and glial cells. In these niches, growth factors and extracellular matrix (ECM) molecules determine the fate of neural stem and progenitor cells (NSPC). However, the precise compounds and the mechanisms that regulate growth factors and other signaling molecules in the niches are unknown. Based on the evidence that NSPCs proliferate next to blood vessels in the dentate gyrus, the concept of a vascular niche for neurogenesis has been initially proposed. In the subventricular zone of the lateral ventricle, the most neurogenic zone in adulthood, we have found that NSPC directly contact a novel type of ECM structure that we have named fractones. Fractones contain heparan sulfate proteoglycans (HSPG) that collect and concentrate the neurogenic growth factor FGF2 at the NSPC surface and likely direct its signaling via tyrosine kinase receptors. Our preliminary results indicate that FGF2 binding to fractone-HSPG is essential for activating FGF2 at the NSPC surface. Moreover, we have found fractones express diverse HSPG at the surface of proliferating NSPC during development, even before the brain vasculature emerges. Therefore, fractones hold considerable promise for promoting growth factors at the stem cell surface to ultimately regulate neurogenesis during development and adulthood.

4.1 Introduction to the Concept of the Neural Stem Cell Niche

Neural stem cells proliferate and differentiate to produce new neurons and glial cells in restricted zones of the adult brain: the subventricular zone (SVZ) of the lateral ventricle, the rostral migratory streams (RMS), the subcallosum zone (SCZ), the olfactory bulb (OB), and the granular layer of the dentate gyrus (DG) (Altman,

F. Mercier (✉), J. Schnack, and M.S.G. Chaumet
Department of Tropical Medicine, Medical Microbiology and Pharmacology, John A, Burns School of Medicine, University of Hawaii, Honolulu, HI 96822, USA
e-mail: fmercier@pbrc.hawaii.edu

T. Seki et al. (eds.), *Neurogenesis in the Adult Brain I: Neurobiology*,
DOI 10.1007/978-4-431-53933-9_4, © Springer 2011

1963, 1969; Altman and Das, 1966; Alvarez-Buylla and Nottebohm, 1988; Seki and Array, 1993; Lois and Alvarez-Buylla, 1994; Doetsch and Alvarez-Buylla, 1996; Seri et al., 2006). In these biological niches, regulation of neural stem and progenitor cell (NSPC) proliferation and differentiation is influenced by growth factors (Palmer et al., 1995; Kuhn et al., 1997; Pencea et al., 2001; Falk and Frisen, 2002; Hienola et al., 2004). However, the physiological mechanism that regulates these growth factors, and therefore NSPC fate, is unknown.

In 2000, Palmer et al. proposed the concept of a vascular niche for neural stem cells, based on the evidence of immediate anatomical associations between proliferating NSPC and blood vessels in the hippocampus. Since their discovery, the authors have focused on the molecular and inflammatory aspects of the niche (Monje et al., 2003; Wexler et al., 2007) without identifying the anatomical and functional vascular elements that serve as a niche for NSPC. Exploring the cellular interactions of the niche (Wurmser et al., 2004) and the extracellular structures that house and that administer growth factors to NSPC is crucial to understanding their proliferation and differentiation in the adult brain.

4.2 The Discovery of Fractones

We recently characterized novel electron-dense structures by transmission electron microscopy in the SVZ of the adult rat (Mercier et al., 2002, 2003). These structures are extracellular, regularly arranged along the ventricle wall, and localized in between the converging processes of NSPC, astrocytes, ependymocytes, and macrophages. Based on their fractal geometry, and in reference to the concept of fractals (Mandelbrot, 1983), these structures have been termed fractones (Mercier et al., 2002). Fractones are small in size (2–5 μm), but their branched morphology allows each fractone to directly contact numerous NSPC and other cells of the ventricle wall (Fig. 4.1a,b; see Fig. 4.6c). By their electron density, fractones resemble basement membranes (basal laminae). Supporting fractone affiliation with basement membranes, we found anatomical connections between fractones and perivascular basement membranes, via elongated projections of electron-dense material, also resembling basement membranes (Mercier et al., 2002, 2003). However, fractones have a unique morphology, and are made of connected fragments that are much thicker (200–300 μm; Fig. 4.1b) than basement membranes, and their arrangement, as a series in the SVZ, is not typical of the conventional sheet-like subepithelial basement membranes (Yurchenko and Schittny, 1990).

After immunolabeling, fractones can be visualized by fluorescence microscopy (Fig. 4.2). Laminin was first used as a marker of fractones (Mercier et al., 2002). Laminin is a basic component of basement membranes and forms, with collagen IV, nidogen, and heparan sulfate proteoglycans (HSPG), a functional lattice of the extracellular matrix (ECM) (Yurchenko and Schyttny, 1990) that intervenes in morphogenesis, cell proliferation, and diverse plasticity processes through organs and tissues of the body (Aviezer et al., 1994; Bernfield et al., 1994; Halfter, 1998; Iozzo, 2005). Immunolabeling with antibody for laminin revealed (laminin+)

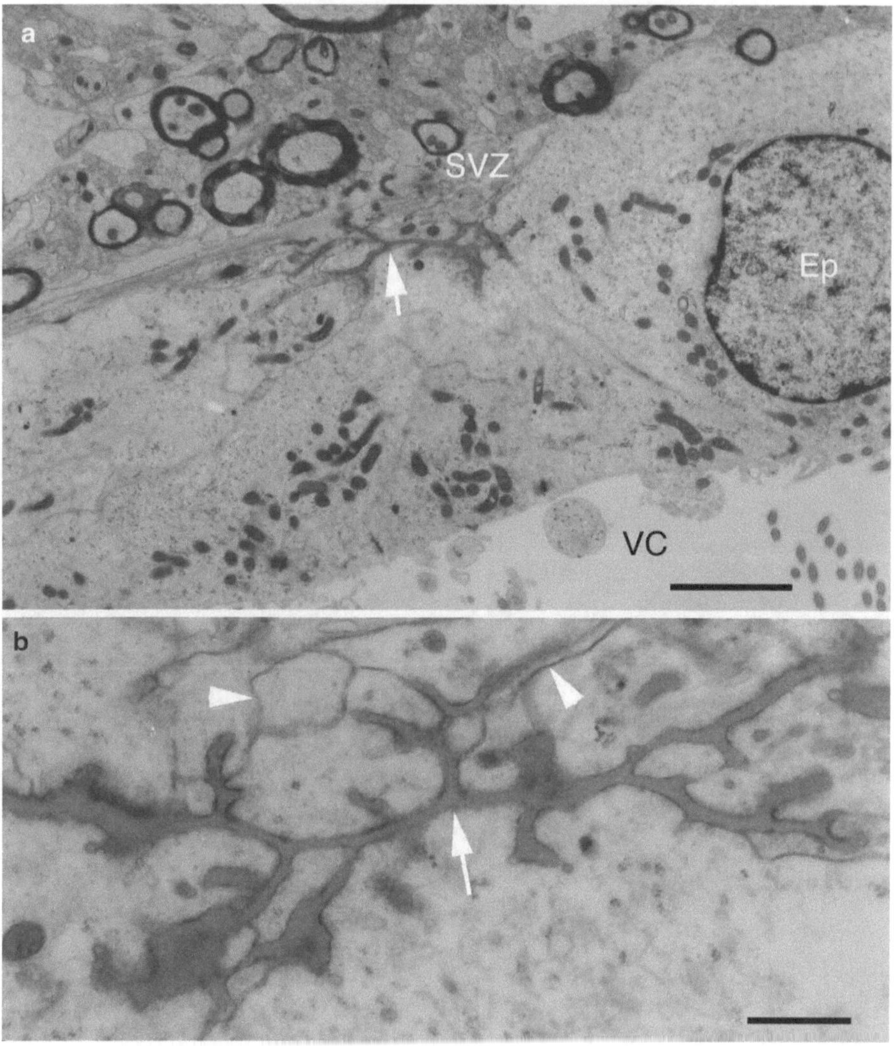

Fig. 4.1 Fractone ultrastructure. Lateral ventricle wall of the adult rat. (**a**) Ultrastructure of the ventricle wall, comprising ependyma (*Ep*) and subventricular zone (*SVZ*). A fractone is visible at the ependyma/SVZ interface (*arrow*). *VC* ventricular cavity. (**b**) Detail of a fractone, visualized as a thick electron-dense extracellular material (*arrow*), with a fractal morphology. *Arrowheads*, plasma membrane of cell processes converging toward the fractone. *Bars* (**a**) 3 μm; (**b**) 1 μm

puncta along the luminal side of the vimentin-positive ependyma in the adult rat, mouse, and human (Mercier et al., 2002; Kerever et al., 2007). The characterization of fractones as laminin+ basement membrane (BM) structures of similar location, distribution, size, and quantity strongly supported the view that the fractones characterized by electron microscopy appear as small puncta by fluorescence microscopy (Mercier and Hatton, 2000). Correlation between the number of fractones visualized by electron microscopy and the number of laminin+ puncta indi-

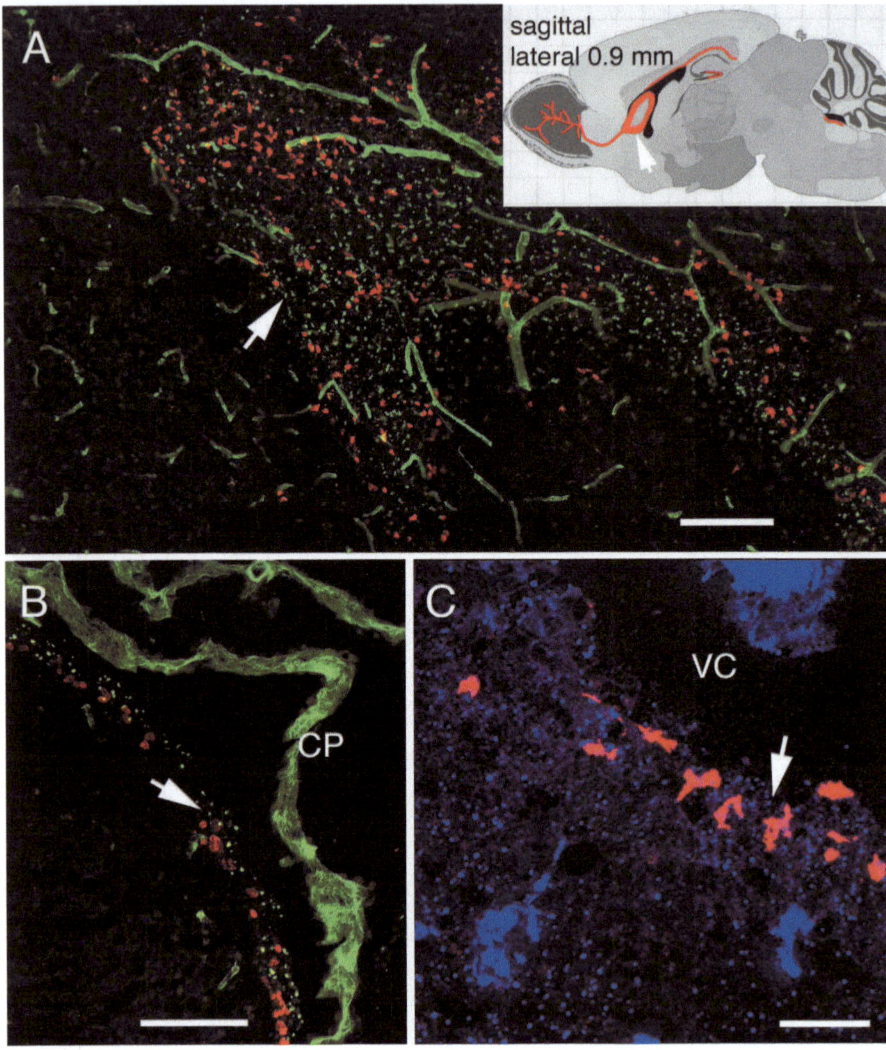

Fig. 4.2 Neural stem and progenitor cells (NSPC) proliferate next to fractones during development and adulthood: adult mouse (**a**, **b**) and mouse in development (**c**). (**a**) After immunolabel for laminin (*green*), fractones appear as puncta (*arrow*) in tangential view of the anterior subventricular zone (SVZa), the most active neurogenic zone during adulthood (the SVZa location is indicated by an *arrow* in the *inset*). The ventricular cavity is not visible in tangential views of the ventricle wall. Proliferating NSPC, visualized by immunolabel for bromodeoxyuridine (BrdU, marker of S-phase mitosis, *red*) 3 h after a single injection of BrdU, are located next to fractones (*arrow*). (**b**) Double immunolabel for heparan sulfate proteoglycan (*green*) and BrdU (*red*), respectively, showing fractones and proliferating NSPC and their association in the SVZ (*arrow*). *CP*, choroid plexus. (**c**) Neuroepithelial cells proliferate next to fractones during early morphogenesis (E11.5 stage). Double immunolabel for laminin (*blue*) and phosphorylated histone 3 (PH3, marker of M-phase mitosis, *red*), respectively, shows fractones and small clusters of cells (*arrow*) that have undergone mitosis. *VC*, ventricular cavity. *Bars* (**a** and **b**) 50 μm; (**c**) 100 μm

cate that all fractones contain laminin. Laminin immunolabel also reveals the meninges, choroid plexus, and the brain blood vessels (Fig. 4.2b), which contain laminin in their basement membranes (Mercier and Hatton, 2000, 2001). The laminin+ puncta of the SVZ are unambiguously distinguished from local capillaries by size, shape, and location. First, the distribution of laminin+ puncta does not suggest a vasculature arrangement (Fig. 4.2a). Second, laminin+ puncta are smaller in size than the smallest capillaries, which are at least 6 μm in diameter (Fig. 4.2a). Third, laminin+ puncta are seen in a single confocal plane whereas the longitudinal axis of capillaries can be followed in the depth of the 25-μm-thick sections.

4.3 Fractone Composition and Distribution in the Adult Brain

Theorizing that fractones represent a form of ECM similar to basement membranes, we investigated fractone components by double immunolabel, using laminin as the reference marker and basement membrane-specific markers on brain frozen sections. We found that all fractones are immunoreactive for the basement membrane components collagen IV, β- and γ-isoforms of laminin, nidogen, and heparan sulfate proteoglycans (HSPG) (Kerever et al., 2007). Most fractones expressed perlecan, a major HSPG that promotes heparin-binding growth factor activity (Aviezer et al., 1994). In the SVZ of the lateral ventricle, a prime neurogenic zone, several fractones were immunoreactive for collagen I (Mercier et al., 2002).

Using laminin or collagen IV immunolabel in serial sections of the whole brain, we found fractones in the SVZ of the whole ventricular system, that is, the lateral ventricle, third ventricle, fourth ventricle, aqueduct of Sylvius, and the spinal canal, in the adult mouse, rat, and human (Mercier et al., 2002, 2003; Kerever et al., 2007). The SVZ of the lateral ventricle, which houses numerous NSPC (Doetsch et al., 1997), is the most neurogenic zone in the adult brain. The SVZ of the other ventricles and spinal cord contain NSPC that produce neurons and glial cells when artificially stimulated by growth factors (Pencea et al., 2001) or in vitro conditions (Weiss et al., 1996). Fractones were not present in the other adult neurogenic zones, that is, in the OB, SCZ, RMS, and DG; those zones lacking ventricles or ventricular extensions in adulthood, therefore, do not possess a SVZ.

4.4 A Highly Organized Fractone and Vascular Niche for Neurogenesis

We used immunolabel for bromodeoxyuridine (BrdU), a marker of DNA incorporation that localizes proliferating cells in S phase, on complete serial sections of the whole brain in adult mice that survived 3 or 6 h after intraperitoneal injection of BrdU to map the neurogenic niches (Saint Georges Chaumet et al., 2008).

In 3 or 6 h, the BrdU-incorporating cells did not have time to significantly migrate and were visualized at the sites of mitosis initiation. We confirmed the existence of a neurogenic niche in the OB, RMS, SVZ of the lateral ventricle, SCZ, and DG, and found no other sites of cell proliferation in the brain parenchyma. The continuum of BrdU+ cells between OB, RMS, SVZ, and SCZ can be interpreted as a single neurogenic niche, centered on the SVZ of the lateral

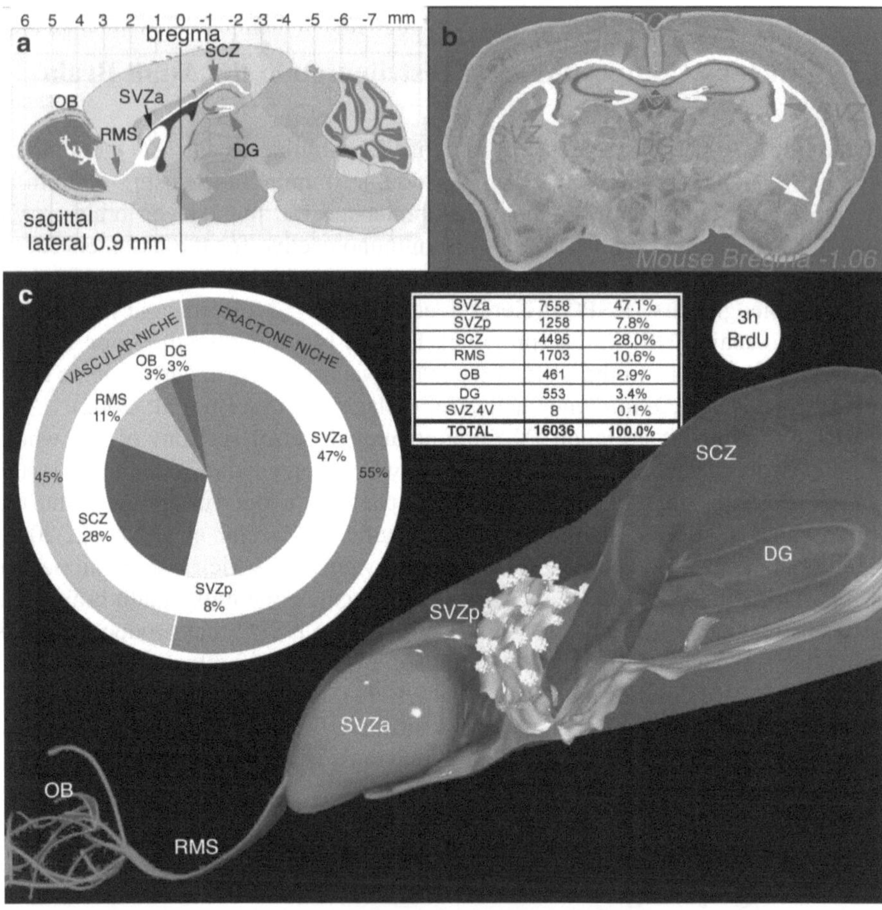

SVZa	7558	47.1%
SVZp	1258	7.8%
SCZ	4495	28.0%
RMS	1703	10.6%
OB	461	2.9%
DG	553	3.4%
SVZ 4V	8	0.1%
TOTAL	16036	100.0%

Fig. 4.3 Maps of the neurogenic niche in the adult mouse brain. (**a**) Sagittal view shows the neurogenic niche (*white*) in the olfactory bulb (*OB*), rostral migratory stream (*RMS*), SVZ of the anterior aspect of the lateral ventricle (*SVZa*), subcallosum zone (*SCZ*), and dentate gyrus (*DG*). (**b**) Coronal view of the neurogenic niche shows the long extensions of the SCZ (*arrow*). (**c**) Proportions of vascular versus fractone niche in the adult mouse brain ("pie chart," *upper left*). The *table* (*upper right*) indicates the number of cells immunoreactive for BrdU, 3 h after injection of BrdU, in the different portions of the neurogenic niche. A three-dimensional graphic (*bottom*) illustrates the organization (sagittal plan) of the neurogenic niche (*gray*) in one brain hemisphere. The choroid plexus of the lateral ventricle is indicated by *White color*

ventricle and dually represented in the left and right hemispheres (Saint Georges Chaumet et al., 2008). The graphics in Fig. 4.3a,c show the anterior portion of the lateral ventricle SVZ (SVZa), which comprises the highest number of proliferating cells in brain, a zone that protrudes from the caudate putamen in the anterior aspect of the lateral ventricle. The RMS, which is also highly neurogenic, is attached to the most anterior extremity of this neurogenic center and is attached to the OB neurogenic zones. Moreover, both posterior aspects of the lateral ventricle (SVZp) and the SCZ are attached to the SVZa. An overview of the neurogenic niche is illustrated in Fig. 4.3a,b (two-dimensional sagittal and coronal representations) and Fig. 4.3c (three-dimensional representation). All BrdU-incorporating cells in the brain parenchyma (excluding meningeal and vascular cells) were counted throughout the brain (Fig. 4.3c). The anterior aspect of the SVZ lateral ventricle, RMS, SCZ, and OB harbored 96.5% of the total number of BrdU-incorporating cells (Saint Georges Chaumet et al., 2008). DG (3%) is apparently disconnected from the main neurogenic centers of the brain, but all are linked to the lamina affixa, a meningeal projection that courses beneath the hippocampus (see Fig. 4.2). Although very restricted in volume, the neurogenic niches form a sheet-like surface that traverses most of the brain from its anterior (OB) to the very lateral portions of the brain via the external capsule and the lateral ventricle (Fig. 4.3b). The anatomy and vascular identity of these niches can be explained by the retraction during development and postnatal life of ventricles between the lateral ventricles and the OB, as well as underneath the corpus callosum. That each of these niches displays different neurogenic abilities suggests that between these neurogenic niches small differences exist in their composition. These differences may arise as a result of anatomical differences and the availability of BM structures, the fractone or vascular niche, or through the cocktail of available soluble factors. Perhaps the distribution of the neurogenic niche throughout the brain minimizes the migratory paths required for new neurons and glial cells to reach their final destination. Anatomically, either directly or via meningeal extensions, the neurogenic zones are interconnected, but what are the specific cues and the physiological organization that control cell proliferation and differentiation in the various NSC niches?

Table 4.1 Basal lamina heparan sulfate proteoglycans (HSPG) (180–467 kDa)

Name	Adult SVZ	Sources	Embryonic (E12.5)	Sources
Perlecan	++++	Fractone[a]/BV/ meninges	++++	Fractone[a]/BV/ meninges
Agrin	?	?	++	Fractone[a]/meninges
Collagen XV	?	?	?	?
Collagen XVIII	?	?	++++	Fractone[a]/meninges

SVZ, subventricular zone; BV, blood vessels

[a] Fractone expression of HSPG is heterogeneous

4.5 Extracellular Matrix and Basement Membranes

The extracellular matrix (ECM) consists of diverse constitutive, functional, and interactive extracellular molecules in all tissues and organs. Although the functions of ECM molecules such as collagens, laminins, fibronectin, HSPGs, and metalloproteinases are not well known in the brain, they participate at virtually all levels of neural tissue functions, providing structural and homeostatic cues. In the brain, ECM molecules are particularly concentrated in the meninges, in the meningeal projections at the interface between brain substructures, in the vasculature, and at the interface of all of these structures with the glia limitans (Mercier et al., 2003; Mercier and Hatton, 2004; Mercier, 2004).

There are a variety of ECM molecules that protect or degrade, focally concentrate, and activate cytogenic growth factors (GFs) and cytokines. These ECM molecules are based at the cell surface or in restricted locations where they are organized as a specialized dense meshwork termed basement membranes (BMs) or basal laminae. It has been shown that ECM molecules contained in BMs and the GFs that are trapped in (or bound to) these BMs are powerful modulators of the interactions that lead to cell proliferation, differentiation, and morphogenesis (Gordon et al., 1987; Roberts et al., 1988).

4.6 Dynamics of Niche-Specific Heparan Sulfate Interactions

4.6.1 Neurogenic Growth Factors Are Heparin-Binding Molecules

It is significant that numerous neurogenic GFs are heparin-binding molecules. As potent regulators of most niches with multiple feedback mechanisms and levels of control, incorporation of HSPG into endogenous signaling pathways is a logical safeguard against overexpression and exacerbation of cellular response. For these reasons, HSPG are a focus of much attention in oncogenesis (Vlodavsky et al., 1990) while interest within the field of neurogenesis has only just begun.

The specific interactions of HSPG and individual GF is highly dependent upon the endogenous HS and the available receptors within the neurogenic niche. The various effects of numerous growth factors that have been determined to require HS to signal participate in a variety of processes from initiation of mitosis, guidance of differentiation, and chemotaxis within the neurogenic niches are summarized in Table 4.2 (Ornitz et al., 1992; Hienola et al., 2006). Fibroblast growth factor 2 (FGF2), the most commonly investigated neurogenic factor, is produced by the choroid plexus, meningeal cells, and astrocytes within the brain and requires HS to bind to its receptor (Guimond et al., 1993; Stopa et al., 2001; Proia et al., 2008) Endogenous sources of the HB growth factors amphiregulin and transforming growth factor-α (TGF-α) have been localized to the neurogenic zones as well as

Table 4.2 Neurogenic heparin-binding growth factors

Fibroblast growth factor (FGF) family	Hematopoietic factors
FGF 1–23	PDGF (AA, BB), GMCSF, platelet factor 4, antithrombin 3
Epidermal growth factor (EGF) family	Angiogenic factors
HB_EGF, amphiregulin, TNF-α	VEGF (1,2), endostatin, endoreppelin
Nerve growth factor (NGF) family	Inflammatory factors
BDNF, heparin-binding growth-associated molecule (HBGAM)	TGF-β (1,2), BMP (2,4), interferon-γ, Noggin, Chordin IL-(1β, 2, 3, 4, 5, 7, 8, 12)
Morphogens	
Wnts (1–15), Notch, ShH (1–2)	

their common receptor (Craig et al., 1996; Falk and Frisen, 2002). Providing additional credence to the numerous models of injury, ischemia and inflammatory models of neurogenesis HS is also a mediator of a number of key inflammatory growth factors and cytokines. A large majority of the TGF-β family and pathways contain HB-factors. TGF-β, a negative modulator of neurogenesis, not only binds to HS but alters its expression in the ECM (Rider, 2006). Bone morphogenetic protein (BMP)-2 and -4 require HS for signaling while their antagonists noggin and chordin require HS to inhibit BMP, suggesting the possibility of a complex feedback network involving these factors (Lim et al., 2000; Paine-Saunders et al., 2002; Jasuja et al., 2004). Interferon-γ demonstrates increased neurogenesis as a result of introduction to NSC populations (Kim et al., 2007). Neurogenesis is a complicated assortment of cellular behaviors that are influenced by a large number of overlapping paracrine factors, most of which rely upon HSPG to coordinate their action within the brain. Unraveling the mystery of the neural stem cell must first begin with discovering the parties involved.

4.6.1.1 Heparan Sulfate Proteoglycans and the Neural Stem cell Niche

HSPG are endogenous molecules of the ECM that interact with paracrine factors via the negatively charged sulfate groups along the HS side chains. They exist as part of the larger glucosaminoglycan (GAG) family that includes keratin sulfate, chondroitin sulfate, dermatin sulfate, and the closest relative of heparan sulfate, heparin. A single HSPG is composed of a single core protein and multiple HS side chain attached at serine residues (Stringer and Gallagher, 1997). HSPG expression and localization in the ECM are determined by the expression of the core protein itself and its interactions with the plasma membrane and other ECM molecules.

HS is composed of a single chain of glucosamine α1,4-glucuronic acid that span 50–200 disaccharides in length and has alternating *N*-acetylated and *N*-sulfated glucosamine regions (Fig. 4.4; Bernfield et al., 1999). The sulfated domains, s-domains, are produced during synthesis by deacetylase/sulfotransferase activity and display large heterogeneity of size and dispersion between individual chains and along the same chain (Unger et al., 1991). The *N*-sulfated s-domains are the regions where

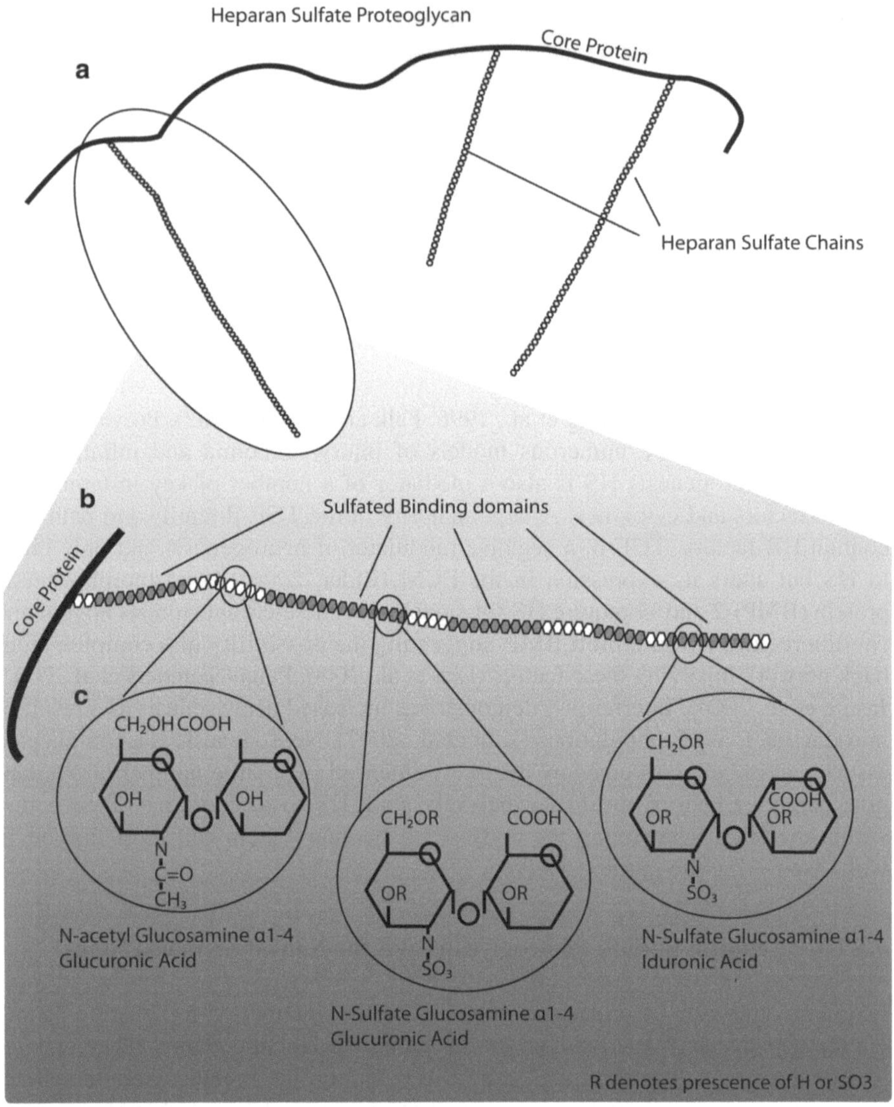

Fig. 4.4 Biochemical structure of heparan sulfate proteoglycan (HSPG) and its HS (GAG) side chains. (**a**) Superstructure of general HSPG with core protein (*black*) and HS side chains. (**b**) Expanded insert of a single HS from HSPG in (**a**) showing unbranched GAG chain. Highly sulfated domains of *N*-sulfated glucosamine (*gray*) are dispersed along the chain by a-domains and are the location for growth factor (GF) interactions. (**c**) Biochemical characterization of HS disaccharides within HS. a-domains are composed of *N*-acetyl glucosamine α1-4 glucuronic acid; s-domains are composed of *N*-sulfated glucosamine α1-4 hexonuric acids. Additional variable sulfation within s-domains (*O*-sulfation) is denoted by "R"

substrates are bound and are also subject to the modification by sulfotransferases and glucoisomerase to increase the substrate specificity and determine the function of the molecule (Stringer and Gallagher, 1997). The charge repulsion from adjacent sulfate groups induces a more rigid form to the s-domains as opposed to the acetylated a-domains, leading to a flexible linear chain with multiple straight regions (Mulloy et al., 1994). As the s-domains are responsible for substrate binding, the entire chain behaves like a number of "sticky rods" linked by flexible chains.

4.6.1.2 Hypervariability of HS and Complex Formation

HS is required for a large number of growth factors and cytokines to form a receptor complex and induce signaling pathways (Gallagher, 2001). The s-domains find positively charged regions on the substrate molecules to facilitate signaling or provide safe haven for the substrate (Saksela et al., 1988). However, not all HS recognize all heparin-binding substrates. Two single HS chains may vary in terms of overall chain length, size, and dispersion of s-domains, and within the s-domains the locations of additional *O*-sulfation and of glucuronic to iduronic acid epimerization. Most modifications occur within the s-domains with the very rare 3-*O*-sulfation of an a-domain (Stringer et al., 1997). The structure of HS is such that in a single short chain segment of eight disaccharides has more than 1 million possible structures by altering the sulfation and epimerization.

The sulfation pattern of the s-domains creates a charge topography that determines substrate specificity. The highly variable nature of the synthetic pathways allow for incredible diversity of the sulfation pattern between individual HS molecules and by association the substrates that they may bind. Determination of HS structure occurs during synthesis and modification in the extracellular space by the coordinated effect of extracellular sulfatases or sulfotransferases (Nakato and Kimata, 2002; Westling and Lindahl, 2002). The mechanism of these modifying enzymes is unknown; however, based upon analysis of HS chain sequence and functional differences it is apparent that their actions are inconsistent, resulting in varied sulfation. The expression of these enzymes by HSPG-producing cells may vary from cell to cell (Lindahl et al., 1998). This structural heterogeneity of HS manifests as a functional heterogeneity that is demonstrated by differential substrate-binding ability in that HS from the same core protein. species may vary between tissues (Smits et al., 2004; Knox et al., 2002), between cellular origin (Whitelock et al., 1996), and on the same singular core protein. Cellular expression of HS-synthesizing and HS-modifying enzymes is differentially expressed and differentially regulated throughout the tissues of the body (Lamanna et al., 2006; Properzi et al., 2008), throughout development (Brickman et al., 1998) and in pathological conditions (Jayson et al., 1998; Nawroth et al., 2007). This finding means that there is no direct correlation between the expression of a core protein and the effects of the HS that it contains. Of particular note is that regions of increased proliferative ability demonstrate a markedly different expression of these HS-modifying enzymes (Grobe et al., 2002; Lindahl et al., 1998; Langsdorf et al., 2007). Despite their presence within all extracellular environments, HS can still provide for unique control of signaling events within the neurogenic niche.

The function of a single HS chain is dependent on its sulfation pattern, with a single HS capable of binding to a number of substrates with varied success. Once bound to a substrate, the HS serves to protect the substrate from proteolytic cleavage (Saksela et al., 1988) until it can be released into the extracellular milieu by heparanase activity (Whitelock et al., 1996), or passed to another HS chain or receptor with higher substrate affinity. In facilitating the signaling event, HS increases receptor affinity for the substrate, which may occur through inducing a conformational change in the substrate, such as with interferon-γ (Lortat-Jacob and Grimaud, 1991), or by stabilizing a dimer structure, such as with FGF2 (Mohammadi et al., 2005). Additional steps that are required for HS to function as a signaling cofactor require additional sequence specificity to allow for receptor and or substrate dimerization (Venkataraman et al., 1996). Differences in the two signaling HS molecules for a single substrate may also affect the mechanism of signaling by induction of altered receptor conformation or by inducing endocytosis (Reiland and Rapraeger, 1993). The involvement of HS in binding to the receptor during complex formation has been demonstrated in some cases and can be altered by the addition or subtraction of a single sulfate group (Goodger et al., 2008).

Individual HS chains may associate with a stable of growth factors with varied affinity and function based upon their specific sulfation pattern (Bernfield et al., 1992; Walker et al., 1994). Within a single niche exist numerous HS chains, each with multiple overlapping functions. Using different portions of the same chain, a single HS may be sufficiently able to facilitate the signaling of tumor necrosis factor-α (TNF-α) and act in a purely storage manner for Wnt while another HS may cause signaling of both Wnt and FGF2. This purpose of this functional redundancy is unknown but may be to allow for feedback control or competition for the HS cofactor within a niche.

The highly variable HS is also subject to significant feedback, largely from molecules with which it normally complexes (Sharma et al., 1998; Kaji et al., 2006). A single growth factor signal may induce expression of specific HS via expression of their core protein (Kobayashi et al., 1992) and upregulate or downregulate HS-modifying enzymes (Narita et al., 2007) to effectively alter the composition of the extracellular HSPG and its ability to interact with growth factors. The importance of understanding niche-specific HS dynamics is crucial to understanding how the growth factors affect cellular behavior within the niche.

4.6.1.3 The HSPG of the Neurogenic Niche

Despite HSPG role in plasticity throughout the body, little has been done to identify those molecules that make up the niche and participate in adult neurogenesis. A large density of both cell-surface and basement membrane HSPG (BM-HSPG) exist within the neurogenic niches (Kerever et al., 2007) but only four, possibly five, have been shown to concentrate within the niche, with more likely to come. Two members of the syndecan family, syndecan-1 and syndecan-3, are found throughout the SVZ-RMS-OB pathway (Kim et al., 1994; Hienola et al., 2006). Syndecan-3 is localized to more mature migrating neuroblasts and has been implicated in heparin-binding growth-associated molecule (HBGAM)-mediated chemotaxis (Hienola et al., 2006), but the origin and function of syndecan-1 in the adult and embryonic brain are unknown. Glypican-1 distribution is diffused throughout the entire brain, but immunoreactivity is strongest within the SVZ (Litwack et al. 1998). Of particular interest is the localization of FGFR2 to astrocytes within the SVZ and DG (Chadashvili et al., 2006). This receptor moonlights as an HSPG (Sakaguchi et al., 1999) and may exist in this form within the niches. Perlecan, an HSPG of the BM variety, was localized to fractones as well as the very interesting HS antibody 10E4 from Seikagaku Corp. (Kerever et al., 2007). In addition to possible roles in growth factor signaling, the HS chains of perlecan have been shown to participate in collagen-1 fibril scaffolding (Yang et al., 2005) and may account for the characterization of some fractones to contain collagen-1 (Mercier et al., 2002). Perlecan-IR structures were not limited to fractones but were a significant component of both the vascular and meningeal basement membranes. Within the fractone population itself, there existed a small percentage of fractone that were void of perlecan immunoreactivity. This compositional heterogeneity of the structures may be the result of an intrinsic nature or fractone turnover caused by matrix metalloproteinase (MMP) and tissue inhibitor of metalloproteinases (TIMP) activity within the highly plastic niche (Whitelock et al., 1996; Fitzgerald et al., 1999). The 10E4 antibody from Seikagaku Corp. is a unique entity in its recognition of a specific sugar sequence of the HS chain itself rather than a traditional peptide epitope. Localization of this antibody in the adult brain was restricted to another subpopulation of fractones and a few specialized meningeal locations such as the choroid plexus (Kerever et al., 2007). Although the specific HS sequence of 10E4 is unknown, it is particularly significant in that it localizes the functional aspect of the HSPG molecule rather than the core protein. The restricted localization of the specialization and its association with FGF2-specific binding capability of fractones bodes well for a model of fractone-mediated neurogenesis.

In the embryo, fractone structures have been identified in the immature brain at numerous times (unpublished results). At embryonic day 12.5 (E12.5), fractone demonstrated immunoreactivity to the additional BM-HSPG agrin and collagen-18. Collagen-18 was localized almost entirely to 10E4, suggesting a possible host for the HS sequence; however, until they can be isolated biochemically it will be difficult to definitively identify the HSPG or multiple HSPG that express the 10E4 epitope. Agrin expression was diminished in comparison to collagen-18 and found within both fractone and meningeal tissue. These two HSPGs also displayed a heterogeneity among the fractone population, mirroring the results seen by perlecan in the adult. Agrin and collagen-18 have yet to be identified within the adult neurogenic zones.

4.6.2 Paracrine Roles of BM-HSPG: Focus on FGF Signaling

From their vantage, BM-HSPG can oversee the various cells within the neurogenic niche and when sought out may present growth factors and cytokines that direct cell fate. This process involves a two-step sequence of growth factor sequestering and storage until presentation to the cell-surface receptor (Fig. 4.5). Sequestration

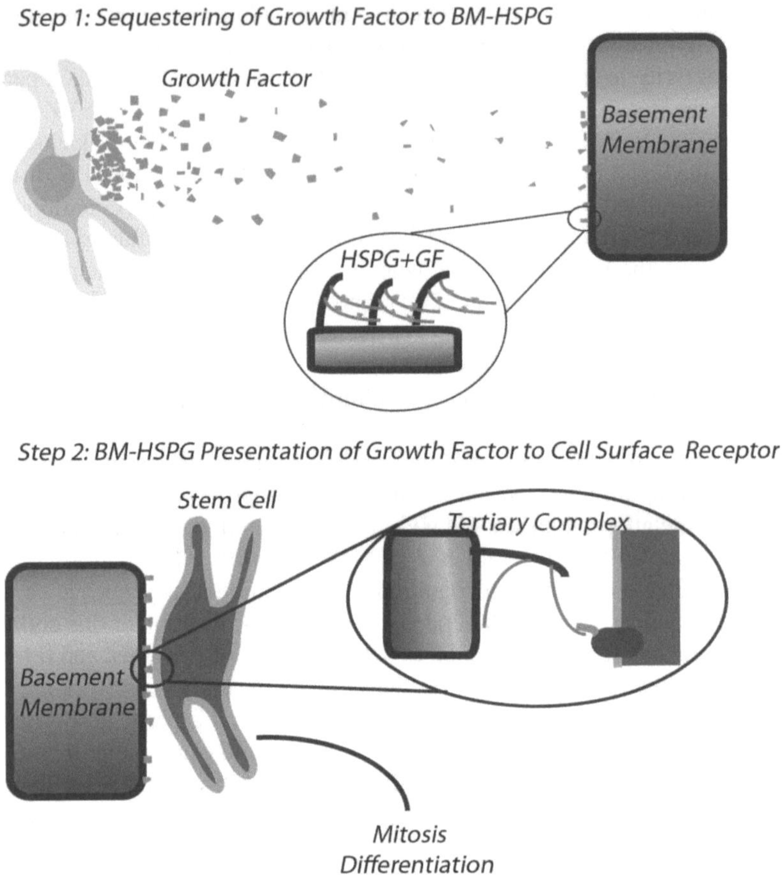

Fig. 4.5 Two-step mechanism for BM-HSPG mediated signaling. *Step 1:* Growth factor (GF) captured by HSPG molecules within the basement membrane (BM). Until presentation of receptor, BM stores GF. *Step 2:* Stem cell contacts BM-HSPG structure to form tertiary complex of HSPG/GF/receptor, resulting in signal transduction and cellular response

among many heparin-binding growth factors implicated in neurogenesis, FGF2 is the most well characterized endogenous signal present in the neurogenic zones. It has been shown that FGF2 requires binding to perlecan to exert its cytogenic activity in the peripheral tissues (Brickman et al., 1995; Aviezer et al., 1994; Yayon et al., 1991; Rapraeger, 1995). Perlecan is a large HSPG that consists of a core protein and several heparan-sulfate sites that represent low-affinity receptors for FGF-2. Yayon and colleagues provided the first demonstration that FGF2 is not biologically active by itself, using COS-7 cells that were manipulated to express a very low level of HSPGs but expressed FGF receptors (Yayon et al., 1991; Aviezer et al., 1994). The authors demonstrated that $HSPG^{-}$ $FGF\text{-}R^{+}$ COS-7 cells cannot proliferate in the presence of FGF2 unless exogenous heparin or perlecan is added to the culture medium. They further demonstrated that FGF-2 binds to the heparan sulfate chains of perlecan and that this binding is necessary for the binding of FGF2 to the high-affinity receptors located at the cell surface. Rapraeger and colleagues further demonstrated that this mechanism physiologically exists in the adult skin, and that two distinct intracellular mechanisms coexist for the biological activation of FGF2. The first mechanism, the perlecan/FGF2 complex, triggers protein kinase cascades that are activated by their association with the cell-surface receptors (FGFR) and ultimately lead to the biological activity of FGF2. Alternatively, the FGF2/perlecan complex enters the cell by endocytosis to reach the cell nucleus where FGF2 acts as a transcription factor (Reiland and Rapraeger, 1993). It is not clear whether one or the other pathway favors cell proliferation versus cell differentiation. Through crystallographic techniques, Goodger et al. (2008) have demonstrated the viability of both the symmetrical and asymmetrical models of HS/FGF2/receptor complex formation based upon the presence or absence of a single sulfate group. Both models have the respective ratio of 1:2:2 of HS:FGF2:FGFR, but the HS of the asymmetrical complex lacks the ability to bind the second receptor although the receptor remains in the complex through its connection to FGF2. It is unclear if both these types occur in vivo; however, this rationale may explain the multiple pathways of FGF2 signaling that Rapraeger et al. have described. In the brain, it is clear that FGF2 induces NSPC proliferation and differentiation both in vivo and in vitro. Although in terms of HS-mediated signaling knowledge of the FGF2 pathway is without doubt the most advanced, little is known of the specific interactions that occur within the restricted niches which account for the widely variable effect of this growth factor upon proliferative processes. The close association of potent mitogenic factors such as FGF2 with perlecan and the other BM-HSPGs within notable mitotic regions such as epidermis and marrow stroma (Klein et al., 1995; Blanpain et al., 2004) highlights the importance of BM-HSPG/GF interactions, especially in light of the discovery of fractone and the vascular niches for neurogenesis.

4.6.3 Fractones Collect and Concentrate Heparin-Binding Growth Factors

Knowing that the ECM, and particularly basement membranes, intervene in the process of growth factor activation to induce cell proliferation and differentiation and ultimately promote cytogenesis (Yayon et al., 1991; Aviezer et al., 1994), we proposed that fractones regulate proliferation, differentiation, and migration in the SVZ (Mercier et al., 2002). To test the potential of fractones for binding FGF2 in vitro, we modified the experimental protocol of Chang et al. (2000) for in situ binding on frozen sections (Chang et al., 2000) and used biotinylated FGF2 as a ligand. Biotinylation allowed visualizing FGF2 together with fractone antigens that were revealed by immunofluorescence in another channel. FGF2 was biotinylated according to the manufacturer (Pierce biotinylation kit) and was incubated on brain frozen sections of the adult mouse and rat before staining with streptavidin Texas red. The sections were also immunostained for laminin or perlecan to determine the exact location of fractones and capillaries in the ventricle wall. After observation by confocal microscopy, we found that biotinylated FGF2 was specifically located in fractones (Kerever et al., 2007). We also demonstrated that fractones could not bind biotinylated FGF2 when the sections were pretreated with heparatinase. Heparatinase effectiveness was assessed by absence of antigenicity for HSPG, using immunolabeling for an antibody specifically directed against N-sulfate HSPG (10E4; Seikagaku). Without N-sulfate domains, fractones cannot bind FGF2. These results indicate that FGF2 binding on fractones depends on HSPG (Kerever et al., 2007).

In other experiments, biotinylated FGF2 was injected into the lateral ventricle of adult rats or mice that were killed at different times after injection to determine whether, and when, FGF2 binding occurs in vivo. Brain sections were incubated with streptavidin Texas red and counterstained for laminin. We found that fractones, and no other structures in brain, sequestered FGF2 in vivo 30 min after cerebral ventricle injection in both rat (Fig. 4.6a,b) and mouse (Kerever et al., 2007). FGF2 was also detected on fractones 5 days after injection, suggesting that fractones store FGF2 after binding (Kerever et al., 2007). Detailed observation of FGF2 binding along the ventricle walls indicated preferential zones in the ventricle walls, with fractones of the corpus callosum and lateral septal nucleus more labeled than those of the caudate putamen in the anterior portion of the lateral ventricle. Most of the fractones of the third ventricle wall did not bind FGF2, even days after injection. Similar results were found with in situ binding of biotinylated FGF2, suggesting that fractones are functionally heterogeneous and have different affinities for GFs according to their location in brain. In similar experiments, we coinjected BrDU with biotinylated FGF2 or native FGF2 (as a control) to determine the neurogenic potential of biotinylated FGF2. After counting BrDU-immunoreactive cells in the lateral ventricle walls, we demonstrated that cerebral ventricle injection of biotinylated FGF2 and native FGF2 induces a similar neurogenic response. This finding suggests that biotinylated FGF2 mimics the effects of native FGF2 and that biotinylated FGF2 can be used as a marker for investigating FGF2/HSPG interactions in fractones.

Based on the potential of HSPG in basement membranes (Figs. 4.4 and 4.5), we hypothesized that fractones promote FGF-2 activity at the NSPC surface to induce NSPC proliferation and differentiation, and to ultimately produce new neurons and glial cells. Figure 4.6c illustrates the ultrastructure of the SVZa niche and its relationship with the ventricular cavity. The potential mechanism that is responsible for neurogenesis is illustrated in Fig. 4.6d. After entering the SVZ via interstitial clefts (Brightman, 2002), and binding to fractone-HSPG, FGF2 would be activated and presented to NSPC surface receptors. Then, intracellular protein kinase cascades would mediate FGF2 neurogenic activity. Experimentally, we studied the effect of preventing FGF2 binding on fractones (via heparanase; Kerever et al., 2007) to determine whether lack of FGF2 binding on fractones results in the loss of FGF2-induced neurogenesis in the SVZa. Our preliminary results indicate that FGF2-induced neurogenesis, assessed by increase of BrdU incorporation in the SVZa, is prevented by heparanase pretreatment. Thus, these preliminary results

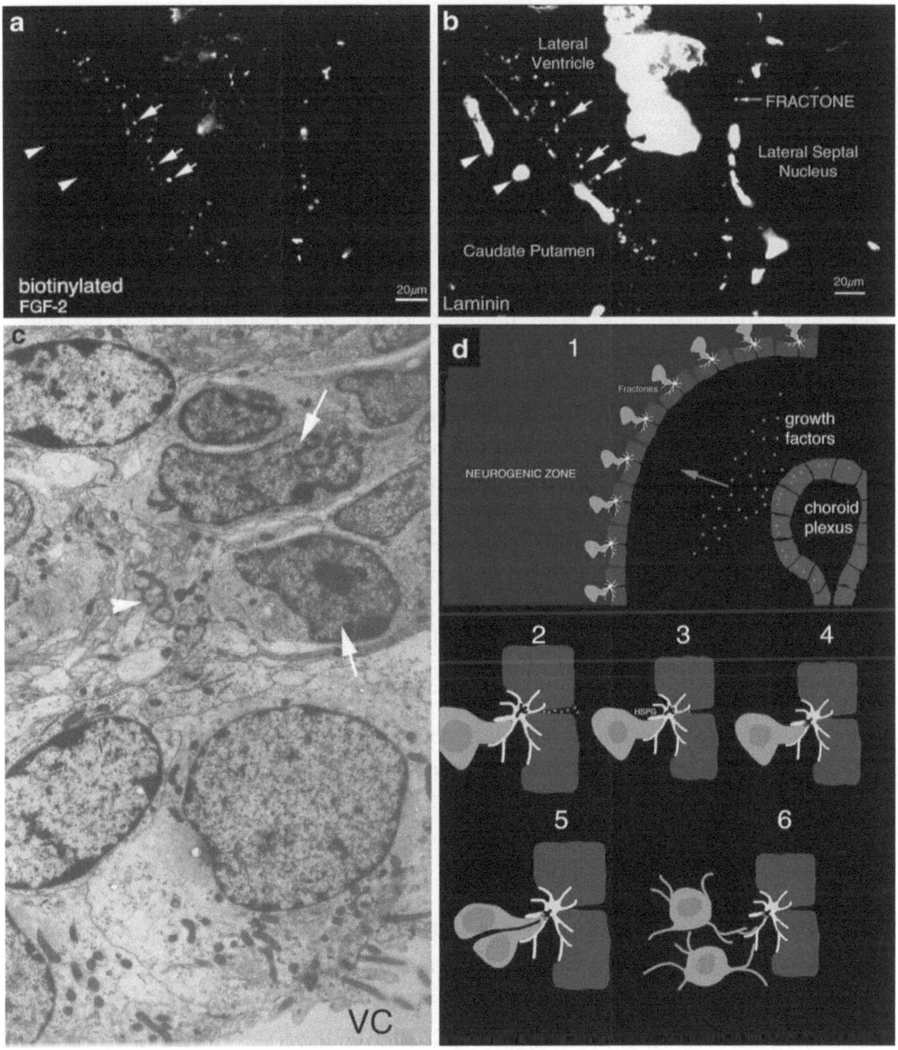

Fig. 4.6 Fractones collect and concentrate fibroblast growth factor 2 (FGF2) via HSPG: hypothesis on fractone function. (**a**) Binding FGF2 to fractones in the adult mouse 30 min after intracerebroventricular injection of biotinylated FGF2. FGF2 binding was revealed by streptavidin Texas red (*arrows*). *Arrowheads* indicate blood vessels that are not bound by FGF2. (**b**) Immunoreactivity for laminin in the section shown in (**a**), indicating the location of fractones (*arrows*) and blood vessels (*arrowheads*). (**c**) Ultrastructure of the SVZa shows neuroblasts (*arrows*) next to a fractone (*arrowhead*). *VC*, ventricular cavity. (**d**) Hypothesis of fractone function in the SVZ of the lateral ventricle. Heparin-binding growth factors that are produced by the choroid plexus are released in the ventricle cavity (*1*), then enter the SVZ via interstitial clefts (*2*) to bind fractone-HSPG (*3*). The HSPG present the growth factor to the NSPC surface receptors (*4*), promoting intracellular protein kinase cascades that trigger proliferation (*5*) and differentiation (*6*)

support the view that HSPG/FGF2 interactions within fractones are responsible for FGF2 activation at the NSPC surface.

4.7 Fractones During Brain Development

To determine whether fractones exist during development, we immunolabeled serial sections of mouse embryos at different stages of development (from E7.5 to E17.5), and postnatal life, with laminin, BrdU, and phosphorylated histone 3 (PH3), a marker of the mitosis M phase. We found numerous laminin+ puncta of size ranging from 0.5 to 3 μm at E7.5 through the growing neuroepithelium. Figure 4.2c shows double immunolabeling for PH3 (revealed by Alexa fluor 546, red) and laminin (revealed by Alexa fluor 647, blue) at the edge of the fourth ventricle. Clusters of postmitotic cells are clearly visible and are surrounded by laminin+ puncta that are similar in size (although additional puncta <1 μm are also visible) and distribution to adult fractones. However, the presence of fractones in the depths of the neuroepithelium, and in perimeningeal locations, clearly indicates that fractone functions may be involved in promoting neuroepithelial cell proliferation during development. Triple immunolabel for laminin, BrdU, and PH3 at all stages of development revealed complex patterns of pre- or postmitotic cell associations with fractones that vary at multiple developmental stages and currently are being investigated. Overall, fractones appear very early during development, entire days before blood vessels, and hold a potential for intervening in the process of cell proliferation in the developing brain.

4.8 Beyond the Neural Stem Cell Niche Function: Fractones Processing Growth Factors and Cytokines

Fractones collect heparin-binding growth factors in the SVZ of the lateral ventricle to provide signaling cues at the stem cell surface. But what is the function of fractones in the SVZ of other ventricles? Neural stem cells, defined by their ability to produce multipotent neurospheres in vitro, that is, by their potential to generate neurons astrocytes and oligodendrocytes, are present throughout the whole SVZ, including the third and fourth ventricles and the spinal canal (Weiss et al., 1996). However, these zones produce either glia (spinal cord) or no new cells at all (third ventricle; Saint Georges Chaumet et al., 2008) under physiological conditions in adulthood. When artificially stimulated by growth factors such as brain-derived neurotrophic factor, these nonneurogenic zones can produce new neurons and glial cells (Pencea et al., 2001; Martens et al., 2002). Together, these data suggest that neural stem cells reside in the SVZ of the whole ventricular system but that the balance of growth factor activity in the nonneurogenic niches prohibits activity.

Therefore, the zones that contain fractones, that is, the neurogenic zone of the lateral ventricle SVZ plus the SVZ of other ventricles (potential neurogenic zones), house neural stem cells. However, vascular neurogenic niches, such as the DG, may not contain neural stem cells, but only progenitors. Supporting this hypothesis, at least one of the vascular niches, the DG, has been demonstrated to contain restricted progenitors, whereas the fractone SVZ niche contains neural stem cells (Seaberg and van der Kooy, 2002). A possible explanation is that growth factor/HSPG interactions provided by fractones in the SVZ lead to an environment that promotes properties in the niche which preserve or generate stem cell fate, whereas vascular niches cannot.

Another important possibility is that fractones collect, concentrate, and promote heparin-binding growth factor activity for other purposes than regulating NSPC behavior in the neurogenic niche. Fractones exist throughout the SVZ of the third ventricle, which is adjacent to neuroendocrine nuclei such as the periventricular nucleus or the suprachiasmatic nucleus. There, fractones and their HSPG may collect heparin-binding growth factors from the cerebrospinal fluid (CSF) or from local astrocytes or microglia to influence neurohormone transcription and release by neurons in the hypothalamus. We have previously reviewed the potential of basement membranes in activating growth factors and cytokines (including numerous heparin-binding molecules) in the neuroendocrine system (Mercier and Hatton, 2004). With a mechanism similar to FGF2 capture in the SVZ of the lateral ventricle, the fractones of the third ventricle may capture, concentrate, and promote heparin-binding growth factors at the astrocytic surface to influence these cells and ultimately modify neuronal function via the numerous dendrites, axons, and neuronal nuclei that converge toward fractones (Mercier et al., 2003).

Knowing the distribution of fractones throughout the SVZ of the whole ventricular system (Mercier et al., 2002, 2003; Kerever et al., 2007), the lack of a tight junction barrier, and the existence of interstitial clefts in the ependyma that allow for the passage of large molecules from the ventricles to the SVZ (Brightman, 1965, 2002), it is conceivable that fractones serve as collectors of heparin-binding growth factors and cytokines for all neural nuclei adjacent to ventricles. The extent of the ventricular system, including recesses (collapsed extensions of the principal cavities) and aqueducts (sylvius) is such that a very important number of neural nuclei are adjacent to an ependyma and SVZ. We have shown that fractones exist in all ventricle cavities and recesses (Mercier et al., 2002, 2003; Kerever et al., 2007). Therefore, numerous neural nuclei may cross-talk via growth factors and cytokines, via the CSF in the ventricles as a carrier for these molecules, and via fractones as collectors/effectors for signaling in the recipient neuronal zone.

The choroid plexus and meninges may also intervene in cross-talk communication within the brain and between brain and blood (thus peripheral organs and tissues). The choroid plexus is an important source of growth factors, including FGF2 (Hayamizu et al., 2001; Stopa et al., 2001), one of the principal neurogenic growth factors in adulthood (Palmer et al., 1995). Another role of the choroid plexus is to provide signaling molecules for NSPC migration (Sawamoto et al., 2006). Meninges also produce growth factors (Mercier and Hatton, 2001, 2004; Wieseler-Frank et al., 2007) that potentially influence neurogenesis and other brain functions. Therefore, choroid plexus and meninges may provide growth factors and other signaling molecules for the NSPC niche and numerous other brain zones.

4.9 Conclusion

Fractones hold a considerable promise as selective reservoirs and processors of heparin-binding growth factors and cytokines in the adult brain. Their location all along the ventricular system allows for capture of these signaling molecules from the OB to the spinal cord, via the CSF playing the role of a carrier of signaling cues through the ventricular system. Choroid plexus is a major source for signaling molecules, but other brain zones facing the ventricles may directly release growth factors and cytokines that could be taken up at every other place of the ventricular system by fractones. Therefore, the ventricles may function as a pathway for these molecules, and fractones may represent the effectors of an intercommunication system within the brain. If this hypothesis is valid, the anatomy of the ventricles would facilitate the exchange of chemical information between distinct neural zones. The choroid plexus and meninges allow for additional routes of communication between the brain and peripheral organs as they are perfectly situated at the interface of the blood and neural tissue proper to permit the exchange of signaling cues. Investigations of this mechanism will be important for understanding the physiology of the neural stem cell niche, offering insights into the interactions that produce, release, capture, and ultimately direct factors and cytokines to the stem cell surface. Analysis of the neurogenic niche will provide links with other proliferative processes and provide a paradigm for investigating stem cell fate. The most promising and significant insight into the function of fractones is that together with the ventricular system and meninges they may form a broad system for administrating cytokines and growth factors in the central nervous system.

References

Altman, J. (1963). Autoradiographic investigation of cell proliferation in the brain of cats and rats. Anat. Rec. 145, 573–591.

Altman, J. (1969). Autoradiographic and histological studies of postnatal neurogenesis. IV. Cell proliferation and migration in the anterior forebrain, with special reference to persisting neurogenesis in the olfactory bulb. J. Comp. Neurol. 137, 433–458.

Altman, J., Das, G.D. (1966). Autoradiographic and histological studies of postnatal neurogenesis. I. A longitudinal investigation of the kinetics, migration and transformation of cells incorporating tritiated thymidine in neonate rats, with special references to postnatal neurogenesis in some brain regions. J. Comp. Neurol. 126, 337–390.

Alvarez-Buylla, A., Nottebohm, F. (1988). Migration of young neurons in adult avian brain. Nature 335, 353–354.

Aviezer, D., Hecht, D., Safran, M., Elsinger, M. David, G., Yayon, A. (1994). Perlecan, basal lamina proteoglycan, promotes basic fibroblast growth factor-receptor binding, mitogenesis, and angiogenesis. Cell 79, 1005–1013.

Bernfield, M., Kokenyesi, R., Kato, M., Hinkes, M.T., Spring, J., Gallo, R.L. (1992). Biology of the syndecans: a family of transmembrane heparan sulfate proteoglycans. Annu. Rev. Cell Biol. 8, 365–93.

Bernfield, M., Banerjee, S.D., Koda, J.E., Rapraeger, A.C. (1994). Remodelling of basement membranes as a mechanism of morphogenetic tissue interactions. In: Trelstadt, R.L. (ed) The role of extracellular matrix in development. New York, Liss, A.R.

Bernfield, M., Gotte, M., Park, P., Reizes, O., Fitzgerald, M., Lincecum, J., Zako, M. (1999). Functions of cell surface heparan sulfate proteoglycans. Annu. Rev. Biochem. 68, 729–777.

Blanpain, C., Lowry, W., Georghegan, A., Polak, L., Fuchs, E. (2004). Self-renewal, multipotency, and the existence of two cell populations within an epithelial stem cell niche. Cell 118, 635–648.

Brickman, Y.G., Ford, M.D., Small, D.H., Bartlett, P.F., Nurcombe, V. (1995). Heparan sulfates mediate the binding of basic fibroblast growth factor to a specific receptor on neural precursor cells. J. Biol. Chem. 270, 24941–24948.

Brickman, Y., Ford, M., Gallagher, J., Nurcombe, V., Bartlett, P., Turnbull, J. (1998). Structural modification of fibroblast growth factor-binding heparan sulfate at a determinative stage of neural development. J. Biol. Chem. 8:5, 4350–4359.

Brightman, M.W. (1965). The distribution within the brain of ferritin injected into cerebrospinal fluid compartments. J. Cell Biol. 26, 99–123.

Brightman, M.W. (2002). The brain's interstitial clefts and their glial walls. J. Neurocytol. 31, 595–603.

Chadashvili, T., Peterson, D. (2006). Cytoarchitecture of fibroblast growth factor receptor 2 (FGFR-2) immunoreactivity in astrocytes of neurogenic and non-neurogenic regions of the young adult and aged rat brain. J. Comp. Neurol. 498:1, 1–15.

Chang, Z., Meyer, K., Rapraeger, A.C., Friedl, A. (2000). Differential ability of heparan sulfate proteoglycans to assemble the fibroblast growth factor receptor complex in situ. FASEB J. 14, 137–144.

Craig, CG, Tropepe, V, Morshead, CM, Reynolds, BA, Weiss, S, van der Kooy, D. (1996). In vivo growth factor expansion of endogenous subependymal neural precursor cell populations in the adult mouse brain. J. Neurosci. 16, 2649–2658.

Doetsch, F., Alvarez-Buylla, A. (1996). Network of tangential pathways for neuronal migration in the adult mammalian brain. Proc. Natl. Acad. Sci. USA 93, 14895–14900.

Doetsch, F., Garcia-Verdugo, J.M., Alvarez-Buylla, A. (1997). Cellular composition and three-dimensional organization in the subventricular germinal zone in the adult mammalian brain. J. Neurosci. 17, 5046–5061.

Falk, A., Frisen, J. (2002). Amphiregulin is a mitogen for adult neural stem cells. J. Neurosci. Res. 69, 757–762.

Fitzgerald, M., Hayward, I.P., Thomas, A.C., Campbell, G.R., Campbell, J.H. (1999). Matrix metalloproteinase can facilitate the heparanase-induced promotion of phenotype change in vascular smooth muscle cells. Atherosclerosis 145:1, 97–106.

Gallagher, J.T. (2001). Heparan sulfate: growth control with a restricted sequence menu. J. Clin. Invest. 108:3, 357–361

Goodger, S., Robinson, C., Murphy, K., Gasiunas, N., Harmer, N., Blundell, T., Pye, D., Gallagher, J. (2008). Evidence that heparin saccharides promote FGF2 mitogenesis through two distinct mechanisms. J. Biol. Chem. 283, 13001–13008.

Gordon, M.Y., Riley, G.P., Watt, S.M., Greaves, M.F. (1987). Compartmentalization of a haematopoeitic growth factor (GM-CSF) by glycosaminoglycans in the bone marrow microenvironment. Nature 326, 403–405.

Grobe, K., Ledin, J., Ringvall, M., Holmborn, K., Forsberg, E., Esko, J.D., Kjellén, L. (2002). Heparan sulfate and development: differential roles of the N-acetylglucosamine N-deacetylase/N-sulfotransferase isozymes. Biochim. Biophys. Acta 1573:3, 209–215.

Guimond, S., Maccarana, M., Olwin, B., Lindahl, U., Rapraeger, A. (1993). Activating and inhibitory heparin sequences for FGF-2 (basic FGF) distinct requirements for FGF-1, FGF-2, and FGF-4. J. Biol. Chem. 268, 23906–23914.

Halfter, W. (1998). Disruption of the retinal basal lamina during early embryonic development leads to a retraction of vitral endfeet, an increase number of ganglion cells, and aberrant axonal outgrowth. J. Comp. Neurol. 397, 99–104.

Hayamizu, T.F., Chan, P.T., Johanson, C.E. (2001). FGF-2 immunoreactivity in adult rat ependyma and choroid plexus: responses to global forebrain ischemia and intraventricular FGF-2. Neurol. Res. 23, 353–358.

Hienola, A., Pekkanen, M., Raulo, E., (2004). HB-GAM inhibits proliferation and enhances differentiation of neural stem cells. Mol. Cell. Neurosci. 26, 75–88.

Hienola, A., Tumova, S., Kulesskiy, E., Rauvala, H. (2006). N-syndecan deficiency impairs neural migration in the brain. J. Cell Biol. 174, 569–580.

Iozzo, R.V. (2005). Basement membrane proteoglycans: from cellar to ceiling. Nat. Rev. Mol. Cell. Biol. 6, 646–656.

Jasuja, R., Allen, B., Pappano, W., Rapraeger, A., Greenspan, D. (2004). Cell-surface heparen sulfate proteoglycans chordin antagonism of bone morphogenetic protein signaling and are necessary for cellular uptake of chordin. J. Biol. Chem. 279, 51289–51297.

Jayson, G.C., Lyon, M., Paraskeva, C., Turnbull, J.E., Deakin, J.A., Gallagher, J.T. (1998). Heparan sulfate undergoes specific structural changes during the progression from human colon adenoma to carcinoma in vitro. J. Biol. Chem. 273:1, 51–57.

Kaji, T., Yamamoto, C., Oh-i, M., Fujiwara, Y., Yamazaki, Y., Morita, T., Plaas, A.H., Wight, T.N. (2006). The vascular endothelial growth factor VEGF165 induces perlecan synthesis via VEGF receptor-2 in cultured human brain microvascular endothelial cells. Biochim. Biophys. Acta 1760:9, 1465–1474.

Kerever, A., Schnack, J., Vellinga, D., Ichikawa, N., Moon, C., Arikawa-Hirasawa, E., Efird, J.T., Mercier, F. (2007). Novel extracellular matrix structures in the neural stem cell niche capture the neurogenic factor FGF-2 from the extracellular milieu. Stem Cells 25, 2146–2157.

Kim, C.W., Goldberger, O.A., Gallo, R.L., Bernfield, M. (1994). Members of the syndecan family of heparan sulfate proteoglycans are expressed in distinct cell-,tissue-, and development-specific patterns. Mol. Biol. Cell 5, 797–805.

Kim, S.J., Son, T.G., Kim, K., Park, H.R., Mattson, M.P., Lee, J. (2007). Interferon-gamma promotes differentiation of neural progenitor cells via the JNK pathway. Neurochem. Res. 32, 1399–1406.

Klein, G., Conzelmann, S., Beck, S., Timpl, R., Müller, C.A. (1995). Perlecan in human bone marrow: a growth-factor-presenting, but anti-adhesive, extracellular matrix component for hematopoietic cells. Matrix Biol. 14, 457–465.

Knox, S., Merry, C., Stringer, S., Melrose, J., Whitelock, J. (2002). Not all perlecans are created equal. J. Biol. Chem. 277, 14657–14665.

Kobayashi, M., Shimada, K., Ozawa, T. (1992). Human platelet-derived transforming growth factor-beta stimulates synthesis of glycosaminoglycans in cultured porcine aortic endothelial cells. Gerontology 38, 36–42.

Kuhn, H.G., Winkler, J., Kempermann, G. Thal, L.J., Gage, F.H. (1997). Epidermal growth factor and fibroblast growth factor-2 have different effects on neural progenitors in the adult rat brain. J. Neurosci. 17, 5820–5827.

Lamanna, W.C., Baldwin, R.J., Padva, M., Kalus, I., Ten Dam, G., van Kuppevelt, T.H., Gallagher, J.T., von Figura, K., Dierks, T., Merry, C.L. (2006). Heparan sulfate 6-O-endosulfatases: discrete in vivo activities and functional co-operativity. Biochem. J. 400, 63–73.

Langsdorf, A., Do, A.T., Kusche-Gullberg, M., Emerson, C.P. Jr, Ai, X. (2007). Sulfs are regulators of growth factor signaling for satellite cell differentiation and muscle regeneration. Dev. Biol. 311:2, 464–477.

Lim, D., Tramontin, A., Trevejo, J., Herrera, D., Garcia-Verdugi, J., Alvarez-Buylla, A. (2000). Noggin antagonizes BMP signaling to create a niche for adult neurogenesis. Neuron 28, 713–726.

Lindahl, U., Kusche-Gullberg, M., Kjellen, L. (1998). Regulated diversity of heparan sulfate. J. Biol. Chem. 273:29, 24979–24982.

Litwack, E.D., Ivins, J.K., Kumbasor, A., Paine-Saunolers, S., Stipp, C.S., Lanoler, A.D. (1998). Expression of the heparan sulfate proteoglycon glypicon-1 in the developing rodent. Dev. Dyn. 211, 72–87.

Lois, C., Alvarez-Buylla, A. (1994). Long distance neuronal migration in the adult mammalian brain. Science 264, 1145–1148.

Lortat-Jacob, H., Grimaud, J.A. (1991). Interferon-gamma C-terminal function: networking hypothesis. Heparan sulfate and heparin, new targets for IFNGamma, protect, relax the cytokine and regulate its activity. Cell Mol. Biol. 37:3, 253–260.

Mandelbrot, B.B. (ed) (1983). The fractal geometry of nature. Freeman, San Francisco

Martens, D.J., Seaber, R.M., van der Kooy, D. (2002). In vivo infusions of exogenous growth factors into the fourth ventricle of the adult mouse brain increase the proliferation of neural progenitors around the fourth ventricle and the central canal of the spinal cord. Eur. J. Neurosci. 16, 1045–1057.

Mercier, F. (2004). Astroglia as a modulation interface between meninges and neurons. In: Hatton, G.I., Parpura, V. (eds) Glial/neuronal signaling. Amsterdam, Kluwer Pub, 125–162.

Mercier, F., Hatton, G.I. (2000). Immunocytochemical basis for a meningeo-glial network. J. Comp. Neurol. 420, 445–465.

Mercier, F., Hatton, G.I. (2001). Connexin 26 and bFGF are primarily expressed in subpial and subependymal layers in adult brain parenchyma: roles in stem cell proliferation and morphological plasticity? J. Comp. Neurol. 431, 88–104.

Mercier, F., Hatton, G.I. (2004). Meninges and perivasculature as mediators of CNS plasticity. In: Bittar, E.E., Hertz, L. (eds.) Non-neuronal cells in the nervous system: function and dysfunction. Elsevier Bioscience, Amsterdam, Adv. Mol. Cell. Biol. 31, 215–253.

Mercier, F., Kitasako, J.T., Hatton, G.I. (2002). Anatomy of the brain neurogenic zones revisited: fractones and the fibroblast/macrophage network. J. Comp. Neurol. 451, 170–188.

Mercier, F., Kitasako, J.T., Hatton, G.I. (2003). Fractones and other basal laminae in the hypothalamus. J. Comp. Neurol. 455, 324–340.

Mohammadi, M., Olsen, S.K., Ibrahimi, O.A. (2005). Structural basis for fibroblast receptor activation. Cyt. Growth Fact. Rev. 16, 107–137.

Monje, M.L., Toda, H., Palmer, T.D. (2003). Inflammatory blockade restores adult hippocampal neurogenesis. Science, 302, 1760–1765.

Mulloy, B., Forster, M.J., Jones, C., Drake, A.F., Johnson, E.A., Davies, D.B. (1994). The effect of variation of substitution on the solution conformation of heparin: a spectroscopic and molecular modelling study. Carbohydr. Res. 255, 1–26.

Nakato, H., Kimata, K. (2002). Heparan sulfate fine structure and specificity of proteoglycan functions. Biochim. Biophys. Acta 1573:3, 312–318.

Narita, K., Chien, J., Mullany, S.A., Staub, J., Qian, X., Lingle, W.L., Shridhar, V. (2007). Loss of HSulf-1 expression enhances autocrine signaling mediated by amphiregulin in breast cancer. J Biol. Chem. 282:19, 14413–14420.

Nawroth, R., Zante, A., Cervantes, S., McManus, M., Helbrok, M., Rosen, S. (2007). Extracellular sulfatases, elements of the WNT signaling pathway, positively regulate growth and tumorigenicity of human pancreatic cancer cells. PLoS One 2:4, 1–11.

Ornitz, D.M., Yayon, A., Flanagan, J.G., Svahn, C.M., Levi, E., Leder, P. (1992). Heparin is required for cell-free binding of basic fibroblastic growth factor to a soluble receptor and for mitogenesis in whole cells. Mol. Cell. Biol. 12, 240–247.

Paine-Saunders, S., Vivano, B., Economides, A., Saunders, S. (2002). Heparan sulfate proteoglycans retain noggins at the cell surface: a potential mechanism for shaping bone morphogenic protein gradients. J. Biol. Chem. 277, 2089–2096.

Palmer, T.D., Ray, J., Gage, F.H. (1995). FGF-2 responsive neuronal progenitors reside in proliferative and quiescent regions of the adult rodent brain. Mol. Cell. Neurosci. 6, 474–486.

Palmer, T.D., Willhoite, A.R., Gage, F.H. (2000). Vascular niche for adult hippocampal neurogenesis. J. Comp. Neurol. 425, 479–494.

Pencea, V., Bingaman, K.D., Wiegand, S.J., Luskin, M.B. (2001). Infusion of brain derived neurotrophic growth factor into the lateral ventricle of the adult rat leads to new neurons in the parenchyma of the striatum, septum, thalamus, and hypothalamus. J. Neurosci. 21, 6706–6717.

Proia, P., Schiera, G., Mineo, M., Ingrassia, A.M., Santoro, G., Savettieri, G., Di Liegro, I. (2008). Astrocytes shed extracellular vesicles that contain fibroblast growth factor-2 and vascular endothelial growth factor. Int. J. Mol. Med. 21, 63–67.

Properzi, F., Lin, R., Kwok, J., Naidu, M., Van Kuppevelt, T., Dam, G., Camargo, L., Raha-Chowdhury, R., Furukawa, Y., Mikami, T., Sugahara, K. (2008). Heparan sulphate proteoglycan in glia and in the normal and injured CNS: expression of sulphotransferases and changes in sulphation. Eur. J. Neurosci. 27, 593–604.

Rapraeger, A. (1995). In the clutches of proteoglycans: how does heparan sulfate regulate FGF binding. Chem. Biol. 2, 137–144.

Reiland, J., Rapraeger, A.C. (1993). Heparan sulfate protoglycan and FGF receptor target basic FGF to different intracellular destinations. J. Cell Sci. 105, 1085–1093.

Rider, C.C. (2006). Heparin/heparan sulphate binding in the TGF-beta cytokine super family. Biochem. Soc. Trans. 34, 458–460.

Roberts, R., Gallagher, J., Spooncer, E., Allen, T.D., Bloomfield, F., Dexter, T.M. (1988). Heparan sulfate bond growth factors: a mechanism for stromal cell mediated haemopoiesis. Nature 332, 376–378.

Sakaguchi, K., Lorenzi, M.V., Bottaro, D.P., Miki, T. (1999). The acidic domain and first immunoglobulin-like loop of fibroblast growth factor receptor 2 modulate downstream signaling through glycosaminoglycan modification. Mol. Cell. Biol. 19, 6754–6764.

Saksela, O., Moscatelli, D., Sommer, A., Rifkin, D.B. (1988). Endothelial cell-derived heparan sulfate binds basic fibroblast growth factor and protects it from proteolytic degradation. J. Cell Biol. 107:2, 743–51.

Sawamoto, K., Wichterle, H., Gonzalez-Perez, O., Choifin, J.A., Yamada, M., Spassky, N., Murcia, N.S., Garcia-Verdugo, J.M., Marin, O., Rubenstein, J.L.M., Tessier-Lavigne, M., Okano, H., Alvarez-Buylla, A. (2006). New neurons follow the flow of cerebrospinal fluid in the adult brain. Science 311, 626–632.

Seaberg, R.M., van der Kooy, D. (2002). Adult rodent neurogenic regions: the ventricular subependyma contains neural stem cells, but the dentate gyrus contains restricted progenitors. J. Neurosci. 22, 1784–1793.

Seki, T., Array, Y. (1993). Highly polysialylated neural cell adhesion molecule (NCAM-H) is expressed by newly generated granule cells in the dentate gyrus of the adult rat. J. Neurosci. 13, 2351–2358.

Seri, B., Herrera, D.G., Gritti, A., Ferron, S., Collado, L., Vescovi, A., Garcia-Verdugo, J.M., Alvarez-Buylla, A. (2006). Composition and organization of the SCZ: a large germinal layer containing neural stem cells in the adult mammalian brain. Cereb. Cortex 16, 103–111.

Sharma, B., Iozzo, R. (1998). Transcriptional silencing of perlecan gene expression by interferon. J. Biol. Chem. 273, 4642–4646.

Smits, N.C., Robbesom, A.A., Versteeg, E.M., van de Westerlo, E.M., Dekhuijzen, P.N., van Kuppevelt, T.H. (2004). Heterogeneity of heparan sulfates in human lung. Am. J. Respir. Cell Mol. Biol. 30, 166–173.

Stopa, E.G., Berzin, T.M., Kim, S., Song, P., Kuo-Leblanc, V., Rodriguez-Wolf, M., Baird, A., Johanson, C.E. (2001). Choroid plexus growth factors: what are the implications for CSF dynamics in Alzheimer's disease? Exp. Neurol. 167, 40–47.

Stringer, S., Gallagher, J. (1997). Heparan sulfate. Int. J. Biochem. Cell Biol. 29:5, 709–714.

Unger, E., Pettersson, I., Eriksson, U.J., Lindahl, U., Kjellén, L. (1991). Decreased activity of the heparan sulfate-modifying enzyme glucosaminyl N-deacetylase in hepatocytes from streptozotocin-diabetic rats. J. Biol. Chem. 266:14, 43–961.

Venkataraman, G., Sasisekharan, V., Herr, A.B., Ornitz, D.M., Waksman, G., Cooney, C.L., Langer, R., Sasisekharan, R. (1996). Preferential self-association of basic fibroblastic growth factor is stabilized by heparin during receptor dimerization and activation. Proc. Natl. Acad. Sci. USA 93, 845–890.

Vlodavsky, I., Korner, G., Ishai-Michaeli, R., Bashkin, P., Bar-Shavit, R., Fuks, Z. (1990). Extracellular matrix-resident growth factors and enzymes: possible involvement in tumor metastasis and angiogenesis. Cancer Metastasis Rev. 32:3, 313–318

Walker, A., Turnbul, J., Gallagher, J. (1994). Specific heparan sulfate saccharides mediate the activity of basic fibroblast growth factor. J. Biol. Chem. 269:2, 931–935.

Weiss, S., Dunne, C., Hewson, J. (1996). Multipotent CNS stem cells are present in the adult mammalian spinal cord and ventricular neuroaxis. J. Neurosci. 16, 7599–7609.
Westling, C., Lindahl, U. (2002). Location of N-unsubstituted glucosamine residues in heparan sulfate. J. Biol. Chem. 277:51, 49247–49255.
Wexler, E.M., Geschwind, D.H., Palmer, T.D. (2007). Lithium regulates adult hippocampal progenitor development through canonical wnt pathway activation. Mol. Psychiatr. 13:285–292.
Whitelock, J., Murdoch, A., Iozzo, R., Underwood, A. (1996). The degradation of human endothelial cell-derived perlecan and release of bound basis fibroblast growth factor by stromelysin collagenase, plasmin, and heparanases. J. Biol. Chem. 271:17, 10079–10086.
Wieseler-Frank, J., Jekich, B.M., Mahoney, J.H., Bland, S.T., Maier, S.F., Watkins, L.R. (2007). A novel immune to CNS communication pathway: cells of the meninges surrounding the spinal cord CSF space produce proinflammatory cytokines in response to an inflammatory cytokine. Brain Behav. Immun. 21, 711–718.
Wurmser, H.M, Palmer, T.D., Gage, F.H. (2004). Neuroscience-Cellular interactions in the stem cell niches. Science 304, 1253–1255.
Yang, W.D., Gomes, R.R. Jr, Alicknavitch, M., Farach-Carson, M.C., Carson, D.D. (2005). Perlecan domain I promotes fibroblast growth factor 2 delivery in collagen I fibril scaffolds. Tissue Eng. 11:1–2, 76–89.
Yayon, A., Klagsbrun, M., Esko, J.D., Leder, P., Ornitz, D.P. (1991). Cell surface, heparin-like molecules are required for binding of basic fibroblast growth factor to its high affinity receptor. Cell 64, 841–848.
Yurchenko, P.D., Schyttny, J.C. (1990). Molecular architecture of basement membranes. FASEB J. 4, 1577–1590.

Part III
Adult Neurogenesis in Non-mammalian Vertebrates

Chapter 5
Adult Neurogenesis in Teleost Fish

Günther K.H. Zupanc

Abstract Adult neurogenesis in the brain of teleost fish has been studied since the 1960s. These investigations have demonstrated mitotic activity in dozens of brain areas and relative rates of cell proliferation 10–100 times higher than in the adult mammalian brain. The source of the new cells are stem cells in distinct proliferation zones, which give rise to both neurons and glial cells. Depending on the brain region, the young cells either reside in close vicinity to the proliferation zone from which they originate, or they migrate long distance to specific target areas. During migration, they are guided by radial glial fibers. Approximately half the young cells undergo apoptotic cell death within the first few weeks of their life; the other half persist long term. This continuous addition of new cells to the population of older cells leads to a permanent growth of the brain. A major function of adult neurogenesis in teleosts appears to be to ensure numerical matching of neurons in the central nervous system and sensory receptors cells or muscle fibers in the periphery. Neurogenesis also plays a central role in brain repair by replacing cells lost to injury by newly generated ones. Large-scale identification of regeneration-associated proteins by proteome analysis has revealed cytoskeletal proteins essential for the formation of new cells, proteins involved in cell proliferation, cellular motility, neuroprotection, and energy metabolism, as well as a zinc finger protein that may act as a transcriptional regulator.

5.1 Introduction

The beginnings of the study of adult neurogenesis in teleost fish date back to the late 1960s and early 1970s, shortly after the pioneering work of Joseph Altman (Altman 1962, 1969; Altman and Das 1965) had suggested the possibility of generation of new neurons in both the intact and the injured mammalian brain.

G.K.H. Zupanc (✉)
Department of Biology, Northeastern University, 134 Mugar Life Science Building, 360 Huntington Avenue, Boston, MA 02115, USA
e-mail: g.zupanc@neu.edu

T. Seki et al. (eds.), *Neurogenesis in the Adult Brain I: Neurobiology*,
DOI 10.1007/978-4-431-53933-9_5, © Springer 2011

Based on classical histological approaches (Kirsche 1967) and [^{3}H]thymidine autoradiography (Rahmann 1968; Kranz and Richter 1970a,b; Richter and Kranz 1970a,b), the early investigations on teleosts revealed mitotic activity in the adult brain and defined some of the major brain areas ("matrix zones") in which these mitoses take place. However, the cell type(s) into which these new cells develop remained elusive at that time.

Detailed analysis of the development of new cells in the adult teleostean brain started in the 1970s and 1980s, particularly by relating cell proliferation in the optic tectum to the continuing formation of new photoreceptor cells in the retina (Meyer 1978; Johns 1982). Most significantly, by ultrastructural examination of the progeny of dividing tectal cells, evidence was obtained that adult-born cells in the teleostean brain are capable of neuronal differentiation (Raymond and Easter 1983), thus firmly establishing the study of adult neurogenesis in teleost fish.

More recent developments in the area of adult neurogenesis in teleost fish have included detailed mappings and quantitative analysis of proliferative activity in the brain of several species (Zupanc and Horschke 1995; Zikopoulos et al. 2000; Ekström et al. 2001; Zupanc et al. 2005; Grandel et al. 2006); demonstration of long-term persistence, neuronal differentiation, and integration of the new nerve cells into existing neural networks (Zupanc and Zupanc 1992; Zupanc et al. 1996, 2005; Ott et al. 1997; Hinsch and Zupanc 2007); isolation and characterization of stem cells as the source of adult-born neurons and glial cells (Hinsch and Zupanc 2006); identification of apoptotic cell death as a regulator of the number of new cells (Soutschek and Zupanc 1995, 1996), and discovery of chromosomal variability as a possible mechanism to regulate gene expression at a genomic level during postembryonic development of brain cells (Rajendran et al. 2007). Moreover, by application of well-defined lesion paradigms, a close link has been demonstrated between adult neurogenesis in the intact brain and the potential of teleost fish to regenerate nervous tissue after injury by formation of new neurons (Hitchcock et al. 1992; Zupanc and Ott 1999). A large number of proteins associated with this process of neuronal regeneration have been identified by employing proteome analysis (Zupanc et al. 2006).

In the following chapter, I first describe the generation and development of new cells in the intact brain of teleost fish. Then, based on a comparative analysis of these findings, I discuss functional and evolutionary aspects of adult neurogenesis in vertebrates. Finally, I review the significance of neurogenesis for brain repair in teleosts, including recent attempts to identify regeneration-associated proteins on a large scale by employment of proteome analysis.

5.2 Adult Neurogenesis

5.2.1 Rate of Cell Proliferation

A distinctive feature of adult neurogenesis in teleost fish is the enormous relative number of new cells produced. Quantitative analysis based on labeling of S-phase cells with the thymidine analogue 5-bromo-2′-deoxyuridine (BrdU) has shown that in *Apteronotus leptorhynchus*, one of the best studied model systems, on average 100,000 cells are generated in the whole brain within any 2-h period (Zupanc and Horschke 1995). In zebrafish, approximately 6,000 cells are born within any 30-min period (Hinsch and Zupanc 2007). These cell numbers correspond to roughly 0.2% and 0.06% of the total population of brain cells in the respective species. Thus, the number of cells that enter the S-phase of mitosis per time unit, relative to the total number of cells in the adult brain, appears to be similar in the two species.

For comparison, in the subventricular zone of adult mice approximately 30,000 cells are formed every day (Lois and Alvarez-Buylla 1994); this corresponds to 0.03% of the estimated 110 million cells in the whole brain of adult mice (Williams 2000). The dentate gyrus of the hippocampus of adult rats gives rise to approximately 9,000 cells every day (Cameron and McKay 2001), corresponding to 0.003% of the estimated 330 million cells in the whole brain of adult rats (Herculano-Houzel and Lent 2005). Thus, the rate of cell proliferation in the adult teleostean brain is at least one, if not two, orders of magnitude higher than in the adult mammalian brain.

5.2.2 *Proliferation Zones*

Another distinctive feature of adult neurogenesis in teleosts is the tremendous number of proliferations zones, small, well-defined areas of the brain in which mitotic cell divisions occur at high rates. Many of these zones are found at or near the surfaces of ventricles or related systems. Although other zones, particularly in the cerebellum, are located in regions distant from any ventricle, at least some of them are thought to be derived from areas located at ventricular surfaces during embryonic stages of development. As a result of the everted development of the fish brain during embryogenesis, the ventricular lumina associated with these areas are obliterated and translocated (Pouwels 1978a,b).

Like the high overall rate of cell proliferation, the large number of proliferation zones in the teleostean brain contrasts with the situation in mammals. In the latter taxon, adult neurogenesis appears to be restricted to two brain areas: first, the anterior part of the subventricular zone of the lateral ventricle from which the immature neurons migrate via a so-called rostral migratory stream into the olfactory bulb and differentiate into granule neurons and periglomerular interneurons (Altman 1969; Luskin 1993; Lois and Alvarez-Buylla 1994; Lois et al. 1996; Pencea et al. 2001; Sanai et al. 2004; Bédard and Parent 2004; Curtis et al. 2007); and second, the subgranular zone of the dentate gyrus, from where the new cells migrate a short distance into the granule cell layer of the hippocampus and develop into mature granule neurons (Altman and Das 1965; Kaplan and Bell 1984; Eriksson et al. 1998; Gould et al. 1999; Kornack and Rakic 1999; Seri et al. 2001).

By contrast, mapping studies have revealed dozens of proliferation zones in the adult teleost fish brain (Figs. 5.1 and 5.2) (Kranz and Richter 1970a,b; Richter and Kranz 1970a,b; Zupanc and Horschke 1995; Zikopoulos et al. 2000; Ekström et al. 2001; Zupanc et al. 2005; Grandel et al. 2006). Quantitative analysis has shown

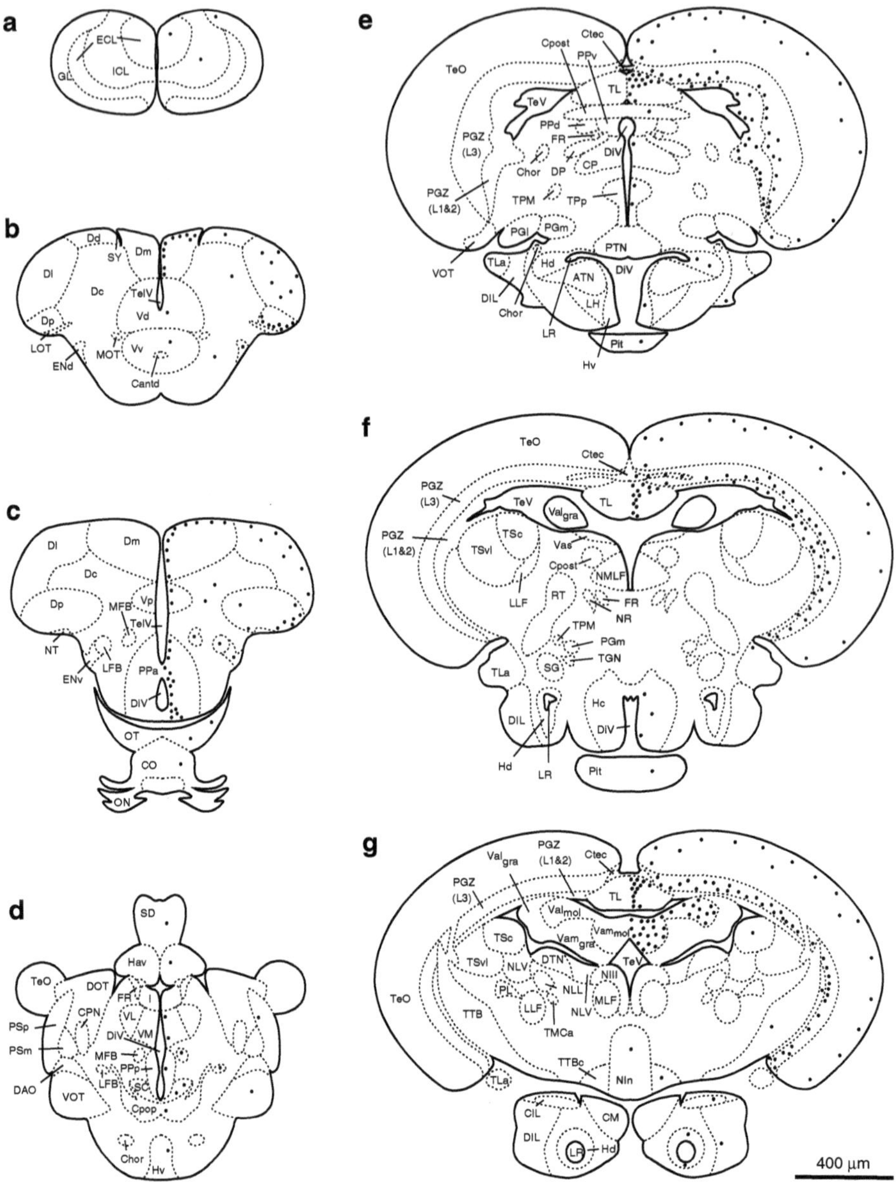

Fig. 5.1 Proliferation zones in rostral divisions of the brain of adult zebrafish. The presence of these zones has been revealed by analysis of 5-bromo-2′-deoxyuridine (BrdU) labeling in transverse brain sections. These sections were obtained through intraperitoneal injection of BrdU, followed by a postadministration survival time of 2 h. Areas of mitotic activity are indicated by *black dots*. Their number correlates with the number of BrdU-labeled cells in a semiquantitative manner. The chartings of the transverse sections correspond roughly to levels 31 (**a**), 85 (**b**), 107 (**c**), 127 (**d**), 153 (**e**), 168 (**f**), and 185 (**g**) of the zebrafish brain atlas (Wullimann et al. 1996). *ATN* nucleus tuberis anterior (anterior tuberal nucleus), *Cantd* commissura anterior, pars dorsalis (anterior commissure, dorsal part), *Chor* commissura horizontalis (horizontal commissure), *CIL* nucleus centralis lobi inferioris hypothalami (central nucleus of the inferior lobe), *CM* corpus mamillare (mammillary body), *CO* chiasma opticum (optic chiasm), *CP* nucleus centralis posterior thalami (central posterior thalamic nucleus), *CPN* nucleus praetectalis centralis (central pretectal nucleus), *Cpop* commissura supraoptica = commissura postoptica (supraoptic commissure = postoptic commissure), *Cpost* commissura posterior (posterior commissure), *Ctec* commissura tecti = commissura intertectalis (tectal

Fig. 5.1 (continued) commissure = intertectal commissure), *Dc* area dorsalis telencephali, zona centralis (central zone of dorsal telencephalic area), *Dd* area dorsalis telencephali, zona dorsalis (dorsal zone of dorsal telencephalic area), *DIL* nucleus diffusus lobi inferioris hypothalami (diffuse nucleus of the inferior lobe), *DiV* ventriculus diencephali (diencephalic ventricle), *Dl* area dorsalis telencephali, zona lateralis (lateral zone of dorsal telencephalic area), *Dm* area dorsalis telencephali, zona medialis (medial zone of dorsal telencephalic area), *DOT* tractus opticus dorsomedialis (dorsomedial optic tract), *Dp* area dorsalis telencephali, zona posterior (posterior zone of dorsal telencephalic area), *DP* nucleus dorsalis posterior thalami (dorsal posterior thalamic nucleus), *DTN* nucleus tegmentalis dorsalis (dorsal tegmental nucleus), *DV* radix descendens nervi trigemini (descending trigeminal root), *ECL* stratum cellulare externum bulbi olfactorii (external cellular layer of olfactory bulb), *EG* eminentia granularis (granular eminence), *ENd* nucleus entopeduncularis, pars dorsalis (entopeduncular nucleus, dorsal part), *ENv* nucleus entopeduncularis, pars ventralis (entopeduncular nucleus, ventral part), *FR* fasciculus retroflexus = tractus habenulointerpeduncularis (habenulo-interpeduncular tract), *GL* stratum glomerulosum bulbi olfactorii (glomerular layer of olfactory bulb), *Hav* nucleus habenularis ventralis (ventral habenular nucleus), *Hc* nucleus periventricularis hypothalami, zona caudalis (caudal zone of periventricular hypothalamus), *Hd* nucleus periventricularis hypothalami, zona dorsalis (dorsal zone of periventricular hypothalamus), *Hv* nucleus periventricularis hypothalami, zona ventralis (ventral zone of periventricular hypothalamus), *I* nucleus intermedius thalami (intermediate thalamic nucleus), *IAF* fibrae arcuatae internae (inner arcuate fibers), *ICL* stratum cellulare internum bulbi olfactorii (internal cellular layer of olfactory bulb), *LFB* fasciculus lateralis telencephali (lateral forebrain bundle), *LH* nucleus lateralis hypothalami (lateral hypothalamic nucleus), *LLF* fasciculus longitudinalis lateralis = lemniscus lateralis (lateral longitudinal fascicle = lateral lemniscus), *LOT* tractus olfactorius lateralis (lateral olfactory tract), *LR* recessus lateralis (lateral recess of diencephalic ventricle), *MFB* fasciculus medialis telencephali (medial forebrain bundle), *MLF* fasciculus longitudinalis medialis (medial longitudinal fascicle), *MOT* tractus olfactorius medialis (medial olfactory tract), *NLL* nucleus lemnisci lateralis (nucleus of the lateral lemniscus), *NLV* nucleus lateralis valvulae (nucleus lateralis valvulae), *NMLF* nucleus fasciculi longitudinalis medialis (nucleus of the medial longitudinal fascicle), *NT* nucleus taeniae (nucleus taeniae), *NIII* nucleus nervi oculomotorii (oculomotor nucleus), *ON* nervus opticus (optic nerve), *OT* tractus opticus (optic tract), *PGl* nucleus praeglomerulosus lateralis (lateral preglomerular nucleus), *PGm* nucleus praeglomerulosus medialis (medial preglomerular nucleus), *PGZ* (L1 & 2) stratum periventriculare tecti optici, laminae 1 and 2 (layers 1 & 2 of periventricular gray zone of optic tectum), *PGZ* (L3) stratum periventriculare tecti optici, lamina 3 (layer 3 of periventricular gray zone of optic tectum), *Pit* hypophysis = glandula pituitaria (hypophysis = pituitary), *PL* nucleus perilemniscularis (perilemniscal nucleus), *PPa* nucleus praeopticus parvocellularis, pars anterior (parvocellular preoptic nucleus, anterior part), *PPd* nucleus praetectalis periventricularis, pars dorsalis (periventricular pretectal nucleus, dorsal part), *PPp* nucleus praeopticus parvocellularis, pars posterior (parvocellular preoptic nucleus, posterior part), *PPv* nucleus praetectalis periventricularis, pars ventralis (periventricular pretectal nucleus, ventral part), *PSm* nucleus praetectalis superficialis, pars magnocellularis (magnocellular superficial pretectal nucleus), *PSp* nucleus praetectalis superficialis, pars parvocellularis (parvocellular superficial pretectal nucleus), *PTN* nucleus tuberis posterior (posterior tuberal nucleus), *SC* nucleus suprachiasmaticus (suprachiasmatic nucleus), *SD* saccus dorsalis (dorsal sac), *SY* sulcus ypsiloniformis (ypsiloniform sulcus), *TelV* ventriculi telencephali (telencephalic ventricles), *TeO* tectum opticum (optic tectum), *TeV* ventriculus mesencephali (tectal ventricle), *TGN* nucleus gustatorius tertius (tertiary gustatory nucleus), *TL* torus longitudinalis (longitudinal torus), *TLa* torus lateralis (lateral torus), *TMCa* tractus mesencephalocerebellaris anterior (anterior mesencephalocerebellar tract), *TPM* tractus praetectomamillaris (pretecto-mammillary tract), *TPp* nucleus periventricularis tuberculi posterioris (periventricular nucleus of posterior tuberculum), *TSc* nucleus centralis tori semicircularis (central nucleus of semicircular torus), *TSvl* nucleus ventrolateralis tori semicircularis (ventrolateral nucleus of semicircular torus), *TTB* tractus tectobulbaris (tecto-bulbar tract), *TTBc* tractus tectobulbaris cruciatus (crossed tecto-bulbar tract), Val_{gra} valvula cerebelli, pars lateralis, stratum granulosum (granular layer of the lateral part of the valvula cerebelli), Val_{mol} valvula cerebelli, pars medialis, stratum moleculare (molecular layer of the lateral part of the valvula cerebelli), Vam_{gra} valvula cerebelli, pars medialis, stratum granulosum (granular layer of the medial part of the valvula cerebelli), Vam_{mol} valvula cerebelli, pars medialis, stratum moleculare (molecular layer of the medial part of the valvula cerebelli), *Vas* lacuna vasculosa areae postremae (vascular lacuna of area postrema), *Vd* area ventralis telencephali, nucleus dorsalis (dorsal nucleus of ventral telencephalic area), *Vl* area ventralis telencepahli, nucleus lateralis (lateral nucleus of ventral telencephalic area), *VL* nucleus ventrolateralis thalami (ventrolateral thalamic nucleus), *VM* nucleus ventromedialis thalami (ventromedial thalamic nucleus), *VOT* tractus opticus ventrolateralis (ventrolateral optic tract), *Vp* area ventralis telencephali, nucleus postcommissuralis (postcommissural nucleus of ventral telencephalic area), *Vv* area ventralis telencephali, nucleus ventralis (ventral nucleus of ventral telencephalic area). (From Zupanc et al. 2005)

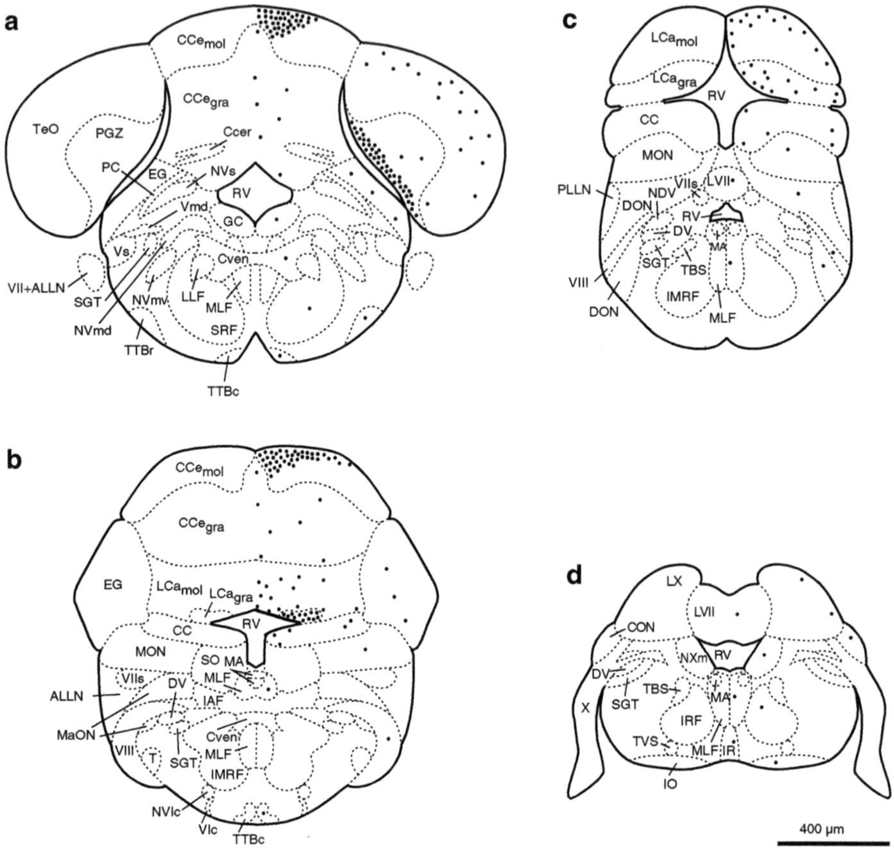

Fig. 5.2 Proliferation zones in caudal divisions of the brain of adult zebrafish. The presence of these zones has been revealed by analysis of BrdU labeling in transverse brain sections at postadministration survival times of 2 h. Areas of mitotic activity are indicated by *black dots*. Their number correlates with the number of BrdU-labeled cells in a semiquantitative manner. The chartings of the transverse sections correspond roughly to levels 219 (**a**), 237 (**b**), 260 (**c**), and 283 (**d**) of the zebrafish brain atlas (Wullimann et al. 1996). *ALLN* nervi lineae lateralis anterioris (anterior lateral line nerves), *CC* crista cerebellaris (cerebellar crest), CCe_{gra} corpus cerebelli, stratum granulosum (granular layer of cerebellar corpus), CCe_{mol} corpus cerebelli, stratum moleculare (molecular layer of cerebellar corpus), *Ccer* commissura cerebelli (cerebellar commissure), *CON* nucleus octavolateralis caudalis (caudal octavolateralis nucleus), *Cven* commissura ventralis rhombencephali (ventral rhombencephalic commissure), *DON* nucleus octavus descendens (descending octaval nucleus), *DV* radix descendens nervi trigemini (descending trigeminal root), *EG* eminentia granularis (granular eminence), *GC* griseum centrale (central gray), *IAF* fibrae arcuatae internae (inner arcuate fibers), *IMRF* formatio reticularis, pars intermedia (intermediate reticular formation), *IO* oliva inferior (inferior olive), *IR* nucleus raphes inferior (inferior raphe), *IRF* formatio reticularis, pars inferior (inferior reticular formation), LCa_{gra} lobus caudalis cerebelli, stratum granulosum (granular layer of caudal lobe of cerebellum), LCa_{mol} lobus caudalis cerebelli, stratum moleculare (molecular layer of caudal lobe of cerebellum), *LLF* fasciculus longitudinalis lateralis = lemniscus lateralis (lateral longitudinal fascicle = lateral lemniscus), *LVII* lobus facialis (facial lobe), *LX* lobus vagus (vagal lobe), *MA* fibra Mauthneri (Mauthner axon), *MaON* nucleus octavus magnocellularis (magnocellular octaval nucleus), *MLF* fasciculus

that the majority of the new brain cells are generated in the cerebellum (Zupanc and Horschke 1995; Hinsch and Zupanc 2007). In two of the three cerebellar subdivisions – the corpus cerebelli and the valvula cerebelli – the proliferation zones are located in specific areas within the molecular layers. In the third cerebellar subdivision, the eminentia granularis, the new cells are generated in a granular layer called the eminentia granularis pars medialis or the lobus caudalis.

The cells that are not produced in the cerebellum arise from various regions of the telencephalon, diencephalon, mesencephalon, and rhombencephalon. Among the latter, two are of particular interest from a comparative point of view: the olfactory bulb and the lateral and posterior zones of the dorsal telencephalon. Based both on neuroanatomical evidence (Northcutt and Braford 1980; Nieuwenhuys and Meek 1990; Braford 1995; Northcutt 1995; Butler 2000; Vargas et al. 2000) and on the results of functional studies (Rodríguez et al. 2002; Portavella et al. 2004), part of the dorsolateral telencephalon is thought to be homologous to the mammalian hippocampus. These results have led to the conclusion that adult neurogenesis in the olfactory bulb and the hippocampus is a conserved vertebrate trait (Zupanc 2006b).

5.2.3 Adult Stem Cells

The new cells in the adult teleostean brain originate from cells with stem cell-like properties. Isolation of such cells from the proliferation zones in the dorsal telencephalon and the cerebellum, followed by examination in culture, have revealed some of their properties (Hinsch and Zupanc 2006). After 3–4 days in culture, these cells give rise to neurospheres that grow through cell proliferation and reach diameters of as much as 140 μm within 3 weeks. Growth of the neurospheres can

Fig. 5.2 (continued) longitudinalis medialis (medial longitudinal fascicle), *MON* nucleus octavolateralis medialis (medial octavolateralis nucleus), *NDV* nucleus descendens nervi trigemini (nucleus of the descending trigeminal root), *NVmd* nucleus motorius nervi trigemini, pars dorsalis (trigeminal motor nucleus, dorsal part), *NVmv* nucleus motorius nervi trigemini, pars ventralis (trigeminal motor nucleus, ventral part), *NVs* nucleus sensorius principalis nervi trigemini (primary sensory trigeminal nucleus), *NVIc* nucleus nervi abducentis, pars caudalis (abducens nucleus, caudal part), *NXm* nucleus motorius nervi vagi (vagal motor nucleus), *PC* tractus cerebellaris posterior (posterior cerebellar tract), *PGZ* stratum periventriculare tecti optici (periventricular gray zone of optic tectum), *PLLN* nervus lineae lateralis posterioris (posterior lateral line nerve), *RV* ventriculus rhombencephali (rhombencephalic ventricle), *SGT* tractus gustatorius secundarius (secondary gustatory tract), *SO* populatio octavia secundaria (secondary octaval population), *SRF* formatio reticularis, pars superior (superior reticular formation), *T* nucleus tangentialis (tangential nucleus), *TBS* tractus bulbospinalis (bulbo-spinal tract), *TeO* tectum opticum (optic tectum), *TTBc* tractus tectobulbaris cruciatus (crossed tecto-bulbar tract), *TTBr* tractus tectobulbaris rectus (uncrossed tecto-bulbar tract), *TVS* tractus vestibulospinalis (vestibulo-spinal tract), *Vmd* radix motoria nervi trigemini, pars dorsalis (dorsal motor root of the trigeminal nerve), *Vs* radix sensoria nervi trigemini (sensory root of the trigeminal nerve), *VIc* radix caudalis nervi abducentis (caudal root of the abducens nerve), *VII* nervus facialis (facial nerve), *VIIs* radix sensoria nervi facialis (sensory root of the facial nerve), *VIII* nervus octavus (octaval nerve), *X* nervus vagus (vagal nerve). (From Zupanc et al. 2005)

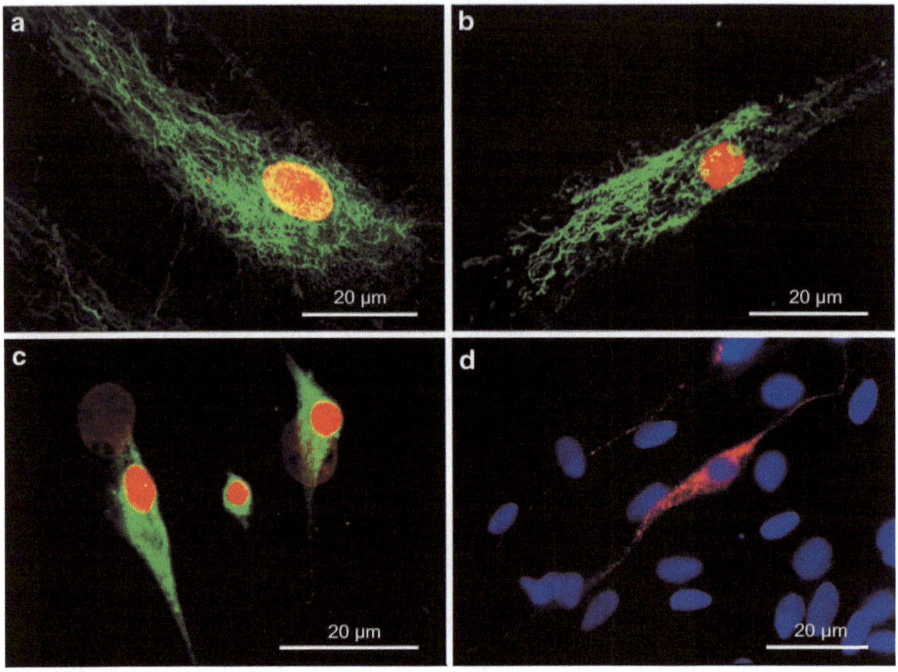

Fig. 5.3 Progeny of adult stem cells isolated from the brain of *Apteronotus leptorhynchus*. Differentiation was induced by plating neurospheres onto a laminin-coated surface and cultivating them in medium containing fetal bovine serum. Approximately 24 h after plating, the cells started to differentiate into various cell types. (**a**) Glial fibrillary acidic protein (GFAP)-immunoreactive astrocyte. (**b**) Vimentin-positive cell. (**c**) Hu-expressing neurons. (**d**) Microtubule-associated protein 2 (MAP2) (2a+2b)-immunopositive neuron. Nuclei were stained with either propidium iodide (*red*) or 4'6-diamidino-2-phenylindole-dihydrochloride (DAPI) (*blue*). Note absence of immunolabeling in some cells in (**c**) and (**d**). (From Hinsch and Zupanc 2006)

be promoted by epidermal growth factor and basic fibroblast growth factor. Differentiation is induced by exposing the neurospheres to fetal bovine serum and laminin as a coating substrate. Under the latter conditions, cells from neurospheres develop into both neurons and glial cells (Fig. 5.3). Because, in addition to this multipotency, the cells isolated from the proliferation zones in the adult brain exhibit the ability for self-renewal, they can be regarded true stem cells.

5.2.4 Migration and Guidance of the Young Cells

Depending on their site of origin, the young cells show either of two major types of behavior during subsequent development. In a number of brain regions, they migrate only a short distance to reside near the area where they were born. In the retina, for example, new cells originate from an annulus of progenitor cells at the junction of the retina and the iris. These progenitors generate neurons that are continuously added appositionally to the margin at the extant retina (Johns and Easter 1977; Meyer 1978; Hagedorn and Fernald 1992; Marcus et al. 1999). Similarly, in the optic tectum – the projection target of the retinal ganglion cells in teleosts – the majority of the new cells are generated at the caudal pole, where they remain during their subsequent development (Raymond and Easter 1983; Mansour-Robaey and Pinganaud 1990; Nguyen et al. 1999; Wullimann and Puelles 1999; Ekström et al. 2001; Candal et al. 2005; Zupanc et al. 2005). As a result, the optic tectum grows asymmetrically by expanding primarily from its caudal end.

By contrast, in other brain areas, the young cells migrate from their proliferation zones over relatively long distances to specific target areas. In the corpus cerebelli and the valvula cerebelli, the new cells are generated in specific proliferation zones in the respective molecular layers. Subsequently, the young cells migrate into the associated granular layers where they spread out evenly (Zupanc et al. 1996, 2005) (Fig. 5.4). Thus, both corpus cerebelli and valvula cerebelli grow in a rather symmetrical fashion.

Radial glial fibers appear to be involved in the long-distance migration by guidance of the young cells, at least in some brain areas (Zupanc and Clint 2003). In both the corpus cerebelli and the valvula cerebelli, such fibers, characterized by their immunoreactivity against vimentin and/or glial fibrillary acidic protein (GFAP), delineate the migratory path taken by the young cerebellar cells. These fibers could provide a scaffolding for the migrating cells, as suggested by double-labeling experiments. A few days after labeling of S-phase cells with BrdU, many of them can be seen in close apposition to GFAP-labeled radial glial fibers. The somata of these cells are distinguished by their elongated morphology (Fig. 5.5), a feature characteristic of migrating cells, as shown during both embryogenesis (Rakic 1971, 1972; Gregory et al. 1988) and adult stages of development (Alvarez-Buylla and Nottebohm 1988).

5.2.5 *Regulation of Cell Numbers by Apoptotic Cell Death*

Quantitative analysis has shown that in the various subdivisions of the cerebellum the number of the new cells drops to half their initial value a few weeks after these cells have arrived at their final target areas, the corresponding granular layers (Zupanc et al. 1996; Ott et al. 1997). Subsequent experiments employing terminal deoxynucleotidyl-transferase-mediated dUTP-biotin nick end-labeling (TUNEL) of 3′-OH ends of DNA demonstrated that this drop is caused by apoptotic cell death (Soutschek and Zupanc 1996). Further, the number of apoptotic cells is manifold larger in the granule cell layers than in the corresponding molecular layers. Taken together, these findings suggest that apoptosis is used as a mechanism to regulate the number of young cells after they have reached their target areas. A similar observation has been made in the teleostean retina during postembryonic development. Apoptotic cell death occurs predominantly in areas where the new cells differentiate and become integrated into visual circuits (Candal et al. 2005).

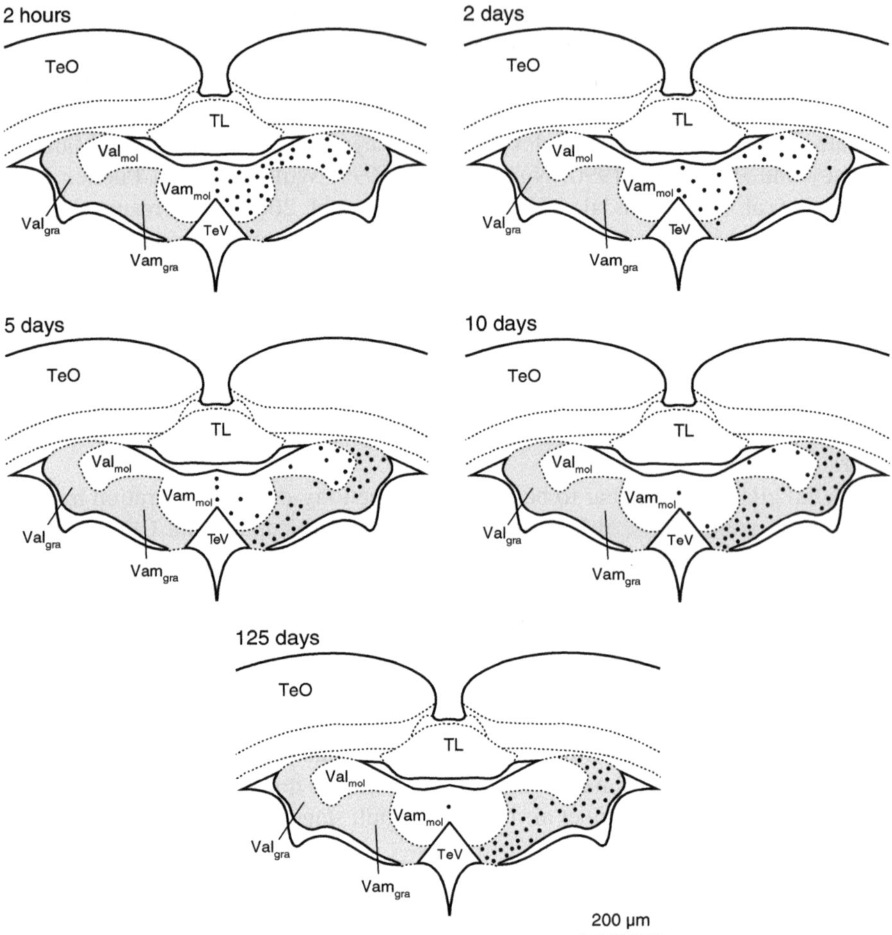

Fig. 5.4 Schematic representation of the distribution of BrdU-labeled cells in the valvula cerebelli of zebrafish after various post-BrdU administration survival times. The location of labeled cells is indicated by *black dots*. The number of dots roughly represents the density of labeled cells in this brain region. *TeO* tectum opticum, *TeV* tectal ventricle, *TL* torus longitudinalis, *Val*$_{gra}$ granule cell layer of the lateral division of the valvula cerebelli, *Val*$_{mol}$ molecular layer of the lateral division of the valvula cerebelli, *Vam*$_{gra}$ granule cell layer of the medial division of the valvula cerebelli, *Vam*$_{mol}$ molecular layer of the medial division of the valvula cerebelli. *Bar* 200 μm. (From Zupanc et al. 2005)

Studies in mammals have shown that during embryonic development apoptosis leads to the elimination of young cells, which, after arrival at the target site, have failed to make proper connections with other neurons and to receive adequate amounts of specific survival factors produced by cells in the target area (Raff 1992; Raff et al. 1993). Although not yet examined in the context of adult neurogenesis, it is possible that a similar mechanism regulates the number of cells born in the adult brain.

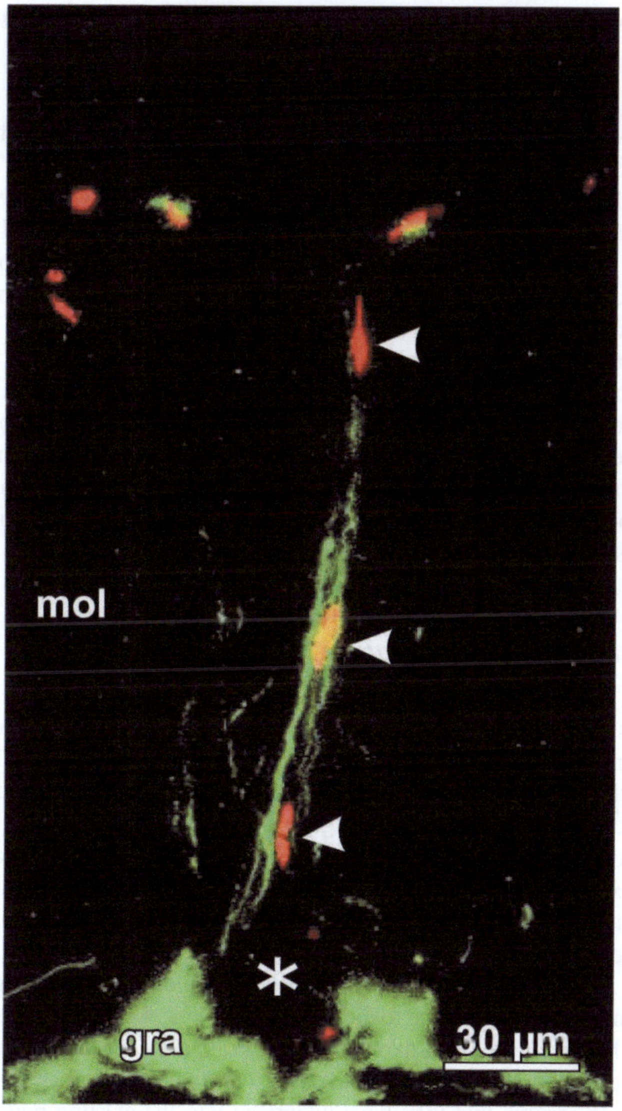

Fig. 5.5 Guidance of young cells by radial glial fibers in the cerebellum of *A. leptorhynchus*. Two days after the administration of BrdU, the fish was killed, and the transverse section through the dorsal portion of the corpus cerebelli was immunostained against both BrdU (*red color*) and GFAP (*green color*). Long GFAP-positive radial glial fascicles run at the midline of the molecular layer (*mol*) through the so-called dorsal tip (*asterisk*) into the granule cell layer (*gra*). Several BrdU-labeled cells, indicated by *arrowheads*, are closely apposed to the radial glial fibers. (From Zupanc 2001)

5.2.6 Long-Term Persistence of New Cells and Permanent Brain Growth

Long-term persistence of the cells generated in the adult teleostean brain has been demonstrated in both *A. leptorhynchus* and zebrafish. In each of the two species, approximately half the initially generated cells survive at least for several hundreds of days and likely for the rest of the fish's life expectancy (Ott et al. 1997; Zupanc et al. 2005; Hinsch and Zupanc 2007). In mammals, persistence of brain cells generated during adulthood has been shown in both the hippocampus and the olfactory bulb (Kaplan et al. 1985; Corotto et al. 1993; Winner et al. 2002; Kempermann et al. 2003).

In *A. leptorhynchus*, quantitative analysis has demonstrated that the long-term survival, together with the continuous production of new cells, leads to a permanent growth of the entire brain, except for very old fish in which the growth rate appears to reach a plateau; this parallels the continuous growth of the body in this species. While the body weight of the fish increases from 1 to 16 g, the total number of brain cells doubles from 5×10^7 to 1×10^8 (Zupanc and Horschke 1995).

5.2.7 Neuronal Differentiation

Experiments in zebrafish have shown that approximately 50% of all adult-born cells that survive long term express the neuron-specific protein Hu (Zupanc et al. 2005; Hinsch and Zupanc 2007). Such neurons are particularly abundant in the dorsal telencephalon, including the region presumably homologous to the mammalian hippocampus, but are also found in other areas of the adult brain.

Further experiments in which BrdU immunohistochemistry was combined with retrograde tract tracing have shown that in the cerebellum of *A. leptorhynchus* and zebrafish at least part of the young cells develop into granule cell neurons (Zupanc et al. 1996, 2005). As in these studies the BrdU-labeled granule cells were retrogradely traced by application of tracer substance to the molecular layers, the experiments also indicate that these cells have developed axons traveling from their somata in the granule cell layers to the corresponding molecular layers to form the parallel fiber network. Thus, it is reasonable to assume that they have integrated into the existing neural network of cerebellar neurons. However, it is unknown whether the axons of the newly generated granule cells make proper synaptic connections with other cells, such as Purkinje cells. It also remains to be examined whether the new granule cells are functional and whether their physiological properties are similar to those of the "older" granule cells.

5.2.8 Genomic Regulation of the Development of the New Cells

How are cellular proliferation and the subsequent development of new brain cells regulated? Traditionally, it was thought that the mechanisms mediating these processes operate on a constant genome and are restricted to alterations at transcriptional and posttranslational levels. However, chromosomal analysis of new cells in the brain of both mammals and teleost fish has provided evidence for somatic genomic alterations through loss and gain of whole chromosomes (Rehen et al. 2001, 2005; Osada et al. 2002; Kaushal et al. 2003; Yurov et al. 2005; Rajendran et al. 2007). Such alterations result in a genetic mosaic of the new cells and, thus, in a likely modification of expression levels of genes residing on the affected chromosomes.

In *A. leptorhynchus*, only 22% of all cells generated in the adult brain display the euploid set of chromosomes, assumed to be 22 (Rajendran et al. 2007) (Fig. 5.6a–d). The remaining cells exhibit numerical chromosomal abnormalities. In most of these aneuploid cells, approximately 84%, loss of chromosomes is observed, whereas chromosome gain is encountered in the remaining 16% of the cells. Not only single chromosomes but also multiple chromosomes may be lost or gained. In the most extreme cases found, up to 16 chromosomes have been lost and up to 12 chromosomes gained, relative to the euploid number of 22 chromosomes. As in other species and tissues, aneuploidy among adult-born brain cells in *A. leptorhynchus* can be accounted for by segregation defects during mitotic division. Such defects include laggards, anaphase bridges, and micronuclei (Rajendran et al. 2007; Zupanc et al. 2009) (Fig. 5.6e–k).

Despite the pronounced variation from the euploid set of chromosomes, aneuploidy in *A. leptorhynchus* does not necessarily result in death of the individual aneuploid cells and in dysfunction of the whole animal. Examination of such cells has shown that a significant number of them can persist long term and develop into neurons (Rajendran et al. 2008; Zupanc et al. 2009). This finding suggests that at least some of the aneuploid cells generated in the adult teleostean brain are competent to develop normally.

5.3 Toward a Functional and Evolutionary Understanding of Adult Neurogenesis in Fish: The Numerical Matching Hypothesis

Why does neurogenesis cease in all but two regions of the mammalian brain at the time of, or shortly after, birth, but persist throughout life in numerous regions of the teleost fish brain? A possible answer is given by a comparative analysis of the differences in the development of motor structures and sensory organs between teleosts and mammals (Zupanc 1999, 2001, 2006a,b, 2008a,b, 2009). Although many species of teleosts and mammals grow during postembryonic stages of development and even throughout life, there is a distinct difference in the growth pattern between these two taxonomic classes. In mammals, postembryonic growth is the result of an increase in size but not in number of individual muscle fibers (Rowe and Goldspink 1969), whereas in fish both the number of muscle fibers and the volume of individual fibers increase (Weatherley and Gill 1985; Fine 1989; Koumans and Akster 1995; Zimmerman and Lowery 1999; for review, see Rowlerson and Veggetti 2001). It is, therefore, possible that the hyperplasia of peripheral motor elements prompts a concomitant increase in the number of central neurons involved in neural control of associated muscle activity (Fig. 5.7).

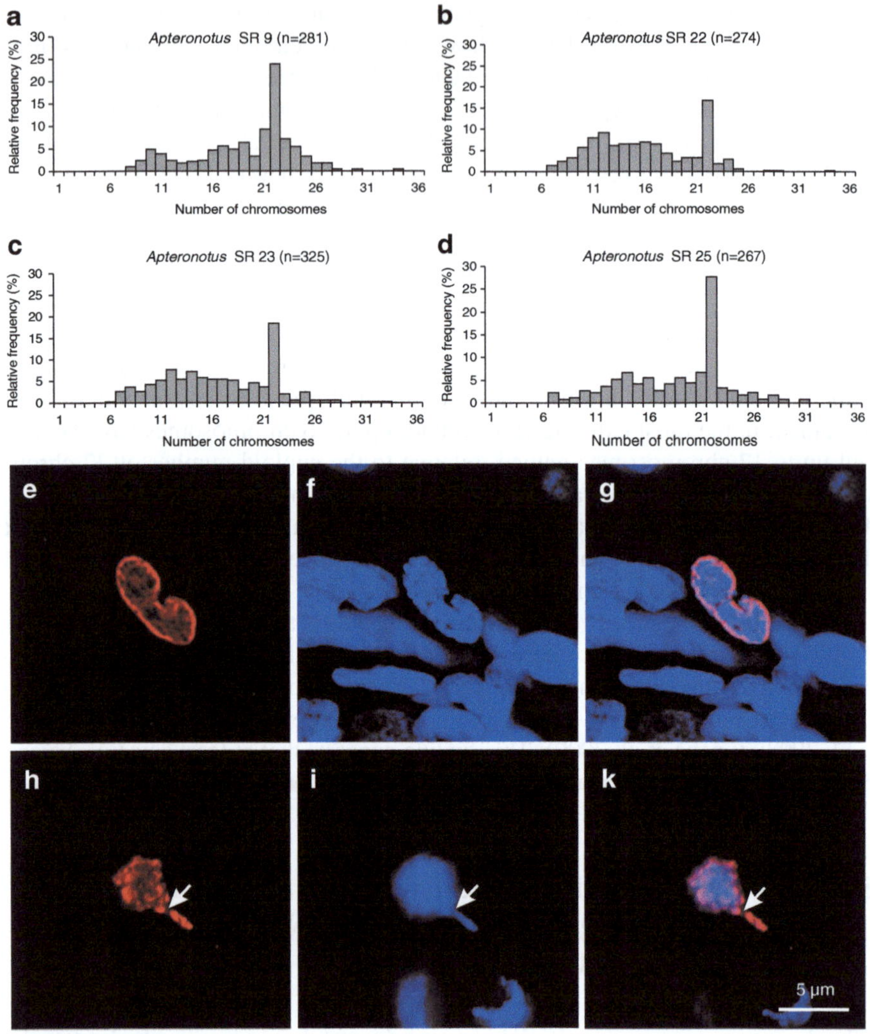

Fig. 5.6 (**a–d**) Chromosome number histograms based on analysis of metaphase chromosome spreads of cells from whole adult brains of four individual fish of *A. leptorhynchus*. Between 17% and 28% of the cells are euploid, having 22 chromosomes. Among the aneuploid cells, the majority lost chromosomes, whereas a rather minor portion gained chromosomes. *n*, number of cells analyzed. (**e–k**) High-power images obtained through deconvolution microscopy of phosphorylated histone-H3-labeled cells (*red*) in the molecular layer of the corpus cerebelli. The nuclei were counterstained with DAPI (*blue*). (**e–g**) Morphologically normal metaphase profile. (**h–k**) Prometaphase/metaphase profile with single laggard. Note the clear gap between the laggard and the daughter nucleus (*arrows*). In addition to the phosphorylated histone-H3 label (**e, h**) and the DAPI counterstain (**f, i**), the overlay (**g, k**) is shown. (After Rajendran et al. 2007)

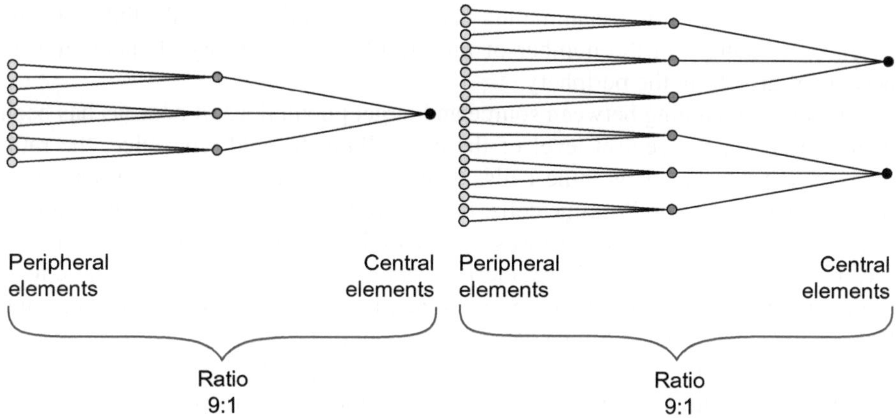

Fig. 5.7 The numerical matching hypothesis. This hypothesis makes predictions about the relationship between peripheral elements, such as muscle fibers or sensory receptor cells, and central elements in the brain involved in motor control of these muscle fibers, or in processing of sensory information received through these receptors, during growth. A number of observations suggest that the ratio of the corresponding peripheral elements and central elements is highly constant and maintained when the number of peripheral elements changes (here, an increase in the number of peripheral elements from 9 to 18 is shown). Hyperplasia of peripheral motor elements, or increase in the number of peripheral sensory elements, therefore, prompts a concomitant increase in the number of the corresponding central neurons. Because such growth through addition of new peripheral elements to the population of existing ones is characteristic of teleost fish, but largely absent in adult mammals, the numerical matching hypothesis can explain the difference in the extent of adult neurogenesis between teleosts and mammals. (From Zupanc 2008b)

Along similar lines, the number of sensory receptor cells, receptor organs, or receptor units in the periphery increases with age, as has been shown in several species of fish. Such a formation of new sensory elements has been demonstrated for sensory hair cells in the inner ear and the lateral line system of various species (Janssen et al. 1987; Corwin 1981; Higgs et al. 2002); for retinal cells in the eyes of goldfish (Johns and Easter 1977); and for electrosensory receptor organs in the gymnotiform fish *Sternopygus dariensis* (Zakon 1984). Thus, similar to the formation of new motor elements, the continuous increase in the number of sensory elements in fish may lead to the generation of new neurons involved in the processing of sensory information, ensuring a numerical matching of central neurons and peripheral sensory elements.

In fish, the matching hypothesis can explain the enormous mitotic activity in the cerebellum. Based on anatomical, physiological, and behavioral experiments, the cerebellum, notably of weakly electric fish, has been proposed to play an important role not only in the well-established function of control and coordination of movements but also in sensory processing. The latter has been demonstrated particularly in the context of the involvement of the cerebellum in tracking movements of objects around the animal and in the generation and subtraction of sensory expectations (for reviews, see Paulin 1993; Bell et al. 1997). On the other hand, the lack of adult neurogenesis in the mammalian cerebellum could be causally linked to the absence of changes in the number of muscle fibers and sensory elements during periods of growth in the periphery.

Numerical matching between source and target populations of neurons has been demonstrated at the central level of the cerebellum. In wild-type mice, the ratio between the number of granule cells, or neurons of the inferior olive, and their postsynaptic target, the Purkinje cells, is remarkably constant. Genetically induced degeneration of Purkinje cells causes a decrease in the number of granule cells and olivary neurons such that the characteristic ratio of the respective presynaptic neuron population and the population of postsynaptic neurons is maintained (Wetts and Herrup 1983; Herrup and Sunter 1987; Herrup et al. 1996).

In fish, there is some experimental evidence supporting the hypothesis that changes in the number of peripheral sensory elements lead to changes in the production of the corresponding central elements. Raymond et al. (1983) found in goldfish that permanent removal of the optic input by enucleation of the eye results in sustained depression of mitotic activity in the tectal proliferation zone on the denervated side compared to the intact one. Temporary denervation by optic nerve crush initially has a similar effect, but upon reinnervation of the tectum by the regenerating optic fibers, proliferation is enhanced on the experimental side compared to the control side.

The core idea of the matching hypothesis, that is, that changes in the number of peripheral motor or sensory elements are causal to alterations in the number of central elements associated with these peripheral elements, can also explain why, in the course of phylogenetic development, adult neurogenesis became restricted to the olfactory bulb and the hippocampus in mammals. These two brain regions are involved in processing of olfactory information perceived by primary sensory cells in the olfactory epithelium. The latter is the only sensory structure in which the primary sensory cells are continuously produced throughout adult life (Graziadei and Graziadei 1979). These new cells replace older olfactory neurons that undergo apoptosis (Magrassi and Graziadei 1995).

The axons of the primary sensory neurons make synaptic contacts with dendrites of mitral and tufted cells within the olfactory bulb. These cells project to the olfactory cortex, including the lateral entorhinal cortex. The latter structure, in turn, sends axons, via the lateral perforant path, to the hippocampus, which synapse on the distal dendrites of the dentate gyrus granule cells (Carlsen et al. 1982; Insausti et al. 2002; van Groen et al. 2002, 2003). Thus, each of the two regions in which neurogenesis occurs within the adult mammalian brain – olfactory bulb and dentate gyrus – is part of the central olfactory pathway. Remarkably, also the entorhinal cortex, in which adult neurogenesis is absent, retains its neurogenic potential during adult life. Neurospheres from this region differentiate into neurons and glia in vitro and after transplantation to the adult dentate gyrus (Fontana et al. 2006).

It is assumed that the primary sensory cells in the olfactory epithelium exhibit just a turnover, instead of a net increase in total number. Nevertheless, it is in agreement with the matching hypothesis that the continuous generation of a subpopulation of these sensory cells leads to a production of new central neurons (directly or indirectly) connected to them, whereas the degeneration of older sensory cells is accompanied by secondary degeneration of central neurons linked to the latter. In such cases, the overall number of peripheral elements and central neurons would vary little, or not at all, over time, although changes in the number of sensory cells or muscle fibers still trigger changes in the number of central neurons so that a specific ratio between the two cellular populations is maintained.

A likely scenario of the phylogenetic development of adult neurogenesis in the vertebrate brain is that this phenomenon constitutes a primitive trait which already existed some 400 million years ago in the common evolutionary ancestors shared by modern mammals, birds, reptiles, amphibians (all of which belong to the taxonomic group of the Sarcopterygii), and ray-finned fishes (Actinopterygii). A major function of this trait has been to ensure a constant ratio of motor (sensory) elements versus central elements within motor (sensory) pathways when the number of peripheral elements grows. However, as soon as the pattern of muscle growth shifted from hyperplasia more toward hypertrophy in the course of the evolution of mammals, the neurogenic potential of brain structures forming part of the motor pathway was reduced in parallel. A similar reduction in the number of neurogenic brain regions and in the rate of neurogenesis occurred when the formation of new sensory cells was more and more abandoned. According to this hypothesis, neurogenesis would have continued only in the olfactory bulb and the hippocampus as a consequence of the continued mitotic activity in the olfactory epithelium.

5.4 Neuronal Regeneration

The phenomenon of adult neurogenesis is closely related to a second phenomenon by which teleosts are distinguished from amniote vertebrates: the ability to replace neurons lost to injury in the adult central nervous system by newly generated ones (Anderson et al. 1983, 1987; Hitchcock et al. 1992; Zupanc and Ott 1999; Cameron 2000; Takeda et al. 2008). This potential, commonly referred to as neuronal regeneration, has been particularly well studied in three neuronal systems: the retina and its primary projection target, the optic tectum (for review, see Otteson and Hitchcock 2003); the spinal cord (for review, see Waxman and Anderson 1986; Sîrbulescu and Zupanc in press); and the cerebellum (for reviews, see Zupanc 1999, 2001, 2006a,b, 2008a,b, 2009; Zupanc and Clint 2003; Zupanc and Zupanc 2006a). In the following, I focus on the latter system, which is based on a series of investigations carried out in *A. leptorhynchus*.

5.4.1 *The Lesion Paradigm*

The study of neuronal regeneration in the cerebellum has greatly benefitted from the availability of a well-defined lesion paradigm in *A. leptorhynchus* (Fig. 5.8) (Zupanc et al. 1998). This paradigm takes advantage of the prominent position of

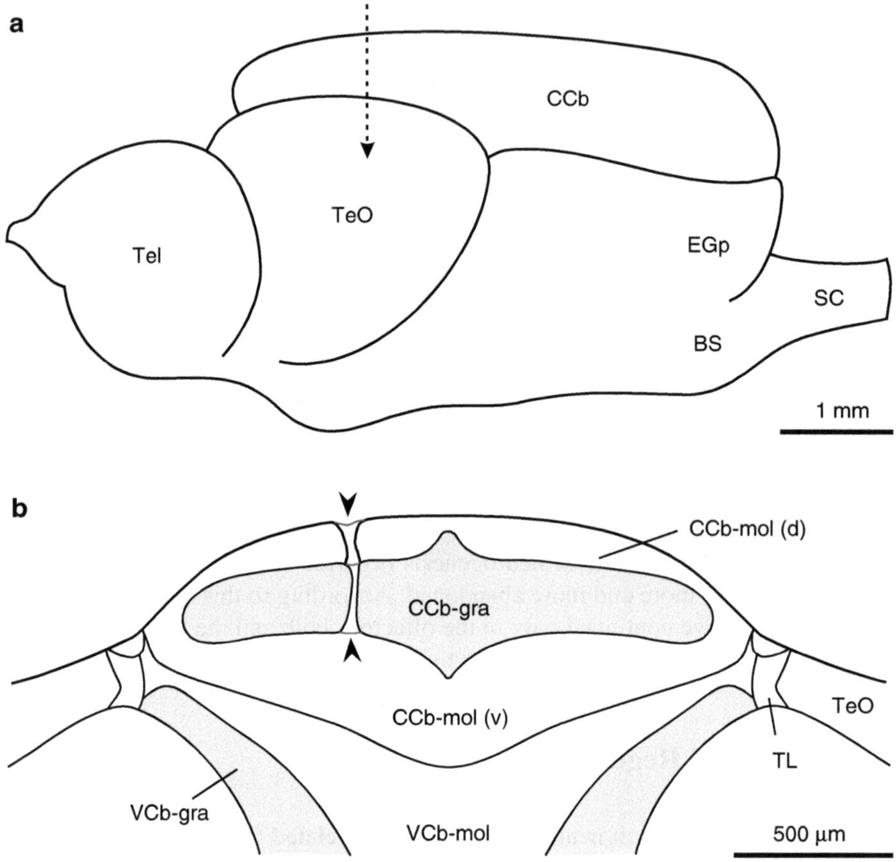

Fig. 5.8 The cerebellar lesion paradigm. (**a**) Side view of the brain of *A. leptorhynchus*. Cerebellar lesions are applied by puncturing the skull of the fish with a surgical blade in a defined location of the head, thus generating a stab wound in one hemisphere in the rostral part of the corpus cerebelli (*CCb*). (**b**) Transverse section through the corpus cerebelli taken at the level indicated by the *arrow* in (**a**). The lesion path (*arrowheads*) travels through the dorsal molecular layer [*CCb-mol(d)*] and the granular layer (*CCb-gra*) roughly halfway between the midline of the brain and the lateral edge of the granular layer. *BS* brainstem, *CCb-mol(v)* ventral molecular layer of corpus cerebelli, *EGp* eminentia granularis pars posterior, *SC* spinal cord, *Tel* telencephalon, *TeO* optic tectum, *TL* torus longitudinalis, *VCb-gra* granule cell layer of valvula cerebelli, *VCb-mol* molecular layer of valvula cerebelli. (From Zupanc and Zupanc 2006a)

the corpus cerebelli, which forms a roof on top of the brain, except over the telencephalon. Cerebellar lesions can, therefore, be made in a defined way by puncturing the skull of the anesthetized fish with a sterile surgical blade, without damaging other parts of the brain. Guided by landmarks on the fish's head, lesions are applied in a defined way, resulting in an approximately 1-mm-deep cut traveling in parasagittal direction and encompassing both the dorsal molecular layer and the granular layer of the corpus cerebelli.

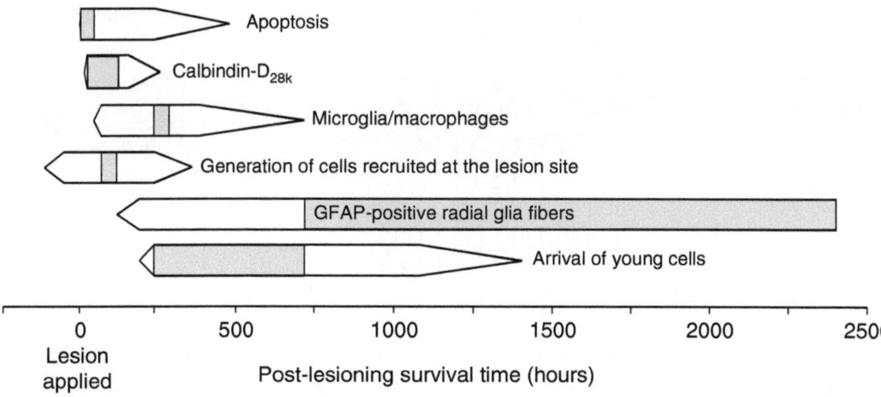

Fig. 5.9 Schematic diagram of the time course of six major processes occurring in response to mechanical lesions applied to the corpus cerebelli of *A. leptorhynchus*. From *top*, these are (1) increase in number of apoptotic cells at and near lesion site (Zupanc et al. 1998); (2) upregulation of calbindin-D_{28k} as a putative survival-promoting factor in the area of the lesion (Zupanc and Zupanc 2006b); (3) increase in density of microglia/macrophages throughout the corpus cerebelli (Zupanc et al. 2003); (4) generation of new cells recruited to the site of lesion 28 days after the injury (Zupanc and Ott 1999) (note that even cells born up to 2 days before the injury participate in the process of neuronal regeneration); (5) increase in density of GFAP-positive radial glia fibers at and near the site of lesion (Clint and Zupanc 2001); and (6) arrival of young cells at site of lesion (Clint and Zupanc 2001). The first slope of each *bar* represents the time period during which the event shows an increase from background levels. The *shaded area* represents the peak period of the event. The second slope represents the time period during which the event decreases to background levels. (From Zupanc 2008b)

Nissl stains of sections taken through the area of such lesions after various postlesion survival times have indicated that neuronal tissue is restored within a few weeks following the injury (Zupanc et al. 1998). This enormous regenerative potential is mediated by a number of well-orchestrated processes, which are discussed in detail below. The time course of several of the major processes is schematically indicated in Fig. 5.9.

5.4.2 Apoptotic Cell Death

The first event observed as part of the overall repair process involves the rapid elimination of damaged cells through apoptosis (Zupanc et al. 1998). The first apoptotic cells, as identified by the TUNEL technique, occur at the lesion site as early as 5 min after application of the stab wound (Fig. 5.10). Thirty minutes after the lesion, the number of cells undergoing apoptosis reaches maximum levels. At 2 days post lesion, the number of apoptotic cells starts to gradually decline until background levels are reached approximately 20 days after the lesion. Over this time period, only a few cells appear to undergo necrotic cell death.

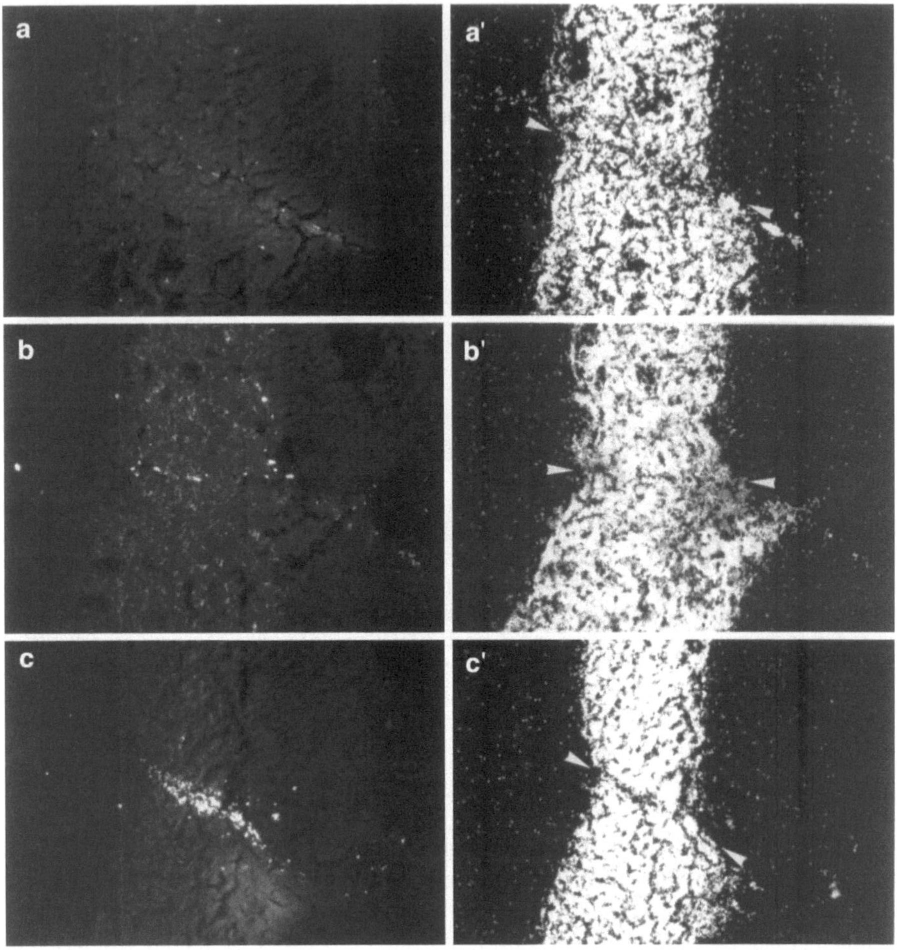

Fig. 5.10 Terminal deoxynucleotidyl transferase dUTP nick end labeling (TUNEL)-positive cells at the site of the lesion in the corpus cerebelli after survival times of 5 min (**a**), 15 min (**b**), 30 min (**c**), 1 day (**d**), 5 days (**e**), and 20 days (**f**). The location of the lesion is indicated by arrowheads in the DAPI counterstains (**a′–f′**). The number of TUNEL-positive cells dramatically increases within the first 30 min, stays at high levels for another day, and then gradually declines to background levels. (After Zupanc et al. 1998)

The elimination of injured cells almost exclusively by apoptosis is remarkable because this contrasts with the type of cell death encountered in mammals. In the latter taxon, necrotic cell death dominates within the core of the injury. Apoptosis occurs in a significant number of cells only in surrounding areas (for reviews, see Beattie et al. 2000; Vajda 2002; Liou et al. 2003). Necrosis often leads to inflammation at the site of the injury (for review, see Kerr et al. 1995). This inflammatory response triggers further necrotic events, thus gradually transforming the site of injury into large cavities devoid of cells (Zhang et al. 1997). These cavities are typically bordered by astrocytic scars that act as mechanical and biochemical barriers preventing the ingrowth of nerve fibers and the migration of cells into the lesion site (for review, see Reier et al. 1983).

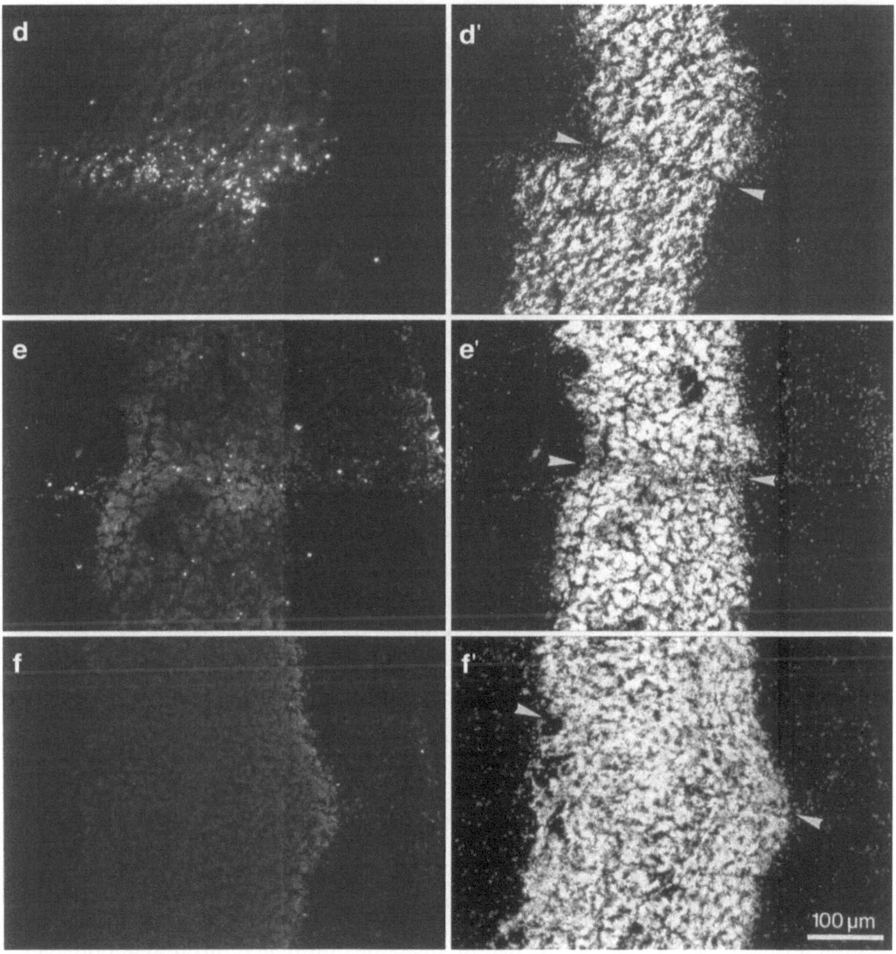

Fig. 5.10 (continued)

In contrast to necrosis, apoptosis is characterized by cell shrinkage, nuclear condensation, and production of membrane-enclosed particles that are digested by other cells. Most significantly, the side effects that accompany necrosis, such as inflammation of the surrounding tissue, are typically absent in apoptosis. Thus, use of this "clean" type of cell death for the elimination of damaged cells appears to be an essential component of the enormous regenerative capability of the adult fish brain.

5.4.3 Removal of Damaged Cells

The cellular debris caused by apoptosis is thought to be removed by the action of microglia/macrophages. Their number is very low in the intact brain but starts to increase at and near the lesion site approximately 3 days after application of the lesion (Zupanc et al. 2003). A similar increase, although less pronounced, is found in the contralateral hemisphere. The areal density of microglia/macrophages reaches maximum levels at approximately 10 days post lesion and returns to background levels approximately 1 month after the lesion. A similar increase in the number of microglia/macrophages in response to central nervous system (CNS) lesions, paralleled by active phagocytic activity, has been found in other systems and species (for reviews, see Kreutzberg 1996; Moore and Thanos 1996; Streit et al. 1999; Nagamoto-Combs et al. 2007).

5.4.4 Neuroprotection

In light of the massive wave of apoptotic cell death after injury, it is likely that survival-promoting factors exist that prevent unrestricted cell death after injuries. Correlative evidence suggests that this might be achieved by calbindin-D_{28k} signaling. The expression of this vitamin D-dependent calcium-binding protein is transiently increased in granular neurons in the lesion path and the immediate vicinity of the lesion in the period between 16 h and 7 days after the lesion (Zupanc and Zupanc 2006b). Calbindin-D_{28k} is likely to exert its neuroprotective function by buffering free intracellular Ca^{2+}, the concentration of which is commonly elevated after brain insults.

5.4.5 Replacement of Damaged Nerve Cells by Newly Generated Neurons

Within a few weeks following the lesion, the cells eliminated through apoptosis are replaced by newly generated ones. Most of these cells are produced 1–10 days after the injury. During this time period, the rate of cell proliferation is increased several fold at the site of the lesion, compared to unlesioned sites of the corpus cerebelli (Zupanc and Ott 1999).

In addition to the cells produced in response to the injury, cells that are born *before* the lesion also contribute to the restoration of damaged tissue. BrdU-labeling experiments have shown that cells generated as early as 2 days before the application of the lesion still participate in the process of regeneration (Zupanc and Ott 1999). This observation suggests a direct relationship between the continuous cell proliferation in the intact brain and neuronal regeneration after injury. The continuous provision of a pool of undifferentiated cells in the intact brain appears to enable fish to recruit new cells much more rapidly in the event of injuries than would be possible by recruiting only cells that are generated in response to such events.

As in the intact cerebellum, the cells destined to replace the ones lost to injury are guided to the lesion site by GFAP-immunopositive radial glial fibers (Clint and

Zupanc 2001). The density of these fibers starts to increase approximately 8 days post lesion. The number of a second population of radial glial fibers, characterized by vimentin immunoreactivity, increases slightly later, about 15 days after the injury (Clint and Zupanc 2002). The function of the vimentin-expressing radial glia is unknown.

At least some, but presumably most, of the new cells recruited for tissue repair develop into granular neurons, the phenotype of most of the neurons that are lost to the injury. Retrograde tracing experiments combined with labeling of S-phase cells with BrdU have further shown that these new granule cell neurons send axons into the associated molecular layer (Zupanc and Ott 1999). This finding suggests that the new granular neurons integrate into the existing neural network of the cerebellum.

5.4.6 *Molecular Identification of Regeneration-Associated Proteins*

What cellular factors mediate the ability of teleost fish to regenerate brain injuries? Considering just the major repair steps described above, it is obvious that a tremendous number of molecular signals must be overall involved in the process of brain regeneration. A large-scale molecular identification of these signals has become feasible with the advent of proteomics (for review, see Zupanc 2007; Zupanc and Zupanc 2011). The applicability of this approach to study brain regeneration in fish has been demonstrated using the regenerating cerebellum of *A. leptorhynchus* as a model system (Fig. 5.11a) (Zupanc et al. 2006). This system combines the cerebellar lesion paradigm, as described above, with differential proteome analysis. Examination of cerebellar tissue collected 3 days after application of lesions (when a marked increase in the generation of new cells recruited for the replacement of the cells lost to injury occurs; Zupanc and Ott 1999) has revealed significant alterations in protein spot intensity on two-dimensional (2D) gels in 53 of nearly 800 protein spots (Fig. 5.11b) (Zupanc et al. 2006). Twenty-four of these differentially regulated proteins could be identified by peptide mass fingerprinting and mass spectrometry/mass spectrometry fragmentation. These proteins can be divided into three groups.

The first group includes cytoskeletal proteins essential for the formation of new cells, such as β-actin and β-tubulin. The transient upregulation of β-actin expression appears to relate to the regeneration of injured axons in the area of the lesion and the development of axons of the newly generated cells, as suggested by immunohistochemical staining of brain sections after application of mechanical lesions to the corpus cerebelli (Zupanc et al. 2006). Such a proposed function is also in line with the results of regeneration studies in mammals, which have demonstrated an upregulation of actin within a few days after axotomy, reflecting the emergence of axonal sprouts and the concomitant needs to lengthen the axon shaft and to support membrane expansion at the growth cone (Tetzlaff et al. 1991; Lund et al. 2002; Lund and McQuarrie 1996). The transient increase in abundance of β-tubulin is closely associated with the appearance of thin capillaries at the site of the lesion several days

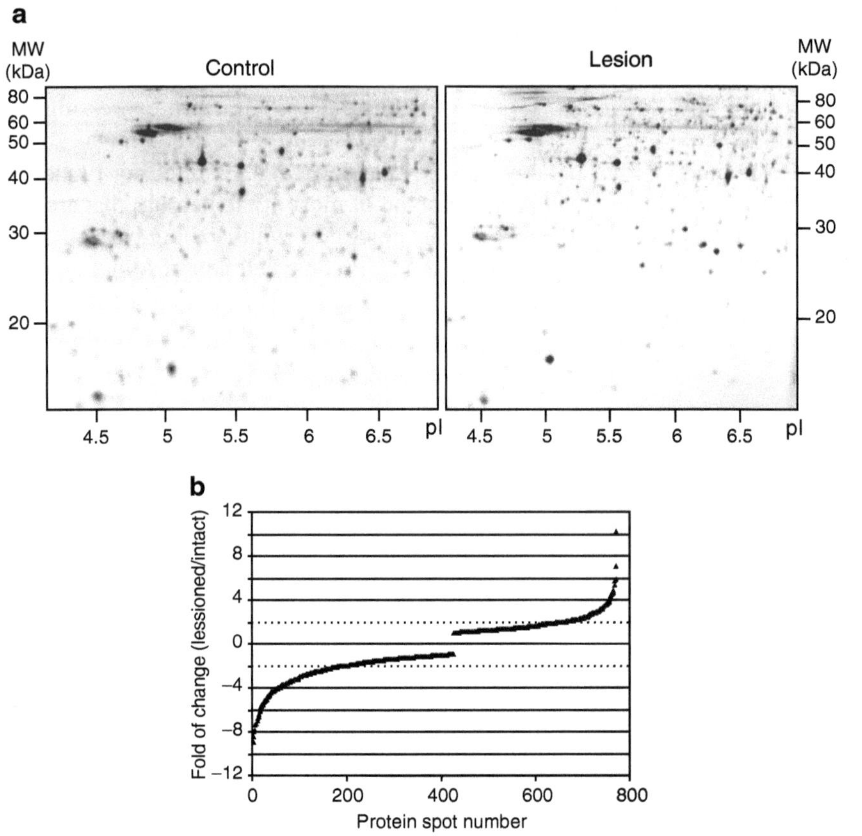

Fig. 5.11 (**a**) Images of two-dimensional (2D) gels of proteins from the intact corpus cerebelli (*Control*) and lesioned corpus cerebelli (*Lesion*) at a postlesion survival time of 3 days in *A. leptorhynchus*. (**b**) Scatter plot of the fold changes of 772 polypeptide spots in the corpus cerebelli of *A. leptorhynchus* 3 days after a lesion compared to unlesioned controls. Decrease in protein abundance is indicated by negative changes, increase by positive changes. Proteins are sorted in ascending order of the spot intensity change. The analysis is based on a comparison of the averaged spot intensities of four 2D gels from lesioned brains and three 2D gels from intact brains. *Dotted lines* indicate the +2-fold and −2-fold threshold. (After Zupanc et al. 2006)

after application of the stab wound, presumably reflecting the generation of new endothelial cells. Also included in this first group are proteins that mediate the correct assembly of these structural proteins. Among the latter are chaperonin containing tailless-complex polypeptide 1, subunit ε, which might mediate chaperoning of cytoskeletal proteins; tropomodulins-3 and -4, which are known to perform capping of slow-growing ends of actin filaments; and bullous pemphigoid antigen 1, which has been proposed to link actin with intermediate filament proteins.

The second group is composed of proteins potentially involved in cell proliferation (such as keratin-10), cellular motility (such as myosin heavy chain), neuroprotection (such as 78,000-Da glucose-regulated protein, glutamine synthetase, and cytosolic aspartate aminotransferase), and energy metabolism (such as α- and β-enolase, as well as F-ATP synthase β-subunit). Remarkably, the abundance of glutamine synthetase is increased after cerebellar lesions in fish (Zupanc et al. 2006), whereas this enzyme is downregulated under traumatic conditions in the mammalian central nervous system (Grosche et al. 1995; Härtig et al. 1995; Lewis et al. 1989, 1994; Oliver et al. 1990; Smith et al. 1991). Glutamine synthetase is a glia-specific enzyme that converts synaptically released glutamate into the nontoxic amino acid glutamine. Under normal conditions, this mechanism prevents the accumulation of neurotoxic amounts of glutamate in neural tissue and thus protects neurons from cell death. However, under traumatic conditions the amount of glutamine synthetase is insufficient to catalyze the excessive amounts of glutamate released by damaged cells. The additional downregulation of glutamine synthetase after brain trauma in mammals aggravates the situation in mammals, thus failing to limit the spread of damage caused by continuous overexcitation of postsynaptic glutamate receptors. On the other hand, the upregulation of glutamine synthetase in fish is likely to provide an important mechanism to reduce the neurodegenerative effect induced by glutamate neurotoxicity. Such differences in the expression pattern of proteins between fish and mammals are particularly interesting because they provide hints about possible cellular correlates for the enormous difference in the regenerative potential between the two vertebrate taxa.

The third group consists of a single protein, bone marrow zinc finger 2. This protein was first identified from a screen of zinc finger proteins in the hematopoietic system (Han et al. 1999). Although nothing is known about its role in the nervous system, its molecular structure and some experimental evidence based on examination of Wilms' tumor, a pediatric kidney cancer, suggest that bone marrow zinc finger 2 acts as a transcriptional regulator (Lee et al. 2002). During regeneration of the teleost cerebellum, it may exert a similar function, that is, regulating the transcription of genes of other proteins.

References

Altman J (1962) Are new neurons formed in the brains of adult mammals? Science 135:1127–1128

Altman J (1969) Autoradiographic and histological studies of postnatal neurogenesis: IV. Cell proliferation and migration in the anterior forebrain, with special reference to persisting neurogenesis in the olfactory bulb. J Comp Neurol 137:433–458

Altman J, Das GD (1965) Autoradiographic and histological evidence of postnatal hippocampal neurogenesis in rats. J Comp Neurol 124:319–336

Alvarez-Buylla A, Nottebohm F (1988) Migration of young neurons in adult avian brain. Nature 335:353–354

Anderson MJ, Waxman SG, Laufer M (1983) Fine structure of regenerated ependyma and spinal cord in *Sternarchus albifrons*. Anat Rec 205:73–83

Anderson MJ, Waxman SG, Lee Y-L, Eng LF (1987) Molecular differentiation of neurons from ependyma-derived cells in tissue cultures of regenerating teleost spinal cord. Mol Brain Res 2:131–136

Beattie MS, Farooqui AA, Bresnahan JC (2000) Review of current evidence for apoptosis after spinal cord injury. J Neurotrauma 17:915–925

Bédard A, Parent A (2004) Evidence of newly generated neurons in the human olfactory bulb. Dev Brain Res 151:159–168
Bell C, Bodznick D, Montgomery J, Bastian J (1997) The generation and subtraction of sensory expectations within cerebellum-like structures. Brain Behav Evol 50 (Suppl. 1):17–31
Braford MR (1995) Comparative aspects of forebrain organization in the ray-finned fishes: touchstones or nor? Brain Behav Evol 46:259–274
Butler AB (2000) Topography and topology of the teleost telencephalon: a paradoxon resolved. Neurosci Lett 293:95–98
Cameron DA (2000) Cellular proliferation and neurogenesis in the injured retina of adult zebrafish. Vis Neurosci 17:789–797
Cameron HA, McKay RD (2001) Adult neurogenesis produces a large pool of new granule cells in the dentate gyrus. J Comp Neurol 435:406–417
Candal E, Anadón R, DeGrip WJ, Rodríguez-Moldes I (2005) Patterns of cell proliferation and cell death in the developing retina and optic tectum of the brown trout. Dev Brain Res 154:101–119
Carlsen J, De Olmos J, Heimer L (1982) Tracing of two-neuron pathways in the olfactory system by the aid of transneuronal degeneration: projections to the amygdaloid body and hippocampal formation. J Comp Neurol 208:196–208
Clint SC, Zupanc GKH (2001) Neuronal regeneration in the cerebellum of adult teleost fish, *Apteronotus leptorhynchus* : guidance of migrating young cells by radial glia. Dev Brain Res 130:15–23
Clint SC, Zupanc GKH (2002) Up-regulation of vimentin expression during regeneration in the adult fish brain. NeuroReport 13:317–320
Corotto FS, Henegar JA, Maruniak JA (1993) Neurogenesis persists in the subependymal layer of the adult mouse brain. Neurosci Lett 149:111–114
Corwin JT (1981) Postembryonic production and aging of inner ear hair cells in sharks. J Comp Neurol 201:541–553
Curtis MA, Kam M, Nannmark U, Anderson MF, Axell MZ, Wikkelso C, Holtas S, van Roon-Mom WMC, Björk-Eriksson T, Nordborg C, Frisén J, Dragunow M, Faull RLM, Eriksson PS (2007) Human neuroblasts migrate to the olfactory bulb via a lateral ventricular extension. Science 315:1243–1249
Ekström P, Johnsson C-M, Ohlin L-M (2001) Ventricular proliferation zones in the brain of an adult teleost fish and their relation to neuromeres and migration (secondary matrix) zones. J Comp Neurol 436:92–110
Eriksson PS, Perfilieva E, Bjork-Eriksson T, Alborn AM, Nordborg C, Peterson DA, Gage FH (1998) Neurogenesis in the adult human hippocampus. Nat Med 4:1313–1317
Fine ML (1989) Embryonic, larval and adult development of the sonic neuromuscular system in the oyster toadfish. Brain Behav Evol 34:13–24
Fontana X, Nacher J, Soriano E, Del Rio JA (2006) Cell proliferation in the adult hippocampal formation of rodents and its modulation by entorhinal and fimbria-fornix afferents. Cereb Cortex 16:301–312
Gould E, Reeves AJ, Fallah M, Tanapat P, Gross CG, Fuchs E (1999) Hippocampal neurogenesis in adult Old World primates. Proc Natl Acad Sci USA 96:5263–5267
Grandel H, Kaslin J, Ganz J, Wenzel I, Brand M (2006) Neural stem cells and neurogenesis in the adult zebrafish brain: origin, proliferation dynamics, migration and cell fate. Dev Biol 295: 263–277
Graziadei PP, Graziadei GA (1979) Neurogenesis and neuron regeneration in the olfactory system of mammals. I. Morphological aspects of differentiation and structural organization of the olfactory sensory neurons. J Neurocytol 8:1–18
Gregory WA, Edmondson JC, Hatten ME, Mason CA (1988) Cytology and neuron-glial apposition of migrating cerebellar granule cells in vitro. J Neurosci 8:1728–1738
Grosche J, Härtig W, Reichenbach A (1995) Expression of glial fibrillary acidic protein (GFAP), glutamine synthetase (GS), and Bcl-2 protooncogene protein by Müller (glial) cells in retinal light damage of rats. Neurosci Lett 185:119–122

Hagedorn M, Fernald RD (1992) Retinal growth and cell addition during embryogenesis in the teleost, *Haplochromis burtoni*. J Comp Neurol 321:193–208
Han Z-G, Zhang Q-H, Ye M, Kan L-X, Gu B-W, He K-L, Shi S-L, Zhou J, Fu G, Mao M, Chen S-J, Yu L, Chen Z (1999) Molecular cloning of six novel Krüppel-like zinc finger genes from hematopoietic cells and identification of a novel transregulatory domain KRNB. J Biol Chem 274:35741–35748
Härtig W, Grosche J, Distler C, Grimm D, el-Hifnawi E, Reichenbach A (1995) Alterations of Müller (glial) cells in dystrophic retinae of RCS rats. J Neurocytol 24:507–517
Herculano-Houzel S, Lent R (2005) Isotropic fractionator: a simple, rapid method for the quantification of total cell and neuron numbers in the brain. J Neurosci 25:2518–2521
Herrup K, Sunter K (1987) Numerical matching during cerebellar development: quantitative analysis of granule cell death in staggerer mouse chimeras. J Neurosci 7:829–836
Herrup K, Shojaeian-Zanjani H, Panzini L, Sunter K, Mariani J (1996) The numerical matching of source and target populations in the CNS: the inferior olive to Purkinje cell projection. Dev Brain Res 96:28–35
Higgs DM, Souza MJ, Wilkins HR, Presson JC, Popper AN (2002) Age- and size-related changes in the inner ear and hearing ability of the adult zebrafish (*Danio rerio*). J Assoc Res Otolaryngol 3:174–184
Hinsch K, Zupanc GKH (2006) Isolation, cultivation, and differentiation of neural stem cells from adult fish brain. J Neurosci Methods 158:75–88
Hinsch K, Zupanc GKH (2007) Generation and long-term persistence of new neurons in the adult zebrafish brain: a quantitative analysis. Neuroscience 146:679–696
Hitchcock PF, Lindsey Myhr KJ, Easter SS, Jr., Mangione-Smith R, Jones DD (1992) Local regeneration in the retina of the goldfish. J Neurobiol 23:187–203
Insausti R, Marcos P, Arroyo-Jimenez MM, Blaizot X, Martinez-Marcos A (2002) Comparative aspects of the olfactory portion of the entorhinal cortex and its projection to the hippocampus in rodents, nonhuman primates, and the human brain. Brain Res Bull 57:557–560
Janssen J, Coombs S, Hoekstra D, Platt C (1987) Anatomy and differential growth of the lateral line system of the mottled sculpin, *Cottus bairdi* (Scorpaeniformes: Cottidae). Brain Behav Evol 30:210–229
Johns PR, Easter SSJ (1977) Growth of the adult goldfish eye: II. Increase in retinal cell number. J Comp Neurol 176:331–342
Johns PR (1982) Formation of photoreceptors in larval and adult goldfish. J Neurosci 2:178–198
Kaplan MS, Bell DH (1984) Mitotic neuroblasts in the 9-day-old and 11-month-old rodent hippocampus. J Neurosci 4:1429–1441
Kaplan MS, McNelly NA, Hinds JW (1985) Population dynamics of adult-formed granule neurons of the rat olfactory bulb. J Comp Neurol 239:117–125
Kaushal D, Contos JJA, Treuner K, Yang AH, Kingsbury MA, Rehen SK, McConnell MJ, Okabe M, Barlow C, Chun J (2003) Alteration of gene expression by chromosome loss in the postnatal mouse brain. J Neurosci 23:5599–5606
Kempermann G, Gast D, Kronenberg G, Yamaguchi M, Gage FH (2003) Early determination and long-term persistence of adult-generated new neurons in the hippocampus of mice. Development 130:391–399
Kerr JFR, Gobé GC, Winterford CM, Harmon BV (1995) Anatomical methods in cell death. In: Schwartz LM, Osborne BA (eds) Cell Death, pp 1–27. Academic Press, San Diego
Kirsche W (1967) Über postembryonale Matrixzonen im Gehirn verschiedener Vertebraten und deren Beziehung zur Hirnbauplanlehre. Z mikrosk-anat Forsch 77:313–406
Kornack DR, Rakic P (1999) Continuation of neurogenesis in the hippocampus of the adult macaque monkey. Proc Natl Acad Sci USA 96:5768–5773
Koumans JTM, Akster HA (1995) Myogenic cells in development and growth of fish. Comp Biochem Physiol 110A:3–20
Kranz D, Richter W (1970a) Autoradiographische Untersuchungen über die Lokalisation der Matrixzonen des Diencephalons von juvenilen und adulten *Lebistes reticulatus* (Teleostei). Z Mikrosk-Anat Forsch 82:42–66

Kranz D, Richter W (1970b) Autoradiographische Untersuchungen zur DNS-Synthese im Cerebellum und in der Medulla oblongata von Teleostiern verschiedenen Lebensalters. Z mikrosk-anat Forsch 82:264–292

Kreutzberg GW (1996) Microglia: a sensor for pathological events in the CNS. Trends Neurosci 19:312–318

Lee TH, Lwu S, Kim J, Pelletier J (2002) Inhibition of Wilms tumor 1 transactivation by bone marrow zinc finger 2, a novel transcriptional repressor. J Biol Chem 277:44826–44837

Lewis GP, Erickson PA, Guerin CJ, Anderson DH, Fisher SK (1989) Changes in the expression of specific Müller cell proteins during long-term retinal detachment. Exp Eye Res 49:93–111

Lewis GP, Guerin CJ, Anderson DH, Matsumoto B, Fisher SK (1994) Rapid changes in the expression of glial cell proteins caused by experimental retinal detachment. Am J Ophthalmol 118:368–376

Liou AKF, Clark RS, Henshall DC, Yin X-M, Chen J (2003) To die or not to die for neurons in ischemia, traumatic brain injury and epilepsy: a review on the stress-activated signaling pathways and apoptotic pathways. Prog Neurobiol 69:103–142

Lois C, Alvarez-Buylla A (1994) Long-distance neuronal migration in the adult mammalian brain. Science 264:1145–1148

Lois C, Garcia-Verdugo JM, Alvarez-Buylla A (1996) Chain migration of neuronal precursors. Science 271:978–981

Lund LM, McQuarrie IG (1996) Axonal regrowth upregulates b-actin and Jun D mRNA expression. J Neurobiol 31:476–486

Lund LM, Machado VM, McQuarrie IG (2002) Increased b-actin and tubulin polymerization in regrowing axons: relationship to the conditioning lesion effect. Exp Neurol 178:306–312

Luskin MB (1993) Restricted proliferation and migration of postnatally generated neurons from the forebrain subventricular zone. Neuron 11:173–189

Magrassi L, Graziadei PP (1995) Cell death in the olfactory epithelium. Anat Embryol 192:77–87

Mansour-Robaey S, Pinganaud G (1990) Quantitative and morphological study of cell proliferation during morphogenesis in the trout visual system. J Hirnforsch 31:495–504

Marcus RC, Delaney CL, Easter SS (1999) Neurogenesis in the visual system of embryonic and adult zebrafish (*Danio rerio*). Vis Neurosci 16:417–424

Meyer RL (1978) Evidence from thymidine labeling for continuing growth of retina and tectum in juvenile goldfish. Exp Neurol 59:99–111

Moore S, Thanos S (1996) The concept of microglia in relation to central nervous system disease and regeneration. Prog Neurobiol 48:441–460

Nagamoto-Combs K, McNeal DW, Morecraft RJ, Combs CK (2007) Prolonged microgliosis in the rhesus monkey central nervous system after traumatic brain injury. J Neurotrauma 24:1719–1742

Nguyen V, Deschet K, Henrich T, Godet E, Joly JS, Wittbrodt J, Chourrout D, Bourrat F (1999) Morphogenesis of the optic tectum in the medaka (*Oryzias latipes*): a morphological and molecular study, with special emphasis on cell proliferation. J Comp Neurol 413:385–404

Nieuwenhuys R, Meek J (1990) The telencephalon of actinopterygian fishes. In: Jones EG, Peters A (eds) Comparative structure and evolution of the cerebral cortex, pp 31–73. Plenum, New York

Northcutt RG (1995) The forebrain of gnathostomes: in search of a morphotype. Brain Behav Evol 46:275–318

Northcutt RG, Braford MR (1980) New observations on the organization and evolution of the telencephalon of actinopterygian fishes. In: Ebbesson SOE (eds) Comparative Neurology of the Telencephalon, pp 41–98. Plenum, New York

Oliver CN, Starke-Reed PE, Stadtman ER, Liu GJ, Carney JM, Floyd RA (1990) Oxidative damage to brain proteins, loss of glutamine synthetase activity, and production of free radicals during ischemia/reperfusion-induced injury to gerbil brain. Proc Natl Acad Sci USA 87:5144–5147

Osada T, Kusakabe H, Akutsu H, Yagi T, Yanagimachi R (2002) Adult murine neurons: their chromatin and chromosome changes and failure to support embryonic development as revealed by nuclear transfer. Cytogenet Genome Res 97:7–12

Ott R, Zupanc GKH, Horschke I (1997) Long-term survival of postembryonically born cells in the cerebellum of gymnotiform fish, *Apteronotus leptorhynchus*. Neurosci Lett 221:185–188

Otteson DC, Hitchcock PF (2003) Stem cells in the teleost retina: persistent neurogenesis and injury-induced regeneration. Vision Res 43:927–936

Paulin MG (1993) The role of the cerebellum in motor control and perception. Brain Behav Evol 41:39–50

Pencea V, Bingaman KD, Freedman LJ, Luskin MB (2001) Neurogenesis in the subventricular zone and rostral migratory stream of the neonatal and adult primate forebrain. Exp Neurol 172:1–16

Portavella M, Torres B, Salas C (2004) Avoidance response in goldfish: emotional and temporal involvement of medial and lateral telencephalic pallium. J Neurosci 24:2335–2342

Pouwels E (1978a) On the development of the cerebellum of the trout, *Salmo gairdneri*: I. Patterns of cell migration. Anat Embryol 152:291–308

Pouwels E (1978b) On the development of the cerebellum of the trout, *Salmo gairdneri*: III. Development of neuronal elements. Anat Embryol 153:37–54

Raff MC (1992) Social controls on cell survival and cell death. Nature 356:397–400

Raff MC, Barres BA, Burne JF, Coles HS, Ishizaki Y, Jacobson MD (1993) Programmed cell death and the control of cell survival: lessons from the nervous system. Science 262:695–700

Rahmann H (1968) Autoradiographische Untersuchungen zum DNS-Stoffwechsel (Mitose-Häufigkeit) im ZNS von *Brachydanio rerio* HAM. BUCH. (Cyprinidae, Pisces). J Hirnforsch 10:279–284

Rajendran RS, Zupanc MM, Lösche A, Westra J, Chun J, Zupanc GKH (2007) Numerical chromosome variation and mitotic segregation defect in the adult brain of teleost fish. Dev Neurobiol 67:1334–1347

Rajendran RS, Wellbrock UM, Zupanc GKH (2011) Apoptotic cell death, long-term persistence, and neuronal differentiation of aneuploid cells generated in the adult brain of teleost fish. Dev Neurobiol 68:1257–1268

Rakic P (1971) Neuron–glia relationship during granule cell migration in developing cerebellar cortex: a Golgi and electronmicroscopic study in *Macacus rhesus*. J Comp Neurol 141: 283–312

Rakic P (1972) Mode of cell migration to the superficial layers of fetal monkey neocortex. J Comp Neurol 145:61–83

Raymond PA, Easter SS, Jr. (1983) Postembryonic growth of the optic tectum in goldfish. I. Location of germinal cells and numbers of neurons produced. J Neurosci 3:1077–1091

Raymond PA, Easter SS, Jr., Burnham JA, Powers MK (1983) Postembryonic growth of the optic tectum in goldfish. II. Modulation of cell proliferation by retinal fiber input. J Neurosci 3: 1092–1099

Rehen SK, McConnell MJ, Kaushal D, Kingsbury MA, Yang AH, Chun J (2001) Chromosomal variation in neurons of the developing and adult mammalian nervous system. Proc Natl Acad Sci USA 98:13361–13366

Rehen SK, Yung YC, McCreight MP, Kaushal D, Yang AH, Almeida BSV, Kingsbury MA, Cabral KMS, McConnell MJ, Anliker B, Fontanoz M, Chun J (2005) Constitutional aneuploidy in the normal human brain. J Neurosci 25:2176–2180

Reier PJ, Stensaas LJ, Guth L (1983) The astrocytic scar as an impediment to regeneration in the central nervous system. In: Kao CC, Bunge RP, Reier PJ (eds) Spinal Cord Reconstruction, pp 163–195. Raven Press, New York

Richter W, Kranz D (1970a) Autoradiographische Untersuchungen über die Abhängigkeit des ^{3}H-Thymidin-Index vom Lebensalter in den Matrixzonen des Telencephalons von *Lebistes reticulatus* (Teleostei). Z mikrosk-anat Forsch 81:530–554

Richter W, Kranz D (1970b) Die Abhängigkeit der DNS-Synthese in den Matrixzonen des Mesencephalons vom Lebensalter der Versuchstiere (*Lebistes reticulatus* – Teleostei): Autoradiographische Untersuchungen. Z mikrosk-anat Forsch 82:76–92

Rodríguez F, López JC, Vargas JP, Gómez Y, Broglio C, Salas C (2002) Conservation of spatial memory function in the pallial forebrain of reptiles and ray-finned fishes. J Neurosci 22:2894–2903

Rowe RWD, Goldspink G (1969) Muscle fibre growth in five different muscles in both sexes of mice. J Anat 104:519–530

Rowlerson A, Veggetti A (2001) Cellular mechanisms of post-embryonic muscle growth in aquaculture species. In: Johnston IA (ed) Muscle Development and Growth, pp 103–140. Academic Press, San Diego

Sanai N, Tramontin AD, Quinones-Hinojosa A, Barbaro NM, Gupta N, Kunwar S, Lawton MT, McDermott MW, Parsa AT, García-Verdugo JM, Berger MS, Alvarez-Buylla A (2004) Unique astrocyte ribbon in adult human brain contains neural stem cells but lacks chain migration. Nature 427:740–744

Seri B, García-Verdugo JM, McEwen BS, Alvarez-Buylla A (2001) Astrocytes give rise to new neurons in the adult mammalian hippocampus. J Neurosci 21:7153–7160

Sîrbulescu RF, Zupanc GKH (in press) Spinal cord repair in regeneration-competent vertebrates: adult teleost fish as a model system. Brain Res Rev

Smith CD, Carney JM, Starke-Reed PE, Oliver CN, Stadtman ER, Floyd RA, Markesbery WR (1991) Excess brain protein oxidation and enzyme dysfunction in normal aging and in Alzheimer disease. Proc Natl Acad Sci USA 88:10540–10543

Soutschek J, Zupanc GKH (1995) Apoptosis as a regulator of cell proliferation in the central posterior/prepacemaker nucleus of adult gymnotiform fish, *Apteronotus leptorhynchus*. Neurosci Lett 202:133–136

Soutschek J, Zupanc GKH (1996) Apoptosis in the cerebellum of adult teleost fish, *Apteronotus leptorhynchus*. Dev Brain Res 97:279–286

Streit WJ, Walter SA, Pennell NA (1999) Reactive microgliosis. Prog Neurobiol 57:563–581

Takeda A, Nakano M, Goris RC, Funakoshi K (2008) Adult neurogenesis with 5-HT expression in lesioned goldfish spinal cord. Neuroscience 151:1132–1141

Tetzlaff W, Alexander SW, Miller FD, Bisby MA (1991) Response of facial and rubrospinal neurons to axotomy: changes in mRNA expression for cytoskeletal proteins and GAP-43. J Neurosci 11:2528–2544

Vajda FJ (2002) Neuroprotection and neurodegenerative disease. J Clin Neurosci 9:4–8

van Groen T, Kadish I, Wyss JM (2002) Species differences in the projections from the entorhinal cortex to the hippocampus. Brain Res Bull 57:553–556

van Groen T, Miettinen P, Kadish I (2003) The entorhinal cortex of the mouse: organization of the projection to the hippocampal formation. Hippocampus 13:133–149

Vargas JP, Rodríguez F, López JC, Arias JL, Salas C (2000) Spatial learning-induced increase in the argyophilic nucleolar organizer region of dorsolateral telencephalic neurons in goldfish. Brain Res 865:77–84

Waxman SG, Anderson MJ (1986) Regeneration of central nervous structures: *Apteronotus* spinal cord as a model system. In: Bullock TH, Heiligenberg W (eds) Electroreception, pp 183–208. John Wiley & Sons, New York

Weatherley AH, Gill HS (1985) Dynamics of increase in muscle fibres in fishes in relation to size and growth. Experientia 41:353–354

Wetts R, Herrup K (1983) Direct correlation between Purkinje and granule cell number in the cerebella of lurcher chimeras and wild-type mice. Brain Res 312:41–47

Williams RW (2000) Mapping genes that modulate brain development: a quantitative genetic approach. In: Goffinet AF, Rakic P (eds) Mouse Brain Development, pp 21–49. Springer-Verlag, New York

Winner B, Cooper-Kuhn CM, Aigner R, Winkler J, Kuhn HG (2002) Long-term survival and cell death of newly generated neurons in the adult rat olfactory bulb. Eur J Neurosci 16:1681–1689

Wullimann MF, Puelles L (1999) Postembryonic neural proliferation in the zebrafish forebrain and its relationship to prosomeric domains. Anat Embryol (Berl) 199:329–348

Wullimann MF, Rupp B, Reichert H (1996) Neuroanatomy of the Zebrafish Brain: A Topological Atlas. Birkhäuser Verlag, Basel/Boston/Berlin

Yurov YB, Iourov IY, Monakhov VV, Soloviev IV, Vostrikov VM, Vorsanova SG (2005) The variation of aneuploidy frequency in the developing and adult human brain revealed by an interphase FISH study. J Histochem Cytochem 53:385–390

Zakon HH (1984) Postembryonic changes in the peripheral electrosensory system of a weakly electric fish: addition of receptor organs with age. J Comp Neurol 228:557–570

Zhang Z, Krebs CJ, Guth L (1997) Experimental analysis of progressive necrosis after spinal cord trauma in the rat: etiological role of the inflammatory response. Exp Neurol 143:141–152

Zikopoulos B, Kentouri M, Dermon CR (2000) Proliferation zones in the adult brain of a sequential hermaphrodite teleost species (*Sparus aurata*). Brain Behav Evol 56:310–322

Zimmerman AM, Lowery MS (1999) Hyperplastic development and hypertrophic growth of muscle fibers in the white seabass (*Atractoscion nobilis*). J Exp Zool 284:299–308

Zupanc GKH (1999) Neurogenesis, cell death and regeneration in the adult gymnotiform brain. J Exp Biol 202:1435–1446

Zupanc GKH (2001) Adult neurogenesis and neuronal regeneration in the central nervous system of teleost fish. Brain Behav Evol 58:250–275

Zupanc GKH (2006a) Neurogenesis and neuronal regeneration in the adult fish brain. J Comp Physiol A 192:649–670

Zupanc GKH (2006b) Adult neurogenesis and neuronal regeneration in the teleost fish brain: implications for the evolution of a primitive vertebrate trait. In: Bullock TH, Rubenstein LR (eds) The Evolution of Nervous Systems in Non-Mammalian Vertebrates, pp 485–520. Academic Press, Oxford

Zupanc GKH (2007) Proteomics of traumatic brain injury and regeneration. Proteomics Clin Appl 1:1362–1372

Zupanc GKH (2008a) Adult neurogenesis in teleost fish. In: Gage FH, Kempermann G, Song H (eds) Adult Neurogenesis, pp 571–592. Cold Spring Harbor Laboratory Press, New York

Zupanc GKH (2008b) Adult neurogenesis and neuronal regeneration in the brain of teleost fish. J Physiol (Paris) 102:357–373

Zupanc GKH (2009) Towards brain repair: insights from teleost fish. Semin Cell Dev Biol 20:683–690

Zupanc GKH, Clint SC (2003) Potential role of radial glia in adult neurogenesis of teleost fish. Glia 43:77–86

Zupanc GKH, Horschke I (1995) Proliferation zones in the brain of adult gymnotiform fish: a quantitative mapping study. J Comp Neurol 353:213–233

Zupanc GKH, Ott R (1999) Cell proliferation after lesions in the cerebellum of adult teleost fish: time course, origin, and type of new cells produced. Exp Neurol 160:78–87

Zupanc GKH, Zupanc MM (1992) Birth and migration of neurons in the central posterior/prepacemaker nucleus during adulthood in weakly electric knifefish (*Eigenmannia* sp.). Proc Natl Acad Sci USA 89:9539–9543

Zupanc GKH, Zupanc MM (2006a) New neurons for the injured brain: mechanisms of neuronal regeneration in adult teleost fish. Regenerative Med 1:207–216

Zupanc MM, Zupanc GKH (2006b) Upregulation of calbindin-D_{28k} expression during regeneration in the adult fish cerebellum. Brain Res 1095:26–34

Zupanc GKH, Zupanc MM (2011) Proteomic analysis of CNS injury and recovery. In: Clelland JD (ed) Genomics, Proteomics, and the Nervous System, pp 511–536. Series: Advances in Neurobiology, Vol 2. Springer-Verlag, New York

Zupanc MM, Horschke I, Ott R, Rascher GB (1996) Postembryonic development of the cerebellum in gymnotiform fish. J Comp Neurol 370:443–464

Zupanc GKH, Kompass KS, Horschke I, Ott R, Schwarz H (1998) Apoptosis after injuries in the cerebellum of adult teleost fish. Exp Neurol 152:221–230

Zupanc GKH, Clint SC, Takimoto N, Hughes ATL, Wellbrock UM, Meissner D (2003) Spatiotemporal distribution of microglia/macrophages during regeneration in the cerebellum of adult teleost fish, *Apteronotus leptorhynchus*: a quantitative analysis. Brain Behav Evol 62:31–42

Zupanc GKH, Hinsch K, Gage FH (2005) Proliferation, migration, neuronal differentiation, and long-term survival of new cells in the adult zebrafish brain. J Comp Neurol 488:290–319

Zupanc MM, Wellbrock UM, Zupanc GKH (2006) Proteome analysis identifies novel protein candidates involved in regeneration of the cerebellum of teleost fish. Proteomics 6:677–696

Zupanc GKH, Wellbrock UM, Sîrbulescu RF, Rajendran RS (2009) Generation, long-term persistence, and neuronal differentiation of cells with nuclear aberrations in the adult zebrafish brain. Neuroscience 159:1338–1348

Chapter 6
Adult Neurogenesis in Reptiles

Susana González-Granero, Melissa Lezameta, and José Manuel García-Verdugo

Abstract Adult neurogenesis in reptiles is a well-documented phenomenon and exists in many telencephalic areas. The newly generated neurons originate along the walls of the lateral ventricles, mainly in the sulci. The putative neural progenitors are radial glial cells. These glial cells give rise to neuroblasts that migrate to their final destination. In general, the new neurons are born in the portion of the ventricular zone (VZ) adjacent to the telencephalic area where they will be recruited and migrate radially through the brain parenchyma along the processes of radial glial cells to their final destination, although migration to the olfactory bulbs (OB) is different. Specifically, it seems that new neurons of the OB are not produced in the ventricular walls of the OB ventricle, but instead in more caudal levels of the lateral ventricles. They then migrate tangentially over the soma of the VZ radial glia until they reach the OB. Adult neurogenesis creates microneurons and projection neurons, and occasionally there is gliogenesis as well.

One last striking feature of the reptilian brain is their ability to regenerate large neuronal populations that have died as a result of physical and chemical injuries.

Abbreviations

3AP	3-Acetylpyridine
^{3}HT	Tritiated thymidine
ADVR	Anterior dorsal ventricular ridge
AOB	Accesory olfactory bulb
AON	Anterior olfactory nucleus

S. González-Granero, M. Lezameta, and J.M. García-Verdugo (✉)
Unidad de Neurobiología Comparada, Instituto Cavanilles de Biodiversidad y Biología Evolutiva, Universidad de Valencia, c/ Catedrático José Beltrán nº2, CP 46980 Paterna, Valencia, Spain
e-mail: j.manuel.garcia@uv.es

J.M. García-Verdugo
Laboratorio de Morfología Celular, Centro de Investigación Príncipe Felipe, Valencia, Spain
and
CIBERNED, Spain

T. Seki et al. (eds.), *Neurogenesis in the Adult Brain I: Neurobiology*,
DOI 10.1007/978-4-431-53933-9_6, © Springer 2011

BrdU	Bromodeoxyuridine
CNS	Central nervous system
CRMP-4	Collapsin response-mediated protein 4
DC	Dorsal cortex
DMC	Dorsomedial cortex
DVR	Dorsal ventricular ridge
GFAP	Glial fibrillary acidic protein
LC	Lateral cortex
MC	Medial cortex
MOB	Main olfactory bulb
NS	Nucleus sphericus
OB	Olfactory bulb
PSA-NCAM	Polysialic neural cell adhesion molecule
sl	Sulcus lateralis
SP	Septum
ssm	Sulcus septomedialis
st	Sulcus terminalis
ST	Striatum
sv	Sulcus ventralis
SVZ	Subventricular zone
VZ	Ventricular zone

6.1 Introduction

Reptiles possess a great capacity to generate neurons during adulthood, both in normal conditions and following an injury (Lopez-Garcia et al. 1988a; Duffy et al. 1992; Perez-Cañellas and Garcia-Verdugo 1996; Font et al. 1997; Perez-Cañellas et al. 1997). The main studies regarding neurogenesis in reptiles have been conducted in various species of lizards (Lopez-Garcia et al. 1988a; Perez-Cañellas and Garcia-Verdugo 1996; Marchioro et al. 2005), but a few can be found in turtles (Perez-Cañellas et al. 1997) and snakes (Holtzman and Halpern 1991).

Adult neurogenesis takes place mainly in the telencephalon. Nevertheless, neurogenesis has also been described in the cerebellum during early postnatal stages, but disappears in the adult (Lopez-Garcia et al. 1990b). Reactive neurogenesis has been studied in the injured spinal cord of lizards, but the data collected support an axon regeneration phenomenon over a truly neurogenic activity (Duffy et al. 1992). Adult neurogenesis has been reported in all the main telencephalic subdivisions: main olfactory bulb and accessory olfactory bulb (MOB and AOB), anterior olfactory nucleus (AON), cerebral cortex (medial, dorsomedial, dorsal and lateral) (MC, DMC, DC, and LC, respectively), dorsal ventricular ridge (DVR), septum (SP), striatum (ST), and nucleus sphericus (NS) (Fig. 6.1). There are striking differences in the intensity of neurogenesis among the different telencephalic areas, and differences between species have also been reported (Font et al. 2001). The newly generated

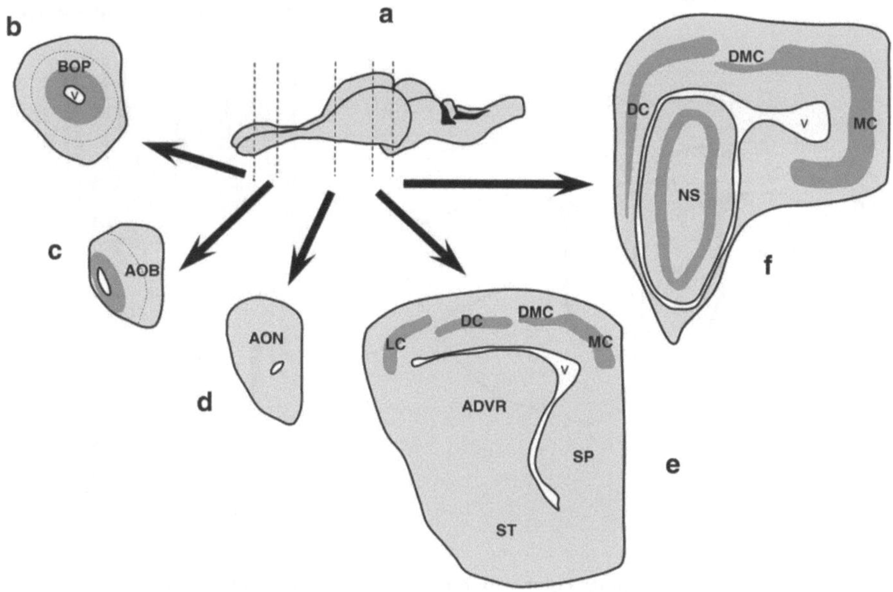

Fig. 6.1 Scheme of different telencephalic levels of the reptilian brain showing reptile brain (**a**), MOB (**b**), AOB (**c**), rostral forebrain (**d**), precomissural level (**e**), and postcomissural level (**f**). *ADVR* anterior dorsal ventricular ridge, *AOB* accessory olfactory bulb, *AON* anterior olfactory nucleus, *DC* dorsal cortex, *DMC* dorsomedial cortex, *LC* lateral cortex, *MC* medial cortex, *MOB* main olfactory bulb, *NS* nucleus sphericus, *SP* septum, *ST* striatum

neurons develop on the walls of the lateral ventricles and migrate radially or tangentially to their target destination.

An interesting aspect of neurogenesis in reptiles is the regenerative potential following an injury. Following chemical or physical injuries, reactive neurogenesis is activated, and large numbers of neuroblasts migrate from the proliferative zones to the injured area to replace the same neuronal types and connections present before the lesion (Font et al. 1991).

Regardless of the large neurogenic potential observed in reptiles, it remains a field of study largely unknown. Most research is conducted in mammals, mainly in rats and mice. Furthermore, little effort has been done to establish a comparative view regarding the data obtained in the different vertebrate groups.

6.2 Proliferative Areas

During adulthood and in normal conditions, neurogenesis in reptiles is restricted to the walls of the lateral ventricles, in the region known as ventricular zone (VZ), as demonstrated by numerous experiments using proliferative markers, such as bromodeoxyuridine (BrdU) or tritiated thymidine (^{3}HT) (Font et al. 2001).

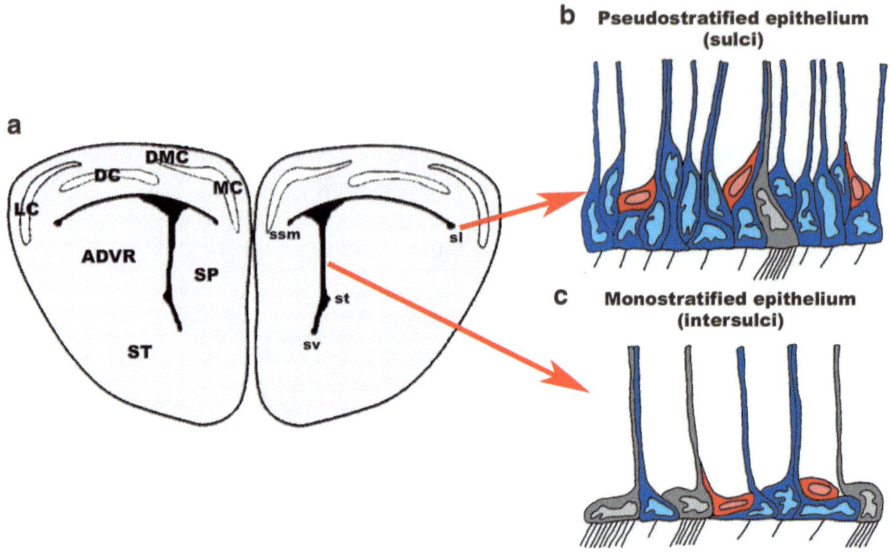

Fig. 6.2 Diagram of the cellular composition in the sulci and intersulci regions in the reptilian brain. (**a**) Transversal section of a precomissural telencephalic level with the main telencephalic areas represented in the left hemisphere and the different sulci shown in the right hemisphere. (**b**) The sulci regions are characterized by a pseudostratified epithelium with columnar cells, mostly composed of B cells (*blue*) and A cells (*red*). (**c**) Intersulci areas, generally a monostratified epithelium that holds mainly flattened radial cells: B cells (*blue*) and E cells (*gray*). *ADVR* anterior dorsal ventricular ridge, *DC* dorsal cortex, *DMC* dorsomedial cortex, *LC* lateral cortex, *MC* medial cortex, *sl* sulcus laterals, *SP* septum, *ssm* sulcus septomedialis, *ST* striatum, *st* sulcus terminalis, *sv* sulcus ventralis

Cells with proliferative potential are found mainly in areas of the VZ known as "sulci" (Kirsche 1983). Sulci are always located in highly folded areas of the ventricular wall and are the remains of the embryonic germinative tissue that generated the telencephalon and which retains its proliferative capacity in the adult (Tineo et al. 1987). Four sulci have been described in reptiles: sulcus septomedialis (ssm), sulcus lateralis (sl), sulcus ventralis (sv), and sulcus terminalis (st) (Fig. 6.2a).

Sulci are present in all reptiles studied to date, but variations in their proliferative capacity have been reported between different species and also between different sulci within the same species. As an example, neurogenic rates in the lizard *Tarentola mauritanica* are much higher in the sulcus septomedialis than in the sulcus lateralis (Perez-Cañellas and Garcia-Verdugo 1996). In contrast, in the turtle *Trachemys scripta*, neurogenesis in the sulcus septomedialis is scarce; most proliferative cells are found in sulci ventralis/terminalis (Perez-Cañellas et al. 1997).

A comparison with other vertebrates reveals basic, possibly ancestral, commonalities in the location of proliferative cells. In birds, proliferative cells are located also in specialized portions of the VZ of the lateral ventricles (Alvarez-Buylla 1990) that are referred to as "hot spots." These are found especially in the dorsal and ventral region of the lateral portion in the ventricular wall (Alvarez-Buylla et al. 1990), near areas of VZ that are also folded. Despite these similarities, no attempt has been made to homologize the proliferative telencephalic areas of birds to the sulci of reptiles and non-amniotic vertebrates. In mammals, adult neurogenesis takes place mainly on the walls of the lateral ventricles, in an area known as the subventricular zone (SVZ) (Doetsch et al. 1999), but in contrast to the situation in reptiles, adult neurogenesis is also found outside the ventricular walls, in the subgranular zone of the hippocampus (Seri et al. 2001). There are no reports of adult neurogenesis in reptiles outside the VZ, except for occasional mitoses in the endothelium and in parenchyma glial cells (Perez-Cañellas and Garcia-Verdugo 1996).

6.3 Cellular Composition of the VZ

The walls of the lateral ventricles are lined by radial glial cells (Stensaas and Stensaas 1968; Garcia-Verdugo et al. 1981), which characteristically have a flattened or cubical soma and a long radial process that often splits into two or more branches. In transverse sections, the radial fiber crosses the entire parenchyma until reaching the pial surface or the surface of blood vessels. Depending on their final destination, the end feet of radial glial cells give rise to the glial-limiting membrane or the hematoencephalic barrier. The cell bodies of radial glial cells are bound to other neighboring glial cells by interdigitations and junctions. Dense junctions joining the fibers of different glial cells are often found in the neuropil. These cells are positive for glial fibrillary acidic protein (GFAP) and glutamine sintetase (Monzon-Mayor et al. 1990).

There is conspicuous regional variation in the organization of the walls of the lateral ventricles, with some areas structured as pseudostratified epithelium holding columnar cells, and other areas organized as a monolayered epithelium of cubic or flattened cells (Yanes-Mendez et al. 1988) (see Fig. 6.2). Proliferative sulci are always found in pseudostratified VZ areas. Most mitoses take place in the sulci, but some can also be observed in areas between sulci.

Of the four sulci that have been described in reptiles, the sulcus septomedialis has been most intensively studied. Three-dimensional reconstructions in the European lizard *Podarcis hispanica* with electron microscopy reveal two separate radial glia populations, known as B and E cells (Garcia-Verdugo et al. 2002) (Fig. 6.2). The ultrastructure of these cells is very similar, both possessing long radial expansions that have abundant intermediate filaments, few microtubules, numerous vacuoles, and elongated mitochondria which usually appear in groups of two or three in their apical portion. The nucleus is located in the apical region, has small chromatin clumps, and its membrane is highly invaginated. The cytoplasm has long dictyosomes, numerous cisternae of rough endoplasmic reticulum, abundant polyribosomes distributed exclusively throughout the cellular soma (absent in the radial expansions), and small packed groups of intermediate filaments, mainly at the base of the radial expansion. The main difference between these cell types is the presence in type B cells of a cilium with a 9+2 structure and a centriole oriented perpendicularly to the basal corpuscle of the cilium, whereas type E cells have 15–20 cilia with a 9+2 structure, without a perpendicular centriole.

Besides these two cells types in contact with the ventricular lumen, a third cell type has been described in the VZ: type A cells, which are not in contact with the ventricular surface (Garcia-Verdugo et al. 2002) (Fig. 6.2). Type A cells are migratory cells, their ultrastructure being similar to that of typical migrating cells found in the neuropil (Garcia-Verdugo et al. 1986a). These cells have an elongated nucleus with a large central nucleolus and lax chromatin, few cytoplasmic organelles, numerous polyribosomes and microtubules, and an internal cilium. Occasionally, fully formed neurons, microglia, and oligodendrocytes are also found near the VZ.

A comparison with the organization of the VZ in other vertebrates, such as birds and mammals, reveals both similarities and differences. Three cell types have also been described in the VZ of birds (Alvarez-Buylla et al. 1998), but there are more differences between the type B and E cells of birds than between the corresponding cell types in reptiles. Besides the presence of one or multiple cilia, there are ultrastructural differences including cell size, thickness of the radial expansion, and the chromatin packing. Mammals have type B, E, and A cells, plus a fourth cell type, called type C (Doetsch et al. 1997), considered as an intermediate cellular type between B and A cells. Type C cells are also referred to as amplifying cells because of their high rates of division. Differences between mammals and reptiles are more pronounced than between birds and reptiles because in mammals B and E cells are not typical radial glial cells. Some basic characteristics are retained, nevertheless, such as the primary cilium and intermediate filaments in B cells, the multiple cilia in E cells, and the numerous ribosomes and microtubules in A cells (Garcia-Verdugo et al. 2002).

6.4 Neural Progenitor Identity

Proliferation markers such as BrdU or ^{3}HT administered to animals killed after short survival times consistently show labeled cells in contact with the ventricular lumen. Moreover, all mitoses found in the VZ are in contact with the ventricular surface and correspond to cells with the ultrastructure of radial glial cells (Font et al. 2001).

There are no precise data considering the nature of the progenitor cell, but considering the different cell types in contact with the lumen already described, it seems likely that radial glial cells are the primary neural precursor for adult neurogenesis in reptiles (Perez-Cañellas and Garcia-Verdugo 1996; Font et al. 2001). Further research is required to establish whether both B and E cells play this role or only one type of these.

In other vertebrate groups, such as birds and mammals, B cells act as neural progenitors (Alvarez-Buylla et al. 1998; Doetsch et al. 1999). In both mammals and birds these cells have a glial phenotype, are GFAP+, and have a single cilium as a special characteristic. Considering this evidence, it seems likely that in reptiles the neural progenitor function is also accomplished by B cells.

6.5 Migration

Newly generated neurons in reptiles are born in the walls of the lateral ventricles and migrate to their final location in the nervous parenchyma, where they differentiate and integrate in existing telencephalic circuits (Lopez-Garcia et al. 1990b).

Although the data are scant, there are differences in the time required for neuroblasts to complete this process. In the tropical lizard *Tropidurus hispidus* it takes 1–2 weeks for newly generated neuroblasts to start migrating, and they do not reach the granular layer of the MC until 15–30 days later (Marchioro et al. 2005). In *Podarcis hispanica*, it takes only 2–4 days for migration to start, and in a week labeled cells can already be seen in the cell layer of the MC (Lopez-Garcia et al. 1988a). Authors of these studies explain the differences based on the presence of GFAP filaments. Radial glial cells in *Tropidurus hispidus* are GFAP–, whereas in *Podarcis hispanica* they are GFAP+ (Lopez-Garcia et al. 1992; Marchioro et al. 2005). Reduced GFAP immunoreactivity is also observed in other lizards living in warm habitats (Ahboucha et al. 2003). However, it is not entirely clear whether this difference in GFAP expression could be responsible for the differences in the timeframe of adult neurogenesis in these two species.

With few exceptions, newly generated neuroblasts follow a radial migration. Cells migrate short distances (500 μm maximum), as their final destiny is usually near the VZ where they are born, along the radial glia expansions (Garcia-Verdugo et al. 1986a; Lopez-Garcia et al. 1988a) and occasionally along blood vessels. Migrating neuroblasts are fusiform, their orientation is perpendicular to the ventricular surface, and they express neuronal migration markers such as collapsin response-mediated protein 4 (CRMP-4) and polysialic neural cell adhesion molecule (PSA-NCAM) (Nacher et al. 2002; Ramirez-Castillejo et al. 2002). This type of migration is characteristic of central nervous system (CNS) embryonic development in vertebrates.

There is one exception to the migration pattern just described, and that concerns migration of newly generated neurons destined for the olfactory bulb (OB). According to studies conducted in different species, most newly generated neurons, if not all, that reach the OB are not formed in situ (despite the presence of the VZ in the OB). Following administration of proliferation markers, no labeled cells are found in the OB ventricle walls after short survival periods. However, after long survival times a large number of labeled neurons are seen in the OB (Peñafiel et al. 1996; Perez-Cañellas and Garcia-Verdugo 1996; Perez-Cañellas et al. 1997). Thus, it seems likely that new neurons in the bulb originate in the telencephalic VZ caudal to the OB and would then migrate tangentially over long distances until reaching their final destination in the OB.

In the lizard *Psammodromus algirus*, OB neurons are generated in the rostral-most telencephalic area: the AON (Peñafiel et al. 1996). In this species, groups of fusiform cells were described along the olfactory peduncles. Such cells show a typical morphology of migrating cells and have a parallel orientation to the ventricular surface. Migration in this case takes place tangentially and along very restricted pathways close to the VZ (Peñafiel et al. 1996; Font et al. 2001). Similar results have been obtained with the turtle *Trachemys scripta*, in which the new cells in the OB originate in the AON and migrate tangentially until they reach the OB (Perez-Cañellas et al. 1997). The origin of OB neurons could be slightly more caudal in the gecko *Tarentola mauritanica*, in which the VZ of the sulci ventralis/terminalis could give rise to neurons destined for the OB. These sulci show an intense proliferating activity that contrasts with the scarce neurons that can be found after long survival times in the surrounding nervous parenchyma (Perez-Cañellas and Garcia-Verdugo 1996). Regardless of the exact origin of the neurons that will be recruited into the OB, it seems the migration is always tangential along the VZ until the cells reach the OB, close to their final target. At that point they can migrate radially, guided by radial glia fibers until they reach their final destination in the granular layer of the MOB or AOB (Peñafiel et al. 1996).

Both types of migration – tangential and radial – are common during early embryonic brain development (Rakic 1971; Austin and Cepko 1990; O'Rourke et al. 1995) and adulthood (Lois and Alvarez-Buylla 1994) in other vertebrate groups. As an example, in mammals, neuroblasts for the OB originate in the SVZ, migrate tangentially along the rostral migratory stream, and then they enter the OB where they migrate radially toward their target layer (Lois and Alvarez-Buylla 1994). Neurons generated in the dentate gyrus, on the other hand, migrate radially along radial glia fibers to their final location in the granular layer (Seri et al. 2001).

6.6 Cell Types Generated

A particularly nagging problem in the study of adult neurogenesis in vertebrates is that of the identity, whether glial or neuronal, of the newly produced cells. Cellular proliferation in the telencephalic VZ could give rise to glial cells or neurons or both. Interspecific differences in the proportion of neurons versus glia have been described, although there are always more neurons than glia generated. The most complete studies have been conduced in the lizards *Podarcis hispanica* and *Tarentola mauritanica* and in the turtle *Trachemys scripta*.

Neurogenesis in lizards produces mostly neurons rather than glia (Lopez-Garcia et al. 1988a; Perez-Cañellas and Garcia-Verdugo 1996). This conclusion is supported by reports indicating that, in the intact lizard brain, free glial cells are very scarce in the telencephalon (Garcia-Verdugo et al. 1981; Font et al. 1995a). In the adult lizard telencephalon, radial glial cells that cover the surface of the ventricles are the predominant astroglial element and provide the functions of astrocytic glia (Garcia-Verdugo et al. 1981).

Proliferation in turtles differs from that in lizards, because in the first both neurogenesis and gliogenesis take place in their adult brain (Perez-Cañellas et al. 1997). Astrocytes and oligodendrocytes labeled with proliferation markers were detected in the turtle brain. However, the data obtained are inconclusive as to whether the newly produced glial cells were generated in situ or whether they were

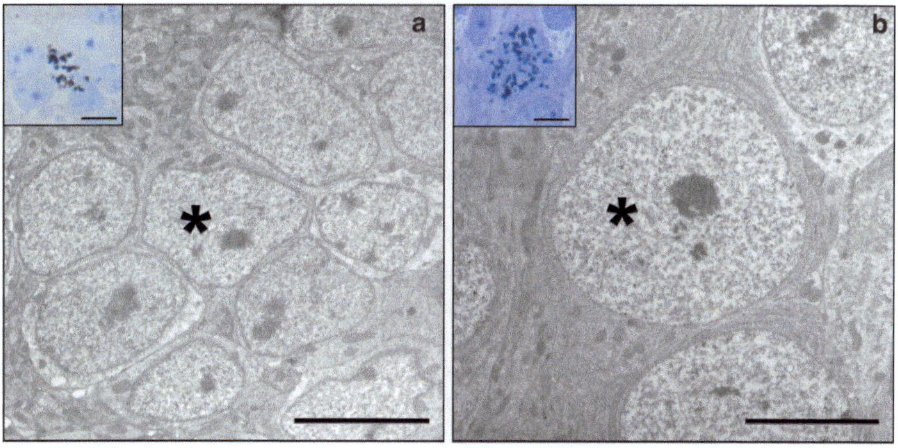

Fig. 6.3 Electron micrographs of ^{3}H-thymidine-labeled neurons in the telencephalon of lizard. The *upper-left insets* show details of the radiolabeled neurons in the autoradiograms from which the ultrathin sections were prepared and examined (*asterisk*). (**a**) Radiolabeled neuron in the olfactory bulbs, a microneuron. (**b**) Radiolabeled neuron in the cell layer of the medial cortex, a projection neuron. *Bars* 5 μm

born in the proliferative VZ. Studies in *Trachemys scripta elegans* suggest that in the VZ situated across the striatum it is mainly glial cells that arise, whereas neurons are produced also in all the other VZ (Perez-Cañellas et al. 1997).

Adult neurogenesis in reptiles produces microneurons (e.g., OB) and projection neurons (e.g., MC) (Peñafiel et al. 1996) (Fig. 6.3). This is a very important difference between reptiles and mammals, in which there is precise documentation only on the generation of microneurons (Kempermann 2006).

A more detailed analysis shows that the cell types originated by adult neurogenesis in the different species of reptiles have different characteristics depending on their final location in the telencephalon. In the OB, organized in a similar way to that described for mammals (Garcia-Verdugo et al. 1986b) (see Fig. 6.1b), newly generated neurons are found mainly in the granular layer and have a rather homogeneous distribution (Garcia-Verdugo et al. 1989; Peñafiel et al. 1996; Perez-Cañellas and Garcia-Verdugo 1996). Occasionally a few cells can be found in the mitral and periglomerular layers, but this occurs in perinatal animals alone (Garcia-Verdugo et al. 1989). The new neurons are similar to those surrounding them and can be characterized by their small spherical nucleus, the lightly packed chromatin, and a prominent nucleolus, as well as the scarce cytoplasm holding numerous ribosomes (Perez-Cañellas and Garcia-Verdugo 1996; Perez-Cañellas et al. 1997) (see Fig. 6.3a).

Adult neurogenesis also occurs in the cortex. Three areas can be distinguished within the cortex in reptiles: the medial, dorsal and the lateral cortices (see Fig. 6.1e). In squamate reptiles, an additional zone called the dorsomedial cortex is has been described. All cortical areas are organized following the same simple three-layered structure: a granular or cellular layer situated in the center, in which most of the neurons are located, and two plexiform layers surrounding the granular layer. The central layer is thus limited dorsally by the external plexiform layer and the meningeal layers and ventrally by the internal plexiform layer and the ependyma that surrounds the ventricle (Ulinski 1990). In a general view, neurogenesis has been observed in all four regions.

There is evidence supporting the homology between MC and the dentate gyrus in the hippocampus in mammals (Butler and Hodos 1996). Both have a series of elements in common: a similar organization and neural population (Luis de la Iglesia and Lopez-Garcia 1997a,b), a postnatal development of the axonal projection containing zinc that has been compared to the hippocampal mossy fiber system in charge of the Timm positive system (Lopez-Garcia and Martinez-Guijarro 1988; Lopez-Garcia et al. 1990a), and the presence of adult neurogenesis (Lopez-Garcia et al. 1988a). Following long survival times, most newly formed neurons in the MC are located in the cellular layer and only a few are found in the inner plexiform layer (Lopez-Garcia et al. 1988a; Perez-Cañellas and Garcia-Verdugo 1996). Two neuronal types can be distinguished in the cellular layer of the MC, referred to as type I and II neurons. Type I are small cells, with packed chromatin and scarce cytoplasm, whereas type II are large cells with a large oval nucleus, lax chromatin, a prominent nucleus, and much cytoplasm, which is rich in organelles (especially Nissl bodies), numerous polyribosomes, and microtubules (Garcia-Verdugo et al. 1984; Davila et al. 1985). Both cell types project to the contralateral hemisphere and are glutamatergic (Lopez-Garcia et al. 1990a; Martinez-Guijarro et al. 1991). There are some interesting observations related to the origin of these two cell types. When newly generated cells are analyzed in this region following treatment with ^{3}HT and intermediate survival terms (1–3 months), most labeled cells correspond to type II neurons (Lopez-Garcia et al. 1988a; Perez-Cañellas and Garcia-Verdugo 1996; Perez-Cañellas et al. 1997) (see Fig. 6.3b). Perinatal animals have mostly type II cells, and these cells are also the only ones that seem to be generated in the adult. It would be expected, as a result, that type II cells should be the predominant neuronal type in the adult brain. Instead, type I cells are more abundant in the adult (Lopez-Garcia et al. 1984), suggesting that type II cells transform into type I cells as a result of specialization associated with a chromatin packing process and a loss in size and number of organelles. To confirm this hypothesis, the fate of labeled neurons should be examined after long survival periods. Labeled cells are also found in the inner plexiform layer. These cells have a fusiform nucleus with a perpendicular orientation regarding the ventricular surface; their ultrastructure is similar to that described for migrating cells in other lizards (Garcia-Verdugo et al. 1986a), and they are CRMP-4+ and PSA-NCAM+ (Nacher et al. 2002; Ramirez-Castillejo et al. 2002). It is thus possible that they are newly produced cells migrating to the granular layer. Occasionally they have spherical nuclei and are situated on an intermediate level (Perez-Cañellas and Garcia-Verdugo 1996).

The DMC region is considered to be homologous to the CA3 in the mammalian hippocampus (Butler and Hodos 1996). Both regions have a contralateral commissural projection and are the main areas receiving input from the zinc-rich mossy fibers of the MC (Martinez-Garcia et al. 1990). A few newly generated neurons have only been described in the inner plexiform layer of the DMC of *Podarcis hispanica* (Font et al. 1997). In some species, such as *Tarentola mauritanica*, there is no neurogenesis in this cortical region (Perez-Cañellas and Garcia-Verdugo 1996), or this telencephalic division may be entirely lacking, as in the turtle *Trachemys scripta* (Perez-Cañellas et al. 1997).

Homologies regarding the DC have been a subject of debate. Certain classic authors argued that it represents the neocortex primordium because it is a polisensorial projection area (Johnston 1916), whereas nowadays homology has been established with the CA1 and the hippocampal subiculum (Martinez-Garcia et al. 1990). There is no neurogenesis in the granular layer of the DC, but new neurons are visible instead in the inner plexiform layer, as a part of neuron clusters as well as isolated (Perez-Cañellas and Garcia-Verdugo 1996). Such cells are small with an irregular nucleus, usually invaginated, packed chromatin, and scarce cytoplasm containing numerous polyribosomes (Perez-Cañellas and Garcia-Verdugo 1996).

The LC is homologous to the olfactory cortex in mammals (piriform and entorhinal cortex) (Butler and Hodos 1996). Both regions receive input from the MOB projections and send projections to the MC/dentate gyrus (Martinez-Garcia et al. 1986; Hoogland and Vermeulen-Vanderzee 1995). Newly generated cells appear in the granular layer and occasionally in the inner plexiform layer of the LC (Perez-Cañellas and Garcia-Verdugo 1996). Those in the granular layer are medium in size with a small spherical nucleus whereas those in the inner plexiform layer have fusiform nuclei and are probably migrating cells. Their ultrastructure is similar to that of migrating cells in the MC (Perez-Cañellas and Garcia-Verdugo 1996), and they are immunoreactive for PSA-NCAM and CRMP-4 (Nacher et al. 2002; Ramirez-Castillejo et al. 2002).

The DVR receives polisensorial inputs and is the main data processing center in reptiles. It is considered homologous to the pallial amygdala and claustropiriform components in mammals (Bruce and Neary 1995; Striedter 1997). Newly generated cells in the DVR are localized in the dorsal region alone, near the VZ, either in characteristic clusters in the region or isolated (Perez-Cañellas and Garcia-Verdugo 1996). They have a large spherical nucleus, lax chromatin, and a prominent nucleolus; the cytoplasm contains numerous organelles, a large rough endoplasmic reticulum, small mitochondria, and numerous dictyosomes and polyribosomes (Perez-Cañellas and Garcia-Verdugo 1996; Perez-Cañellas et al. 1997).

Newly generated cells in the striatum do not seem to be restricted to certain areas; instead, they are spread widely and homogeneously (Perez-Cañellas and Garcia-Verdugo 1996). They show a small round nucleus, slightly packed chromatin, and a prominent nucleolus. Cytoplasm is scarce but rich in organelles, with small dark mitochondria, a large Golgi system, an extensive rough endoplasmic reticulum, numerous polyribosomes, and dense secretion granules (Perez-Cañellas et al. 1997).

The septum (SP) is another telencephalic region in which adult neurogenesis has not been observed in all species studied. This is the case in the lizard *Tarentola mauritanica*, which lacks adult neurogenesis in this region, whereas in *Trachemys scripta* new neurons are originated with a very similar appearance to that described for the striatum, except for the absence of secretion granules and a smaller rough endoplasmic reticulum (Perez-Cañellas et al. 1997).

The nucleus sphericus (NS) is a characteristic region in most reptiles; it includes the most posterior area of the dorsal ventricular ridge (DVR) and is considered to be homologous to the amygdaline complex in mammals (Ulinski and Kanarek 1973; Butler and Hodos 1996; Martinez-Garcia et al. 2002). It is composed of four concentric layers, which from the outermost inward are referred to as ependymal, marginal, mural, and hilus, respectively (see Fig. 6.1f). Newly formed neurons are localized mainly in external mural layer, near the VZ, often in pairs (Perez-Sanchez et al. 1989; Perez-Cañellas and Garcia-Verdugo 1996). They have a small nucleus and a similar ultrastructure to those in the DC (Perez-Cañellas and Garcia-Verdugo 1996). In *Trachemys scripta* this telencephalic subdivision is absent also (Perez-Cañellas et al. 1997).

6.7 Interspecific Differences in Adult Neurogenesis

Neurogenesis has been described in several species of reptiles in separate regions, although always in telencephalic areas (Table 6.1). The entire telencephalon has been analyzed for neurogenic activity in only two species of lizards, *Podarcis hispanica* and *Tarentola mauritanica*, and one species of turtles, *Trachemys scripta* (Lopez-Garcia et al. 1988a; Garcia-Verdugo et al. 1989; Perez-Sanchez et al. 1989; Lopez-Garcia et al. 1990b; Perez-Cañellas and Garcia-Verdugo 1996; Font et al. 1997; Perez-Cañellas et al. 1997). Studies involving other species have evaluated more precise regions, studying one or more telencephalic areas (Garcia-Verdugo et al. 1986a; Duffy et al. 1992; Peñafiel et al. 1996). There is also evidence for neurogenesis in newborn natricine snakes, after which the authors suggest that some postnatal neurogenesis may occur in the accessory and main olfactory bulbs and in the NS (Holtzman and Halpern 1991; Holtzman 1993).

There are studies supporting adult neurogenesis in the main divisions of the lacertid telencephalon, including the main olfactory bulb (MOB), accessory olfactory bulb (AOB), anterior olfactory nucleus (AON), the four cortical areas, septum (SP), dorsal ventricular ridge (DVR), striatum, and nucleus sphericus (NS) (Lopez-Garcia et al. 1988a,b, 1990b; Garcia-Verdugo et al. 1989; Perez-Sanchez et al. 1989; Font et al. 1997). Specifically, it has been observed for *Podarcis hispanica* that neurogenesis is more intense in the NS, although still occurring in the MOB, AOB, and MC. Neurogenic rates in the DMC and SP are considerably low (Font et al. 1997).

Within the family Gekonidae there are only data available for the species *Tarentola mauritanica* (Perez-Cañellas and Garcia-Verdugo 1996). Adult neurogenesis occurs in all the separate telencephalic regions except in one of the cortical regions, the DMC, and in the SP. The highest neurogenesis rates in *Tarentola* occur in the MC and the OB, whereas the lowest involve the lateral cortex (LC) and the striatum.

Table 6.1 Reptile species in which adult neurogenesis has been studied

Species	Region	Technique	Reference
Anolis carolinensis	MC, spinal cord	^{3}HT	Duffy et al. (1990)
Emys orbicularis *Lacerta trilineata*	Telencephalon	^{3}HT	Goffinet et al. (1986)
Gallotia galloti	MC	EM	Garcia-Verdugo et al. (1986a)
Gallotia galloti	Telencephalon	BrdU	Delgado-Gonzalez et al. (2008)
Lacerta viridis	DC	^{3}HT	Minelli et al. (1977)
Lampropholis	Spinal cord	^{3}HT, EM	Alibardi (1994)
Podarcis hispanica	OB	^{3}HT	Garcia-Verdugo et al. (1989)
Podarcis hispanica	MC	^{3}HT, EM	Lopez-Garcia et al. (1988a), Font et al. (1991), Molowny et al. (1995)
Podarcis hispanica	MC	Timm+, EM, ^{3}HT, GFAP	Lopez-Garcia et al. (1988b), Lopez-Garcia et al. (1992), Lopez-Garcia (1993)
Podarcis hispanica	MC	HRP, ^{3}HT, EM	Lopez-Garcia et al. (1990a)
Podarcis hispanica	MC	^{3}HT, PCNA	Ramirez et al. (1997)
Podarcis hispanica	NS	^{3}HT	Perez-Sanchez et al. (1989)
Podarcis hispanica	Telencephalon	^{3}HT, EM, HRP	Lopez-Garcia et al. (1990a,b)
Podarcis hispanica	Telencephalon	^{3}HT, silver impregnation Golgi staining, EM	Font et al. (1995b), Font et al. (1997)
Podarcis hispanica	Telencephalon	BrdU, CRMP-4	Nacher et al. (2002)
Podarcis hispanica	Telencephalon	PSA-NCAM, BrdU	Ramirez-Castillejo et al. (2002)
Psammodromus algirus	OB	^{3}HT	Peñafiel et al. (1996)
Tarentola mauritanica	Telencephalon	BrdU, ^{3}HT, EM	Perez-Cañellas and Garcia-Verdugo (1996)
Trachemys scripta	Telencephalon	^{3}HT, EM, BrdU	Perez-Cañellas et al. (1997)
Triturus cristatus	Optic tectum	^{3}HT, AChE	Minelli et al. (1987)
Tropidurus hispidus	MC	BrdU	Marchioro et al. (2005)

Brain areas of different species of reptiles in which adult neurogenesis has been studied. The techniques used are detailed in each case. *AChE* acetylcholinesterase, *BrdU* bromodeoxyuridine, *DC* dorsal cortex, *EM* electron microscopy, *GFAP* glial fibrillary acid protein, *HRP* horseradish peroxidase, *MC* medial cortex, *OB* olfactory bulb, *PCNA* proliferating cell nuclear antigen, *PSA-NCAM* polysialic acid-neural cell adhesion molecule, *^{3}HT* tritiated thymidine

Adult neurogenesis can also be found in turtles in all telencephalic regions, although it has to be taken into account that they lack the DMC and NS (Perez-Cañellas et al. 1997). Adult neurogenesis in *Trachemys scripta* is more intense in the OB and barely exists in the MC and SP (Perez-Cañellas et al. 1997). Except for the OB, it seems that in turtles, in contrast to lizards, postnatal neurogenesis in the telencephalon is a reminiscent process because only small numbers of neuroblasts integrate into the neuronal telencephalic populations (Font et al. 2001). Neurogenesis has also been described outside the telencephalon, in the cerebellum of *Podarcis hispanica*, although it must be considered that the animals analyzed were perinatal (Lopez-Garcia et al. 1990b). Labeled cells have also been seen in the cerebellum of *Trachemys scripta* (Perez-Cañellas et al. 1997) that are morphologically similar to neighboring granular cells, suggesting such cells are neurons instead of glia. However, an ultrastructural analysis is needed to confirm the presence of adult neurogenesis in the cerebellum in both species.

Routine light and electron microscopic techniques, as well as degeneration-sensitive silver impregnation procedures, have failed to detect neuronal death in the telencephalon of adult lizards and turtles (Lopez-Garcia et al. 1990a; Font et al. 1995b, 1997). This suggests that the neurons born in the brain of adult reptiles represent a net addition to already existing populations, rather than a replacement for older neurons. Furthermore, neurons labeled with ^{3}HT or BrdU survive for long periods of time in their target areas (Perez-Cañellas et al. 1997; Font et al. 2001). From this finding we conclude that the permanent addition of new neurons, together with their long-term survival, forms the basis for the continuous growth of most telencephalic areas in adult reptiles.

6.8 Regenerative Potential in the Central Nervous System

It is well established that reptiles have a significant regenerative ability in their CNS. Several studies in different species of lizards support such regeneration in the adult brain following an injury, whether it is chemical or physical (Font et al. 1997; Romero-Aleman et al. 2004).

With respect to physical (i.e., mechanical) injuries, the extent of regeneration depends on the size of the injury. For example, a small incision on the surface of the DC in the lizard *Gallotia galloti* will require 240 days to fully recover (Romero-Aleman et al. 2004). In contrast, experiments in which a deeper injury affected part of the DC and DMC in the lizard *Lacerta viridis* show that at 260 days following the injury the area has not yet recovered (Minelli et al. 1977). In the latter experiment, the injured area recovers its width but lacks its characteristic cellular layer. According to the authors, the regeneration is still progressing 240 days after the injury, and a complete regeneration would probably be possible allowing a longer term of survival.

Chemical injuries have been studied mostly by administration of the neurotoxin 3-acetylpyridine (3AP) in *Podarcis hispanica* (Font et al. 1991, 1995b, 1997;

Lopez-Garcia et al. 1992; Desfilis et al. 1993; Lopez-Garcia 1993; Molowny et al. 1995). 3AP causes neuronal degeneration of several telencephalic areas including MC, DC, LC, anterior dorsal ventricular ridge (ADVR), NS, and the lateral amygdaloid nucleus (Font et al. 1991; Font et al. 1997). Within these regions, the largest neuronal damage affects the cellular layer of the MC, in which 97% of neurons degenerate by pyknosis 7 days following administration of the neurotoxin (Fig. 6.4a). The level of injury varies greatly in all other regions of the telencephalon and usually affects less than 50% of their respective cell populations (Font et al. 1997). Neurons in the DMC are also damaged by the toxin, but damage to this area is delayed and takes place later than the initial injury, suggesting that it may be caused by denervation as a consequence of the loss of neurons of their main projection from the MC. Six to eight weeks following the injury, most areas are completely regenerated (Font et al. 1991) (Fig. 6.4c).

Regardless of the type of injury, two main phenomena take place during recovery: (1) an increase of microglia in the injured area and (2) an increase in the rate of neurogenesis in the VZ, especially in the area beneath the damaged region (Font et al. 2001; Romero-Aleman et al. 2004). Microglia increases in number during the first few days (Font et al. 1995b; Romero-Aleman et al. 2004), and, together with radial glia cells, they eliminate cellular debris (Lopez-Garcia et al. 1994; Font et al. 1995b; Nacher et al. 1999a,b). In addition to an increase in the number of microglia, the radial glia fibers simultaneously elongate and show pseudopodial prolongations surrounding pyknotic nuclei and other cellular debris. Later on, numerous dense bodies can be seen inside the radial glia cells, which suggests transportation of debris

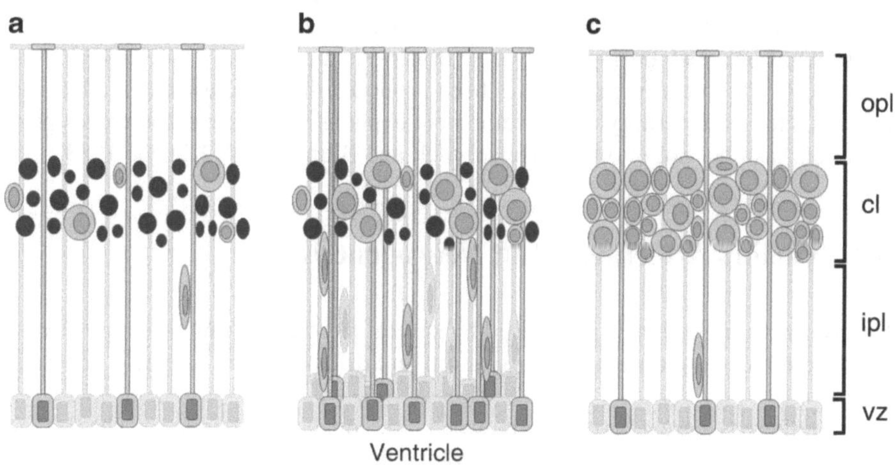

Fig. 6.4 Scheme of the evolution of a 3-acetylpiridine (3AP) lesion in the medial cortex (MC). (**a**) Initially, death affects most of the neurons in the soma layer of the MC by pyknosis. (**b**) Afterward, there is a hypertrophy of the ventricular zone (VZ), resulting in the generation of large numbers of neuroblasts. (**c**) Two months following the injury, the cell layer of the MC is completely regenerated. *cl* cellular layer, *ipl* inner plexiform layer, *opl* outer plexiform layer, *VZ* ventricular zone

from the degenerative sites to the ependyma and possibly to the ventricular system (Font et al. 1995b). The second phenomenon involves a hypertrophy of the VZ consequent to a process of reactive neurogenesis, which generates new neurons to replace those lost by the injury (Fig. 6.4b). It appears that astrocytes are also generated, probably in an effort to help discard debris. This finding suggests that in contrast to normal neurogenesis, reactive proliferation in the telencephalon of lizards gives rise to both neurons and glial cells (Font et al. 1995b).

Radial glial cells therefore play an essential role in the regeneration process, eliminating cellular debris, acting as multipotent stem cells that give rise to neurons and glia, and guiding newly formed neurons to their final location (Alibardi 1994; Font et al. 1995b; Chernoff 1996).

One possible explanation for the absence of complete regeneration in the injured DC of *Lacerta viridis* following a physical lesion is that the insult also damages radial glia cells, which appear to be essential for proper regeneration.

There is a different type of regeneration in the CNS of lizards, which is not driven by neuronal generation, but instead by regrowth of sectioned fibers. Evidence of this process has been found following the section of the caudal spinal cord in *Anolis carolinensis* (Simpson and Duffy 1994) and of the optic nerve in *Ctenophorus ornatus* (Beazley et al. 1997) and *Gallotia galloti* (Lang et al. 1998, 2002).

Neurogenesis can be induced experimentally in other groups of vertebrates such as fish (Zupanc and Ott 1999), amphibians (Kirsche 1983; Minelli et al. 1987; Del Grande et al. 1990; Margotta and Morelli 1996), songbirds (Scharff et al. 2000), and mammals (Gould and Tanapat 1997). Nevertheless, lizards are still unique in their ability to regenerate large portions of their cerebral cortex bilaterally during adulthood, which may be a result of their ability to generate new neurons in the cortex under normal conditions (Font et al. 1991). Because the cellular mechanisms that are brought into play during reactive neurogenesis are basically the same as those involved in adult neurogenesis, it has been proposed that the presence of adult neurogenesis long after embryonic development explains why regeneration takes place in certain taxa but not in others (Hulsebosch and Bittner 1980; Kirsche 1983).

6.9 Factors Affecting Adult Neurogenesis

Neurogenesis in reptiles is influenced by several factors that can have a positive or negative effect on proliferation, migration, and differentiation of newly generated neurons.

First, there is a great variability between individuals of the same species (Lopez-Garcia et al. 1990b; Marchioro et al. 2005), probably the result of differences in sex, age, and/or social status. Generally, studies do not take sex into consideration, and as a result there are no studies demonstrating that sex can influence neurogenesis. Age has been shown to affect neurogenesis, and as a result differences can be seen during development that are more or less obvious depending on the species. In the European lizard *Podarcis hispanica*, neurogenesis decreases with age, and as a consequence the rate of neurogenesis is higher in young animals than in adults, and higher in the latter than in aged specimens (Lopez-Garcia et al. 1988a; Molowny et al. 1995). In the tropical lizard *Tropidurus hispidus*, on the other hand, neurogenesis is more intense in adults than in juvenile and aged animals (Marchioro et al. 2005). It is possible that these differences in proliferative neurogenetic activity are related to the physical or social environment experienced by the experimental subjects during the survival period preceding their death. Perhaps younger and nondominant animals are subject to considerable stress in their holding terraria because of the impossibility of escaping from aggressive adults and dominant lizards. It is well known that stress suppresses postnatal neurogenesis in the mammalian hippocampus (Gould and Tanapat 1999), and we ought to consider a similar possibility in lizards. In this sense a recent study conducted in *Gallotia galloti* shows a decrease in neurogenesis in captivity in comparison to wild animals (Delgado-Gonzalez et al. 2008).

Seasonality may also modulate adult neurogenesis in reptiles, as there is evidence that neurogenesis increases with long photoperiods and decreases with short ones (Molowny et al. 1995; Delgado-Gonzalez et al. 2008). A more detailed analysis, comparing neurogenic rates during different seasons in *Gallotia galloti*, shows that spring is the season with the highest neurogenesis, followed by summer, winter, and autumn (Delgado-Gonzalez et al. 2008). Such variations could be caused by changes associated with the spring reproduction season. The relationship between temperature and neurogenesis has been analyzed in other studies in which it was observed that under low temperatures, migration, but not proliferation, is reduced (Ramirez et al. 1997).

Morphological variations related to seasonal changes can be observed by analyzing the VZ. The width of the VZ in the sulci varies seasonally and possibly also in relationship to changes in proliferation activity (Garcia-Verdugo et al. 2002).

Acknowledgments Special thanks to Enrique Font, from the Instituto Cavanilles de Biodiversidad y Biologia Evolutiva, and Nader Sanai, MD, at the University of California San Francisco, for their comments and suggestions on the manuscript. This review was supported by CIBERNED and TerCel.

References

Ahboucha S, Laalaoui A, Didier-Bazes M, Montange M, Cooper HM, Gamrani H (2003) Differential patterns of glial fibrillary acidic protein-immunolabeling in the brain of adult lizards. J Comp Neurol 464:159–171.

Alibardi L (1994) H3-thymidine labeled cerebrospinal fluid contacting cells in the regenerating caudal spinal cord of the lizard Lampropholis. Ann Anat 176:347–356.

Alvarez-Buylla A (1990) Mechanism of neurogenesis in adult avian brain. Experientia 46:948–955.

Alvarez-Buylla A, Theelen M, Nottebohm F (1990) Proliferation "hot spots" in adult avian ventricular zone reveal radial cell division. Neuron 5:101–109.

Alvarez-Buylla A, Garcia-Verdugo JM, Mateo AS, Merchant-Larios H (1998) Primary neural precursors and intermitotic nuclear migration in the ventricular zone of adult canaries. J Neurosci 18:1020–1037.
Austin CP, Cepko CL (1990) Cellular migration patterns in the developing mouse cerebral cortex. Development 110:713–732.
Beazley LD, Sheard PW, Tennant M, Starac D, Dunlop SA (1997) Optic nerve regenerates but does not restore topographic projections in the lizard Ctenophorus ornatus. J Comp Neurol 377:105–120.
Bruce LL, Neary TJ (1995) The limbic system of tetrapods: a comparative analysis of cortical and amygdalar populations. Brain Behav Evol 46:224–234.
Butler AB, Hodos W (1996) Comparative vertebrate neuroanatomy: Evolution and adaptation. Wiley-Liss, New York.
Chernoff EA (1996) Spinal cord regeneration: a phenomenon unique to urodeles? Int J Dev Biol 40:823–831.
Davila JC, Guirado S, De la Calle A, Marin-Giron F (1985) Electron microscopy of the medial cortex in the lizard *Psammodromus algirus*. J Morphol 185:327–338.
Del Grande P, Franceschini V, Minelli G, Ciani F (1990) Mitotic activity of the telencephalic matrix areas following optic tectum or pallial cortex lesion in newt. Z Mikrosk Anat Forsch 104:617–624.
Delgado-Gonzalez FJ, Alonso-Fuentes A, Delgado-Fumero A, Garcia-Verdugo JM, Gonzalez-Granero S, Trujillo-Trujillo CM, Damas-Hernandez MC (2008) Seasonal differences in ventricular proliferation of adult Gallotia galloti lizards. Brain Res 1191:39–46.
Desfilis E, Garcia-Verdugo JM, Font E (1993) Regeneration in the adult lizard brain: further evidence from 3TH autoradiography. Eur J Neurosci S6:290.
Doetsch F, Garcia-Verdugo JM, Alvarez-Buylla A (1997) Cellular composition and three-dimensional organization of the subventricular germinal zone in the adult mammalian brain. J Neurosci 17:5046–5061.
Doetsch F, Caille I, Lim DA, Garcia-Verdugo JM, Alvarez-Buylla A (1999) Subventricular zone astrocytes are neural stem cells in the adult mammalian brain. Cell 97:703–716.
Duffy MT, Simpson SB, Liebich DR, Davis BM (1990) Origin of spinal cord axons in the lizard regenerated tail: supernormal projections from local spinal neurons. J Comp Neurol 293:208–222.
Duffy MT, Liebich DR, Garner LK, Hawrych A, Simpson SB, Jr., Davis BM (1992) Axonal sprouting and frank regeneration in the lizard tail spinal cord: correlation between changes in synaptic circuitry and axonal growth. J Comp Neurol 316:363–374.
Font E, Garcia-Verdugo JM, Alcantara S, Lopez-Garcia C (1991) Neuron regeneration reverses 3-acetylpyridine-induced cell loss in the cerebral cortex of adult lizards. Brain Res 551:230–235.
Font C, Hoogland PV, Vermeulen van der Zee E, Perez-Clausell J, Martinez-Garcia F (1995a) The septal complex of the telencephalon of the lizard Podarcis hispanica. I. Chemoarchitectonical organization. J Comp Neurol 359:117–130.
Font E, Garcia-Verdugo JM, Desfilis E, Perez-Cañellas M (1995b) Neuron-glia interrelations during 3-acetylpyridine-induced degeneration and regeneration in the adult lizard brain. In: Vernadakis A, Roots B (eds) Neuron-glia interrelations during phylogeny: II. Plasticity and regeneration. Humana press Inc, Totowa.
Font E, Desfilis E, Perez-Cañellas M, Alcantara S, Garcia-Verdugo JM (1997) 3-Acetylpyridine-induced degeneration and regeneration in the adult lizard brain: a qualitative and quantitative analysis. Brain Res 754:245–259.
Font E, Desfilis E, Perez-Cañellas MM, Garcia-Verdugo JM (2001) Neurogenesis and neuronal regeneration in the adult reptilian brain. Brain Behav Evol 58:276–295.
Garcia-Verdugo JM, Berbel PJ, Lopez-Garcia C (1981) [Golgi and electron microscopy study of cerebral ependymocytes of the lizard Lacerta galloti]. Trab Inst Cajal 72:269–278.
Garcia-Verdugo JM, Berbel-Navarro P, Regidor-Garcia J, Lopez-Garcia C (1984) Ultrastructure of neuronal cell bodies in the medial cortex of Lacerta galloti. J Hirnforsch 25:187–196.

Garcia-Verdugo JM, Fariñas I, Molowny A, Lopez-Garcia C (1986a) Ultrastructure of putative migrating cells in the cerebral cortex of Lacerta galloti. J Morphol 189:189–197.

Garcia-Verdugo JM, Llahi S, Fariñas I, Martin V (1986b) Laminar organization of the main olfactory bulb of Podarcis hispanica: an electron microscopic and Golgi study. J Hirnforsch 27:87–100.

Garcia-Verdugo JM, Llahi S, Ferrer I, Lopez-Garcia C (1989) Postnatal neurogenesis in the olfactory bulbs of a lizard. A tritiated thymidine autoradiographic study. Neurosci Lett 98:247–252.

Garcia-Verdugo JM, Ferron S, Flames N, Collado L, Desfilis E, Font E (2002) The proliferative ventricular zone in adult vertebrates: a comparative study using reptiles, birds, and mammals. Brain Res Bull 57:765–775.

Goffinet AM, Daumerie C, Langerwerf B, Pieau C (1986) Neurogenesis in reptilian cortical structures: 3H-thymidine autoradiographic analysis. J Comp Neurol 243:106–116.

Gould E, Tanapat P (1997) Lesion-induced proliferation of neuronal progenitors in the dentate gyrus of the adult rat. Neuroscience 80:427–436.

Gould E, Tanapat P (1999) Stress and hippocampal neurogenesis. Biol Psychiatry 46:1472–1479.

Holtzman DA (1993) The ontogeny of nasal chemical senses in garter snakes. Brain Behav Evol 41:163–170.

Holtzman DA, Halpern M (1991) Incorporation of 3H-thymidine in telencephalic structures of the vomeronasal and olfactory systems of embryonic garter snakes. J Comp Neurol 304: 450–466.

Hoogland PV, Vermeulen-Vanderzee E (1995) Efferent connections of the lateral cortex of the lizard Gekko gecko: evidence for separate origins of medial and lateral pathways from the lateral cortex to the hypothalamus. J Comp Neurol 352:469–480.

Hulsebosch CE, Bittner GD (1980) Evolution of abilities to regenerate neurons in central nervous system. Am Nat 115:276–284.

Johnston JB (1916) Evidence of a motor pallium in the forebrain of reptiles. J Comp Neurol 26:469–480.

Kempermann G (2006) Adult neurogenesis: stem cells and neuronal development in the adult brain. Oxford, New York.

Kirsche W (1983) The significance of matrix zones for brain regeneration and brain transplantation with special consideration of lower vertebrates. In: Wallace RB, Das GD (eds) Neural tissue transplantation research. Springer, Berlin.

Lang DM, Monzon-Mayor M, Bandtlow CE, Stuermer CA (1998) Retinal axon regeneration in the lizard Gallotia galloti in the presence of CNS myelin and oligodendrocytes. Glia 23:61–74.

Lang DM, Romero-Aleman MM, Arbelo-Galvan JF, Stuermer CA, Monzon-Mayor M (2002) Regeneration of retinal axons in the lizard Gallotia galloti is not linked to generation of new retinal ganglion cells. J Neurobiol 52:322–335.

Lois C, Alvarez-Buylla A (1994) Long-distance neuronal migration in the adult mammalian brain. Science 264:1145–1148.

Lopez-Garcia C (1993) Postnatal neurogenesis and regeneration in the lizard cerebral cortex. In: Cuello AC (ed) Restorative neurology. Elsevier, Amsterdam.

Lopez-Garcia C, Martinez-Guijarro FJ (1988) Neurons in the medial cortex give rise to Timm-positive boutons in the cerebral cortex of lizards. Brain Res 463:205–217.

Lopez-Garcia C, Tineo PL, Del Corral J (1984) Increase of the neuron number in some cerebral cortical areas of a lizard, Podarcis hispanica, (Steind., 1870), during postnatal periods of life. J Hirnforsch 25:255–259.

Lopez-Garcia C, Molowny A, Garcia-Verdugo JM, Ferrer I (1988a) Delayed postnatal neurogenesis in the cerebral cortex of lizards. Brain Res 471:167–174.

Lopez-Garcia C, Molowny A, Rodriguez-Serna R, Garcia Verdugo JM, Martinez-Guijarro FJ (1988b) Postnatal development of neurons in the telencephalic cortex of lizards. In: Schwerdtfeger WK, Smeets WJAJ (eds) The forebrain of reptiles: Current concepts of structure and function. Karger, Basel.

Lopez-Garcia C, Molowny A, Garcia-Verdugo JM, Martinez-Guijarro FJ, Bernabeu A (1990a) Late generated neurons in the medial cortex of adult lizards send axons that reach the Timm-reactive zones. Brain Res Dev Brain Res 57:249–254.
Lopez-Garcia C, Molowny A, Garcia-Verdugo JM, Perez-Sanchez F, Martinez-Guijarro FJ (1990b) Postnatal neurogenesis in the brain of the lizard Podarcis hispanica. In: Schwerdtfeger WK, Germroth P (eds) The forebrain in nonmammals: New aspects of structure and development. Springer, Berlin.
Lopez-Garcia C, Molowny A, Martinez-Guijarro FJ, Blasco-Ibanez JM, Luis de la Iglesia JA, Bernabeu A, Garcia-Verdugo JM (1992) Lesion and regeneration in the medial cerebral cortex of lizards. Histol Histopathol 7:725–746.
Lopez-Garcia C, Nacher J, Castellano B, Luis de la Iglesia JA, Molowny A (1994) Transitory disappearance of microglia during the regeneration of the lizard medial cortex. Glia 12:52–61.
Luis de la Iglesia JA, Lopez-Garcia C (1997a) A Golgi study of the short-axon interneurons of the cell layer and inner plexiform layer of the medial cortex of the lizard Podarcis hispanica. J Comp Neurol 385:565–598.
Luis de la Iglesia JA, Lopez-Garcia C (1997b) A Golgi study of the principal projection neurons of the medial cortex of the lizard Podarcis hispanica. J Comp Neurol 385:528–564.
Marchioro M, Nunes JM, Ramalho AM, Molowny A, Perez-Martinez E, Ponsoda X, Lopez-Garcia C (2005) Postnatal neurogenesis in the medial cortex of the tropical lizard Tropidurus hispidus. Neuroscience 134:407–413.
Margotta V, Morelli A (1996) Encephalic matrix areas and post-natal neurogenesis under natural and experimental conditions. Anim Biol 5:117–131.
Martinez-Garcia F, Amiguet M, Olucha F, Lopez-Garcia C (1986) Connections of the lateral cortex in the lizard Podarcis hispanica. Neurosci Lett 63:39–44.
Martinez-Garcia F, Amiguet M, Schwerdfeger WK (1990) Interhemispheric connections through the pallial commissures in the brain of *Podarcis hispanica* and *Gallotia-Stehlinii* (Reptilia, Lacertidae). J Morphol 205:17–31.
Martinez-Garcia F, Martinez-Marcos A, Lanuza E (2002) The pallial amygdala of amniote vertebrates: evolution of the concept, evolution of the structure. Brain Res Bull 57:463–469.
Martinez-Guijarro FJ, Soriano E, Del Rio JA, Lopez-Garcia C (1991) Zinc-positive boutons in the cerebral cortex of lizards show glutamate immunoreactivity. J Neurocytol 20:834–843.
Minelli G, del Grande P, Mambelli MC (1977) Preliminary study of the regenerative processes of the dorsal cortex of the telencephalon of Lacerta viridis. Z Mikrosk Anat Forsch 91:241–256.
Minelli G, Franceschini V, Del Grande P, Ciani F (1987) Newly-formed neurons in the regenerating optic tectum of Triturus cristatus carnifex. Basic Appl Histochem 31:43–52.
Molowny A, Nacher J, Lopez-Garcia C (1995) Reactive neurogenesis during regeneration of the lesioned medial cerebral cortex of lizards. Neuroscience 68:823–836.
Monzon-Mayor M, Yanes C, Tholey G, De Barry J, Gombos G (1990) Immunohistochemical localization of glutamine synthetase in mesencephalon and telencephalon of the lizard Gallotia galloti during ontogeny. Glia 3:81–97.
Nacher J, Ramirez C, Palop JJ, Molowny A, Luis de la Iglesia JA, Lopez-Garcia C (1999a) Radial glia and cell debris removal during lesion-regeneration of the lizard medial cortex. Histol Histopathol 14:89–101.
Nacher J, Ramirez C, Palop JJ, Artal P, Molowny A, Lopez-Garcia C (1999b) Microglial cells during the lesion-regeneration of the lizard medial cortex. Histol Histopathol 14:103–117.
Nacher J, Soriano S, Varea E, Molowny A, Ponsoda X, Lopez-Garcia C (2002) CRMP-4 expression in the adult cerebral cortex and other telencephalic areas of the lizard Podarcis hispanica. Brain Res Dev Brain Res 139:285–294.
O'Rourke NA, Sullivan DP, Kaznowski CE, Jacobs AA, McConnell SK (1995) Tangential migration of neurons in the developing cerebral cortex. Development 121:2165–2176.
Peñafiel A, Gutierrez A, Martin R, Perez-Cañellas MM, de la Calle A (1996) A tangential neuronal migration in the olfactory bulbs of adult lizards. Neuroreport 7:1257–1260.

Perez-Cañellas MM, Garcia-Verdugo JM (1996) Adult neurogenesis in the telencephalon of a lizard: a [3H]thymidine autoradiographic and bromodeoxyuridine immunocytochemical study. Brain Res Dev Brain Res 93:49–61.

Perez-Cañellas MM, Font E, Garcia-Verdugo JM (1997) Postnatal neurogenesis in the telencephalon of turtles: evidence for nonradial migration of new neurons from distant proliferative ventricular zones to the olfactory bulbs. Brain Res Dev Brain Res 101:125–137.

Perez-Sanchez F, Molowny A, Garcia-Verdugo JM, Lopez-Garcia C (1989) Postnatal neurogenesis in the nucleus sphericus of the lizard, Podarcis hispanica. Neurosci Lett 106:71–75.

Rakic P (1971) Guidance of neurons migrating to the fetal monkey neocortex. Brain Res 33: 471–476.

Ramirez C, Nacher J, Molowny A, Sanchez-Sanchez F, Irurzun A, Lopez-Garcia C (1997) Photoperiod-temperature and neuroblast proliferation-migration in the adult lizard cortex. Neuroreport 8:2337–2342.

Ramirez-Castillejo C, Nacher J, Molowny A, Ponsoda X, Lopez-Garcia C (2002) PSA-NCAM immunocytochemistry in the cerebral cortex and other telencephalic areas of the lizard Podarcis hispanica: differential expression during medial cortex neuronal regeneration. J Comp Neurol 453:145–156.

Romero-Aleman MM, Monzon-Mayor M, Yanes C, Lang D (2004) Radial glial cells, proliferating periventricular cells, and microglia might contribute to successful structural repair in the cerebral cortex of the lizard Gallotia galloti. Exp Neurol 188:74–85.

Scharff C, Kirn JR, Grossman M, Macklis JD, Nottebohm F (2000) Targeted neuronal death affects neuronal replacement and vocal behavior in adult songbirds. Neuron 25:481–492.

Seri B, Garcia-Verdugo JM, McEwen BS, Alvarez-Buylla A (2001) Astrocytes give rise to new neurons in the adult mammalian hippocampus. J Neurosci 21:7153–7160.

Simpson SB, Jr., Duffy MT (1994) The lizard spinal cord: a model system for the study of spinal cord injury and repair. Prog Brain Res 103:229–241.

Stensaas LJ, Stensaas SS (1968) Light microscopy of glial cells in turtles and birds. Z Zellforsch Mikrosk Anat 91:315–340.

Striedter GF (1997) The telencephalon of tetrapods in evolution. Brain Behav Evol 49:179–213.

Tineo PL, Planelles MD, Del-Corral J (1987) Modifications in cortical ependyma of the lizard, Podarcis hispanica, during postnatal development. J Hirnforsch 28:485–489.

Ulinski PS (1990) The cerebral cortex of reptiles. In: Jones EG, Peters A (eds) Cerebral cortex. Plenum, New York.

Ulinski PS, Kanarek DA (1973) Cytoarchitecture of nucleus sphericus in the common boa, Constrictor constrictor. J Comp Neurol 151:159–174.

Yanes-Mendez C, Martin-Trujillo JM, Perez-Batista MA, Mozon-Mayor M, Marrero A (1988) Ependymogenesis of the lizard basal areas. I. Ependymal zones. Z Mikrosk Anat Forsch 102:555–572.

Zupanc GK, Ott R (1999) Cell proliferation after lesions in the cerebellum of adult teleost fish: time course, origin, and type of new cells produced. Exp Neurol 160:78–87.

Part IV
Adult Neurogenesis in the Hippocampus

Chapter 7
From Embryonic to Adult Neurogenesis in the Dentate Gyrus

Tatsunori Seki

Abstract Adult neurogenesis is now a very popular phenomenon in neuroscience. It is widely accepted that neurons continue to be generated in special regions of the adult brain such as the subventricular zone (SVZ) of the forebrain and the subgranular zone (SGZ) of the dentate granule cell layer. Adult neurogenesis, however, is not a special type of neurogenesis that occurs only in the adult stage, but is a part of the persistent neurogenesis that continues from embryonic to adult stages. Here, we describe the similarities and differences in the properties of progenitors and cellular architecture of proliferative zones from embryonic to adult periods and discuss the elements required for persistent neurogenesis.

7.1 Discovery of New Neurons and Their Progenitors in the Adult Brain

Historically, neurogenesis has been explored exclusively in the embryonic and early postnatal brain, and for a long time there was almost no study of adult neurogenesis (Jacobson 1991; Gross 2000; Kempermann 2006). In the early stage of research on embryonic neurogenesis, mitotic cells and cell migration were analyzed by conventional staining methods such as Nissl and Golgi methods. In the early 1960s, adult neurogenesis was discovered by Altman, who used [^{3}H]thymidine autoradiography and found newborn neurons in the subgranular zone (SGZ) of the adult rat dentate gyrus (Altman and Das 1965; Gross 2000; Kempermann 2006). Thereafter, study of adult neurogenesis was undertaken by a few groups of scientists using the same methods (Kaplan and Hinds 1977; Goldman and Nottbohm 1983; Kaplan 2001; Alvarez-Buylla and Nottebohm 1988; Gould et al. 1992;

T. Seki (✉)
Department of Histology and Neuroanatomy, Tokyo Medical University, Tokyo 160-8402, Japan
e-mail: sekit@tokyo-med.ac.jp

T. Seki et al. (eds.), *Neurogenesis in the Adult Brain I: Neurobiology*,
DOI 10.1007/978-4-431-53933-9_7, © Springer 2011

Cameron et al. 1993; Gould and Gross 2002). However, it took about 40 years before adult neurogenesis became very popular in the field of neuroscience.

One of the reasons why adult neurogenesis was not widely accepted is that autoradiography is time consuming because it requires a few weeks or sometimes more than 1 month to detect [^{3}H]thymidine-labeled newborn cells, and also only nuclei are visualized in the proliferating cells that incorporate [^{3}H]thymidine during a short period of time after the [^{3}H]thymidine injection. Researchers could not obtain complete images of all the developing neurons including their dendrites and axons.

In the early 1990s, two new techniques were introduced in this field. In 1991, immunostaining with an antibody against polysialic acid (PSA), a carbohydrate portion of neural cell adhesion molecules (NCAM), demonstrated that only granule cells located in the innermost region of the granule cell layer (GCL) express PSA–NCAM (Seki and Arai 1991). Judging from their distribution and morphology, it was concluded that the positive cells are newly generated granule cells, implying that PSA is a useful marker for newborn neurons in the adult hippocampus. Another technique is the labeling of dividing cells with bromodeoxyuridine (BrdU), a thymidine analogue that was first used in the embryonic and early postnatal hippocampus by Nowakowski's group (Miller and Nowakowski 1988), but not in adults. In 1993, fate analysis using a combination of PSA immunohistochemistry and BrdU labeling clearly demonstrated that newly generated granule cells express PSA transiently in young adult rats (Seki and Arai 1993b) and also in aged rats (Seki and Arai 1995). Immunohistochemistry for PSA is the first method to visualize the entire and detailed shape of all developing neurons in a short time, including the dendrites and axons, in the adult hippocampus and the subventricular zone (SVZ) of the adult forebrain (Seki and Arai 1993a,b, 1995). Thereafter, the combination of PSA immunostaining and BrdU labeling came into use by Alvarez-Buylla's and Gage's groups (Rousselot et al. 1995; Kuhn et al. 1996), which greatly contributed to the progress of this field.

In the late 1990s, multiple immunostainings using various cell markers for stem cells, immature neurons and glia with BrdU labeling became a standard method to explore adult neurogenesis (Gross 2000; Kempermann 2006). Additionally, a retrovirus vector bearing the green fluorescence protein (GFP) gene and transgenic mice such as Nestin-GFP, doublecortin (DCX)-GFP, and proopiomelanocortin (POMC)-GFP are now available to label newly generated cells and for time-lapse imaging and electrophysiology in slices of hippocampal tissues (Yamaguchi et al. 2000; Fukuda et al. 2003; Zhao 2007; Enikolopov and Overstreet-Wadiche 2008).

Despite much progress in the techniques, the identity of the neural stem cells or primary progenitors that give rise to neurons in the adult brain remained unsolved until 1999. An unexpected answer to this issue was determined by Alvarez-Buylla's group (Doetsch et al. 1999; Seri et al. 2001). The stem cells were found to have features of astrocytes, including glial fibrillary acidic protein (GFAP), brain lipid-binding protein (BLBP), glutamate aspartate transporter (GLAST), and electrophysiological properties of astrocytes; this allowed us to use GFAP-GFP transgenic mice to identify stem cells (Steiner et al. 2006; Liu et al. 2010).

These techniques have enabled progress in many aspects of the study of the unique neurogenesis in the adult brain. However, to fully understand the mechanism and function of adult neurogenesis, it is also important to reveal how neurogenesis continues from embryonic to adult stages or how prospective stem cell-like progenitors are incorporated into the postnatal GCL during the perinatal period and how the profile of stem cells is maintained during the postnatal period.

7.2 Embryonic Neurogenesis

7.2.1 Morphogenesis of the Hippocampus

The hippocampal formation is derived from a caudomedial portion of the cortical primordium, called the cortical hem, whose ventral part makes connections to the choroid plexus (Fig. 7.1) (Li and Pleasure 2007; Hebert and Fishell 2008; O'Leary and Sahara 2008; Subramanian and Tole 2009). The progenitors of both pyramidal cells and granule cells are generated in the ventricular proliferative zone of the caudomedial pallium formed by a pseudostratified, columnar epithelium. According to Altman and Bayer (Altman and Bayer 1990a,b,c), the neurogenic hippocampal neuroepithelium is divided into three discrete parts that are arranged in order from dorsal to ventral: the ammonic neuroepithelium, the primary dentate neuroepithelium, and the fimbrial glioepithelium.

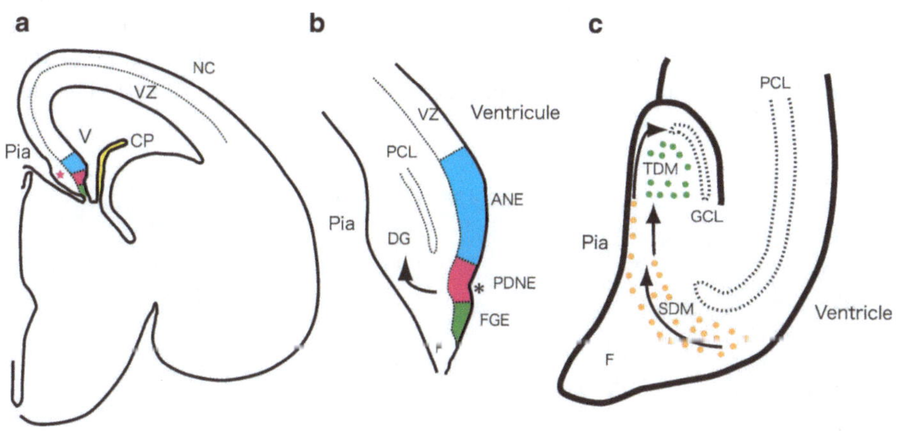

Fig. 7.1 Schematic drawings of hippocampal development. (**a**) Coronal section of the forebrain in embryonic day 16–18 (E16–18) rat embryo. The hippocampus is derived from a caudomedial portion of the cortical primordium, the cortical hem (*colored region*). *Magenta asterisk*, prospective dentate region. (**b**) Hippocampal primordium at E16–18. The ventricular zone (VZ) is divided into three regions: ammonic neuroepithelium (*ANE*), primary dentate neuroepithelium (*PDNE*), and fimbrial glioepithelium (*FGE*). *Black asterisk* dentate notch, *arrow* dentate migration. (**c**) Proliferative cells in the dentate migration (*arrows*) form a secondary dentate matrix (SDM, *orange circles*) around E21. Then, one dentate migration follows the subpial route to form the granule cell layer (*GCL*), and another reaches the hilus to form a tertiary dentate matrix (*TDM*, *green circles*). *CP* choroid plexus, *DG* dentate gyrus, *F* fimbria, *NC* neocortex, *PCL* pyramidal cell layer, *V* ventricle (Prepared on the basis of Altaman and Bayer 1993a)

The developmental processes of the pyramidal layer are roughly similar to those of the neocortical layer (Altman and Bayer 1990a,b,c). The progenitors of the pyramidal cells are produced in the ventricular zone, migrate radially as does the neocortex (Kriegstein and Noctor 2004; Rakic 2007), and reach their final position to form the pyramidal cell layer (PCL) during the embryonic period. However, the morphogenesis of the dentate GCL is fundamentally different from that of the hippocampal PCL (Angevine 1965; Stensaas 1967; Schlessinger et al. 1975; Bayer 1980; Eckenhoff and Rakic 1984a; Rickmann et al. 1987b; Altman and Bayer 1990b; Sievers et al. 1992a; Li and Pleasure 2005; Nakahira and Yuasa 2005).

According to studies by Altman and Bayer using [^{3}H]thymidine autoradiography (Altman and Bayer 1990a,b,c), the putative stem cells of the granule cells initially appear on embryonic day 16 (day E16) in rats in the primary dentate neuroepithelium, which is located around a ventricular indentation, or the dentate notch at the dorsal edge of the fimbria (Fig. 7.1a,b). The progenitors generated in the primary dentate neuroepithelium do not simply radially migrate, but travel through the dorsal portion of the fimbria to the pial side where the dentate gyrus is formed. This migratory stream is called dentate migration. These migrating cells maintain their proliferative capacity and form a secondary dentate matrix along their migratory stream, including suprafimbrial, subpial, and prospective dentate regions (Fig. 7.1c). Gradually, the dentate migration divides into two routes. The first dentate migration follows the subpial route and forms the GCL (the outer shell), whereas the secondary dentate migration reaches the prospective hilus. During the perinatal and early postnatal periods, the hilus becomes the main proliferative zone, called the tertiary dentate matrix, which establishes the inner shell of the GCL while the secondary matrix is diminished (Fig. 7.2a,d). The dual

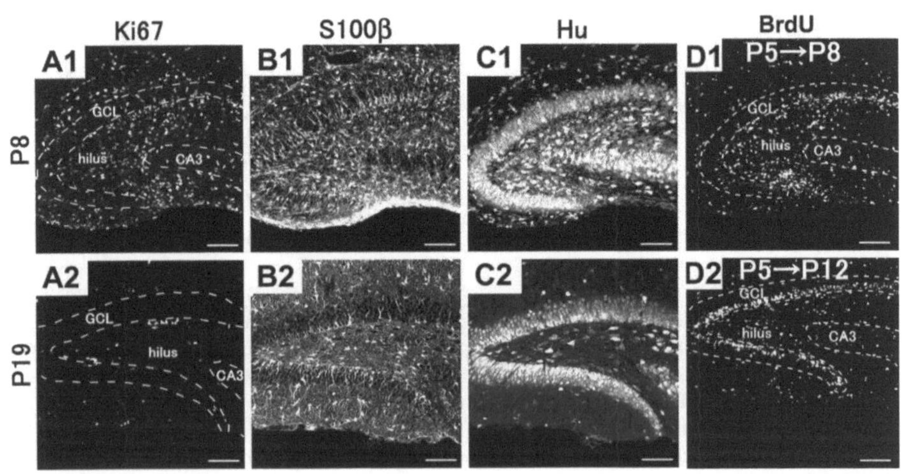

Fig. 7.2 (**a–c**) Comparison of the distribution patterns of the Ki67-, S100β-, and Hu-expressing cells in the dentate gyrus between postnatal day 8 (P8) and P19. (**d**) Change in the distribution of bromodeoxyuridine (BrdU)-labeled cells in rats injected with BrdU on P5 and fixed at 3 (P8) and 7 (P12) days after the injection. *CA3* CA3 field of the pyramidal cell layer, *GCL* granule cell layer. *Bars* 50 μm. (Modified, with permission, from Namba et al. 2005)

formation of the outer and inner shells of the GCL has been confirmed by recent mouse aggregation chimera analysis (Martin et al. 2002). In rats, only 15% of granule cells are created during the embryonic period and 85% are generated after birth (Angevine 1965; Schlessinger et al. 1975; Bayer 1980).

7.2.2 Molecules Regulating Development of the Hippocampus

Modern molecular biological approaches have shed light on the regulatory mechanism underlying the development of the hippocampus or, more restrictedly, the dentate gyrus (Table 7.1). The initial development of the hippocampus is regulated by morphogens that are secreted by one of the patterning centers of the cortex, the cortical hem, which is thought to be a putative organizer for the hippocampus (Li and Pleasure 2005; O'Leary et al. 2007; Hebert and Fishell 2008; Subramanian et al. 2009). In the dorsal telencephalon, two patterning centers are mainly implicated

Table 7.1 Molecules regulating neurogenesis in embryonic/early postnatal and adult periods

	Embryonic/early postnatal period	Adult period
Shh/Gli	Han et al. (2008)	Lai et al. (2003)
	Breunig et al. (2008)	Machold et al. (2003)
Wnt	Lee et al. (2000)	Lie et al. (2005)
	Zhou et al. (2004)	
	Machon et al. (2007)	
	Galceran et al. (2000)	
	Solberg et al. (2008)	
	Roelink (2000)	
Lhx	Mangale et al. (2008)	
EMX2	Pellegrini et al. (1996)	Hong et al. (2007)
	Yoshida et al. (1997)	
	Oldekamp et al. (2004)	
	Zhao et al. (2006)	
SDF-1 (CXCl12)/CXCR4	Lu et al. (2002)	Kolodziej et al. (2008)
	Bagri et al. (2002)	
	Li et al. (2009)	
Notch	Breunig et al. (2007)	Salama-Cohen et al. (2006)
Reelin	Forster et al. (2002)	Gong et al. (2007)
	Forster et al. (2006)	Muller et al. (2009)
	Li et al. (2009)	
	Sibbe et al. (2009)	
Mash1 (Ascl1)		Kim et al. (2007)
		Jessberger et al. (2008)
Neurogenin	Galichet et al. (2008)	Ozen et al. (2007)
NeuroD	Miyata et al. (1999)	Gao et al. (2009)
	Schwab et al. (2000)	
	Liu et al. (2000)	
DISC1	Enomoto et al. (2009)	Kim et al. (2009)
	Meyer and Morris (2009)	Duan et al. (2007)

in the arealization: the commissural plate (anterior neural ridge), which expresses fibroblast growth factors (FGFs), and the cortical hem, a caudomedial portion connected to the choroid plexus, which expresses bone morphogenetic proteins (BMPs) and Wingless/Int proteins (Wnts) (O'Leary et al. 2007; Hebert and Fishell 2008; Subramanian et al. 2009). BMPs are required for the formation of the choroid plexus (Hebert et al. 2002). Wnt3A has been shown to be crucial for the development of the entire hippocampus (Lee et al. 2000; Roelink 2000). Defects in signaling molecules in the Wnt/beta-catenin pathway, such as LRP6, a coreceptor (Zhou et al. 2004), beta-catenin, a cytoplasmic mediator (Machon et al. 2003), and Lef1, a nuclear mediator (Galceran et al. 2000; Roelink 2000), are associated with decreased dentate granule cell production or failure of development of the dentate gyrus. These results suggest that canonical Wnt signaling is essential in the generation of the granule cells. Intracellular effectors of BMP and Wnt signaling are reported to regulate the expression of a basic helix-loop-helix (bHLH) transcription factor, empty spiracles homologue 2 (Emx2) (Theil et al. 2002; Shinozaki et al. 2004). Deficiency in Emx2 causes a lack of the dentate GCL (Pellegrini et al. 1996; Yoshida et al. 1997). Additionally, experiments with chimeric mice in which Lhx2-null cells (Lhx2 is a lim-homeodomain transcription factor regulating neocortical cell fate) are surrounded by wild-type cortical neuroepithelium have shown that ectopic hem tissue can induce an ectopic hippocampus, suggesting that Lhx2 suppresses the fate of the cortical hem, which is an organizer for the hippocampus (Mangale et al. 2008).

The formation of the GCL has recently been reported to be regulated by Shh signal, which is well known to antagonize BMP/Wnt signals in the spinal cord, and the antagonism controls the dorsoventral polarity and cell differentiation of the spinal cord. In the developing hippocampus, Shh signaling regulates expansion and differentiation of neural stem cells through the primary cilia, which are necessary for vertebrate hedgehog signaling (Davis et al. 2006; Han et al. 2008).

The migration of the dentate granule precursor cells is controlled by factors produced by components around the pia mater, such as the meninges and Cajal-Retzius cells. A pioneering study by Hartmann et al. showed that the destruction of meningeal cells by the injection of 6-hydroxydopamine into the interhemispheric fissure prevents the formation of the infrapyramidal blade of the GCL, suggesting that meningeal cells are involved in the development of the GCL (Hartmann et al. 1992). Meninges around the hippocampus from the hippocampal fissure to the fimbria is reported to express stromal cell-derived factor 1 (SDF-1)/CXCl12, a chemokine that is the only known ligand for CXCR4 chemokine receptor (Lu et al. 2002). CXCR4 mRNA is expressed by dentate progenitor cells, particularly in secondary and tertiary dentate matrix including the suprafimbrial, subpial, and hilar regions. In mice lacking the CXCR4 receptor, precursors of the granule cells fail to reach the GCL and remain in the hilus, suggesting that SDF-1/CXCR4 chemokine signaling regulates migration of dentate granule cells (Bagri et al. 2002; Lu et al. 2002).

Cajal-Retizus cells are located in the marginal zone of the forebrain, just below the pia mater, and secrete the extracellular molecule reelin, which has been shown to be required for proper migration of granule cells (Forster et al. 2002, 2006; Sibbe et al. 2009). The reelin binds to the very low density lipoprotein receptor and apolipoprotein E receptor 2, the cytoplasmic domains of both of which bind to adaptor protein disabled 1 (Dab1) in neurons and glia cells, and reelin also binds to $beta_1$-class integrin. In mutant mice that are deficient for reelin (*reeler*), Dab1 (*scrambler*), and $beta_1$-integrins, the migration of the granule cells is defective (Forster et al. 2002). A recent study indicated that reelin signaling is required for the migration from the subpial pool to the SGZ, whereas SDF-1 (CXCl12)/CXCR4 signaling is essential for the initial formation of the subpial dentate precursor pool (Li et al. 2009). These molecules are suggested to affect the formation of radial glia that serve as a scaffold for the migration of the granule cell precursors.

As for the granule cell migration, many papers suggest that, during the perinatal period, granule cell precursors migrate along radial glial fibers in a manner similar to that in the neocortex (Eckenhoff and Rakic 1984b; Rickmann et al. 1987a; Sievers et al. 1992b; Yuasa 2001; Nakahira and Yuasa 2005; Barry et al. 2008). The radial glial cells are disorganized in mutants that show malformation of the dentate gyrus, probably because of the deficient migration of granule cells. Findings for mutant mice with defects in SDF-1/CXCR4 (Li et al. 2009), reelin, Da1, $beta_1$-integrins (Forster et al. 2002), and nuclear factor Ib (Barry et al. 2008) suggest that these molecules are required for radial glial formation and the proper development of the GCL. However, it should also be noted that the migration pattern of dentate granule cells is not simply identical to radial glia-dependent migration of neocortical neuroblasts (Altman and Bayer 1990a). Recently, the cell migration of dentate granule cells has been shown to be disturbed by knockdown of *disrupted-in-schizophrenial* (*DISC1*), a gene related to major psychiatric disorders (Meyer and Morris 2009), and a molecule that interacts with DISC1, namely, girdin (Enomoto et al. 2009).

The production of the dentate granule cells is regulated by molecules that specify neural cell fate and neuronal differentiation. A study of gain-of-function and loss-of-function of Notch1 showed that Notch1 signaling regulates a variety of events including proliferation, cell-cycle exit, differentiation of progenitors, and dendritic morphology of newborn granule cells (Breunig et al. 2007). Neurogenin-2, one of the bHLH proneural proteins, is reported to have an essential role in the proliferation and differentiation of dentate progenitors (Galichet et al. 2008). Another bHLH protein, NeuroD, is needed for the differentiation of progenitors into granule neurons (Miyata et al. 1999; Liu et al. 2000; Schwab et al. 2000). These findings suggest that the neuronal differentiation-related bHLH proteins are required for the production of dentate granule cells in the developing hippocampus.

7.2.3 Future Problems

Despite the detailed description and molecular analysis of the morphogenesis of the dentate gyrus, important questions remain regarding the properties of progenitors in the primary dentate neuroepithelium and secondary proliferative zone, for example, if they have astroglial characteristics such as neural stem cells in the adult hippocampus, how they differentiate into neurons, and which progenitor cells become future adult stem cells or granule neurons. Some reports on studies with *GFAP-Cre* mice suggest that, during the perinatal period, GFAP-expressing neural progenitors may produce granule cells (Garcia et al. 2004; Ganat et al. 2006). However, additional work that focuses on individual cells is required to answer these questions.

7.3 Early Postnatal Neurogenesis

7.3.1 Early Postnatal Development of the Dentate Gyrus

Early postnatal neurogenesis is important to understand the mechanism of persistent neurogenesis that expands from embryonic to adult periods, because it constitutes a transition state between embryonic and adult neurogenesis. In the early postnatal period, there are two types of progenitor derived from secondary dentate matrix: one type of progenitor migrates below the pial surface and forms the outer shell of the GCL by postnatal day 5 (P5) in rats, and the other type migrates to the prospective hilus or tertiary dentate matrix (see Fig. 7.1c). The newborn cells generated in the hilus are added to the inner part of the GCL and form the inner shell of the GCL (Altman and Bayer 1990a,b,c; Martin et al. 2002). As postnatal time proceeds, the hilar proliferative zone is confined to the SGZ, the adult proliferative zone. Therefore, the hilar proliferative zone or the tertiary dentate matrix is the main source of granule cells during the early postnatal period.

Despite many studies on the embryonic development of the dentate gyrus (Li and Pleasure 2005) and the properties of adult progenitors (Kriegstein and Alvarez-Buylla 2009; Johnson et al. 2009; Suh et al. 2009), the exact features of the early dentate progenitor cells in the hilus were unknown until recently.

7.3.2 Properties of Progenitor Cells

A recent immunohistochemical study showed that the hilus during the first postnatal week is filled with a dense population of small cells that differ in morphology from adult types of hilar cell, and express S100β, an astrocyte marker, or Hu, a neuronal marker, but almost disappear by postnatal week 3, which suggests neural differentiation of progenitor cells in the transiently formed tertiary dentate matrix (see Fig. 7.2) (Namba et al. 2005). Analyses based on the expression of Ki67, a cell-cycle marker, and BrdU labeling have shown that a majority of proliferating cells in the early postnatal hilus express astrocytic and radial glial markers including GFAP, S100β, GLAST, and nestin (Fig. 7.3a). Other proliferating cells in the postnatal hilus express neuronal markers such as Hu and MASH-1. Furthermore, fate analysis using BrdU labeling revealed that a majority of the S100β+ astrocyte-like

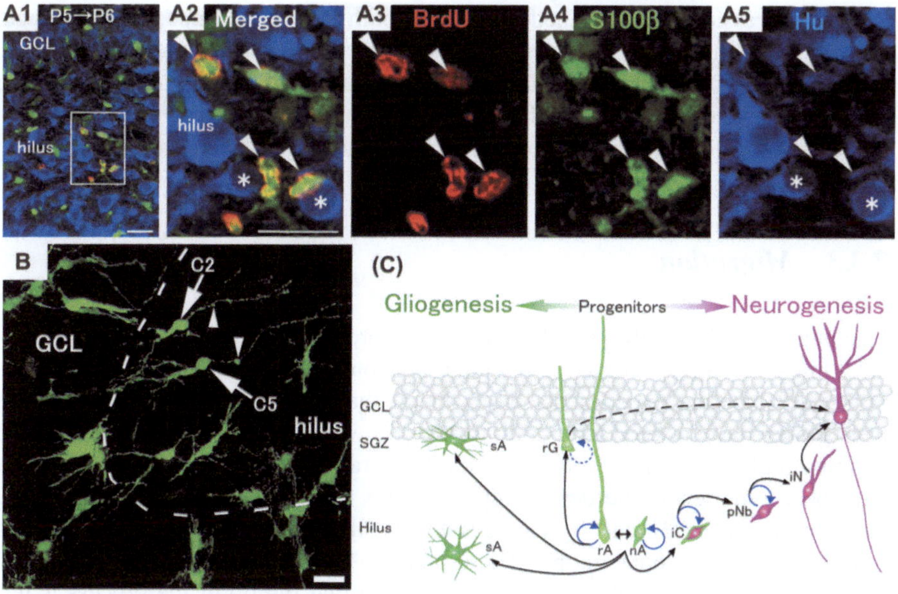

Fig. 7.3 (**a**) BrdU-labeled cells in the hilus express S100β in a rat injected with BrdU at P5 and fixed 1 day later (P6). (**b**) A retroviral vector bearing the green fluorescence protein (GFP) gene was injected into the dentate gyrus of a rat on P5; the rat was fixed on P8. (**c**) A model of postnatal neurogenesis. The hilus contains nonradial astrocyte-like cells (*nA*) and radial astrocyte-like cells (*rA*). They proliferate and produce three types of cells: granule neurons (*gN*), star-shaped astrocytes (*sA*), and radial glia-like cells (*rG*). A majority of proliferative astrocyte-like cells (nA or rA) differentiate into immature neurons (*iN*) via proliferative neuroblasts (*pNb*), which express neuronal marker (*green*), and/or transient intermediate cells (*iC*), which express both neuronal and astrocytic markers (*red*). A minority of proliferative astrocyte-like cells (nA or rA) differentiate into star-shaped astrocytes (sA) and radial glia-like cells (rG). The latter cells are assumed to become neural progenitor cells. *GCL* granule cell layer, *SGZ* subgranular zone. (Modified, with permission, from Namba et al. 2005)

hilar proliferating cells at postnatal day 5 (P5) differentiate into Hu+ immature neurons at P19. These results indicate that the majority of the GFAP+ hilar proliferating cells or tertiary germinal cells are neural progenitors that produce neuroblasts (Fig. 7.3c) (Namba et al. 2005). This finding is supported by in vitro experiments showing that GFAP-expressing cells taken from postnatal hippocampus give rise to neurons in culture (Laywell et al. 2000; Malatesta et al. 2000; Imura et al. 2003). Additionally, our recent time-lapse experiments suggest that GFAP+ primary cells are intermediate progenitor-like cells that undergo symmetrical cell division to generate a pair of immature neurons or neuronal precursors (Namba et al. 2007; Seki et al. 2009).

S100β expression in GFAP+ neural progenitor cells seems to differ between early postnatal rats and mice, and even in rats between early postnatal and adult stages. In contrast to the findings for early postnatal rats, our preliminary experiments showed that astrocytic proliferating cells in early postnatal mice exhibit no

S100β expression (unpublished data). In both rats and mice of adult age, GFAP-expressing neural progenitors are S100β negative (Filippov et al. 2003; Kempermann 2006; Seki et al. 2007). Therefore, in rats, S100β expression may be one of the traits of early postnatal GFAP+ progenitors and could be a useful molecular marker to distinguish early postnatal progenitors from adult ones.

7.3.3 Migration

Fate analysis using BrdU has also shown that proliferating cells in the early postnatal hilus eventually migrate to the inner portion of the GCL. Furthermore, retrovirus-GFP labeling experiments have shown that Hu+/GFP+ positive migrating neuroblasts have a very long trailing process with varicosities (see Fig. 7.3b) (Namba et al. 2005). Interestingly, similar long trailing processes of migrating cells have been reported in the developing neocortex (Schwartz et al. 1991; Hatanaka and Murakami 2002), and it is suggested that the long trailing processes of the migrating cells become axons. Therefore, Hu+/GFP+ hilar neuroblasts could migrate, extending a trailing process or axon so that the tip of the process is left near the CA3c pyramidal cells, which are a target of the granule cell axons.

Contact of the migrating neuroblasts with radial processes of radial glia cells in the postnatal hilus has been previously reported (Eckenhoff and Rakic 1984a; Rickmann et al. 1987b). The radial glial processes are proposed to act as a scaffold for the migrating neuroblasts (Eckenhoff and Rakic 1984a; Rickmann et al. 1987b; Forster et al. 2002). Recent retrovirus-GFP experiments show that relatively small immature neurons, with characteristic signs of moving cells, make contact with both astrocytic processes and their cell bodies (Namba et al. 2005). Although the astrocytic processes could play a role as a scaffold for the migrating cells, as considered previously, another possibility is that the processes contribute to cell-to-cell interaction between radial astrocytes and immature neurons as has been suggested in the developing neocortex (Seki and Arai 1999; Alonso 2001; Seki 2003; Ever and Gaiano 2005).

7.3.4 Molecules Regulating Postnatal Development of the GCL

As the GCL continuously develops throughout the perinatal period, the early postnatal development of the GCL shares similar molecular mechanisms with the embryonic one (see Table 7.1). Attention should be particularly paid to molecules that regulate the expansion and differentiation of progenitors, such as Shh (Han et al. 2008) and neurogenin 2 (Galichet et al. 2008), and migration such as SDF-1/CXCl12 secreted by the meninges (Bagri et al. 2002; Lu et al. 2002) and reelin by Cajal-Retzius cells (Forster et al. 2002, 2006; Sibbe et al. 2009), both of which affect the arrangement of the radial glial scaffold.

7.3.5 Future Problems

Several previous studies have described the presence of GFAP-positive proliferative glioblasts in the postnatal hilar region that are integrated into the SGZ and become radial glial cells (Eckenhoff and Rakic 1984a; Rickmann et al. 1987b; Sievers et al. 1992a; Yuasa 2001). It is likely that a small population of the hilar astrocytic proliferating cells differentiate into radial glial cells in the SGZ, although recent BrdU- and retrovirus-eGFP-labeling experiments have demonstrated that a majority of the proliferative cells similar to glioblasts give rise to neurons(Namba et al. 2005). Because radial glial cells in the adult GCL are considered to be adult neural stem cells, it is hypothesized that the radial glial cells derived from the hilar proliferative zone and integrated into the GCL become adult neural stem cells, implying that the hilar proliferative zone is a source of the adult neural progenitors as well as granule cells(Namba et al. 2005). Further experiments should be conducted to distinguish between the two types of GFAP+ cells in the early postnatal hilus: one to produce neurons and the other to become adult neural stem cells.

7.4 Adult Neurogenesis

7.4.1 Adult Neurogenic Region

Neurogenesis in the adult hippocampus occurs in the SGZ, a narrow zone between the hilus and GCL (Fig. 7.4a). The SGZ is proposed to have a special microenvironment, called a niche, which possesses the ability to maintain the properties of stem cells and regulate their differentiation. The presence of a niche in the dentate gyrus was initially indicated by transplantation experiments. When neural stem cells or progenitors from neurogenic or nonneurogenic regions are transplanted into the dentate gyrus and the transplanted cells finally settle down in the SGZ, they can differentiate into neurons (Suhonen et al. 1996; Shihabuddin et al. 2000). However, when they migrate to CA1 and CA3 PCLs, they become glial cells (Shihabuddin et al. 2000). Furthermore, when the neural stem or progenitor cells from neurogenic regions are transplanted into nonneurogenic regions such as the cerebellum and spinal cord, they only differentiate into glia (Suhonen et al. 1996; Shihabuddin et al. 2000).

The architecture of niches has been intensively investigated in the SVZ of the adult forebrain that lies adjacent to the ependymal cell layer lining the lateral ventricular wall and contains stem cells which contact the ventricle by a thin cellular protrusion (Lathia et al. 2007; Riquelme et al. 2008; Kriegstein and Alvarez-Buylla 2009; Miller and Gauthier-Fisher 2009; Suh et al. 2009). The SVZ niche is composed of stem cells, intermediate progenitors, neuroblasts, and blood vessels with specialized basal lamina called fractones (Andreu-Agullo et al. 2009).

Similarly, the SGZ niche is composed of stem cells, intermediate progenitors, neuroblasts, and blood vessels. However, the niche of the SGZ is not related to the

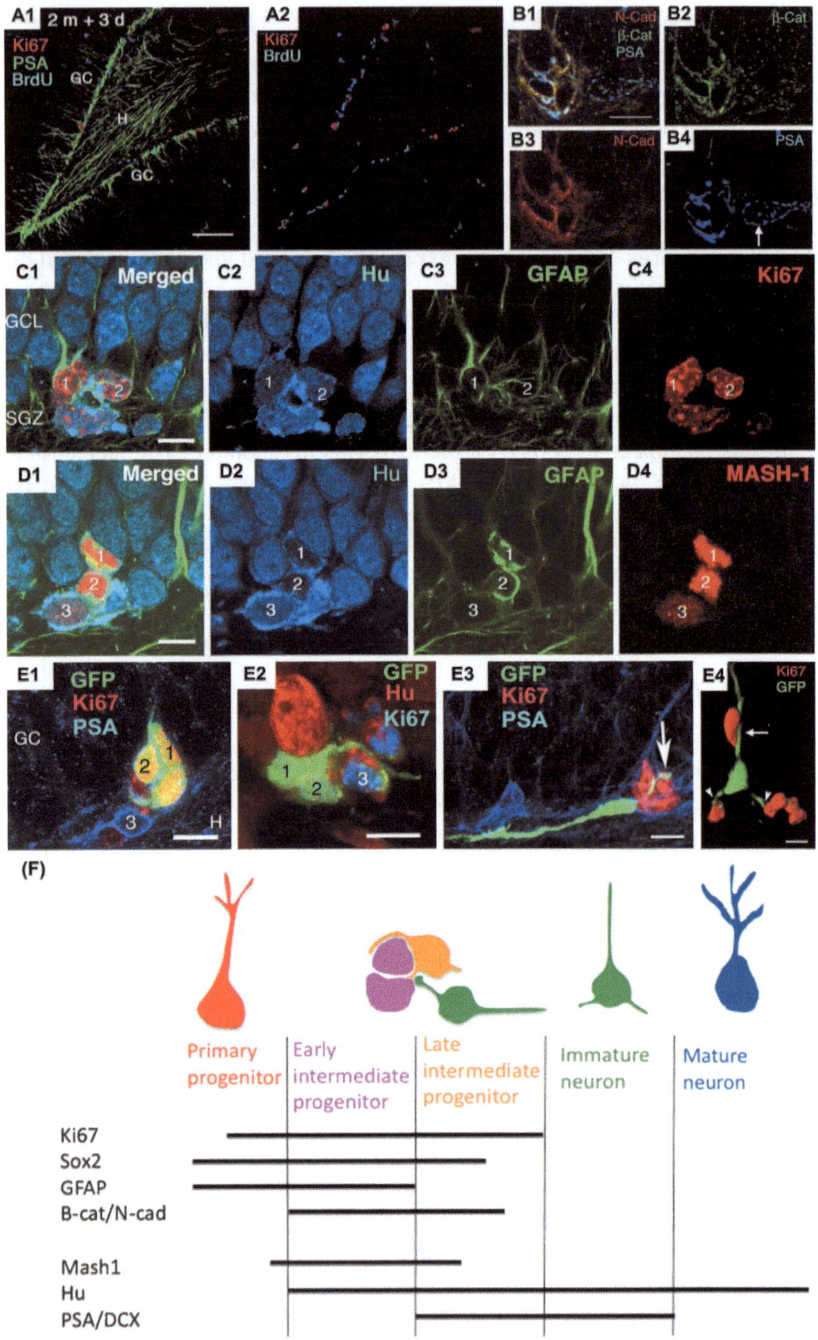

Fig. 7.4 (**a**) Polysialic acid-positive (PSA+) immature neurons, Ki67+ proliferating cells, and BrdU+ 3-day-old newborn neurons in the hippocampus of 2-month-old rats. (**b**) β-Catenin and N-cadherin exhibit identical distributions on the plasma membrane of cluster cells. Some cells also express PSA.

ventricle and does not contain ventricle-contacting stem cells. It is not known if the fractons are included in the SGZ niches. These differences imply that the SGZ niches may lack some of the properties exhibited by the SVZ niches.

The stem cells of both SVZ and SGZ are GFAP+ astrocyte-like cells (Doetsch et al. 1999; Seri et al. 2001; Kriegstein and Alvarez-Buylla 2009). In the early process of adult neurogenesis, GFAP+ primary progenitors, neuronal marker-positive intermediate progenitors, and postmitotic neuroblasts make clusters (Fig. 7.4) (Palmer et al. 2000; Seki 2002; Seri et al. 2004; Seki et al. 2007). In the clusters, the proliferating cells and GFAP+ nonproliferating cells sometimes engulf neighboring cells by their short processes (Fig. 7.4e) (Palmer et al. 2000; Seki 2002; Seri et al. 2004; Shapiro et al. 2005; Seki et al. 2007). Cellular components of the clusters appear to serve as neurogenic niches. Neural stem cells cocultured on astrocytes from the hippocampus are induced to differentiate neurons, suggesting that hippocampal astrocytes regulate the cell fate of neural stem cells (Song et al. 2002). Blood vessels sometimes run along the border between the GCL and hilus or the neurogenic SGZ. The clusters of proliferating cells make contact with blood capillaries, particularly their branching points (Palmer et al. 2000; Seki 2003), although not all the clusters are associated with blood capillaries. It has been proposed that signals from cellular components of blood vessels regulate neurogenesis in the SGZ. One of the candidates of the signals is vascular endothelial growth factor (VEGF), which induces both hippocampal neurogenesis and angiogenesis (Cao et al. 2004).

It is thought that there are signaling systems via soluble factors secreted by niche cells and via cell-to-cell contact between cellular components of the niches, both of which allow stem cells to self-renew or differentiate into neuron-committed cells. These regulatory molecules are discussed next.

7.4.2 Properties of Adult Progenitors

Stem cell-like primary progenitors of the adult dentate neurogenic regions possess astrocytic features and express GFAP (Seri et al. 2001). The primary progenitors develop gradually into neurons through certain steps. According to the models proposed by Kempermann and Alvarez-Buylla (Kempermann et al. 2004; Kriegstein and Alvarez-Buylla 2009), the primary progenitors, termed type 1 cells or B cells, exhibit radial morphology, express some astrocytic and radial glial markers, such

Fig. 7.4 (continued) (**c**) Cluster cells (*1, 2*) in the SGZ express glial fibrillary acidic protein (GFAP), Hu, and Ki67 simultaneously. (**d**) Cluster cells (*1, 2*) in the SGZ express GFAP, Hu, and Mash1 simultaneously. (**e**) Retrovirus-GFP-labeled 3-day-old cells (E1–3) and 5-day-old cells (E4). Ki67+ proliferating cells (*1, 2*) form a cluster, and a Ki67– cell is strongly positive for PSA (E1). Note the cells with short (E2), long (E3), or basal processes (E4) that are closely associated with Ki67+ proliferating cells. (**f**) Cell clustering, migration, neurite formation, and expression of specific markers in adult hippocampal neurogenesis. (Modified, with permission, from Seki et al. 2007) as GFAP, GLAST, and BLBP, and divide slowly. The next progenitors with tangentially oriented processes, called type 2–3 or type D cells, express some neuronal markers including PSA, doublecortin, and Tbr2 (Steiner et al. 2006; Seki et al. 2007; Hodge et al. 2008). They proliferate actively and serve as amplifying progenitors or intermediate progenitors. The primary and intermediate progenitors have different electrophysiological properties (Filippov et al. 2003; Fukuda et al. 2003).

In this regard, recent studies using a neuronal marker Hu and proneural protein Mash1 (Ascl1) indicate that the GFAP+ proliferating cells are not a single group but are divided into two groups: GFAP+ cells with or without neuronal marker expression (Fig. 7.4c,d,f) (Seki et al. 2007; Liu et al. 2010). In adult rat experiments with a proliferating cell marker, Ki67, about one-third of proliferating cells in the SGZ exhibit GFAP expression. The majority of GFAP+ proliferating cells also express a neuronal marker, Hu, and the half proneural protein Mash1 (Ascl1), suggesting that the majority of GFAP+ proliferative precursor cells are committed to the neuronal cell lineage. Furthermore, some of the Hu+/GFAP+ and MASH-1+/GFAP+ cells have a radial process, although GFAP+ radial cells (or radial glial cells) were previously considered to be primary precursor cells. These results suggest that at least the subpopulation of GFAP+ radial cells is already committed to the neuronal cell lineage and the identity of the GFAP+ radial cells should be reconsidered. It is assumed that the Hu+/GFAP+ and MASH-1+/GFAP+ proliferating cells could represent an intermediate state or early intermediate progenitors when proliferative GFAP+ cells differentiate into proliferative Hu+ neuronal precursor cells.

The progeny or daughter cells of GFAP+ primary progenitors can be analyzed in transgenic mice that express enhanced green fluorescent protein (eGFP) under the control of the mouse GFAP promoter (mGFAPp-eGFP Tg mice) (Suzuki et al. 2003), because eGFP persists in the progeny for a short time after the GFAP immunoreactivity disappears as a result of the high stability of exogenous eGFP compared with that of endogenous GFAP (Corish and Tyler-Smith 1999; Verkhusha et al. 2003; Ozen et al. 2007; Hodge et al. 2008; Kim et al. 2008). In the mGFAPp-eGFP Tg mice, proliferating eGFP+ cells consist of GFAP+ and GFAP– cells, both of which express a stem cell marker, Sox2 (Liu et al. 2010). The subpopulation of proliferating GFAP+/eGFP+ cells express proneural protein Mash1 and neuronal marker Hu, whereas the proliferating GFAP–/eGFP+ cells express additional immature neuronal markers, such as polysialic acid–neural cell adhesion molecule (PSA–NCAM) and doublecortin. BrdU-labeling analysis suggests that GFAP–/eGFP+ cells are progenitors that transiently appear during neuronal differentiation of GFAP+/eGFP+ cells. Taken together, these studies indicate the existence of early intermediate progenitors that express both GFAP and neuronal markers such as Hu and Mash1(Liu et al. 2010). These adult intermediate progenitors may divide symmetrically to produce neurons, such as GFAP+ progenitors in early postnatal dentate gyrus (Namba et al. 2007; Seki et al. 2009).

This model can explain interesting previously reported results that showed a transient increase in GFAP+ progenitors during adult neurogenesis in normal and voluntarily exercised mice (Steiner et al. 2004). In this respect, if the GFAP+ cells are neural stem cells that divide symmetrically to produce a stem cell and a neuron, the GFAP+ cells never decrease in number after they have increased once. Actually, it was concluded that the transient increase is caused by gliogenesis. However, if the GFAP+ cells are intermediate cells that divide symmetrically to produce a pair of neurons, they exhibit a transient increase in GFAP+ intermediate progenitors and then an increase in neurons. This latter model can explain an increase in neuron number with voluntary wheel running.

7.4.3 Migration

Although progenitor cells initially make clusters, immature neurons eventually line up along the SGZ separately, suggesting that they may migrate tangentially from clusters through the SGZ (Fig. 7.4a). In fact, BrdU/Ki67 immunohistochemistry has shown that about 80% of cell division occurs in the clusters, and within a week 75% of the cells generated inside the clusters migrate to the outside (Seki et al. 2007). Furthermore, retrovirus-GFP labeling indicates that proliferating cells inside clusters have short tangential processes, and when postmitotic cells withdraw from the proliferative site and migrate tangentially, they extend long tangential processes (Seki et al. 2007). In agreement with these results, several studies have shown short tangential processes in type 2 neuron-like progenitors (van Praag et al. 2002; Fukuda et al. 2003; Alvarez-Buylla and Lim 2004; Kempermann et al. 2004; Esposito et al. 2005; Tozuka et al. 2005). Furthermore, several reports have described long tangential processes of immature granule cells in the adult dentate GCL in normal and epileptic rats (Seress and Mrzljak 1987; Spigelman et al. 1998; Nacher et al. 2001; Dashtipour et al. 2002; Seki 2002; Jones et al. 2003; Ribak et al. 2004; Seri et al. 2004; Esposito et al. 2005; Shapiro and Ribak 2005; Seki et al. 2007), although the function of these processes remains unknown.

During the tangential migration, the migrating cells leave a tangential process behind in the clusters containing proliferating progenitors (Fig. 7.4e,f) (Seki et al. 2007). In this regard, it is worth noting that similar intercellular relationships have already been well studied in the developing neocortex (Rakic 1971; Marin-Padilla 1998; Miyata et al. 2001; Tamamaki et al. 2001; Gaiano and Fishell 2002; Fujita 2003; Rakic 2003; Tabata and Nakajima 2003; Noctor et al. 2004; Ever and Gaiano 2005), where the migrating neuroblasts make contact with the processes of radial glial cells that have been demonstrated to be neural progenitor cells (Rakic 1971; Marin-Padilla 1998; Miyata et al. 2001; Tamamaki et al. 2001; Fujita 2003; Rakic 2003; Tabata and Nakajima 2003; Noctor et al. 2004); this work suggests that embryonic and adult neurogenesis may share similar mechanisms of neuronal development.

7.4.4 Molecules Regulating Hippocampal Adult Neurogenesis

The initial stage of adult neurogenesis is thought to be regulated by signaling molecules that are expressed or secreted by cellular components of neurogenic niches (Johnson et al. 2009; Suh et al. 2009). Although the later stage of adult neurogenesis is controlled by neurotransmitters and neural activity (Johnson et al. 2009; Suh et al. 2009), this discussion focuses on the regulation of the initial stage because of limited space in this review.

Despite the difference in neurogenic niches among embryonic, early postnatal, and adult dentate gyrus, common signals are reported to contribute to the regulation of adult neurogenesis. Wnt3, a key regulator during the early morphogenesis of the hippocampus, is expressed in the SGZ of the adult hippocampus (Shimogori et al. 2004). A study using overexpression of Wnt3 and blocking of Wnt/beta-catenin signaling suggested that Wnt signaling controls neuronal differentiation of primary progenitors and the proliferation of neuron-committed progenitors (Lie et al. 2005). Furthermore, strong β-catenin expression is found to be confined to the plasma membrane of clustering cells of the SGZ in the adult hippocampus, and the distribution of the β-catenin expression on the plasma membrane exactly coincides with N-cadherin expression (Seki et al. 2007). β-Catenin is a multifunctional molecule that serves as a linker protein of cell adhesion molecules, cadherins, and a mediator of Wnt signaling pathway (Nelson and Nusse 2004). As Wnt/β-catenin signaling can control cell proliferation in adult hippocampal neural progenitors, β-catenin in the clusters could function in Wnt signaling pathway to regulate adult neurogenesis (Madsen et al. 2003; Lie et al. 2005). However, the coincidence of β-catenin and N-cadherin in the clusters suggests another possibility that β-catenin works in a signaling system mediated by N-cadherin-mediated cell-to-cell contact. Because the intercellular interaction by N-cadherin/β-catenin is reported to regulate proliferation and differentiation of neural stem cells in the developing neocortex (Noles and Chenn 2007), the N-cadherin/β-catenin in the adult hippocampus may contribute to control the proliferation of adult neural progenitors.

Shh is reported to be required for maintenance of stemness in dentate progenitor cells (Machold et al. 2003). Long-term mapping experiments show that Shh-responding cells in the GCL have the property of quiescent neural stem cells that can self-renew and produce neurons for more than a year (Ahn and Joyner 2005). The upregulation of Shh by injection of adeno-associated viral vector-Shh into the dentate gyrus leads to increases in hippocampal neural progenitors (Lai et al. 2003). The removal of Shh signaling in Shh- or smoothened-deficient mice results in a dramatic decrease in the number of dentate progenitor cells (Machold et al. 2003). Notch signaling can control proliferation, cell fate of progenitors, and dendritic morphology of newborn granule cells during the postnatal period, but in the adult hippocampus this molecule is involved in controlling dendritic development (Salama-Cohen et al. 2006). Reelin is shown to regulate migration of neuroblasts in normal and epileptic hippocampus (Gong et al. 2007; Muller et al. 2009). Prolonged seizures decrease reelin expression in the adult dentate gyrus and lead to ectopic migration and aberrant integration of newly generated granule cells (Gong et al. 2007). In agreement with the findings of in vivo experiments, reelin was found to cause detachment of chain-migrating neuroblasts in explant cultures of dentate gyrus.

Although our understanding of the molecular mechanisms underlying regulation of adult neurogenesis has recently advanced, exact information is needed as to which cells express or produce the regulatory molecules and which cells are affected by them.

7.4.5 Perspective

Finally, I would like to compare the development of the GCL with that of other brain regions (Fig. 7.5). In the development of all brain regions, initial neural stem cells are located in the VZ. In the developing neocortex, stem cells with radial morphology are situated in the VZ, and their radial processes extend throughout the entire cortical layer to the pial surface and make direct contact with the basal lamina. The stem cells produce intermediate progenitors that are detached from the ventricular surface; the intermediate progenitors proliferate in the SVZ and give rise to neurons during the embryonic period but disappear soon after birth.

In the development of the SVZ stem cells that produce olfactory interneurons, embryonic stem cells change their initial position from the VZ to the SVZ during the postnatal period, but adult stem cells maintain contact with the ventricular surface with a thin process.

In the development of the GCL of the cerebellum, ventricular stem cells migrate to the external GCL just below the pial surface and relatively close to the choroid plexus, proliferate there, then migrate further to the internal GCL and become cerebellar granule cells. The cerebellar neurogenesis ceases by postnatal week 3 in rodents.

Interestingly, the dentate gyrus seems to have some similarities in morphogenesis with the other three brain regions. For example, the hippocampus and cerebellum are derived from neuroepithelium around the edge (cortical hem and rhombic lip) of the dorsal portions of telencephalon and metencephalon, respectively, both of which are attached to the choroid plexus. Both progenitors are generated in the ventricular

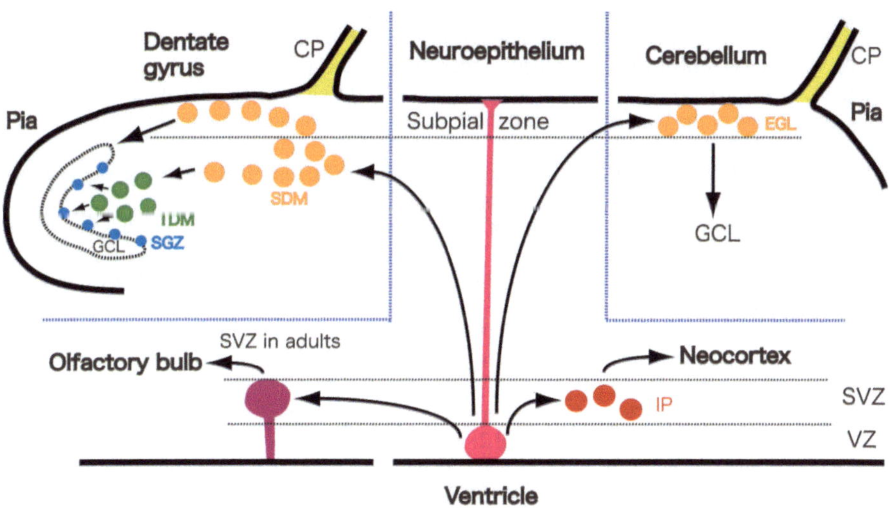

Fig. 7.5 Comparison of neurogenesis among the neocortex, dentate gyrus, subventricular zone of the forebrain, and cerebellum. *CP* choroid plexus, *EGL* external granular layer, *IP* intermediate progenitors, *GCL* granule cell layer, *SDM* secondary dentate matrix, *SGZ* subgranular zone, *SVZ* subventricular zone, *TDM* tertiary dentate matrix, *VZ* ventricular zone

zone and migrate to the subpial portion in which secondary proliferative zones (or secondary matrix) are formed. The progenitors in the secondary proliferative zone give rise to small interneurons (granule cells). Both neuronal productions in the subpial regions continue into the early postnatal period to form the cerebellar and dentate GCLs. There may be a similar mechanism in early postnatal neurogenesis in the subpial regions of the dentate and cerebellum. Furthermore, in the dentate gyrus, progenitor cells of the secondary dentate matrix migrate to the hilus and establish the tertiary dentate matrix. During the late postnatal period, the progenitors of the tertiary dentate matrix are gradually confined to the SGZ, which contains adult stem cell-like cells. Therefore, the prospective adult stem cell-like cells may migrate from the ventricular zone to the SGZ via the secondary and tertiary dentate matrix. To determine how dentate neurogenesis continues from embryonic to adult stages, further studies are needed to show how dentate progenitors are influenced by each niche including the ventricular zone, subpial region, hilus, and SGZ.

Acknowledgments I thank Dr. Takashi Namba for careful reading of the manuscript and valuable comments.

References

Ahn, S and Joyner, A L (2005) In vivo analysis of quiescent adult neural stem cells responding to Sonic hedgehog. Nature 437: 894–897

Alonso, G (2001) Proliferation of progenitor cells in the adult rat brain correlates with the presence of vimentin-expressing astrocytes. Glia 34: 253–266

Altman, J and Bayer, S A (1990a) Migration and distribution of two populations of hippocampal granule cell precursors during the perinatal and postnatal periods. J Comp Neurol 301: 365–381

Altman, J, and Bayer, S A (1990b) Mosaic organization of the hippocampal neuroepithelium and the multiple germinal sources of dentate granule cells. J Comp Neurol 301: 325–342

Altman, J and Bayer, S A (1990c) Prolonged sojourn of developing pyramidal cells in the intermediate zone of the hippocampus and their settling in the stratum pyramidale. J Comp Neurol 301: 343–364

Altman, J and Das, G D (1965) Autoradiographic and histological evidence of postnatal hippocampal neurogenesis in rats. J Comp Neurol 124: 319–335

Alvarez-Buylla, A and Lim, D A (2004) For the long run: maintaining germinal niches in the adult brain. Neuron 41: 683–686

Alvarez-Buylla, A and Nottebohm, F (1988) Migration of young neurons in adult avian brain. Nature 335: 353–354

Andreu-Agullo, C, Morante-Redolat, J M, Delgado, A C et al (2009) Vascular niche factor PEDF modulates Notch-dependent stemness in the adult subependymal zone. Nat Neurosci 12: 1514–1523

Angevine, J B, Jr. (1965) Time of neuron origin in the hippocampal region. An autoradiographic study in the mouse. Exp Neurol: Suppl 2: 1–70

Bagri, A, Gurney, T, He, X et al (2002) The chemokine SDF1 regulates migration of dentate granule cells. Development 129: 4249–4260

Barry, G, Piper, M, Lindwall, C et al (2008) Specific glial populations regulate hippocampal morphogenesis. J Neurosci 28: 12328–12340

Bayer, S A (1980) Development of the hippocampal region in the rat. I. Neurogenesis examined with 3H-thymidine autoradiography. J Comp Neurol 190: 87–114

Breunig, J J, Silbereis, J, Vaccarino, F M et al (2007) Notch regulates cell fate and dendrite morphology of newborn neurons in the postnatal dentate gyrus. Proc Natl Acad Sci USA 104: 20558–20563

Breunig, J J, Sarkisian, M R, Arellano, J I et al (2008) Primary cilia regulate hippocampal neurogenesis by mediating sonic hedgehog signaling. Proc Natl Acad Sci USA 105: 13127–13132

Cameron H A, Woolley C S, McEwen B S et al (1993) Differentiation of newly born neurons and glia in the dentate gyrus of the adult rat. Neuroscience 56: 337–344

Cao, L, Jiao, X, Zuzga, D S et al (2004) VEGF links hippocampal activity with neurogenesis, learning and memory. Nat Genet 36: 827–835

Corish, P and Tyler-Smith, C (1999) Attenuation of green fluorescent protein half-life in mammalian cells. Protein Eng 12: 1035–1040

Dashtipour, K, Yan, X X, Dinh, T T et al (2002) Quantitative and morphological analysis of dentate granule cells with recurrent basal dendrites from normal and epileptic rats. Hippocampus 12: 235–244

Davis, E E, Brueckner, M and Katsanis, N (2006) The emerging complexity of the vertebrate cilium: new functional roles for an ancient organelle. Dev Cell 11: 9–19

Doetsch, F, Caille, I, Lim, D A et al (1999) Subventricular zone astrocytes are neural stem cells in the adult mammalian brain. Cell 97: 703–716

Duan, X, Chang, J H, Ge, S et al (2007) Disrupted-in-Schizophrenia 1 regulates integration of newly generated neurons in the adult brain. Cell 130: 1146–1158

Eckenhoff, M F and Rakic, P (1984) Radial organization of the hippocampal dentate gyrus: a Golgi, ultrastructural, and immunocytochemical analysis in the developing rhesus monkey. J Comp Neurol 223: 1–21

Enikolopov, G and Overstreet-Wadiche, L (2008) The use of reporter mouse lines to study adult neurogenesis. In: Gage, F. H. et al. (eds.), Adult neurogenesis. Cold Spring Harbor Laboratory Press, New York

Enomoto, A, Asai, N, Namba, T et al (2009) Roles of disrupted-in-schizophrenia 1-interacting protein girdin in postnatal development of the dentate gyrus. Neuron 63: 774–787

Esposito, M S, Piatti, V C, Laplagne, D A et al (2005) Neuronal differentiation in the adult hippocampus recapitulates embryonic development. J Neurosci 25: 10074–10086

Ever, L and Gaiano, N (2005) Radial 'glial' progenitors: neurogenesis and signaling. Curr Opin Neurobiol 15: 29–33

Filippov, V, Kronenberg, G, Pivneva, T et al (2003) Subpopulation of nestin-expressing progenitor cells in the adult murine hippocampus shows electrophysiological and morphological characteristics of astrocytes. Mol Cell Neurosci 23: 373–382

Forster, E, Tielsch, A, Saum, B et al (2002) Reelin, disabled 1, and beta 1 integrins are required for the formation of the radial glial scaffold in the hippocampus. Proc Natl Acad Sci USA 99: 13178–13183

Forster, E, Jossin, Y, Zhao, S et al (2006) Recent progress in understanding the role of Reelin in radial neuronal migration, with specific emphasis on the dentate gyrus. Eur J Neurosci 23: 901–909

Fujita, S (2003) The discovery of the matrix cell, the identification of the multipotent neural stem cell and the development of the central nervous system. Cell Struct Funct 28: 205–228

Fukuda, S, Kato, F, Tozuka, Y et al (2003) Two distinct subpopulations of nestin-positive cells in adult mouse dentate gyrus. J Neurosci 23: 9357–9366

Gaiano, N and Fishell, G (2002) The role of notch in promoting glial and neural stem cell fates. Annu Rev Neurosci 25: 471–490

Galceran, J, Miyashita-Lin, E M, Devaney, E et al (2000) Hippocampus development and generation of dentate gyrus granule cells is regulated by LEF1. Development 127: 469–482

Galichet, C, Guillemot, F, Parras, C M (2008) Neurogenin 2 has an essential role in development of the dentate gyrus. Development 135: 2031–2041

Ganat, Y M, Silbereis, J, Cave, C et al (2006) Early postnatal astroglial cells produce multilineage precursors and neural stem cells in vivo. J Neurosci 26: 8609–8621

Gao, Z, Ure, K, Ables, J L et al (2009) Neurod1 is essential for the survival and maturation of adult-born neurons. Nat Neurosci 12: 1090–1092

Garcia, A D, Doan, N B, Imura, T et al (2004) GFAP-expressing progenitors are the principal source of constitutive neurogenesis in adult mouse forebrain. Nat Neurosci 7: 1233–1241
Goldman, S A and Nottbohm, F (1983) Neuronal production, migration, and differentiation in a vocal control nucleus of the adult female canary brain. Proc Natl Acad Sci USA 80: 2390–2394
Gong, C, Wang, T W, Huang, H S et al (2007) Reelin regulates neuronal progenitor migration in intact and epileptic hippocampus. J Neurosci 27: 1803–1811
Gould, E and Gross, C G (2002) Neurogenesis in adult mammals: some progress and problems. J Neurosci 22: 619–623
Gould, E, Cameron, H A, Daniels, D C et al (1992) Adrenal hormones suppress cell division in the adult rat dentate gyrus. J Neurosci 12: 3642–3650
Gross, C G (2000) Neurogenesis in the adult brain: death of a dogma. Nat Rev Neurosci 1: 67–73
Han, Y G, Spassky, N, Romaguera-Ros, M et al (2008) Hedgehog signaling and primary cilia are required for the formation of adult neural stem cells. Nat Neurosci 11: 277–284
Hartmann, D, Sievers, J, Pehlemann, F W et al (1992) Destruction of meningeal cells over the medial cerebral hemisphere of newborn hamsters prevents the formation of the infrapyramidal blade of the dentate gyrus. J Comp Neurol 320: 33–61
Hatanaka, Y and Murakami, F (2002) In vitro analysis of the origin, migratory behavior, and maturation of cortical pyramidal cells. J Comp Neurol 454: 1–14
Hebert, J M and Fishell, G (2008) The genetics of early telencephalon patterning: some assembly required. Nat Rev Neurosci 9: 678–685
Hebert, J M, Mishina, Y and McConnell, S K (2002) BMP signaling is required locally to pattern the dorsal telencephalic midline. Neuron 35: 1029–1041
Hodge, R D, Kowalczyk, T D, Wolf, S A et al (2008) Intermediate progenitors in adult hippocampal neurogenesis: Tbr2 expression and coordinate regulation of neuronal output. J Neurosci 28: 3707–3717
Hong, S M, Liu, Z, Fan, Y et al (2007) Reduced hippocampal neurogenesis and skill reaching performance in adult Emx1 mutant mice. Exp Neurol 206: 24–32
Imura, T, Kornblum, H I and Sofroniew, M V (2003) The predominant neural stem cell isolated from postnatal and adult forebrain but not early embryonic forebrain expresses GFAP. J Neurosci 23: 2824–2832
Jacobson, M (1991) Developmental neurobiology. Plenum Press, New York
Jessberger, S, Toni, N, Clemenson, G D, Jr. et al (2008) Directed differentiation of hippocampal stem/progenitor cells in the adult brain. Nat Neurosci 11: 888–893
Johnson, M A, Ables, J L and Eisch, A J (2009) Cell-intrinsic signals that regulate adult neurogenesis in vivo: insights from inducible approaches. BMB Rep 42: 245–259
Jones, S P, Rahimi, O, O'Boyle, M P et al (2003) Maturation of granule cell dendrites after mossy fiber arrival in hippocampal field CA3. Hippocampus 13: 413–427
Kaplan, M S (2001) Environment complexity stimulates visual cortex neurogenesis: death of a dogma and a research career. Trends Neurosci 24: 617–620
Kaplan, M S and Hinds, J W (1977) Neurogenesis in the adult rat: electron microscopic analysis of light radioautographs. Science 197: 1092–1094
Kempermann, G (2006) Adult neurogenesis. Oxford University Press, New York
Kempermann, G, Jessberger, S, Steiner, B et al (2004) Milestones of neuronal development in the adult hippocampus. Trends Neurosci 27: 447–452
Kim, E J, Leung, C T, Reed, R R et al (2007) In vivo analysis of Ascl1 defined progenitors reveals distinct developmental dynamics during adult neurogenesis and gliogenesis. J Neurosci 27: 12764–12774
Kim, E J, Battiste, J, Nakagawa, Y et al (2008) Ascl1 (Mash1) lineage cells contribute to discrete cell populations in CNS architecture. Mol Cell Neurosci 38: 595–606
Kim, J Y, Duan, X, Liu, C Y et al (2009) DISC1 regulates new neuron development in the adult brain via modulation of AKT-mTOR signaling through KIAA1212. Neuron 63: 761–773
Kolodziej, A, Schulz, S, Guyon, A et al (2008) Tonic activation of CXC chemokine receptor 4 in immature granule cells supports neurogenesis in the adult dentate gyrus. J Neurosci 28: 4488–4500

Kriegstein, A and Alvarez-Buylla, A (2009) The glial nature of embryonic and adult neural stem cells. Annu Rev Neurosci 32: 149–184

Kriegstein, A R and Noctor, S C (2004) Patterns of neuronal migration in the embryonic cortex. Trends Neurosci 27: 392–399

Kuhn, H G, Dickinson-Anson, H and Gage, F H (1996) Neurogenesis in the dentate gyrus of the adult rat: age-related decrease of neuronal progenitor proliferation. J Neurosci 16: 2027–2033

Lai, K, Kaspar, B K, Gage, F H et al (2003) Sonic hedgehog regulates adult neural progenitor proliferation in vitro and in vivo. Nat Neurosci 6: 21–27

Lathia, J D, Rao, M S, Mattson, M P et al (2007) The microenvironment of the embryonic neural stem cell: lessons from adult niches? Dev Dyn 236: 3267–3282

Laywell, E D, Rakic, P, Kukekov, V G et al (2000) Identification of a multipotent astrocytic stem cell in the immature and adult mouse brain. Proc Natl Acad Sci USA 97: 13883–13888

Lee, S M, Tole, S, Grove, E et al (2000) A local Wnt-3a signal is required for development of the mammalian hippocampus. Development 127: 457–467

Li, G and Pleasure, S J (2005) Morphogenesis of the dentate gyrus: what we are learning from mouse mutants. Dev Neurosci 27: 93–99

Li, G and Pleasure, S J (2007) Genetic regulation of dentate gyrus morphogenesis. Prog Brain Res 163: 143–152

Li, G, Kataoka, H, Coughlin, S R et al (2009) Identification of a transient subpial neurogenic zone in the developing dentate gyrus and its regulation by Cxcl12 and reelin signaling. Development 136: 327–335

Lie, D C, Colamarino, S A, Song, H J et al (2005) Wnt signalling regulates adult hippocampal neurogenesis. Nature 437: 1370–1375

Liu, M, Pleasure, S J, Collins, A E et al (2000) Loss of BETA2/NeuroD leads to malformation of the dentate gyrus and epilepsy. Proc Natl Acad Sci USA 97: 865–870

Liu, Y, Namba, T, Liu, J et al (2010) Glial fibrillary acidic protein-expressing neural progenitors give rise to immature neurons via early intermediate progenitors expressing both glial fibrillary acidic protein and neuronal markers in the adult hippocampus. Neuroscience 166: 241–251

Lu, M, Grove, E A and Miller, R J (2002) Abnormal development of the hippocampal dentate gyrus in mice lacking the CXCR4 chemokine receptor. Proc Natl Acad Sci USA 99: 7090–7095

Machold, R, Hayashi, S, Rutlin, M et al (2003) Sonic hedgehog is required for progenitor cell maintenance in telencephalic stem cell niches. Neuron 39: 937–950

Machon, O, van den Bout, C J, Backman, M et al (2003) Role of beta-catenin in the developing cortical and hippocampal neuroepithelium. Neuroscience 122: 129–143

Machon, O, Backman, M, Machonova, O et al (2007) A dynamic gradient of Wnt signaling controls initiation of neurogenesis in the mammalian cortex and cellular specification in the hippocampus. Dev Biol 311: 223–237

Madsen, T M, Newton, S S, Eaton, M E et al (2003) Chronic electroconvulsive seizure up-regulates beta-catenin expression in rat hippocampus: role in adult neurogenesis. Biol Psychiatry 54: 1006–1014

Malatesta, P, Hartfuss, E and Gotz, M (2000) Isolation of radial glial cells by fluorescent-activated cell sorting reveals a neuronal lineage. Development 127: 5253–5263

Mangale, V S, Hirokawa, K E, Satyaki, P R et al (2008) Lhx2 selector activity specifies cortical identity and suppresses hippocampal organizer fate. Science 319: 304–309

Marin-Padilla, M (1998) Cajal-Retzius cells and the development of the neocortex. Trends Neurosci 21: 64–71

Martin, L A, Tan, S S and Goldowitz, D (2002) Clonal architecture of the mouse hippocampus. J Neurosci 22: 3520–3530

Meyer, K D and Morris, J A (2009) Disc1 regulates granule cell migration in the developing hippocampus. Hum Mol Genet 18: 3286–3297

Miller, F D and Gauthier-Fisher, A (2009) Home at last: neural stem cell niches defined. Cell Stem Cell 4: 507–510

Miller, M W and Nowakowski, R S (1988) Use of bromodeoxyuridine-immunohistochemistry to examine the proliferation, migration and time of origin of cells in the central nervous system. Brain Res 457: 44–52

Miyata, T, Maeda, T and Lee, J E (1999) NeuroD is required for differentiation of the granule cells in the cerebellum and hippocampus. Genes Dev 13: 1647–1652

Miyata, T, Kawaguchi, A, Okano, H et al (2001) Asymmetric inheritance of radial glial fibers by cortical neurons. Neuron 31: 727–741

Muller, M C, Osswald, M, Tinnes, S et al (2009) Exogenous reelin prevents granule cell dispersion in experimental epilepsy. Exp Neurol 216: 390–397

Nacher, J, Crespo, C and McEwen, B S (2001) Doublecortin expression in the adult rat telencephalon. Eur J Neurosci 14: 629–644

Nakahira, E and Yuasa, S (2005) Neuronal generation, migration, and differentiation in the mouse hippocampal primoridium as revealed by enhanced green fluorescent protein gene transfer by means of in utero electroporation. J Comp Neurol 483: 329–340

Namba, T, Mochizuki, H, Onodera, M et al (2005) The fate of neural progenitor cells expressing astrocytic and radial glial markers in the postnatal rat dentate gyrus. Eur J Neurosci 22: 1928–1941

Namba, T, Namiki, H and Seki, T (2007) Direct evidence for neuronal differentiation of astrocyte-like progenitors in the postnatal hippocampus. Neurosci Res 58 Suppl 1: S209

Nelson, W J and Nusse, R (2004) Convergence of Wnt, beta-catenin, and cadherin pathways. Science 303: 1483–1487

Noctor, S C, Martinez-Cerdeno, V, Ivic, L et al (2004) Cortical neurons arise in symmetric and asymmetric division zones and migrate through specific phases. Nat Neurosci 7: 136–144

Noles, S R and Chenn, A (2007) Cadherin inhibition of beta-catenin signaling regulates the proliferation and differentiation of neural precursor cells. Mol Cell Neurosci 35: 549–558

O'Leary, D D and Sahara, S (2008) Genetic regulation of arealization of the neocortex. Curr Opin Neurobiol 18: 90–100

O'Leary, D D, Chou, S J and Sahara, S (2007) Area patterning of the mammalian cortex. Neuron 56: 252–269

Oldekamp, J, Kraemer, N, Alvarez-Bolado, G et al (2004) bHLH gene expression in the Emx2-deficient dentate gyrus reveals defective granule cells and absence of migrating precursors. Cereb Cortex 14: 1045–1058

Ozen, I, Galichet, C, Watts, C et al (2007) Proliferating neuronal progenitors in the postnatal hippocampus transiently express the proneural gene Ngn2. Eur J Neurosci 25: 2591–2603

Palmer, T D, Willhoite, A R and Gage, F H (2000) Vascular niche for adult hippocampal neurogenesis. J Comp Neurol 425: 479–494

Pellegrini, M, Mansouri, A, Simeone, A et al (1996) Dentate gyrus formation requires Emx2. Development 122: 3893–3898

Rakic, P (1971) Guidance of neurons migrating to the fetal monkey neocortex. Brain Res 33: 471–476

Rakic, P (2003) Developmental and evolutionary adaptations of cortical radial glia. Cereb Cortex 13: 541–549

Rakic, P (2007) The radial edifice of cortical architecture: from neuronal silhouettes to genetic engineering. Brain Res Rev 55: 204–219

Ribak, C E, Korn, M J, Shan, Z et al (2004) Dendritic growth cones and recurrent basal dendrites are typical features of newly generated dentate granule cells in the adult hippocampus. Brain Res 1000: 195–199

Rickmann, M, Amaral, D G and Cowan, W M (1987) Organization of radial glial cells during the development of the rat dentate gyrus. J Comp Neurol 264: 449–479

Riquelme, P A, Drapeau, E and Doetsch, F (2008) Brain micro-ecologies: neural stem cell niches in the adult mammalian brain. Philos Trans R Soc Lond B Biol Sci 363: 123–137

Roelink, H (2000) Hippocampus formation: an intriguing collaboration. Curr Biol 10: R279–R281

Rousselot, P, Lois, C and Alvarez-Buylla, A D (1995) Embryonic (PSA) N-CAM reveals chains of migrating neuroblasts between the lateral ventricle and the olfactory bulb of adult mice. J Comp Neurol 351: 51–61

Salama-Cohen, P, Arevalo, M A, Grantyn, R et al (2006) Notch and NGF/p75NTR control dendrite morphology and the balance of excitatory/inhibitory synaptic input to hippocampal neurones through Neurogenin 3. J Neurochem 97: 1269–1278

Schlessinger, A R, Cowan, W M and Gottlieb, D I (1975) An autoradiographic study of the time of origin and the pattern of granule cell migration in the dentate gyrus of the rat. J Comp Neurol 159: 149–175

Schwab, M H, Bartholomae, A, Heimrich, B et al (2000) Neuronal basic helix-loop-helix proteins (NEX and BETA2/Neuro D) regulate terminal granule cell differentiation in the hippocampus. J Neurosci 20: 3714–3724

Schwartz, M L, Rakic, P and Goldman-Rakic, P S (1991) Early phenotype expression of cortical neurons: evidence that a subclass of migrating neurons have callosal axons. Proc Natl Acad Sci USA 88: 1354–1358

Seki, T (2002) Hippocampal adult neurogenesis occurs in a microenvironment provided by PSA-NCAM-expressing immature neurons. J Neurosci Res 69: 772–783

Seki, T (2003) Microenvironmental elements supporting adult hippocampal neurogenesis. Anat Sci Int 78: 69–78

Seki, T and Arai, Y (1991) The persistent expression of a highly polysialylated NCAM in the dentate gyrus of the adult rat. Neurosci Res 12: 503–513

Seki, T and Arai, Y (1993a) Distribution and possible roles of the highly polysialylated neural cell adhesion molecule (NCAM-H) in the developing and adult central nervous system. Neurosci Res 17: 265–290

Seki, T and Arai, Y (1993b) Highly polysialylated neural cell adhesion molecule (NCAM-H) is expressed by newly generated granule cells in the dentate gyrus of the adult rat. J Neurosci 13: 2351–2358

Seki, T and Arai, Y (1995) Age-related production of new granule cells in the adult dentate gyrus. Neuroreport 6: 2479–2482

Seki, T and Arai, Y (1999) Temporal and spacial relationships between PSA-NCAM-expressing, newly generated granule cells, and radial glia-like cells in the adult dentate gyrus. J Comp Neurol 410: 503–513

Seki, T, Namba, T, Mochizuki, H et al (2007) Clustering, migration, and neurite formation of neural precursor cells in the adult rat hippocampus. J Comp Neurol 502: 275–290

Seki, T, Namba, T, Liu, Y et al (2009) How do GFAP-expressing neural progenitors divide and generate neuron-committed progeny in the postnatal hippocampus? Neuroscience Meeting Planner. Chicago, IL: Society for Neuroscience Program No. 506.14

Seress, L and Mrzljak, L (1987) Basal dendrites of granule cells are normal features of the fetal and adult dentate gyrus of both monkey and human hippocampal formations. Brain Res 405: 169–174

Seri, B, Garcia-Verdugo, J M, McEwen, B S et al (2001) Astrocytes give rise to new neurons in the adult mammalian hippocampus. J Neurosci 21: 7153–7160

Seri, B, Garcia-Verdugo, J M, Collado-Morente, L et al (2004) Cell types, lineage, and architecture of the germinal zone in the adult dentate gyrus. J Comp Neurol 478: 359–378

Shapiro, L A and Ribak, C E (2005) Integration of newly born dentate granule cells into adult brains: hypotheses based on normal and epileptic rodents. Brain Res Brain Res Rev 48: 43–56

Shapiro, L A, Korn, M J, Shan, Z et al (2005) GFAP-expressing radial glia-like cell bodies are involved in a one-to-one relationship with doublecortin-immunolabeled newborn neurons in the adult dentate gyrus. Brain Res 1040: 81–91

Shihabuddin, L S, Horner, P J, Ray, J et al (2000) Adult spinal cord stem cells generate neurons after transplantation in the adult dentate gyrus. J Neurosci 20: 8727–8735

Shimogori, T, VanSant, J, Paik, E et al (2004) Members of the Wnt, Fz, and Frp gene families expressed in postnatal mouse cerebral cortex. J Comp Neurol 473: 496–510

Shinozaki, K, Yoshida, M, Nakamura, M et al (2004) Emx1 and Emx2 cooperate in initial phase of archipallium development. Mech Dev 121: 475–489

Sibbe, M, Forster, E, Basak, O et al (2009) Reelin and Notch1 cooperate in the development of the dentate gyrus. J Neurosci 29: 8578–8585
Sievers, J, Hartmann, D, Pehlemann, F W et al (1992) Development of astroglial cells in the proliferative matrices, the granule cell layer, and the hippocampal fissure of the hamster dentate gyrus. J Comp Neurol 320: 1–32
Solberg, N, Machon, O and Krauss, S (2008) Effect of canonical Wnt inhibition in the neurogenic cortex, hippocampus, and premigratory dentate gyrus progenitor pool. Dev Dyn 237: 1799–1811
Song, H, Stevens, C F and Gage, F H (2002) Astroglia induce neurogenesis from adult neural stem cells. Nature 417: 39–44
Spigelman, I, Yan, X X, Obenaus, A et al (1998) Dentate granule cells form novel basal dendrites in a rat model of temporal lobe epilepsy. Neuroscience 86: 109–120
Steiner, B, Kronenberg, G, Jessberger, S et al (2004) Differential regulation of gliogenesis in the context of adult hippocampal neurogenesis in mice. Glia 46: 41–52
Steiner, B, Klempin, F, Wang, L et al (2006) Type-2 cells as link between glial and neuronal lineage in adult hippocampal neurogenesis. Glia 54: 805–814
Stensaas, L J (1967) The development of hippocampal and dorsolateral pallial region of the cerebral hemisphere in fetal rabbits. V. Sixty millimeter stage, glial cell morphology. J Comp Neurol 131: 423–436
Subramanian, L and Tole, S (2009) Mechanisms underlying the specification, positional regulation, and function of the cortical hem. Cereb Cortex 19 Suppl 1: i90–i95
Subramanian, L, Remedios, R, Shetty, A et al (2009) Signals from the edges: the cortical hem and antihem in telencephalic development. Semin Cell Dev Biol 20: 712–718
Suh, H, Deng, W and Gage, F H (2009) Signaling in adult neurogenesis. Annu Rev Cell Dev Biol 25: 253–275
Suhonen, J O, Peterson, D A, Ray, J et al (1996) Differentiation of adult hippocampus-derived progenitors into olfactory neurons in vivo. Nature 383: 624–627
Suzuki, R, Watanabe, J, Arata, S et al (2003) A transgenic mouse model for the detailed morphological study of astrocytes. Neurosci Res 47: 451–454
Tabata, H and Nakajima, K (2003) Multipolar migration: the third mode of radial neuronal migration in the developing cerebral cortex. J Neurosci 23: 9996–10001
Tamamaki, N, Nakamura, K, Okamoto, K et al (2001) Radial glia is a progenitor of neocortical neurons in the developing cerebral cortex. Neurosci Res 41: 51–60
Theil, T, Aydin, S, Koch, S et al (2002) Wnt and Bmp signalling cooperatively regulate graded Emx2 expression in the dorsal telencephalon. Development 129: 3045–3054
Tozuka, Y, Fukuda, S, Namba, T et al (2005) GABAergic excitation promotes neuronal differentiation in adult hippocampal progenitor cells. Neuron 47: 803–815
van Praag, H, Schinder, A F, Christie, B R et al (2002) Functional neurogenesis in the adult hippocampus. Nature 415: 1030–1034
Verkhusha, V V, Kuznetsova, I M, Stepanenko, O V et al (2003) High stability of Discosoma DsRed as compared to Aequorea EGFP. Biochemistry 42: 7879–7884
Yamaguchi, M, Saito, H, Suzuki, M et al (2000) Visualization of neurogenesis in the central nervous system using nestin promoter-GFP transgenic mice [In Process Citation]. Neuroreport 11: 1991–1996
Yoshida, M, Suda, Y, Matsuo, I et al (1997) Emx1 and Emx2 functions in development of dorsal telencephalon. Development 124: 101–111
Yuasa, S (2001) Development of astrocytes in the mouse hippocampus as tracked by tenascin-C gene expression. Arch Histol Cytol 64: 149–158
Zhao, C (2007) Retrovirus-mediated cell labeling. In: Gage, F. H. et al. (eds.), Adult neurogenesis. Cold Spring Harbor Laboratory Press, New York
Zhao, T, Kraemer, N, Oldekamp, J et al (2006) Emx2 in the developing hippocampal fissure region. Eur J Neurosci 23: 2895–2907
Zhou, C J, Zhao, C and Pleasure, S J (2004) Wnt signaling mutants have decreased dentate granule cell production and radial glial scaffolding abnormalities. J Neurosci 24: 121–126

Chapter 8
Activity-Dependent Regulation of the Early Phase of Adult Hippocampal Neurogenesis

Tatsuhiro Hisatsune, Yoko Ide, and Rokuya Nochi

Abstract New neurons are generated from neural stem cells through the neuronal progenitor cell stage at the dentate gyrus of the hippocampal formation throughout life, and contribute to higher cognitive function, such as learning and memory. Physiological conditions, such as voluntary exercise or enriched environment, or pathophysiological conditions, such as stroke or epilepsy, upregulate the generation of new neurons, but the mechanism for governing the promotion of adult neurogenesis has not been fully clarified. It has been widely accepted that the neural activities of the hippocampal circuitry regulate the generation of hippocampal new neurons. Actually, a number of neurotransmitter systems – serotoninergic, GABAergic, cholinergic, purinergic, and glutamatergic systems – are involved in the regulation of adult hippocampal neurogenesis process. In this chapter, we propose a possible involvement of activity-dependent regulation of the fate of neural stem/progenitor cells in enhanced hippocampal neurogenesis. In a series of studies, we and others have reported how hippocampal activity affects the generation of new neurons, even at the early phase of adult hippocampal neurogenesis. These findings may also contribute to the pharmaceutical enhancement of adult neurogenesis for treating neurodegenerative and psychological disorders.

8.1 Overview

New neurons are continuously generated in the hippocampal circuit throughout life (Altman and Das, 1965, 1967; Kaplan and Hinds, 1977; Kaplan and Bell, 1984; Gould et al., 1998; Eriksson et al., 1998; Kornack and Rakic, 1999), integrated into hippocampal circuit (van Praag et al., 2002; Toni et al., 2007), and contribute to cognitive functions of nervous system, such as learning and memory (Shors et al., 2001; Saxe et al., 2006; Zhang et al., 2008; Dupret et al., 2008).

T. Hisatsune (✉), Y. Ide, and R. Nochi
Department of Integrated Biosciences, The University of Tokyo, Kashiwa, Chiba 277-8562, Japan
e-mail: hisatsune@k.u-tokyo.ac.jp

T. Seki et al. (eds.), *Neurogenesis in the Adult Brain I: Neurobiology*,
DOI 10.1007/978-4-431-53933-9_8, © Springer 2011

At the subgranular zone of the dentate gyrus, new neurons develop from the neural stem cell through the neuronal progenitor cell stage (Seri et al., 2001; Fukuda et al., 2003; Filippov et al., 2003; Kempermann et al., 2004). A number of physiological and pathological conditions, including voluntary exercise (van Praag et al., 1999; Fabel et al., 2003; Pereira et al., 2007) and ischemic stroke (Jin et al., 2001; Yoshimura et al., 2001; Tonchev et al., 2003; Koketsu et al., 2006), upregulate the generation of new neurons. Although the mechanism for governing the promotion of adult neurogenesis has not yet been clarified, the notion that enhanced neurogenesis is mediated through the regulation of the process of neurogenesis by neural activity is widely accepted (Cameron et al., 1995; Deisseroth et al., 2004; Tozuka et al., 2005; Bruel-Jungerman et al., 2006; Chun et al., 2006).

In this chapter, we focus on how activity controls the process of adult neurogenesis from neural stem cells, to neuronal progenitor cells, to early postmitotic immature new neurons. Actually, a number of neurotransmitter systems, γ-aminobutyric acid (GABA)-ergic (Tozuka et al., 2005; Ge et al., 2006), serotoninergic (Santarelli et al., 2003), cholinergic (Mohapel et al., 2005; Kaneko et al., 2006; Itou et al., 2011), and glutamatergic (Cameron et al., 1995), are involved in the regulation of the early phase of adult hippocampal neurogenesis. These findings may also contribute to the pharmaceutical enhancement of adult neurogenesis for treating neurodegenerative and psychological disorders.

8.2 Hippocampal Circuits in the Dentate Gyrus

The dentate gyrus is a hippocampal region where new neurons are continuously generated from neural stem cells. After neuronal maturation, new neurons are integrated as young granule cells at the granular layer of the dentate gyrus. As tri-synaptic transmissions in the hippocampal circuit, granule cells receive inputs from the entorhinal cortex as the perforant pathway, and extend axonal projections as mossy fibers toward the CA3 pyramidal neurons (Fig. 8.1). CA3 pyramidal neurons extend axonal projections as Shaffer collateral toward the CA1 pyramidal neurons. Then, CA1 pyramidal neurons send a neural signal back to neurons at the entorhinal cortex. At the dentate gyrus, both the excitatory interneurons (mossy cells) and the inhibitory interneurons (for example, basket cells) are interconnected with the tri-synaptic hippocampal network, and these interneurons may have direct interactions with neural stem cells and neuronal progenitor cells.

Mossy cells receive inputs basically from mossy fibers of granule cells. Some mossy cells extend their dendrites toward the outer molecular layer and may receive inputs from the entorhinal cortex via the perforant pathway. Mossy cells extend their axons widely to the inner molecular layer of the dentate gyrus even toward the contralateral side of the hippocampus, in addition to the ipsilateral side. As neural stem cells extend their cellular processes toward the inner molecular layers, therefore, it would be highly likely that neural stem cells could contact with the axonal fiber from mossy cells as a source for the glutamatergic neurotransmitter system.

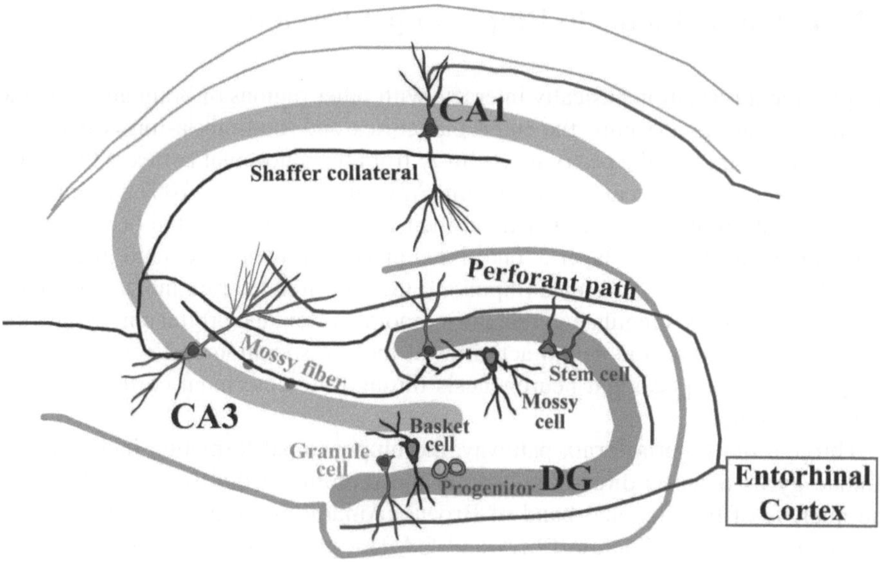

Fig. 8.1 The hippocampal circuit comprises tri-synaptic transmissions in the dentate gyrus (*DG*), *CA3*, and *CA1* regions. In addition to the feed-forward tri-synaptic pathway, there might be a feedback pathway from CA3 to mossy cells

In the dentate gyrus, at least six types of GABAergic interneurons are involved: axo-axonic cells, basket 1 cells, basket 2 cells, hilar commissural-associational pathway-related (HICAP) cells, hilar perforant path-associated (HIPP) cells, and molecular layer perforant path-associated (MOPP) cells contribute to the neural activity in the dentate gyrus (Freund and Buzsáki, 1996). In detail, axo-axonic cells and basket 1 cells, which are parvalbumin (PV) positive, locate in all layers of the dentate gyrus, and innervate both granule cell layer and hilus. Basket 2 cells, which are cholecystokinin (CCK) positive, locate within or just below the granule cell layer bordering the hilus and are involved in local circuits. They receive inputs from both the perforant path and from the local mossy fiber collaterals, whereas their axons form synapses with granule cells in the granule cell layer. HICAP cells receive both symmetrical and asymmetrical synapses from all layers, and their axons mainly spread just above the granular layer. Therefore, they are thought to be involved in both feed-forward and feed-back circuits. HIPP cells have dendritic trees limited to the hilus that perfectly coincide with mossy fiber collaterals and extend axons to the outer two-thirds of the molecular layer. Thus, HIPP cells may mediate feedback inhibition of granule cells. Finally, MOPP cells make axonal and dendritic trees only to the outer two-thirds of the dentate molecular layer, although they span widely in transverse length. This interneuron type is likely to be driven in a feed-forward manner, largely by entorhinal afferents. In a previous study, we found that neuronal progenitor cells have direct neural inputs from a hippocampal GABAergic network (Tozuka et al., 2005).

8.3 Extrinsic Inputs to Hippocampal Circuits

Hippocampal formation basically interacts with other regions of brain areas via the two major gateway systems: the entorhinal cortex and the fimbria-fornix pathway. As already described, the perforant pathway from the entorhinal cortex is the major input gateway for hippocampal formation (Amaral and Lavenex, 2007). From layer 2, entorhinal projection neurons extend their axons toward both the dentate gyrus and the CA3 areas. From layer 3, entorhinal projection neurons extend their axons toward the CA1 areas. Afferents (inputs) to the entorhinal cortex include the association cortex of the cingulate gyrus, the temporal cortex, the orbital cortex, and the amygdala. Therefore, the neural activities at these association cortices may affect the process of hippocampal neurogenesis through the entorhinal cortex-perforant pathway system.

Through the fimbria-fornix pathway, the hippocampal formation including the dentate gyrus receives direct inputs from the septum (medial septal nucleus and the nucleus of the diagonal band of Broca) (Amaral and Lavenex, 2007). From the septum, both cholinergic fibers and GABAergic fibers penetrate toward the granule cell layers. Therefore, it would be highly likely that the neural activities of the septum may involve the regulation of hippocampal neurogenesis through the direct influences against neural stem cells or neuronal progenitor cells. The fimbria-fornix pathway also sends the glutamatergic inputs to hippocampal formations from the contralateral hippocampus and the hypothalamus (the supramammillary area), as another glutamatergic input other than the entorhinal cortex-perforant pathway system. These excitatory inputs perhaps reach the subgranular zone of the dentate gyrus, directly or indirectly, where both neural stem cells and neuronal progenitor cells reside.

In addition, the hippocampal formation receives monoaminergic ascending pathways (serotoninergic and noradrenergic) (Amaral and Lavenex, 2007). The penetrations of both serotoninergic and noradrenergic fibers are observed at the subgranular zone of the dentate gyrus. Santarelli et al. (2003) have elegantly demonstrated that these monoaminergic systems affect the process of hippocampal neurogenesis.

8.4 Theta-Rhythm as a Typical Hippocampal Circuit Activity

In the hippocampus of freely moving rodents, six prominent EEG patterns have been identified: four rhythmical and two nonrhythmical (O'Keefe, 2007). Rhythmical patterns include theta (4–12 Hz), beta (12–30 Hz), gamma (30–100 Hz), and ripple (100–400 Hz) waves. Nonrhythmical patterns include large irregular amplitude activity (LIA) and small irregular amplitude activity (SIA). Some patterns can co-occur (e.g., theta with gamma, LIA with ripples), whereas others appear to be mutually exclusive (e.g., theta, LIA, and SIA). The latter three waveforms appear to correspond to mutually exclusive states of hippocampal functioning. Hippocampal EEG activities in the theta range are described next in detail. LIA with ripples occur during quiet sitting and slow-wave sleep, sometimes in a slow rhythm (delta range at 0.5–4 Hz). SIA is a very low amplitude pattern and occurs only occasionally.

Theta-rhythm (also called theta oscillations) is a typical EEG activity in the hippocampal circuit. It consists of theta waves in combination with gamma spikes. Although theta-rhythm was initially observed by Jung and Kornmullar (1938), its function has remained undefined. Theta-rhythm may represent a hippocampal arousal pattern (Green and Arduini, 1954; Vanderwolf, 1969; Bland, 1986), which could be related to learning and memory (Kahana et al., 1999; Cornwell et al., 2008). Actually, the hippocampal region is critically involved in memory, as revealed in both animal and human studies. In addition to learning and memory, elevation of theta-rhythm is observed during voluntary exercise, such as running, walking, jumping, and swimming (Vanderwolf, 1969).

Theta-rhythm are often classified into two subtypes on the basis of behavioral correlation and pharmacological sensitivity. One is a-theta or type 2 theta (for arousal/attention theta; 4–8 Hz), which is sensitive to anticholinergic drugs such as atropine, and the other is t-theta or type 1 theta (intention to move, for translation movement theta; 8–12 Hz), which may be also dependent on serotoninergic. Theta-rhythm in hippocampal circuit is mediated by several types of neurons (Fig. 8.2). At the dentate gyrus, both projection neurons of granule cells and interneurons (excitatory mossy cells and inhibitory basket cells) are involved in the theta circuit. These cells receive neural inputs from layer 2 of the entorhinal cortex, medial septum (MS), and feedback projections from CA3 pyramidal neurons. Theta-rhythm runs through the tri-synaptic hippocampal circuit from the dentate gyrus→CA3→CA1, then back to the entorhinal cortex (Buzsáki, 2002, 2006). Theta-rhythm is a synchronized neural firing of these neurons in the hippocampus.

Lesions of the medial septal nucleus eliminate both types of theta activity. Inactivation by intraseptal injections of the local anesthetic procaine or GABAergic drug muscimol eliminates all theta from the hippocampus. Intraseptal injection of cholinergic antagonists (such as atropine) also blocks both thetas if the animal sits quietly, but leaves movement-related t-theta intact (Lawson and Bland, 1993). In contrast, lesions of the entorhinal cortex eliminate t-theta but leave a-theta intact (Kramis et al., 1975). These results suggest two pacemakers for theta generation: septum and entorhinal cortex.

For considering theta-rhythm as the driving force for the promotion of adult neurogenesis as a physiological elevation of hippocampal network activity, we shall take the two types of theta-rhythm into account. Type 1 theta from the entorhinal cortex occurs during voluntary movements whereas type 2 theta from the septum occurs for arousal/attention. For the direct activation of neural stem cells and/or neuronal progenitor cells by type 1 or type 2 theta-rhythm, it is possible to selectively deliver electrical stimulation at the perforant pathway or fimbria-fornix pathway, respectively.

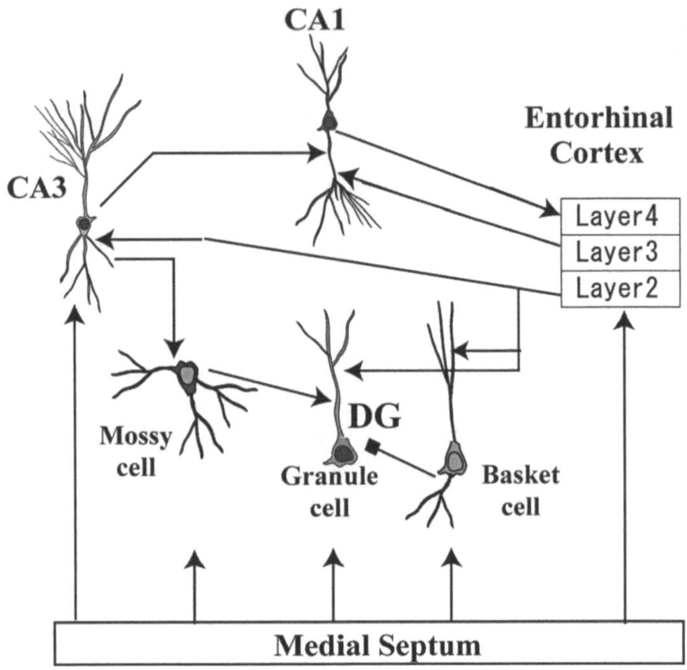

Fig. 8.2 The hippocampal circuit mediates theta-rhythm. Theta-rhythm is regulated by several types of neurons, such as basket cells, granule cells, mossy cells, CA3 pyramidal cells, and CA1 pyramidal cells. These cells receive neural inputs from the entorhinal cortex (layers 2 and 3) and the medial septum (MS). Theta-rhythm is induced by synchronized neural firing of these neurons in the hippocampus

8.5 Process of Adult Hippocampal Neurogenesis

In the dentate gyrus of the hippocampus, new neurons are generated from neural stem cells through neuronal progenitor cells. Although the nomenclature of these precursor cells is still relatively complicated, in this chapter we simply use the two terms neural stem cells and neuronal progenitor cells (Fig. 8.3). After the terminal differentiation, postmitotic new neurons survive and become granule cells with 12 weeks of neuronal maturation (van Praag et al., 2002). Each differentiation step is strictly regulated by genetic mechanisms (for review, see Abrous et al., 2005), as is fully explained by other chapter of this book (for example; Chaps. 7).

Neural stem cells reside at the subgranular zone of dentate gyrus. In fresh hippocampal slices from nestin-promoter-GFP (NestinP-GFP) transgenic mice (Yamaguchi et al., 2000), we visualize both neural stem cells and neuronal progenitor cells (Fukuda et al., 2003). By means of patch-clamp recordings, we observed many glial characteristics in neural stem cells. The notable membrane property of neural stem cells is their low input resistance (IR) value (<500 MΩ;

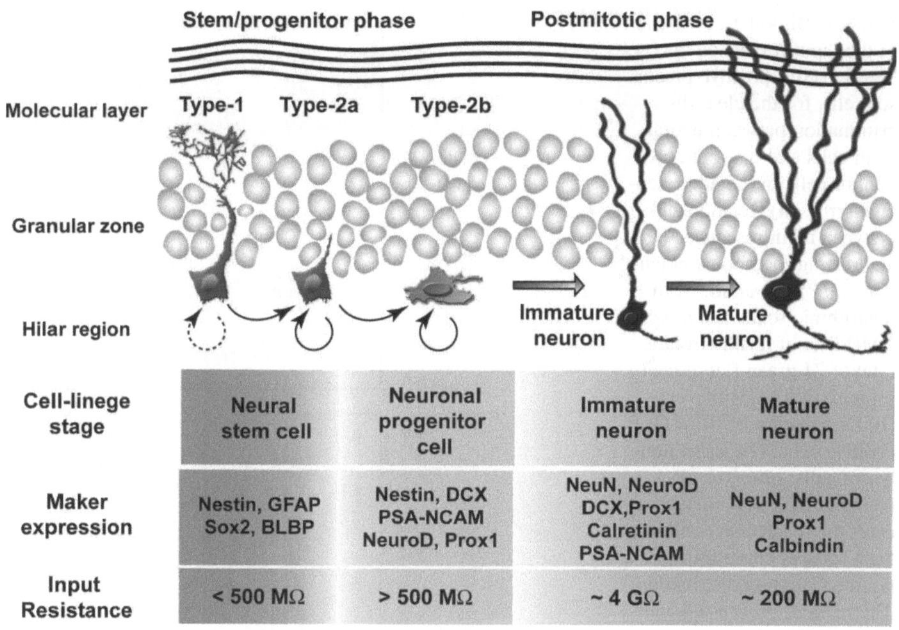

Fig. 8.3 Process of adult hippocampal neurogenesis. Neural stem cells give rise to mature new neurons through the neuronal progenitor cell stage. The value of the input resistance, one of the membrane properties, clearly differentiates neural stem cells from neuronal progenitor cells

Fig. 8.4). From this value, we clearly discriminate between neural stem cells and neuronal progenitor cells because neuronal progenitor cells have a higher IR value (>500 MΩ). In addition, we did not detect any inward currents through the voltage-gated Na channel in neural stem cells. By immunohistochemical staining, we detected numbers of glial proteins such as glial fibrillary acid protein (GFAP) and glutamate aspartate transporter (GLAST) in neural stem cells.

Morphologically, neural stem cells could be further differentiated into the two subtypes. Radial glia-like neural stem cells have thick radial processes toward the molecular layer of the dentate gyrus and behave as quiescent stem cells. We seldom observe the division of radial glia-like stem cells; therefore, dividing stem cells may not possess the thick radial process. After careful confocal observation, however, we can still trace the thin process from these round-type stem cells. We also detect GFAP expression at the initial segment of the proximal process. In the developing nervous system, most of the dividing stem cells manifest such morphology and extend their fine processes toward the pial surface area. Gage and colleagues named quiescent stem cells and dividing stem cells, respectively, by using the Sox2-promoter driven GFP mouse (Suh et al., 2007). Kempermann and colleagues have pointed out that the expression of Sox2 is limited to type 1 and type 2a cells. Recently, they realized the type2a cell is a type of glial cell (Steiner et al., 2006). In our previous publication, we had called neural stem cells (with glial phenotype)

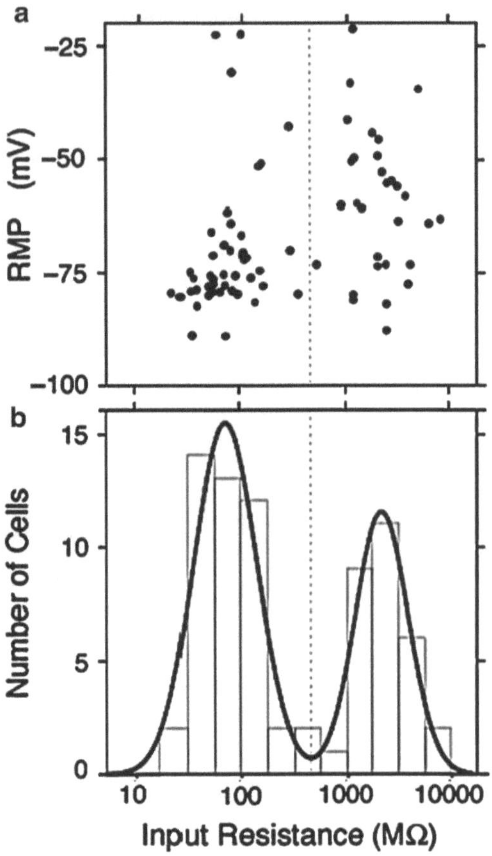

Fig. 8.4 Bimodal distribution of the input resistance in NestinP-GFP-positive precursor cells, for the clear discrimination between neural stem cells and neuronal progenitor cells in the adult dentate gyrus. (**a**) Relationship between input resistance (IR) and resting membrane potential (*RMP*). *Abscissa*, IR in logarithmic scale; *ordinate*, RMP. Plot of results from a total of 74 green fluorescent protein (GFP)-positive cells. (**b**) Distribution of IR of GFP-positive cells. *Ordinate*, number of cells; *abscissa*, IR [same as (**a**)]. The *curve* indicates the result of a nonlinear, least squares curve fitting of a double Gaussian distribution. The *vertical broken line* indicates the trough between two Gaussian peaks at ~500 MΩ. (For more details, see Fukuda et al., 2003)

a type I cell. Because of this confusion, in this chapter, we propose simply calling quiescent stem cells and dividing stem cells (Gage's nomenclature), type 1 and type 2a cells (Kempermann's nomenclature), B cells (Arturo Alvarez-Buylla's nomenclature), and type I cells (our initial nomenclature) as neural stem cells.

The next step of differentiation toward new neurons is neuronal progenitor cells. These progenitor cells still divide, and begin to possess several characteristics: physiologically, they have high IR (>500 MΩ) and immunohistochemically, GFAP expression is downregulated, and the processes toward the granule cell layer disappear. Kempermann and colleagues have termed these neuronal progenitor cells in the course of the neuronal differentiation process as type 2b and type 3 cells. The discrimination between type 2b and type 3 is the detection of nestin-promoter-driven GFP molecules. We detect the expression of GFP molecules in type 2b cells but not in type 3 cells. The fine subclassification of differentiating cells sometimes contributes to understanding the molecular mechanism for neuronal differentiation but simultaneously leads to too much information, especially for readers unfamiliar with this particular area. Therefore, to simplify, We propose calling type 2b and type 3 after Kempermann's nomenclature, D cells (after Arturo Alvarez-Buylla), and type II cells (our initial nomenclature) as neuronal progenitor cells.

8.6 Activity-Dependent Regulation of Adult Neurogenesis

A number of physiological environments and situations regulate the genesis of adult-born neurons. Among them, enriched environments and physical exercise are two popular stimuli that promote adult neurogenesis. In 1997, Kempermann et al. reported that exposure to an enriched environment has a survival-promoting effect on newly generated cells in the adult dentate gyrus (Kempermann et al., 1997), whereas physical activity may induce precursor cell proliferation (van Praag et al., 1999; Suh et al., 2007). A comparison of different strains of mice showed that an enriched environment affects hippocampal neurogenesis differently according to the genetic background. In C57BL/6 mice that have a high baseline level of neurogenesis, environmental enrichment increases cell survival but not cell proliferation, whereas in 129/SvJ mice, which have low levels of neurogenesis, it also increases cell proliferation (Kempermann et al., 1997, 1998).

Physical stimuli, including voluntary exercise, motivated behavior, novelty environments, exploratory behavior, and learning, enhance the rate of hippocampal neurogenesis. This spectrum of behavior is related to the elevation of theta-rhythm activities in hippocampal circuits. Therefore, the theta-rhythm would be the direct force to drive activation of neural stem cells and/or neuronal progenitor cells, which give rise to adult-born neurons.

In addition to physiological stimuli, several pathophysiological conditions, such as epilepsy or ischemic stroke, also promote the process of hippocampal neurogenesis (Abrous et al., 2005). In terms of hippocampal EEG, these pathological conditions also elevate the level of hippocampal theta-band activities. Thus, elevation of the theta-rhythm would be a general mechanism for activating hippocampal neurogenesis.

On the other hand, several physiological and pathological conditions suppress the genesis of new neurons: psychological stress, sleep deprivation, Alzheimer's disease, and aging (Abrous et al., 2005). The mechanism for the attenuation of neurogenesis remains to be elucidated. The modulation of hippocampal EEG activities may be involved in the downregulation of neurogenesis in these conditions.

As mentioned before, type 2 theta (a-theta) activity is from the septum and is mediated by both cholinergic and GABAergic transmission. These fibers directly penetrate into the subgranular zone of the dentate gyrus. In this area, cholinergic fibers basically form synapses onto GABAergic interneurons (basket cells) and hilar excitatory interneurons (mossy cells). Physical activity-related theta (type 1 theta or t-theta) is mediated through the perforant pathway from the entorhinal cortex. Because the entorhinal cortex is a main gate from most of the cortical areas toward the hippocampal circuit, type 1 theta therefore could be modulated by sensory inputs and the intention of movement.

In combination with the theta-rhythm circuit, catecholaminergic (serotoninergic, noradrenergic, and dopaminergic) systems may modulate the rate of neurogenesis. Both noradrenergic and serotoninergic fibers penetrate directly toward the neurogenic area; thus, it is possible to speculate on the influence of these systems on activity at the dentate gyrus. Very recently, Airan et al. have suggested that the serotoninergic system may upregulate the power of EEG activities at the dentate gyrus (Airan et al., 2007). Taken together, type 2 theta would be dependent on serotoninergic neurotransmission, and neurogenesis enhanced by treatment with serotonin-specific reuptake inhibitor (SSRI) (Santarelli et al., 2003) might be the result of upregulation of hippocampal EEG activities. If the doperminergic system would also modulate neurogenesis, this effect could be caused by an indirect effect through the modification of function of septal neurons, where we observed the robust enervation of dopaminergic fibers.

8.7 Neural Inputs and Regulations on Neural Stem Cells

Neural stem cells residing within the subgranular zone in the adult dentate gyrus are the source of new neurons because they give rise to neuronal progenitor cells, which finally differentiate to mature new neurons (Fukuda et al., 2003; Kempermann et al., 2004). The neural stem cells include the radial glia-like cells with GFAP-positive long processes (type 1) and weakly GFAP-expressing progenitor cells with short and thin processes (type 2a) (Itou et al., 2011). Neural stem cells have the ability for both self-renewal and generation of progenitor cells by symmetrical and asymmetrical division.

In the adult dentate gyrus, neural stem cells are clearly identified by using the NestinP-GFP transgenic mouse (Yamaguchi et al., 2000), in which enhanced EGF (EGFP) is driven by nestin second intron enhancers. Neural stem (type 1 and 2a) cells have characteristic electrophysiological properties: they have a small IR (<500 MΩ) whereas type 2b cells present a larger IR (>500 MΩ) (Fig. 8.4). This range of neural stem cells substantially overlaps with that for recordings from the dentate gyrus astrocytes in hippocampal slices of adult rats (mean, 175 MΩ; range, 119–278 MΩ; Bordey and Sontheimer, 2000), although, in the developmental stage, the IR of glial cells is relatively high (Kressin et al., 1995; Matthias et al., 2003). Additionally, neural stem cells were characterized by their properties: the significantly higher resting membrane potential (RMP) and the lack of voltage-dependent inward currents upon depolarization.

In the adult dentate gyrus, neural stem cells reach their radial processes toward the molecular layer, where the perforant path from the entorhinal cortex or fibers from hippocampal mossy cells run through, and their radial processes locate close to the asymmetrical synapses between the mossy cells and granule cells in the inner molecular layer. Therefore, neural stem cells are integrated into the hippocampal local circuit through which theta-rhythm activities run. Interestingly, in the hippocampal slice, stimulation of the perforant path with a high theta-burst protocol induced strong calcium responses in neural stem cells (Fig. 8.5). Therefore, neural

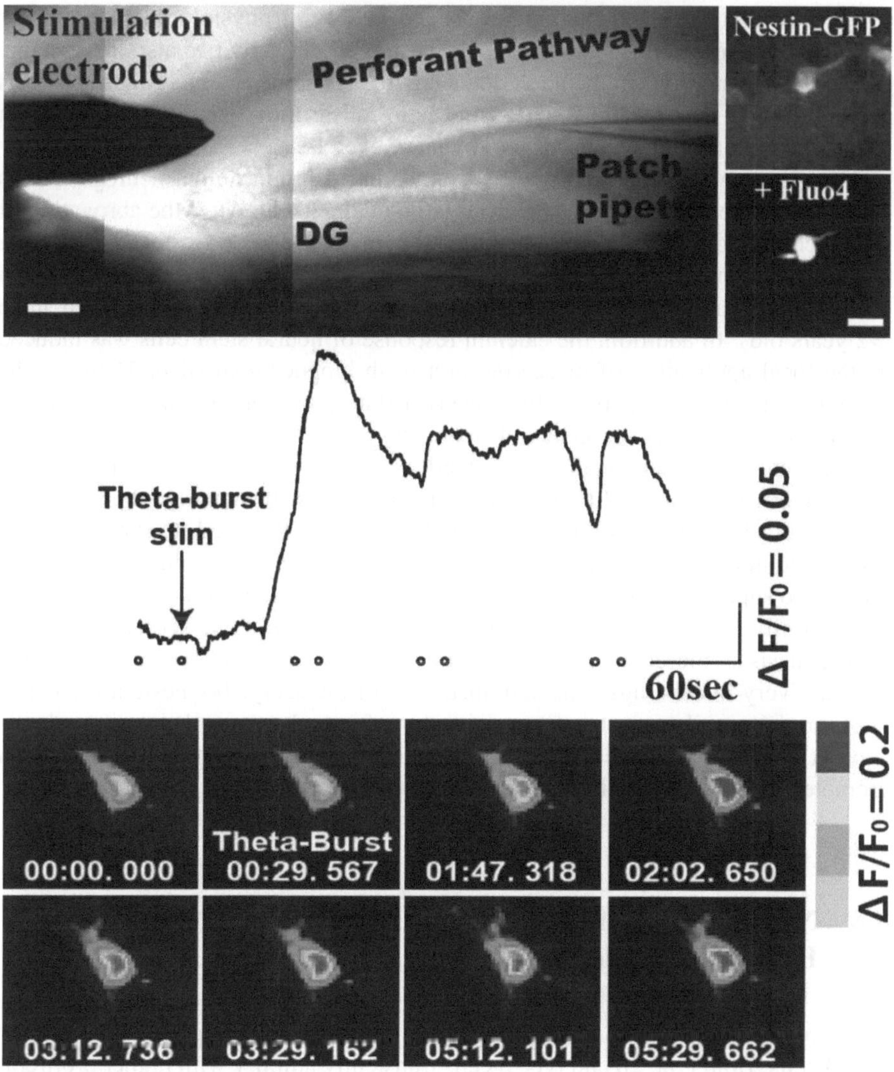

Fig. 8.5 Theta-burst stimulation activates neural stem cells in a hippocampal slice. A calcium-sensitive dye, Fluo-4, was injected through the patch pipette into neural stem cells, which bear the NestinP-GFP signal. After theta-burst at the perforant pathway, we observed a rapid increase in the fluorescence signal from fluo-4-loaded neural stem cells. Note that most of calcium was released from the endoplasmic reticulum around the cell nucleus. (From Nochi et al., unpublished data)

stem cells could be integrated into the hippocampal local circuit and have potential to receive neural input. Accumulating evidence shows that the hippocampal circuit activity mediates the proliferation and differentiation rate of neural stem/progenitor cells (Deisseroth et al., 2004; Tozuka et al., 2005; Bruel-Jungerman et al., 2006;

Chun et al., 2006). Theta-rhythm is elevated in several physiological conditions, such as voluntary exercise and learning. These physical activities also increase the genesis of new neurons in the adult dentate gyrus. There is an emerging hypothesis that elevation in the cell division of neural stem cells is the reason for the promotion of the generation of new neurons by voluntary exercise.

Very recently, we reported that exercise-induced hippocampal neurogenesis is dependent on septal cholinergic neurons (Itou et al., 2011). After the abrogation of septal cholinergic neurons by the injection of p75-SAP (Advanced Targeting Systems), the enhancement effect of wheel running exercise on neural stem cell proliferation was completely deleted. This effect was also found in aged mice (>2 years old). In addition, the calcium response of neural stem cells was induced by the focal application of muscarine in a fresh hippocampal slice. These results strongly suggest that acetylcholine released during wheel running may activate muscarinic-ACh receptors on neural stem cells.

The genesis of new neurons gradually decreases during the aging process (Hattiangady and Shetty, 2008); however, neural stem cells maintain their numbers in the adult dentate gyrus throughout the life, even in the aged. Our recent study using aged nonhuman primates (*Macaca fascicularis* >20 years old; Aizawa et al., 2011) also supports the notion that the number of neural stem cells is not decreased substantially in aged animals. Therefore, the reason for the decrease of adult neurogenesis is the change of proliferating activity of neural stem cells in aged animals. Very interestingly, the administration of the acetylcholinesterase (AChE) inhibitor donepezil promotes the proliferation of neural stem cells in aged animals (Itou et al., 2011). In aged animals, voluntary exercise or cholinergic activation promotes the generation of adult hippocampal new neurons.

8.8 Neural Inputs and Regulations on Neuronal Progenitor Cells

As mentioned previously, neural progenitor cells can be sorted into type 2b and type 3 cells. Both type 2b and type 3 cells transiently amplify with higher proliferative activity compared with neural stem cells (type 1 and type 2a) (Fukuda et al., 2003), but exit the cell cycle within 1–3 days after several divisions. Type 2b cells can be identified by their expression of both nestin and polysialated neural cell adhesion molecule (PSA-NCAM) or extremely high IR (>500 MΩ), whereas type 3 cells are negative with nestin. However, studies to characterize type 3 cells electrophysiologically are lacking at the present time.

From electron microscopy observation, we identified the synaptic inputs toward neuronal progenitor cells (Fig. 8.6). Neuronal progenitor cells are D cells in another nomenclature, which means dense nucleus in electron microscopy. On the other hand, neural stem cells have a bright nucleus and extend their processes toward the molecular layer. By GFP-targeted recording, we found that neuronal progenitor (type 2b) cells receive active direct neural inputs from

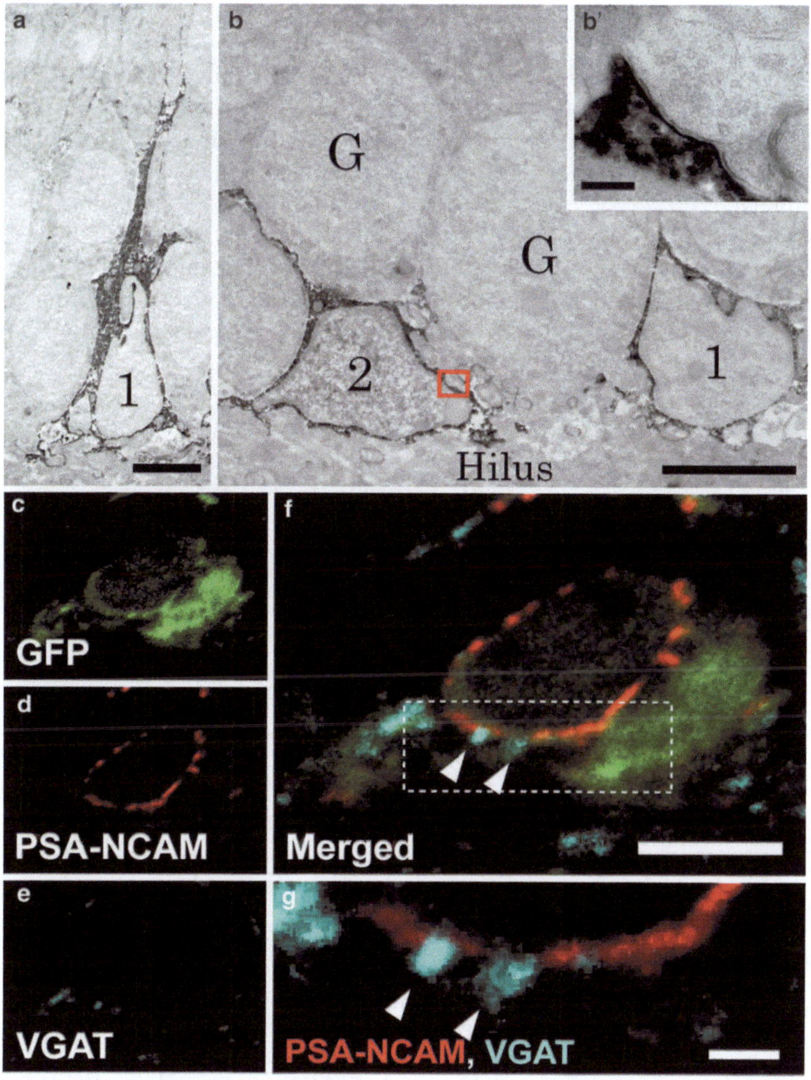

Fig. 8.6 Neuronal progenitor cells receive γ-aminobutyric acid (GABA)-ergic inputs. (**a, b**) Immunoelectron microscopy images of neural stem cells and neuronal progenitor cells (type 2b cells). GFP was visualized with electron-dense DAB precipitation after anti-GFP staining. (**a**) Image showing neural stem cells with thick radial process (type 1 cell). (**b**) Image showing neuronal progenitor cells (type 2b cell) with dark nuclei and scant cytoplasm (*C*). Cell marked with *1* might be classified into the current nomenclature of type 2a, which has a bright nucleus, but does not extend a thick radial process. (**b′**) Enlarged view of *boxed area* in (**b**). A terminal-like structure apposed to the plasma membrane of a neuronal progenitor cell. *1*, neural stem cell (type 1 and type 2a); *2*, neuronal progenitor cells (type 2b); *G*, granule cell. (**c–g**) Confocal images of the subgranular zone of adult mice. VGAT (vesicular GABA transporter)-expressing terminal-like structures in close contact with neuronal progenitor cell (type 2b cell). (**c**) Nestin-GFP expression. (**d**) Polysialated neural cell adhesion molecule (PSA-NCAM) staining. (**e**) VGAT staining. (**f**) Merged. (**g**) Enlarged view of *boxed area* in (**f**), focusing on the VGAT+ nerve terminals on the neuronal progenitor cell (type 2b cell) at high magnification. Note GABAergic terminals making contact on a neuronal progenitor cell (type 2b cell). *Bars* (**a**,**b**) 4 μm; (**b′**) 200 nm; (**f**) 5 μm; (**g**) 1 μm. (For more details, see Tozuka et al., 2005)

the hippocampal circuitry (Tozuka et al., 2005): this input was GABAergic but not glutamatergic. Neuronal progenitor cells that were visualized by NestinP-GFP and PSA-NCAM staining have GABAergic neural input, which was identified by the staining of vesicular-type GABA transporters located at the axonal terminals.

One of the most interesting characteristics of the type 2b cells is the neuronal differentiation induced by GABAergic input. The systemic injection of GABAergic agonists caused 50% increase in the number of new neurons (Fig. 8.7; from Tozuka et al., 2005). Because neuronal progenitor cells (type 2b cells) have high chloride

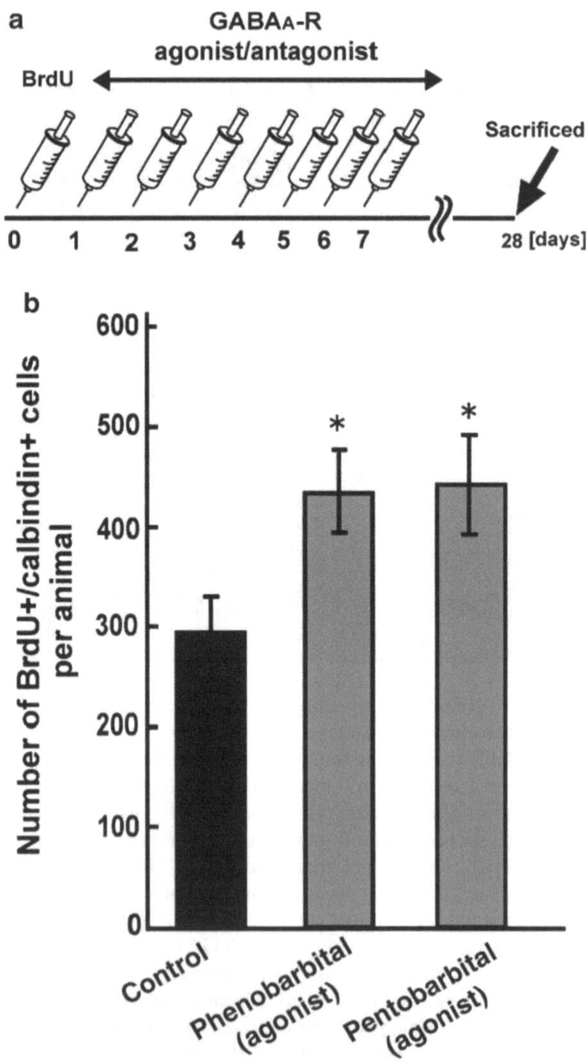

Fig. 8.7 GABAergic treatment promotes the neuronal differentiation of neuronal progenitor (type 2b) cells. (**a**) Experimental paradigm for assessing the neuronal differentiation of neuronal progenitor cells (type 2b cells). (**b**) Number of bromodeoxyuridine (BrdU)-labeled calbindin+ cells in each animal. Note the significant increase of BrdU+ new neurons in the $GABA_A$-R agonist-treated groups. *Error bars* indicate SEM $*P<0.05$, Student's *t* test for unpaired samples. (For more details, see Tozuka et al., 2005)

content from the expression of Na^+-K^+-$2Cl^-$ transporter (NKCC1), they depolarize by activation of $GABA_A$ receptors (Tozuka et al., 2005). This depolarization elicits an elevation of intracellular Ca^{2+} influx through the activation of voltage-gated Ca^{2+} channels such as the L-type Ca^{2+} channel. In consequence, expression of NeuroD, a positive regulator of neuronal differentiation, will be induced by the Ca^{2+} influx.

The $GABA_A$ receptors of type 2b cells can be activated by release of GABA from hippocampal GABAergic interneurons such as basket cells or axo-axonic cells (Tozuka et al., 2005; Ge et al., 2006). Several neural pathways are thought to influence hippocampal GABAergic interneurons. GABAergic interneurons are widely spread in hippocampus, but those neurons in the hilus receive excitatory inputs from the entorhinal cortex, which innervate the molecular layer through the perforant path. GABAergic interneurons in hilus also receive acetylcholinergic input from the septum, so septal excitation can also affect type 2b cells indirectly.

N-Methyl-D-aspartic acid (NMDA) receptor (NMDA-R) blockade in vivo also robustly increases proliferation of progenitor cells (Cameron et al., 1995; Nacher et al., 2001). This proliferation might suggest a direct influence of NMDA antagonists on adult progenitor cells; however, by means of targeted recording we have detected no significant functional expression of NMDA-Rs on either stem cells or progenitor cells in the adult dentate gyrus (Wang et al., 2005; Tozuka et al., 2005). It is possible that this phenomenon results from indirect effects mediated via GABAergic inputs to type 2b cells. GABA release from hippocampal interneurons requires NMDA-R activation (Matsuyama et al., 1997).

8.9 Neural Inputs and Regulation on Postmitotic Immature New Neurons

After the final mitosis of neuronal progenitor cells, newborn cells start to mature as neurons (Zhao et al., 2006). Control in adult hippocampal neurogenesis is also prominent at the stage of early postmitotic neurons (before 2 weeks; Gould et al., 1999; Döbrössy et al., 2003; Dupret et al., 2007). Because glutamatergic inputs are seen in the new neurons from 2 to 4 weeks after birth, and start to act as mature granule neurons from 4 weeks, it has been expected that there are other sources of neural input (GABAergic or cholinergic) toward immature new neurons before 2 weeks after mitosis for the selection of new neurons.

Interestingly, the depolarizing response to GABA input remains until 2 weeks after birth, for their high chloride content, although it switches to hyperpolarizing response because the K^+-Cl^- cotransporter (KCC2) is upregulated and decreases the chloride concentration (Esposito et al., 2005; Ge et al., 2006). In addition, immature neurons distinguished by PSA-NCAM may also receive acetylcholinergic inputs (Kaneko et al., 2006). We also observed both GABAergic and cholinergic inputs to new neurons at least 7 days after mitosis (Ide et al., unpublished observations). To our surprise, new neurons have already grown their axonal fibers, which is called the "mossy fiber,"

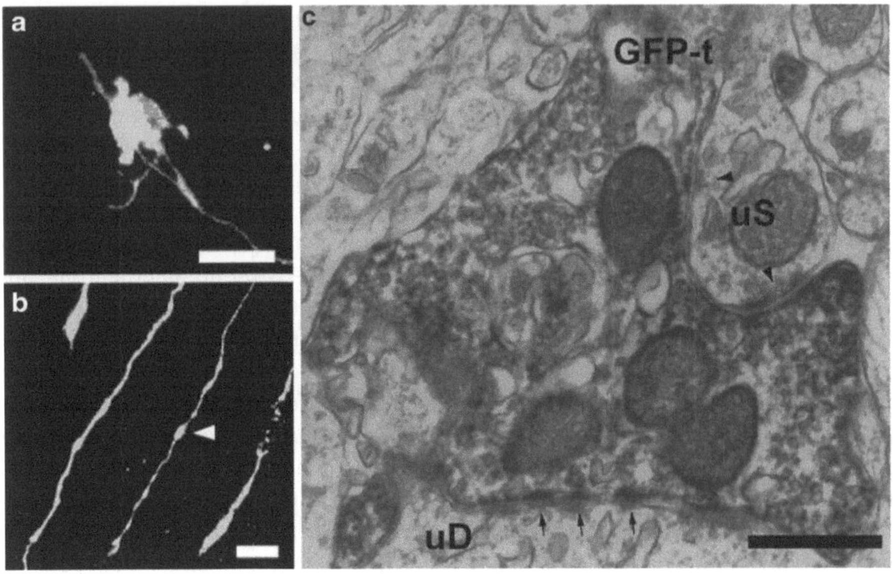

Fig. 8.8 Mossy fiber terminals of immature new neuron. New neurons labeled with enhanced green fluorescent protein (*GFP*) tagged with the membrane-targeting GAP43 palmitoylation site (GGFP) form mossy fiber terminals toward CA3 within a week after final mitosis. Both large bouton (**a**) and small terminal (**b**; *arrowhead*) are innervated to hilus and CA3. (**c**) GGFP-labeled large boutons showed asymmetrical synapse with unlabeled spines (*uS*; *arrowhead*), and puncta adherens-like junctions with unlabeled dendritic shaft (*uD*; *arrow*). GGFP expression to newborn cells was designed with adenovirus injection to Nestin-Cre transgenic mouse. *Bars* (**a**,**b**) 5 μm; (**c**) 500 nm. (For more details, see Ide et al., 2008)

and form synapses with neurons in the hilus and CA3 areas just 7 days after their final mitosis (Fig. 8.8; Ide et al., 2008). This finding suggests that new neurons are already integrated in the hippocampal network within a week after birth. Although 1-week-old neurons might be still too early to functionally recruit into spatial memory networks (Kee et al., 2007), it is suggested that 1- to 2-week-old newborn neurons are involved in particular learning tasks (Shors et al., 2001; Olariu et al., 2005).

Therefore, the survival and the maturation of newborn neurons can be controlled by many kinds of neural activities, GABAergic and/or cholinergic excitation. Knockdown of the NKCC1 by short-hairpin RNA reveals that depolarization by GABAergic input plays a critical role in the development for dendrites and formation of synapses (Ge et al., 2006). In addition, the cholinergic system also supports the survival of new neurons (Mohapel et al., 2005; Kaneko et al., 2006). Furthermore, it can be assumed that the postsynaptic neurons may deliver some signals for cell survival or maturation of immature new neurons about 1–2 weeks after final mitosis. Although glutamatergic inputs to immature neurons are seen after 2 weeks, earlier integration of immature new neurons before 2 weeks may also explain why the survival of the new neurons is influenced by experience, such as spatial or learning memory (Gould et al., 1999; Döbrössy et al., 2003; Tashiro et al., 2006; Dupret et al., 2007; Kee et al., 2007).

8.10 Summary and Perspectives

Adult hippocampal neurogenesis provides hope for ameliorating the decline in cognitive functions in normal aging and mild cognitive impairment in psychological illness, vascular dementia, or Alzheimer's disease. Hundreds of recent reports have indicated that many environmental stimuli, such as exercise, exploration, enriched environments, and learning and memory, improve the rate of the generation of new neurons; however, the mechanism governing the enhanced neurogenesis has remained elusive. Although endothelial components may be involved in enhanced neurogenesis, hippocampal circuits would play pivotal roles in the promotion of adult neurogenesis by environmental stimuli. Other reports have demonstrated that several pharmaceutical treatments also improve the rate of neurogenesis at the adult dentate gyrus. It can be reasonably assumed that hippocampal circuits mediate the effect of these pharmaceutical treatments on adult hippocampal neurogenesis.

Theta-rhythm is a typical network activity in hippocampal circuits during actively behaving animals and humans. In addition, the power of theta-rhythm is affected by the tone of any neurotransmitter systems. Therefore, we could propose that regulation of the genesis of hippocampal new neurons would be mediated by the dynamics of theta-rhythm in hippocampal circuits, for example, duration and power, which could be tuned by environmental stimuli or pharmaceutical treatment.

Although from a conventional aspect it had been assumed that theta-rhythm might affect only mature neurons fully integrated in the hippocampal tri-synaptic circuitry, in this chapter we have shown evidence that well supports the view that even dividing precursor cells (neural stem cells and neuronal progenitor cells), as well as early postmitotic new neurons, can sense the activities of hippocampal circuits, especially the power of theta-rhythm activity, as regulation of hippocampal neurogenesis would occur at its early stage, from neural stem cells, to neuronal progenitor cells, to early postmitotic new neurons (before 3 weeks). These lines of research may contribute to develop therapeutic interventions for age-related cognitive decline.

Acknowledgments We thank Drs. Satoshi Fukuda and Yusuke Tozuka for their discussion, comments, and the images.

References

Abrous DN, Koehl M, Le Moal M (2005) Adult neurogenesis: from precursors to network and physiology. Physiol Rev 85:523–569

Airan RD, Meltzer LA, Roy M et al (2007) High-speed imaging reveals neurophysiological links to behavior in an animal model of depression. Science 317:819–823

Aizawa K, Ageyama N, Terao K, Hisatsune T (2011) Primate-specific alterations in neural stem/progenitor cells in the aged hippocampus. Neurobiol Aging, doi: 10.1016/j.neurobiolaging.32:140–150 (2011).12.011, online

Altman J, Das GD (1965) Autoradiographic and histological evidence of postnatal hippocampal neurogenesis in rats. J Comp Neurol 124:319–335

Altman J, Das GD (1967) Postnatal neurogenesis in the guinea-pig. Nature 214:1098–1101

Amaral D, Lavenex P (2007) Hippocampal neuroanatomy. In: Anderson P, Morris R, Amaral D, Bliss T, O'Keefe J (eds) The hippocampus book. Oxford University Press, New York, p 37–114

Bland BH (1986) The physiology and pharmacology of hippocampal formation theta rhythms. Prog Neurobiol 26:1–54

Bordey A, Sontheimer H (2000) Ion channel expression by astrocytes in situ: comparison of different CNS regions. Glia 30:27–38

Bruel-Jungerman E, Davis S, Rampon C et al (2006) Long-term potentiation enhances neurogenesis in the adult dentate gyrus. J Neurosci 26:5888–5893

Buzsáki G (2002) Theta oscillations in the hippocampus. Neuron 33:325–340

Buzsáki G (2006) Rhythms of the brain. Oxford University Press, New York

Cameron HA, McEwen BS, Gould E (1995) Regulation of adult neurogenesis by excitatory input and NMDA receptor activation in the dentate gyrus. J Neurosci 15:4687–4692

Chun SK, Sun W, Park, JJ et al (2006) Enhanced proliferation of progenitor cells following long-term potentiation induction in the rat dentate gyrus. Neurobiol Learn Mem 86:322–329

Cornwell BR, Johnson LL, Holroyd T et al (2008) Human hippocampal and parahippocampal theta during goal-directed spatial navigation predicts performance on a virtual Morris water maze. J Neurosci 28:5983–5990

Deisseroth K, Singla S, Toda H et al (2004) Excitation-neurogenesis coupling in adult neural stem/progenitor cells. Neuron 42:535–552

Döbrössy MD, Drapeau E, Aurousseau C et al (2003) Differential effects of learning on neurogenesis: learning increases or decreases the number of newly born cells depending on their birth date. Mol Psychiatry 8:974–982

Dupret D, Fabre A, Döbrössy MD et al (2007) Spatial learning depends on both the addition and removal of new hippocampal neurons. PLoS Biol 5:e214

Dupret D, Revest JM, Koehl M et al (2008) Spatial relational memory requires hippocampal adult neurogenesis. PLoS One 3:e1959

Eriksson PS, Perfilieva E, Bjork-Eriksson T et al (1998) Neurogenesis in the adult human hippocampus. Nat Med 4:1313–1317

Esposito MS, Piatti VC, Laplagne DA et al (2005) Neuronal differentiation in the adult hippocampus recapitulates embryonic development. J Neurosci 25:10074–10086

Fabel K, Tam B, Kaufer D et al (2003) VEGF is necessary for exercise-induced adult hippocampal neurogenesis. Eur J Neurosci 18:2803–2812

Filippov V, Kronenberg G, Pivneva T et al (2003) Subpopulation of nestin-expressing progenitor cells in the adult murine hippocampus shows electrophysiological and morphological characteristics of astrocytes. Mol Cell Neurosci 23:373–382

Freund TF, Buzsáki G (1996) Interneurons of the hippocampus. Hippocampus 6:347–470

Fukuda S, Kato F, Tozuka Y et al (2003) Two distinct subpopulations of nestin-positive cells in adult mouse dentate gyrus. J Neurosci 23:9357–9366

Ge S, Goh EL, Sailor KA et al (2006) GABA regulates synaptic integration of newly generated neurons in the adult brain. Nature 439:589–593

Gould E, Tanapat P, McEwen BS et al (1998) Proliferation of granule cell precursors in the dentate gyrus of adult monkeys is diminished by stress. Proc Natl Acad Sci USA 95:3168–3171

Gould E, Beylin A, Tanapat P et al (1999) Learning enhances adult neurogenesis in the hippocampal formation. Nat Neurosci 2:260–265

Green JD, Arduini A (1954) Hippocampal electrical activity in arousal. J Neurophysiol 17:533–557

Hattiangady B, Shetty AK (2008) Aging does not alter the number or phenotype of putative stem/progenitor cells in the neurogenic region of the hippocampus. Neurobiol Aging 29:129–147

Ide Y, Fujiyama F, Okamoto-Furuta K, Tamamaki N, Kaneko T, Hisatsune T (2008) Rapid integration of young newborn dentate gyrus granule cells in the adult hippocampal circuitry. Eur J Neurosci 28:2381–2392

Itou Y, Nochi R, Kuribayashi H, Saito Y, Hisatsune T (2011) Cholinergic activation of hippocampal neural stem cells in aged dentate gyrus. Hippocampus 21:446–459

Jin K, Minami M, Lan JQ et al (2001) Neurogenesis in dentate subgranular zone and rostral subventricular zone after focal cerebral ischemia in the rat. Proc Natl Acad Sci USA 98:4710–4715

Jung R, Kornmullar A (1938) Eine methodik der ableitung lokalisierter potential schwankingen aus subcorticalen hirnyebieten. Arch Psychiat Neruenkr 109:1–30

Kahana MJ, Sekuler R, Caplan JB et al (1999) Human theta oscillations exhibit task dependence during virtual maze navigation. Nature 24:781–784

Kaneko N, Okano H, Sawamoto K (2006) Role of the cholinergic system in regulating survival of newborn neurons in the adult mouse dentate gyrus and olfactory bulb. Genes Cells 11:1145–1159

Kaplan MS, Bell DH (1984) Mitotic neuroblasts in the 9-day-old and 11-month-old rodent hippocampus. J Neurosci 4:1429–1441

Kaplan MS, Hinds JW (1977) Neurogenesis in the adult rat: electron microscopic analysis of light radioautographs. Science 197:1092–1094

Kee N, Teixeira CM, Wang AH et al (2007) Preferential incorporation of adult-generated granule cells into spatial memory networks in the dentate gyrus. Nat Neurosci 10:355–362

Kempermann G, Kuhn HG, Gage FH (1997) Genetic influence on neurogenesis in the dentate gyrus of adult mice. Proc Natl Acad Sci USA 94:10409–10414

Kempermann G, Brandon EP, Gage FH (1998) Environmental stimulation of 129/SvJ mice causes increased cell proliferation and neurogenesis in the adult dentate gyrus. Curr Biol 8:939–942

Kempermann G, Jessberger S, Steiner B et al (2004) Milestones of neuronal development in the adult hippocampus. Trends Neurosci 27:447–452

Koketsu D, Furuichi Y, Maeda, M et al (2006) Increased number of new neurons in the olfactory bulb and hippocampus of adult non-human primates after focal ischemia. Exp Neurol 199:92–102

Kornack DR, Rakic P (1999) Continuation of neurogenesis in the hippocampus of the adult macaque monkey. Proc Natl Acad Sci USA 96:5768–5773

Kramis R, Vanderwolf CH, Bland BH (1975) Two types of hippocampal rhythmical slow activity in both the rabbit and the rat: relations to behavior and effects of atropine, diethyl ether, urethane, and pentobarbital. Exp Neurol 49:58–85

Kressin K, Kuprijanova E, Jabs R et al (1995) Developmental regulation of Na+ and K+ conductances in glial cells of mouse hippocampal brain slices. Glia 15:173–187

Lawson VH, Bland BH (1993) The role of the septohippocampal pathway in the regulation of hippocampal field activity and behavior: analysis by the intraseptal microinfusion of carbachol, atropine, and procaine. Exp Neurol 120:132–144

Matsuyama S, Nei K, Tanaka C (1997) Regulation of GABA release via NMDA and 5-HT1A receptors in guinea pig dentate gyrus. Brain Res 761:105–112

Matthias K, Kirchhoff F, Seifert G et al (2003) Segregated expression of AMPA-type glutamate receptors and glutamate transporters defines distinct astrocyte populations in the mouse hippocampus. J Neurosci 23:1750–1758

Mohapel P, Leanza G, Kokaia M et al (2005) Forebrain acetylcholine regulates adult hippocampal neurogenesis and learning. Neurobiol Aging 26:939–946

Nacher J, Rosell DR, Alonso-Llosa G et al (2001) NMDA receptor antagonist treatment induces a long-lasting increase in the number of proliferating cells, PSA-NCAM-immunoreactive granule neurons and radial glia in the adult rat dentate gyrus. Eur J Neurosci 13:512–520

O'Keefe (2007) Hippocampal neurophysiology in the behaving animal. In: Andersen P, Morris R, Amaral D, et al (eds) The hippocampus book. Oxford University Press, New York

Olariu A, Cleaver KM, Shore LE et al (2005) A natural form of learning can increase and decrease the survival of new neurons in the dentate gyrus. Hippocampus 15:750–762

Pereira AC, Huddleston DE, Brickman AM et al (2007) An in vivo correlate of exercise-induced neurogenesis in the adult dentate gyrus. Proc Natl Acad Sci USA 104:5638–5643

Santarelli L, Saxe M, Gross C et al (2003) Requirement of hippocampal neurogenesis for the behavioral effects of antidepressants. Science 301:805–809

Saxe MD, Battaglia F, Wang J et al (2006) Ablation of hippocampal neurogenesis impairs contextual fear conditioning and synaptic plasticity in the dentate gyrus. Proc Natl Acad Sci USA 103:17501–17506

Seri B, Garcia-Verdugo JM, McEwen BS et al (2001) Astrocytes give rise to new neurons in the adult mammalian hippocampus. J Neurosci 21:7153–7160
Shors TJ, Miesegaes G, Beylin A et al (2001) Neurogenesis in the adult is involved in the formation of trace memories. Nature 410:372–376
Steiner B, Klempin F, Wang L et al (2006) Type-2 cells as link between glial and neuronal lineage in adult hippocampal neurogenesis. Glia 54:805–814
Suh H, Consiglio A, Ray J et al (2007) In vivo fate analysis reveals the multipotent and self-renewal capacities of Sox2(+) neural stem cells in the adult hippocampus. Cell Stem Cell 1:515–528
Tashiro A, Sandler VM, Toni N et al (2006) NMDA-receptor-mediated, cell-specific integration of new neurons in adult dentate gyrus. Nature 442:929–933
Tonchev AB, Yamashima T, Zhao L et al (2003) Proliferation of neural and neuronal progenitors after global brain ischemia in young adult macaque monkeys. Mol Cell Neurosci 23:292–301
Toni N, Teng EM, Bushong EA et al (2007) Synapse formation on neurons born in the adult hippocampus. Nat Neurosci 10:727–734
Tozuka Y, Fukuda S, Namba T et al (2005) GABAergic excitation promotes neuronal differentiation in adult hippocampal progenitor cells. Neuron 47:803–815
van Praag H, Kempermann G, Gage FH (1999) Running increases cell proliferation and neurogenesis in the adult mouse dentate gyrus. Nat Neurosci 2:266–270
van Praag H, Schinder AF, Christie BR et al (2002) Functional neurogenesis in the adult hippocampus. Nature 415:1030–1034
Vanderwolf CH (1969) Hippocampal electrical activity and voluntary movement in the rat. Electroencephalogr Clin Neurophysiol 26:407–418
Wang LP, Kempermann G, Kettenmann H (2005) A subpopulation of precursor cells in the mouse dentate gyrus receives synaptic GABAergic input. Mol Cell Neurosci 29:181–189
Yamaguchi M, Saito H, Suzuki M et al (2000) Visualization of neurogenesis in the central nervous system using nestin promoter-GFP transgenic mice. Neuroreport 26:1991–1996
Yoshimura S, Takagi Y, Harada J et al (2001) FGF-2 regulation of neurogenesis in adult hippocampus after brain injury. Proc Natl Acad Sci USA 98:5874–5879
Zhang CL, Zou Y, He W et al (2008) A role for adult TLX-positive neural stem cells in learning and behaviour. Nature 451:1004–1007
Zhao C, Teng EM, Summers RG Jr., et al (2006) Distinct morphological stages of dentate granule neuron maturation in the adult mouse hippocampus. J Neurosci 26:3–11

Chapter 9
Integration of New Neurons into the Adult Hippocampus

Wei Deng, Chunmei Zhao, and Fred H. Gage

Abstract The hippocampus is one of the regions where continuous addition of new neurons occurs in adult mammalian brains. In this review, we describe the process of integration of newborn neurons into the neural circuitry of the hippocampus in adult rodents and discuss how hippocampal neurogenesis is regulated by the neural network activity. We further address the functional implications of adult hippocampal neurogenesis in learning and memory.

Abbreviations

AMPA	α-Amino-3-hydroxy-5-methyl-4-isoxazolepropionic acid
BLBP	Brain lipid-binding protein
BrdU	Bromodeoxyuridine
CNS	Central nervous system
Dcx	Doublecortin
DG	Dentate gyrus
GFAP	Glial fibrillary acidic protein
IEGs	Immediate early genes
LTP	Long-term potentiation
MAM	Methylazoxymethanol acetate
MWM	Morris water maze
NeuN	Neuronal nuclei
NKCC1	Na-K-2Cl cotransporter
NMDA	*N*-methyl-D-aspartic acid
POMC	Pro-opiomelanocortin
PSA-NCAM	Polysialated form of neural cell adhesion molecule
SGZ	Subgranular zone
Tk	Thymidine kinase

W. Deng, C. Zhao, and F.H. Gage (✉)
Laboratory of Genetics, Salk Institute for Biological Studies,
10010 North Torrey Pines Road, La Jolla, CA 92037, USA
e-mail: gage@salk.edu

T. Seki et al. (eds.), *Neurogenesis in the Adult Brain I: Neurobiology*,
DOI 10.1007/978-4-431-53933-9_9, © Springer 2011

9.1 Introduction

The hippocampus is one of the brain regions where new neurons are continuously generated in adult animals, including humans. In the hippocampus, the generation of new neurons starts with the proliferation of progenitor cells, which are located in a special region named the subgranular zone (SGZ), lying at the interface between the granule cell layer and the hilus of the dentate gyrus (DG) (Fig. 9.1a,d). Two types of progenitors can be identified according to their distinct morphologies and expression of molecular markers. Type 1 cells, expressing molecular markers such as glial fibrillary acidic protein (GFAP), Sry-related transcriptional factor Sox2, Nestin, and brain lipid-binding protein (BLBP), have a radial process spanning the granule cell layer that ramifies in the inner molecular layer. Type 2 cells have only short processes and do not express GFAP or BLBP. These progenitors reside in a microenvironment that allows the majority of their daughter cells to differentiate into granule cells of the DG.

Existing in the microenvironment of the DG is a complex neural network (Fig. 9.1a; Table 9.1) (Amaral and Lavenex 2007). The dentate granule cells, with elaborate dendritic arborizations in the molecular layer, receive excitatory inputs from the entorhinal cortex through the perforant pathway and from the hilar mossy cells through the commissural-associational pathway. Granule cells also receive inhibitory inputs from the local interneurons, which are distributed randomly in the DG and can be modulated by mossy cells. The neurons in the hilus are major targets for the inputs from other brain regions, such as the cholinergic and GABAergic inputs from the medial septum and the basal forebrain, the serotonergic inputs from the raphe nuclei, the norepinephrinergic inputs from the locus coeruleus, and the dopaminergic inputs from the ventral tegmental area. Some of these input fibers also project into the molecular layer of the DG and other areas of the hippocampus. The major efferents from the granule cells are the mossy fiber pathways to the pyramidal neurons of CA3. The mossy fibers also innervate the mossy cells and interneurons in the hilus, forming both positive and negative feedback loops. In addition to neurons, numerous local glia cells, such as astrocytes, oligodendrocytes, and microglia, and the local vasculature play supportive roles in the neural network. Therefore, the activity of the neural network and of all cells within this network can exert direct and indirect influences on the process of hippocampal neurogenesis.

The phenomenon of adding new neurons in the adult hippocampus circuitry raises interesting questions. How are new neurons produced and integrated into the neural circuits of the adult hippocampus? How does the surrounding neural network activity influence the process of adult hippocampal neurogenesis? How does the incorporation of new neurons into the existing neural networks influence the function of the hippocampus? In this chapter, we first review the basic process of neurogenesis in the hippocampus. Hippocampal neurogenesis is subject to regulation by both intrinsic and extrinsic factors in the neurogenic microenvironment of the DG. Several recent reviews have comprehensively summarized updated findings about this subject (Abrous et al. 2005; Zhao et al. 2008). Here, we focus our discussion

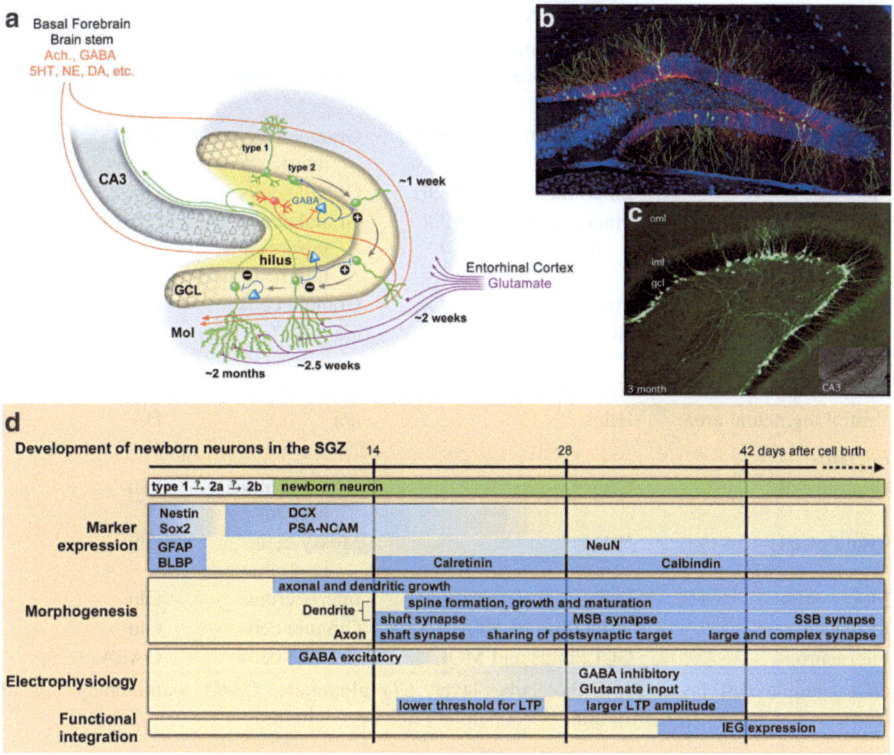

Fig. 9.1 Hippocampal neurogenesis. (**a**) Adult neurogenesis and the neural circuits in the dentate gyrus. Type 1 and type 2 progenitors are located in the subgranular zone (SGZ) and are morphologically distinct. Type 2 progenitors are positively regulated by tonic GABA, which is released by local interneurons (*blue cells*). *Green cells* in the granule cell layer (GCL) represent newborn neurons at progressive stages of neuronal maturation, with their dendrites elaborating in the molecular layer (Mol) and their axons extending towards CA3. GABA is an excitatory (*plus signs*) neurotransmitter for newborn neurons younger than 2 weeks old (*blue lines*). During the third week after cell birth, GABAergic inputs into newborn neurons change from excitatory to inhibitory (*minus signs*), which coincides with the growth of dendritic spines and the receiving of excitatory glutamatergic inputs from entorhinal cortex in these neurons (*purple lines*). The *red cell* in the hilus represents a mossy cell, which receives inputs from granule cells and innervates local interneurons and granule cells (*red lines*). The granule cells and hilar interneurons also receive inputs from the basal forebrain and the brain stem (*orange lines*) mediated by a variety of neurotransmitters, such as acetylcholine (Ach), serotonin (5HT), norepinephrine (NE), dopamine (DA) and GABA. (**b** and **c**) Labeling of adult-born neurons in the dentate gyrus using various approaches. In (**b**) *green cells* represent 8-week-old adult-born granule cells labeled by a GFP expressing retrovirus; *red cells* represent immature adult-born neurons labeled by Dcx; *blue cells*, labeled by NeuN, represent mature neurons. A horizontal section of the dentate gyrus from a 3-month-old *POMC-EGFP* transgenic mouse is shown in (**c**). *Labeled cells* represent immature granule cells. *oml* outer molecular layer, *iml* inner molecular layer, *gcl* granule cell layer. (**d**) Timeline for the development of newborn neurons in the SGZ, which is characterized by the expression of specific molecular markers, distinct morphologic and electrophysiological properties, and functional integration. *MSB* multiple-synapse boutons, *SSB* single-synapse boutons, *LTP* long-term potentiation, *IEG* immediate early gene. (**a**, **b** and **d**) Modified, with permission, from Zhao et al. (2008). (**c**) modified, with permission, from Overstreet et al. (2004)

Table 9.1 The neural network in the dentate gyrus

Anatomical origins	Target regions in hippocampus	Target neurons	Neurotransmitters
Afferents			
Entorhinal cortex	Outer two thirds of MOL	Granule cells, interneurons	Glu
Mossy cells	Inner one third of MOL	Granule cells	Glu
Medial septal nucleus, nucleus of diagonal band of Broca	Inner one third of MOL, hilus (SGZ)	Granule cells, mossy cells, interneurons	GABA, Ach
Supermammillary body	Inner one third of MOL	Granule cells	Glu, Substance P
Pontine nucleus of locus coeruleus	Hilus	n/a	NE
Median raphe nucleus	Hilus (SGZ)	Interneurons	5HT
Ventral tegmental area	Hilus	n/a	DA
Efferents			
Granule cells	CA3	CA3 pyramidal neurons	Glu
Granule cells	Hilus	Mossy cells, interneurons	Glu
Mossy cells	Hilus	Interneurons	Glu
Mossy cells	Inner third of MOL	Granule cells	Glu
Interneurons	GCL, hilus and MOL	Granule cells	GABA

GCL granule cell layer, *MOL* molecular layer, *Glu* glutamate, *GABA* γ-aminobutyric acid, *NE* norepinephrine, *5HT* serotonin, *DA* dopamine, *Ach* acetylcholine

on how neurogenesis can be regulated by neural network activity under both physiological and pathological conditions. In the last section, we summarize studies suggesting the potential functions of hippocampal neurogenesis.

9.2 Morphogenesis of Newborn Neurons in the Adult Hippocampus

Three different methods have been developed to study the morphogenesis of newborn neurons in the adult hippocampus: (1) the combination of Bromodeoxyuridine (BrdU) birthdating and certain immature neuron markers such as Doublecortin (Dcx) and the polysialated form of Neural Cell Adhesion Molecule (PSA-NCAM) revealed by immunohistochemistry (Fig. 9.1b); (2) the combination of BrdU birthdating and transgenic mouse lines in which fluorescent protein markers are specifically expressed in immature neurons. Two examples for such transgenic mice are POMC-GFP and Dcx-dsRed, in which reporter molecules GFP and dsRed are expressed from the *pro-opiomelanocortin* (*POMC*) and *dcx* gene promoters, respectively. (Brown et al. 2003; Overstreet et al. 2004) (Fig. 9.1c); and (3) retroviral labeling (van Praag et al. 2002) (Fig. 9.1b). Unlike the first two approaches, retrovirus-mediated cell labeling allows the birthdating and the direct visualization of cell morphology simultaneously. Therefore, it has been one of the major tools used in recent years to examine the morphological and electrophysiological properties of newborn neurons.

Newborn cells are frequently found to be clustered shortly after labeling by BrdU or retrovirus. They may migrate tangentially from the proliferating cell clusters before the formation of the primary radial process – the presumptive apical dendrites (Seki et al. 2007). Radial processes can be found in newborn cells as early as 2 days after BrdU labeling (Shapiro et al. 2007). The apical dendrites of newborn cells reach the inner, middle, and outer third of the molecular layer around 10, 14, and 17 days after cell birth, respectively (Esposito et al. 2005; Ge et al. 2006; Zhao et al. 2006). The dendrites of most of the newborn cells reach the hippocampal fissure by 21 days after cell birth, and the extent to which these processes continue to grow has not been characterized (Fig. 9.1a).

Dendritic spines are the major sites of excitatory synapses on granule cells and are first found on newborn cells during the third week after cell birth, shortly after the apical dendrites reach the middle third of the molecular layer (Zhao et al. 2006). The mechanism of initiation of spine growth is unclear, but it probably involves interaction with putative afferent connections. The growth of dendritic spines reaches a plateau around 56 days after cell birth; however, spine morphology is still subject to change. For example, the density of mushroom spines continues to increase after 56 days (Zhao et al. 2006). Electron microscopic studies have shown that newborn dentate granule cells form synapses on both dendritic shafts and spines (Shapiro et al. 2007; Toni et al. 2007). Interestingly, newborn cells are more likely to form synapses with multiple-synapse boutons when they are 3–4 weeks old than when they are more mature (Toni et al. 2007). It has been proposed that spines from new neurons may compete with existing spines for presynaptic terminals (Fig. 9.1a,d).

Similar to mature granule cells, newborn cells project their axons into the hilus and CA3 area through mossy fibers. The axons of newborn granule cells can be found in the CA3 area proximal to the DG as early as 10 days after cell labeling, and they rapidly project to the distal CA3 area (Zhao et al. 2006). The mossy fiber axons of the newborn cells reach the distal CA3 area as early as 16 days after cell birth. Anatomical analyses at the electromicroscopic level showed that these mossy fiber axons had already formed synapses with interneurons in the hilus and pyramidal cells in the CA3 at 17 days after cell birth. The size and complexity of these efferent synapses increase during the maturation of the cell, and they are indistinguishable from synapses formed by unlabeled mossy fiber axons at 75 days after cell birth (Toni et al. 2008) (Fig. 9.1a,d).

In summary, the morphogenesis of newborn granule cells is a prolonged process, involving the establishment of cell polarity and the coordinated development of dendrites and axons. Dendritic development includes several stages: establishment of the primary apical dendrite, growth and branching of dendritic processes, formation and modification of dendritic spines, and formation and maturation of afferent synapses. Similar to dendritic growth, the morphogenesis of mossy fiber axons includes an initial rapid growth of axonal fibers and the formation and maturation of efferent synapses.

9.3 Physiological Characteristics of Newborn Neurons in the Adult Hippocampus

The course of newborn neuron maturation can be roughly divided into five stages with distinct physiological properties at each stage (Fig. 9.1d). The first stage is characterized by the trophic effect of γ-aminobutyric acid (GABA), which promotes neuronal differentiation of the type 2 hippocampal progenitors (Tozuka et al. 2005). The second stage involves the depolarization of young immature neurons by GABA because of their higher intracellular Cl^- concentration (Esposito et al. 2005; Overstreet-Wadiche et al. 2005; Ge et al. 2006). The GABAergic excitatory response in immature cells appears to be critical for the maturation and integration of newborn neurons (Ge et al. 2006). The third stage marks the transition of GABA depolarization to hyperpolarization in newborn cells, as well as the formation of glutamate excitatory input (Esposito et al. 2005; Ge et al. 2007). This transition coincides with the initiation of dendritic spine formation (Zhao et al. 2006). The fourth stage is characterized by enhanced long-term potentiation (LTP) in newborn cells, which is reflected in the low threshold for LTP induction when the cells are 3–4 weeks old and the higher LTP amplitude when the cells are 4–6 weeks old (Schmidt-Hieber et al. 2004; Ge et al. 2007). The fifth stage involves the modulation and maturation of synaptic input of newborn cells. At this stage, newborn neurons eventually display similar physiological properties to neurons born during embryonic and postnatal neurogenesis (van Praag et al. 2002; Laplagne et al. 2006, 2007).

The integration of newborn neurons can also be studied by examining the expression of immediate early genes (IEGs), such as *c-fos*, *zif268*, *arc*, *homer1a*, etc., that are rapidly induced by behavioral experience and can serve as activity reporters for neurons (reviewed by Kubik et al. 2007). Zif268 expression can be detected in neurons as young as 2 weeks old, when LTP is induced in the perforant path electrically in vivo (Bruel-Jungerman et al. 2006). This observation suggests that newborn neurons at 2 weeks of age are able to display an activity-dependent response despite their immature membrane properties. However, the activity-dependent expression of IEGs (c-Fos, Zif268 and Homer1A) has not been detected in 15-day-old newborn neurons following kainic acid- or pentylenetetrazol-induced seizures (Jessberger and Kempermann 2003). Furthermore, the ratio of newborn neurons expressing IEGs is comparable to that of older mature neurons only when the cells are more than 5 weeks old (Jessberger and Kempermann 2003). In addition, normal hippocampal activity, such as the acquisition of Morris water maze (MWM) and enriched environment exposure, is able to induce the expression of activity-dependent IEGs in 6-week-old newborn neurons (Jessberger and Kempermann 2003; Kee et al. 2007; Tashiro et al. 2007).

Although the expression of IEGs has not been detected in newborn neurons at immature stages (0–3 weeks old) in normal behaving animals, experiences of mice during this time window can modulate the activity of these newborn neurons in an experience-dependent manner when they become older. Using BrdU labeling to birthdate newborn neurons, Tashiro et al. (2007) exposed the mice to either an enriched environment or home cage when the BrdU-labeled cells were 2 weeks old. When these cells reached 6 weeks of age, the mice were re-exposed to the same enriched environment and increased numbers of active newborn neurons were found in the group of mice with previous exposure to the enriched environment. However, when they trained these mice in MWM instead of re-exposure to the enriched environment, no such increases in activity were detected. This type of experience-dependent modulation of newborn neuron activity also occurs in adult-born olfactory granule neurons (Magavi et al. 2005).

Analysis of the expression of IEGs further reveals that adult-born neurons in the DG may be preferentially incorporated into the neural network supporting spatial memory. Kee et al. (2007) birthdated newborn cells with BrdU, trained mice in MWM at various time points after labeling and examined the expressions of c-Fos and Arc in 10-week-old BrdU-labeled neurons after a series of memory retrieval tests. They found that the proportion of c-Fos- or Arc-positive cells was higher in BrdU-labeled newborn neurons than in the Neuronal Nuclei (NeuN)-labeled mature neurons when the training occurred 4 or more weeks after BrdU labeling. In another study, 5-month-old adult-born neurons were preferentially activated when rats were allowed to explore a novel environment (Ramirez-Amaya et al. 2006). However, Jessberger and Kempermann (2003) were not able to detect this preferential activation of newborn neurons in mice trained in MWM or in mice enduring chemical-induced seizures. The preferential incorporation of young adult-born neurons (age 4–8 weeks) is consistent with their enhanced plasticity, as discussed above. It is not clear why the 5-month-old adult-born neurons maintained this preference, given that adult-born neurons at this age are not morphologically or physiologically different from mature granule cells.

Together, the electrophysiological properties and the expression of the IEGs suggest the functional integration of adult-born neurons into hippocampal circuits. Moreover, the integrated newborn neurons are subject to experience-dependent activation and are preferentially incorporated into spatial memory networks, suggesting the involvement of these adult-born neurons in hippocampal information processing.

9.4 Regulation of Hippocampal Neurogenesis by Neural Circuit Activity

Adult neurogenesis occurs in a microenvironment containing complex neural networks. Thus, neural circuit activity in the hippocampus is a key factor in both the proliferation of neural progenitor cells and the subsequent survival/integration of newborn neurons. As discussed above, the granule cells receive afferents from many brain regions (Fig. 9.1a). The putative roles of these pathways in hippocampal neurogenesis have been examined (Table 9.2).

To examine the modulation of neurogenesis by inputs from the entorhinal cortex, Bruel-Jungerman et al. (2006) and Chun et al. (2006) induced LTP in adult rats

Table 9.2 Regulation of hippocampal neurogenesis by neural circuit activity

Experimental manipulations	Affected afferents/circuits	Effects on neurogenesis
LTP induction of MPP	Stimulation of perforant pathway	Increases proliferation and newborn cell survival
Ibotenic acid injection to entorhinal cortex	Lesion of entorhinal cortex	Increases proliferation
Entorhinal axotomy	Lesion of perforant pathway	No effect on newborn cell survival
Fornix/fimbria lesion	Disruption of septo-hippocampus pathway	Decreases proliferation
NMDA injection to medial septum	Lesion of septal GABAergic neurons	Decreases newborn cell survival
Icv infusion of 192 IgG-saporin	Lesion of forebrain cholinergic neurons	Decreases proliferation and neurogenesis
Short term pilocarpine-induced seizure	Lesion of mossy cells	Slows down newborn neuron maturation
5,7-DHT injection to raphe nuclei	Lesion of raphe nucleus	Decreases proliferation and neurogenesis
Fetal raphe nuclei grafting	Serotonergic innervations	Increases proliferation and neurogenesis
Systemic DSP-4 treatment	Depletion of norepinephrine	Decreases proliferation and no effect on survival
Systemic MTPT treatment	Dopaminergic dennervations	Decreases proliferation
Electroconvulsive treatment	Brief seizures	Increases proliferation and neurogenesis
Prolonged pilocarpine-induced seizures	Induction of status epilepticus	Increases proliferation and neurogenesis, ectopic newborn neuron migration, accelerated newborn neuron maturation
Prolonged perforant path stimulation	Induction of focal hippocampal seizures	Increases proliferation
Amygdala kindling	Induction of seizures	Increases proliferation
Kainic acid treatment	Induction of convulsive seizures	Increases neurogenesis, persistence of basal dendrites, aberrant synapse formation

LTP long-term potentiation, *MPP* medial perforant path, *icv* intracerebroventricular, *5,7-DHT* 5,7-dihydroxytryptamine, *DSP-4* *N*-(2-chloroethyl)-*N*-ethyl-2-bromobenzylamine hydrochloride, *MTPT* 1-methyl-4-phenyl-1,2,3,6-tetrahydropyridine

by the tetanic stimulation of medial perforant path via a unilaterally implanted electrode. This manipulation resulted in the stimulation of neural progenitor proliferation in the ipsilateral but not contralateral side of the hippocampus. LTP induction also enhances the survival of 1- to 2-week-old newborn neurons, which readily respond to LTP stimulation by expressing immediate early gene *zif268*. However, lesion of the entorhinal cortex leads to increased progenitor cell proliferation in the DG (Cameron et al. 1995), and disruption of perforant path projection has no effect on the survival of newborn neurons (Fontana et al. 2006). Given the inconsistency of these observations, it is currently unclear how the entorhinal inputs affect hippocampal neurogenesis.

Disruption of the septal/basal forebrain projection by fornix/fimbria lesion causes a decrease in progenitor cell proliferation (Fontana et al. 2006). Infusion of *N*-methyl-D-aspartic acid (NMDA) into the medial septum strongly decreases GABAergic inputs into the hippocampus and results in a severe reduction in newborn cell survival (Van der Borght et al. 2005). Through the septo-hippocampal pathway, the hippocampus also receives cholinergic inputs, which have a major impact on hippocampus-dependent learning and memory. Immunotoxic lesion of cholinergic neurons in the medial septum leads to decreased proliferation of hippocampal progenitors and reduced neurogenesis in SGZ (Cooper-Kuhn et al. 2004; Mohapel et al. 2005).

The DG receives the serotonergic inputs from raphe nuclei. Neurotoxin lesion of serotonergic neurons in raphe nuclei leads to decreased cell proliferation and neurogenesis, which can be reversed by serotonergic reinnervation (Brezun and Daszuta 1999, 2000a,b). The role of norepinephrine in the regulation of SGZ neurogenesis has been examined by selective neurotoxic depletion of norepinephrine, which results in decreased SGZ cell proliferation (Kulkarni et al. 2002). Dopaminergic fibers have been detected in the SGZ of the hippocampus. Neurotoxic depletion of dopaminergic innervations in adult brain results in decreased cell proliferation in SGZ, which can be restored by dopaminergic reinnervation (Hoglinger et al. 2004). The involvement of these circuitries in hippocampal neurogenesis is also suggested by pharmacological and genetic studies of the corresponding receptors for monoamines, which are summarized by Zhao et al. (2008).

To investigate the role of the associational/commissural afferents in hippocampal neurogenesis, Marques-Mari et al. (2007) reported that destroying the mossy cells by pilocarpine-induced seizure impeded the maturation of newborn neurons. Although it is difficult to reach definitive conclusions from these studies due to the side effects of experimental manipulations, these observations revealed the important role of neural circuit activity in the regulation of hippocampal neurogenesis.

Hippocampal neurogenesis is also influenced by aberrant neural circuit activities under pathological or therapeutic conditions. Electroconvulsive treatment, which induces a brief seizure in the brain and has therapeutic effects on many emotional disorders, increases the proliferation of neural progenitors and neurogenesis in the hippocampus (Madsen et al. 2000; Malberg et al. 2000). Prolonged seizures can cause various abnormalities in hippocampal neurogenesis: (1) pilocarpine- or electrically induced seizures increase neural progenitor proliferation and neurogenesis (Parent et al. 1997, 1998); (2) seizures cause abnormal morphogenesis of adult-born neurons. The neurons born shortly after seizure induction ectopically extend spiny dendrites into the hilus, suggesting the aberrant integration of these neurons (Jessberger et al. 2007). Another study reported that immature neurons born 1–3 weeks before seizure also possessed basal dendrites (Walter et al. 2007), which were not detected by Jessberger and colleagues. As also seen in mature granule cells, seizure induced the sprouting of mossy fibers in 4-week-old newborn neurons

(Jessberger et al. 2007); (3) ectopic migration is often detected in seizure-induced newborn neurons, which are abnormally located in the hilus and the molecular layer (Parent et al. 2006). Further studies suggested that the loss of Reelin secreted by hilar interneurons was responsible for this ectopic migration (Gong et al. 2007); and (4) neurons born into an epileptic environment survive for a long time after seizure. They aberrantly integrate into the neural circuit, given the synaptic formation on their basal dendrites and their expression of c-Fos after handling-induced seizures (Jessberger et al. 2007). In addition, seizures accelerate the integration of newborn neurons, as indicated by both morphological and electrophysiological parameters (Overstreet-Wadiche et al. 2006). Together, these studies suggest abnormal circuit activity leads to aberrant adult neurogenesis in the hippocampus.

The involvement of neural activity in hippocampal neurogenesis has been demonstrated in animals with behavioral experiences. Learning a hippocampus-dependent task but not a hippocampus-independent task increases the survival of immature newborn neurons (Gould et al. 1999). Further studies have revealed that it was learning but not training that increased newborn neuron survival (Leuner et al. 2006; Sisti et al. 2007). In addition, this survival-enhancing effect is only limited to newborn neurons at particular neuronal maturation stages (Dupret et al. 2007). Moreover, living in an environmental environment, which provides numerous opportunities for hippocampal stimulation, enhances the survival of newborn neuron (Kempermann et al. 1997a). Finally, as discussed above, abnormal hippocampal activities, such as pilocarpine-induced seizures, also have a great impact on the production and functional integration of newborn neurons.

Recent studies have demonstrated that there is a critical time window for the experience-dependent enhancement of newborn neurons' survival. Tashiro et al. (2007) labeled newborn cells with BrdU and exposed mice to an enriched environment for a week at various time points after BrdU labeling. They found that the environmental enrichment only enhanced the survival of newborn neurons of 0- to 3-week-old but not that of 4-week-old neurons. Similarly, MWM learning only enhanced the survival of newborn neurons younger than 4 weeks old (Kee et al. 2007). Moreover, the survival of 1-to 2-week-old newborn neurons was promoted by LTP induction in the medial perforant path (Bruel-Jungerman et al. 2006). All these observations suggest that the activity of hippocampal circuits has a pro-survival effect on newborn neurons within a critical time window during adult hippocampal neurogenesis. In the olfactory bulb, activity-dependent regulation for the survival of the newborn neurons has also been observed in a critical time period (Yamaguchi and Mori 2005).

Neurotransmitters are among the key molecular factors that may mediate the regulation of hippocampal neurogenesis by circuit activity. Among several classes of neurotransmitters involved in regulating hippocampal neurogenesis, glutamate and GABA have been extensively studied. In the brain, the excitatory inputs are mainly mediated by glutamate. There are four major types of glutamate receptors, including NMDA receptor, α-amino-3-hydroxy-5-methyl-4-isoxazolepropionic acid (AMPA) receptor, Kainate receptor and metabotrophic glutamate receptors. Of these, the effect of NMDA signaling on hippocampal neurogenesis has been examined in most detail, given its key role in LTP (Nacher and McEwen 2006). Activation of the NMDA receptor by systemic administration of NMDA results in decreased cell proliferation in DG (Cameron et al. 1995). In contrast, inhibiting NMDA signaling by antagonists of NMDA receptor leads to a rapid increase in neural progenitor proliferation (Cameron et al. 1995; Okuyama et al. 2004). This effect can last for a few days and occurs in both young and aged animals (Nacher et al. 2001, 2003; Okuyama et al. 2004). Consistently, decreased expression of NR1, the essential subunit of NMDA receptor, correlates with increased neural progenitor proliferation/maturation in mice (Bursztajn et al. 2007). However, whether this increased proliferation results in an increased production of adult-born neuron remains controversial (Cameron et al. 1995; Joo et al. 2007). Similarly, blocking the AMPA/Kainate receptors or group II metabotrophic glutamate receptors leads to an up-regulation of progenitor proliferation and/or maturation (Bernabeu and Sharp 2000; Yoshimizu and Chaki 2004). It is not clear whether glutamate regulates the proliferation of neural progenitors in a cell-autonomous manner. The expression of NMDA receptors in neural progenitors and immature adult-born neurons has been suggested by both in vitro and in vivo studies (Joo et al. 2007; Nacher et al. 2007). However, hippocampal progenitors do not respond to NMDA or AMPA stimulation in electrophysiological recordings (Tozuka et al. 2005), suggesting that NMDA signaling could have an indirect impact on the proliferation of hippocampal neural progenitors. In contrast, the survival of newborn neurons is partially dependent on the NMDA receptor in a cell-autonomous manner during a critical time window when these neurons start to receive glutamatergic inputs, as revealed by the retrovirus-mediated single cell knockout of NR1 (Tashiro et al. 2006). In addition, the survival of NR1-deficient newborn neurons is slightly increased by the global inhibition of NMDA receptor in the critical time window, which does not affect the survival of wildtype adult-born neurons. Thus, it is proposed that newborn neurons may compete for certain factors for their survival via the activation of NMDA receptors, similar to activity-dependent survival of neurons during development.

GABA is another important regulator of progenitor cell proliferation and differentiation as well as the maturation of newborn neurons in hippocampal neurogenesis. GABA, which is likely released from local hilar interneurons and septal inputs, is an excitatory stimulus for the type 2 neural progenitors and immature neurons younger than 2 weeks old (Tozuka et al. 2005; Ge et al. 2006). Morphological studies suggest that there may be direct synaptic connections between hilar interneurons and type 2 cells, though GABA could elicit its effects on neural progenitors or immature neurons through tonic activation. The GABAergic input into progenitor cells promotes the differentiation of these progenitors into neurons via the induction of proneural gene NeuroD. Consistently, GABA agonists lead to decreased numbers of dividing type 2 progenitors and increased numbers of newborn neurons. Furthermore, disruption of the Na-K-2Cl cotransporter (NKCC1) decreases the intracellular Cl^- ion concentration, converts the GABA input from depolarization to hyperpolarization and significantly impairs the maturation and synaptic integration of newborn neurons. Finally, depolarizing GABAergic network activity is stronger in postnatal mice compared to adult animals, correlating with an increased maturation rate of newborn neurons in postnatal DG (Overstreet-Wadiche et al. 2006).

In summary, the local circuit activity in the adult hippocampus plays an important role in the production and subsequent integration of newborn neurons during adult neurogenesis, with the neurotransmitters as key mediators.

9.5 Functional Implications of Hippocampal Neurogenesis

The integration of adult-born neurons into the existing circuitry of the hippocampus indicates their involvement in hippocampus-dependent brain functions. This finding has led to extensive investigation of the functional significance of hippocampal neurogenesis in cognition and emotionality. The level of hippocampal neurogenesis is correlated with emotional status: hippocampal neurogenesis is decreased in stressed animals and is increased by anti-depressant treatments. Furthermore, the effectiveness of anti-depressants depends on hippocampal neurogenesis. The involvement of hippocampal neurogenesis in affective behavior will be discussed in detail in other chapters. The potential roles of hippocampal neurogenesis in learning and memory have been examined using both experimental approaches and computational simulation. The computation modeling of hippocampal neurogenesis has been recently reviewed by Aimone and Wiskott (2007). We will summarize the current experimental evidence that suggests the functional implications of hippocampal neurogenesis in learning and memory, using both correlative and reductionistic approaches.

9.5.1 Correlational Evidence

The genetic background of an animal has a major impact on hippocampal neurogenesis (Kempermann et al. 1997b). An animal's performance in MWM correlates with the level of hippocampal neurogenesis (Kempermann and Gage 2002). However, this observation could be complicated by the fact that the locomotor activity of individuals is also subject to the regulation by genetic makeup. This correlation between hippocampal neurogenesis and learning and memory has also been found in mutant mice that have a deficiency in individual molecules. For example, *mbd1*$^{-/-}$ mice with reduced hippocampal neurogenesis have defects in DG LTP and are defective in both acquisition and retention of MWM (Zhao et al. 2003). Similarly, mice deficient in FGFR1 in the central nervous system (CNS) have reduced cell proliferation in SGZ and are defective in spatial memory consolidation in the MWM test (Zhao et al. 2007). However, deletion of these genes during CNS development prevents us from bridging the causal gap between adult neurogenesis and cognition.

The experiences of animals have a powerful influence on hippocampal neurogenesis. Voluntary exercise increases progenitor proliferation in SGZ and selectively enhances LTP in DG, which correlates with improved performance in MWM in both young and aged mice (van Praag et al. 1999, 2005). As mentioned above, exposure to an environmental environment promotes the survival of newborn neurons and leads to improved performance in MWM in both young and aged mice (Kempermann et al. 1997a, 1998, 2002). Stress and aging are two major negative regulators of hippocampal neurogenesis. Although some studies suggest that stressed animals are defective in learning and memory, the relationship between stress and cognition is rather complicated (Shors 2004). However, it is well documented that a progressive decline in neurogenesis occurs with aging in a variety of species ranging from rodent to primates (reviewed by Klempin and Kempermann 2007). Taking advantage of spontaneous individual difference in hippocampal neurogenesis in aged rats, a correlation between hippocampal neurogenesis and the performance in MWM has been revealed in some but not other studies.

In general, experience-regulated hippocampal neurogenesis is correlated with the animal's cognitive performance. However, these environmental factors also influence structural plasticity and molecular regulations in the adult brain. It is not clear whether the changes in hippocampal neurogenesis are the main causes for the changed cognitive performance. It was recently reported that the beneficial effects of the environmental enrichment on spatial memory were not dependent on SGZ neurogenesis, which was ablated by X-ray irradiation (Meshi et al. 2006). However, another study suggested that the improvement in object recognition memory induced by the enriched environment required adult neurogenesis (Bruel-Jungerman et al. 2005). In a third study using *presenilin1*$^{-/-}$ mice, the enhancement of hippocampal neurogenesis by the environmental enrichment was attenuated (Feng et al. 2001). The *presenilin1*$^{-/-}$ mice formed stronger contextual fear memory when they were exposed to an enriched environment after the acquisition of fear conditioning. Therefore, it remains to be clarified whether hippocampal neurogenesis is a major causal factor for the changes in cognitive behavior under these conditions.

9.5.2 *Causal Evidence*

To more directly examine the function of hippocampal neurogenesis, several experimental methods have been developed to reduce adult neurogenesis. (1) Systemic administration of anti-mitotic drugs, such as methylazoxymethanol acetate (MAM), is able to effectively reduce hippocampal neurogenesis (Shors et al. 2001). However, high levels of MAM are toxic to animals and may have deleterious effects on animals' health (Dupret et al. 2005). (2) Low-dose irradiation of either whole brain or restricted brain regions almost completely abolishes hippocampal neurogenesis (Wojtowicz 2006). However, irradiation induced inflammation not only has a negative effect on neurogenesis but also harmfully influences many other processes in the brain (Monje et al. 2003). (3) Genetically engineered mouse models are able to allow more specific ablating of hippocampal neurogenesis. For example, in *GFAP-tk* transgenic mice, the thymidine kinase (tk) from *herpes simplex virus* is expressed in neural progenitors and astrocytes driven by the promoter of *gfap* gene (Garcia et al. 2004). The proliferating tk-expressing neural progenitors are ablated when animals are administered the nucleotide analog, ganciclovir, which disrupts chain elongation during DNA replication. A recent study reported that the inducible deletion of *tlx* (a gene encoding an orphan nuclear receptor) in adult mice led to a reduction in hippocampal neurogenesis, though all cells in adult animals could be affected and additional controls had to be included (Zhang et al. 2008).

Spatial learning and memory have been evaluated in animals with reduced neurogenesis, given the critical role of the hippocampus in this process. Although most studies failed to detect any defects in spatial learning, impaired long-term retention of spatial memory was detected in irradiated rats (Snyder et al. 2005). Furthermore, during both training and reverse training phases, defective learning was detected in the conditional $tlx^{-/-}$ mice when the gene deletion was induced in adulthood (Zhang et al. 2008). In addition, X-ray irradiation of 2-month-old mice led to impaired spatial learning and memory in the Barnes maze but not in the MWM (Raber et al. 2004). However, this observation likely depends on the age of the animal. When 3-week-old mice were irradiated, they displayed impaired spatial learning and memory in MWM but not in the Barnes maze (Rola et al. 2004). More intriguingly, it is reported that ablating neurogenesis, using either irradiation or *GFAP-tk* mice, led to an improvement in hippocampus-dependent working memory in the 8-armed radial maze (Saxe et al. 2007).

To examine the role of hippocampal neurogenesis in hippocampus-dependent conditioning, Shors et al. (2001) systemically treated rats with MAM and found that these rats failed to form conditioned responses in trace eyeblink conditioning or trace fear conditioning. However, MAM treatment did not affect contextual fear conditioning, which is also dependent on hippocampus (Shors et al. 2002). Similarly, no contextual fear conditioning defects were detected in the inducible conditional $tlx^{-/-}$ mice (Zhang et al. 2008). In contrast, irradiated rats or mice and ganciclovir-treated *GFAP-tk* mice were defective in contextual fear conditioning (Saxe et al. 2006; Winocur et al. 2006).

Place and object recognition memories have also been examined in animals with ablated neurogenesis. In a delayed, non-matching-to-sample test conducted in a water maze, Winocur et al. (2006) reported that irradiated rats did not perform as well as sham irradiated controls when the time interval between training and test trials became long enough. However, object recognition memory was not affected in irradiated or MAM-treated rats when recognition memory was examined in a simpler context (Bruel-Jungerman et al. 2005). Nevertheless, a place recognition defect was detected 1 or 3 weeks after irradiation in rats (Madsen et al. 2003). In addition, as mentioned above, MAM treatment prevented the enhancement of object recognition memory by enriched environment (Bruel-Jungerman et al. 2005).

In summary, these studies suggest that hippocampal neurogenesis could influence the cognitive behavior of animals. However, the contradictory nature of the results from these studies makes it difficult to reach any definitive conclusions about the exact functions of hippocampal neurogenesis in cognition. These discrepancies are probably due to differences in animal species and strains, the detailed behavior procedures and the different knockdown strategies with undesired side effects. Moreover, current studies mainly use established behavioral tests, which are based on whole hippocampal lesion models. These methods may not be sensitive enough to reveal the function of hippocampal neurogenesis, since newborn neurons only constitute a small part of the anatomical structure of the hippocampus. Therefore, to definitively demonstrate the function of SGZ neurogenesis, selective ablation approaches with few side effects and specific behavioral tests aimed at challenging the newborn neurons need to be developed in the future.

9.6 Concluding Remarks

Intensive research in past decades revealed many of the detailed processes of hippocampal neurogenesis, from the proliferation of neural stem/progenitor cells, to the differentiation of the newborn cells, to the maturation and migration of newborn neurons and to the final integration of adult-born neurons into existing neural circuits. The process of hippocampal neurogenesis is subject to extensive regulation by both neural circuit activities and extrinsic/intrinsic molecular factors. However, the mechanisms of these various types of regulation are largely unknown. The integration of newborn neurons into adult hippocampal circuits has a consequence in modulating animals' behavior, although the exact functions of adult neurogenesis in hippocampus-related information processing remain elusive. As discussed in other chapters of this book, it has become clear that hippocampal neurogenesis readily responds to pathological conditions in adult brains, which provides opportunities for new therapeutic development. Further investigations addressing these issues will help us not only to better understand the basic biology of hippocampal neurogenesis but also to develop new therapies to combat neurological disorders in the adult brain.

Acknowledgments The authors thank Jamie Simon for illustrations and Mary Lynn Gage for helpful comments. This work is supported by National Institutes of Health, Picower Foundation, Lookout Foundation, and J.S. McDonnell Foundations.

References

Abrous DN, Koehl M, Le Moal M (2005). Adult neurogenesis: from precursors to network and physiology. Physiol Rev 85: 523–569.

Aimone JB, Wiskott L (2007). Computational modeling of adult neurogenesis. In: Gage FH, Kempermann G, Song H (eds.) Adult neurogenesis. Cold Spring Harbor Laboratory Press, Cold Spring Harbor.

Amaral D, Lavenex P (2007). Hippocampal neuroanatomy. In: Anderson P, Morris R, Amaral D, Bliss T, O'Keefe J (eds.) The hippocampus book. Oxford University Press, Oxford.
Bernabeu R, Sharp FR (2000). NMDA and AMPA/kainate glutamate receptors modulate dentate neurogenesis and CA3 synapsin-I in normal and ischemic hippocampus. J Cereb Blood Flow Metab 20: 1669–1680.
Brezun JM, Daszuta A (1999). Depletion in serotonin decreases neurogenesis in the dentate gyrus and the subventricular zone of adult rats. Neuroscience 89: 999–1002.
Brezun JM, Daszuta A (2000a). Serotonin may stimulate granule cell proliferation in the adult hippocampus, as observed in rats grafted with foetal raphe neurons. Eur J Neurosci 12: 391–396.
Brezun JM, Daszuta A (2000b). Serotonergic reinnervation reverses lesion-induced decreases in PSA-NCAM labeling and proliferation of hippocampal cells in adult rats. Hippocampus 10: 37–46.
Brown JP, Couillard-Despres S, Cooper-Kuhn CM et al (2003). Transient expression of doublecortin during adult neurogenesis. J Comp Neurol 467: 1–10.
Bruel-Jungerman E, Laroche S, Rampon C (2005). New neurons in the dentate gyrus are involved in the expression of enhanced long-term memory following environmental enrichment. Eur J Neurosci 21: 513–521.
Bruel-Jungerman E, Davis S, Rampon C et al (2006). Long-term potentiation enhances neurogenesis in the adult dentate gyrus. J Neurosci 26: 5888–5893.
Bursztajn S, Falls WA, Berman SA et al (2007). Cell proliferation in the brains of NMDAR NR1 transgenic mice. Brain Res 1172: 10–20.
Cameron HA, McEwen BS, Gould E (1995). Regulation of adult neurogenesis by excitatory input and NMDA receptor activation in the dentate gyrus. J Neurosci 15: 4687–4692.
Chun SK, Sun W, Park JJ et al (2006). Enhanced proliferation of progenitor cells following long-term potentiation induction in the rat dentate gyrus. Neurobiol Learn Mem 86: 322–329.
Cooper-Kuhn CM, Winkler J, Kuhn HG (2004). Decreased neurogenesis after cholinergic forebrain lesion in the adult rat. J Neurosci Res 77: 155–165.
Dupret D, Montaron MF, Drapeau E et al (2005). Methylazoxymethanol acetate does not fully block cell genesis in the young and aged dentate gyrus. Eur J Neurosci 22: 778–783.
Dupret D, Fabre A, Dobrossy MD et al (2007). Spatial learning depends on both the addition and removal of new hippocampal neurons. PLoS Biol 5: e214.
Esposito MS, Piatti VC, Laplagne DA et al (2005). Neuronal differentiation in the adult hippocampus recapitulates embryonic development. J Neurosci 25: 10074–10086.
Feng R, Rampon C, Tang YP et al (2001). Deficient neurogenesis in forebrain-specific presenilin-1 knockout mice is associated with reduced clearance of hippocampal memory traces. Neuron 32: 911–926.
Fontana X, Nacher J, Soriano E et al (2006). Cell proliferation in the adult hippocampal formation of rodents and its modulation by entorhinal and fimbria-fornix afferents. Cereb Cortex 16: 301–312.
Garcia AD, Doan NB, Imura T et al (2004). GFAP-expressing progenitors are the principal source of constitutive neurogenesis in adult mouse forebrain. Nat Neurosci 7: 1233–1241.
Ge S, Goh EL, Sailor KA et al (2006). GABA regulates synaptic integration of newly generated neurons in the adult brain. Nature 439: 589–593.
Ge S, Yang CH, Hsu KS et al (2007). A critical period for enhanced synaptic plasticity in newly generated neurons of the adult brain. Neuron 54: 559–566.
Gong C, Wang TW, Huang HS et al (2007). Reelin regulates neuronal progenitor migration in intact and epileptic hippocampus. J Neurosci 27: 1803–1811.
Gould E, Beylin A, Tanapat P et al (1999). Learning enhances adult neurogenesis in the hippocampal formation. Nat Neurosci 2: 260–265.
Hoglinger GU, Rizk P, Muriel MP et al (2004). Dopamine depletion impairs precursor cell proliferation in Parkinson disease. Nat Neurosci 7: 726–735.
Jessberger S, Kempermann G (2003). Adult-born hippocampal neurons mature into activity-dependent responsiveness. Eur J Neurosci 18: 2707–2712.

Jessberger S, Zhao C, Toni N et al (2007). Seizure-associated, aberrant neurogenesis in adult rats characterized with retrovirus-mediated cell labeling. J Neurosci 27: 9400–9407.

Joo JY, Kim BW, Lee JS et al (2007). Activation of NMDA receptors increases proliferation and differentiation of hippocampal neural progenitor cells. J Cell Sci 120: 1358–1370.

Kee N, Teixeira CM, Wang AH et al (2007). Preferential incorporation of adult-generated granule cells into spatial memory networks in the dentate gyrus. Nat Neurosci 10: 355–362.

Kempermann G, Gage FH (2002). Genetic determinants of adult hippocampal neurogenesis correlate with acquisition, but not probe trial performance, in the water maze task. Eur J Neurosci 16: 129–136.

Kempermann G, Kuhn HG, Gage FH (1997a). More hippocampal neurons in adult mice living in an enriched environment. Nature 386: 493–495.

Kempermann G, Kuhn HG, Gage FH (1997b). Genetic influence on neurogenesis in the dentate gyrus of adult mice. Proc Natl Acad Sci USA 94: 10409–10414.

Kempermann G, Kuhn HG, Gage FH (1998). Experience-induced neurogenesis in the senescent dentate gyrus. J Neurosci 18: 3206–3212.

Kempermann G, Gast D, Gage FH (2002). Neuroplasticity in old age: sustained fivefold induction of hippocampal neurogenesis by long-term environmental enrichment. Ann Neurol 52: 135–143.

Klempin F, Kempermann G (2007). Adult hippocampal neurogenesis and aging. Eur Arch Psychiatry Clin Neurosci 257: 271–280.

Kubik S, Miyashita T, Guzowski JF (2007). Using immediate-early genes to map hippocampal subregional functions. Learn Mem 14: 758–770.

Kulkarni VA, Jha S, Vaidya VA (2002). Depletion of norepinephrine decreases the proliferation, but does not influence the survival and differentiation, of granule cell progenitors in the adult rat hippocampus. Eur J Neurosci 16: 2008–2012.

Laplagne DA, Esposito MS, Piatti VC et al (2006). Functional convergence of neurons generated in the developing and adult hippocampus. PLoS Biol 4: e409.

Laplagne DA, Kamienkowski JE, Esposito MS et al (2007). Similar GABAergic inputs in dentate granule cells born during embryonic and adult neurogenesis. Eur J Neurosci 25: 2973–2981.

Leuner B, Waddell J, Gould E et al (2006). Temporal discontiguity is neither necessary nor sufficient for learning-induced effects on adult neurogenesis. J Neurosci 26: 13437–13442.

Madsen TM, Treschow A, Bengzon J et al (2000). Increased neurogenesis in a model of electroconvulsive therapy. Biol Psychiatry 47: 1043–1049.

Madsen TM, Kristjansen PE, Bolwig TG et al (2003). Arrested neuronal proliferation and impaired hippocampal function following fractionated brain irradiation in the adult rat. Neuroscience 119: 635–642.

Magavi SS, Mitchell BD, Szentirmai O et al (2005). Adult-born and preexisting olfactory granule neurons undergo distinct experience-dependent modifications of their olfactory responses in vivo. J Neurosci 25: 10729–10739.

Malberg JE, Eisch AJ, Nestler EJ et al (2000). Chronic antidepressant treatment increases neurogenesis in adult rat hippocampus. J Neurosci 20: 9104–9110.

Marques-Mari AI, Nacher J, Crespo C et al (2007). Loss of input from the mossy cells blocks maturation of newly generated granule cells. Hippocampus 17: 510–524.

Meshi D, Drew MR, Saxe M et al (2006). Hippocampal neurogenesis is not required for behavioral effects of environmental enrichment. Nat Neurosci 9: 729–731.

Mohapel P, Leanza G, Kokaia M et al (2005). Forebrain acetylcholine regulates adult hippocampal neurogenesis and learning. Neurobiol Aging 26: 939–946.

Monje ML, Toda H, Palmer TD (2003). Inflammatory blockade restores adult hippocampal neurogenesis. Science 302: 1760–1765.

Nacher J, McEwen BS (2006). The role of N-methyl-D-asparate receptors in neurogenesis. Hippocampus 16: 267–270.

Nacher J, Rosell DR, Alonso-Llosa G et al (2001). NMDA receptor antagonist treatment induces a long-lasting increase in the number of proliferating cells, PSA-NCAM-immunoreactive granule neurons and radial glia in the adult rat dentate gyrus. Eur J Neurosci 13: 512–520.

Nacher J, Alonso-Llosa G, Rosell DR et al (2003). NMDA receptor antagonist treatment increases the production of new neurons in the aged rat hippocampus. Neurobiol Aging 24: 273–284.
Nacher J, Varea E, Miguel Blasco-Ibanez J et al (2007). N-methyl-d-aspartate receptor expression during adult neurogenesis in the rat dentate gyrus. Neuroscience 144: 855–864.
Okuyama N, Takagi N, Kawai T et al (2004). Phosphorylation of extracellular-regulating kinase in NMDA receptor antagonist-induced newly generated neurons in the adult rat dentate gyrus. J Neurochem 88: 717–725.
Overstreet LS, Hentges ST, Bumaschny VF et al (2004). A transgenic marker for newly born granule cells in dentate gyrus. J Neurosci 24: 3251–3259.
Overstreet Wadiche L, Bromberg DA, Bensen AL et al (2005). GABAergic signaling to newborn neurons in dentate gyrus. J Neurophysiol 94: 4528–4532.
Overstreet-Wadiche LS, Bromberg DA, Bensen AL et al (2006). Seizures accelerate functional integration of adult-generated granule cells. J Neurosci 26: 4095–4103.
Parent JM, Yu TW, Leibowitz RT et al (1997). Dentate granule cell neurogenesis is increased by seizures and contributes to aberrant network reorganization in the adult rat hippocampus. J Neurosci 17: 3727–3738.
Parent JM, Janumpalli S, McNamara JO et al (1998). Increased dentate granule cell neurogenesis following amygdala kindling in the adult rat. Neurosci Lett 247: 9–12.
Parent JM, Elliott RC, Pleasure SJ et al (2006). Aberrant seizure-induced neurogenesis in experimental temporal lobe epilepsy. Ann Neurol 59: 81–91.
Raber J, Rola R, LeFevour A et al (2004). Radiation-induced cognitive impairments are associated with changes in indicators of hippocampal neurogenesis. Radiat Res 162: 39–47.
Ramirez-Amaya V, Marrone DF, Gage FH et al (2006). Integration of new neurons into functional neural networks. J Neurosci 26: 12237–12241.
Rola R, Raber J, Rizk A et al (2004). Radiation-induced impairment of hippocampal neurogenesis is associated with cognitive deficits in young mice. Exp Neurol 188: 316–330.
Saxe MD, Battaglia F, Wang JW et al (2006). Ablation of hippocampal neurogenesis impairs contextual fear conditioning and synaptic plasticity in the dentate gyrus. Proc Natl Acad Sci USA 103: 17501–17506.
Saxe MD, Malleret G, Vronskaya S et al (2007). Paradoxical influence of hippocampal neurogenesis on working memory. Proc Natl Acad Sci USA 104: 4642–4646.
Schmidt-Hieber C, Jonas P, Bischofberger J (2004). Enhanced synaptic plasticity in newly generated granule cells of the adult hippocampus. Nature 429: 184–187.
Seki T, Namba T, Mochizuki H et al (2007). Clustering, migration, and neurite formation of neural precursor cells in the adult rat hippocampus. J Comp Neurol 502: 275–290.
Shapiro LA, Upadhyaya P, Ribak CE (2007). Spatiotemporal profile of dendritic outgrowth from newly born granule cells in the adult rat dentate gyrus. Brain Res 1149: 30–37.
Shors TJ (2004). Learning during stressful times. Learn Mem 11: 137–144.
Shors TJ, Miesegaes G, Beylin A et al (2001). Neurogenesis in the adult is involved in the formation of trace memories. Nature 410: 372–376.
Shors TJ, Townsend DA, Zhao M et al (2002). Neurogenesis may relate to some but not all types of hippocampal-dependent learning. Hippocampus 12: 578–584.
Sisti HM, Glass AL, Shors TJ (2007). Neurogenesis and the spacing effect: learning over time enhances memory and the survival of new neurons. Learn Mem 14: 368–375.
Snyder JS, Hong NS, McDonald RJ et al (2005). A role for adult neurogenesis in spatial long-term memory. Neuroscience 130: 843–852.
Tashiro A, Zhao C, Gage FH (2006). Retrovirus-mediated single-cell gene knockout technique in adult newborn neurons in vivo. Nat Protoc 1: 3049–3055.
Tashiro A, Makino H, Gage FH (2007). Experience-specific functional modification of the dentate gyrus through adult neurogenesis: a critical period during an immature stage. J Neurosci 27: 3252–3259.
Toni N, Teng EM, Bushong EA et al (2007). Synapse formation on neurons born in the adult hippocampus. Nat Neurosci 10: 727–734.

Toni N, Laplagne DA, Zhao C et al (2008). Neurons born in the adult dentate gyrus form functional synapses with target cells. Nat Neurosci 11: 901–907.
Tozuka Y, Fukuda S, Namba T et al (2005). GABAergic excitation promotes neuronal differentiation in adult hippocampal progenitor cells. Neuron 47: 803–815.
Van der Borght K, Mulder J, Keijser JN et al (2005). Input from the medial septum regulates adult hippocampal neurogenesis. Brain Res Bull 67: 117–125.
van Praag H, Christie BR, Sejnowski TJ et al (1999). Running enhances neurogenesis, learning, and long-term potentiation in mice. Proc Natl Acad Sci USA 96: 13427–13431.
van Praag H, Schinder AF, Christie BR et al (2002). Functional neurogenesis in the adult hippocampus. Nature 415: 1030–1034.
van Praag H, Shubert T, Zhao C et al (2005). Exercise enhances learning and hippocampal neurogenesis in aged mice. J Neurosci 25: 8680–8685.
Walter C, Murphy BL, Pun RY et al (2007). Pilocarpine-induced seizures cause selective time-dependent changes to adult-generated hippocampal dentate granule cells. J Neurosci 27: 7541–7552.
Winocur G, Wojtowicz JM, Sekeres M et al (2006). Inhibition of neurogenesis interferes with hippocampus-dependent memory function. Hippocampus 16: 296–304.
Wojtowicz JM (2006). Irradiation as an experimental tool in studies of adult neurogenesis. Hippocampus 16: 261–266.
Yamaguchi M, Mori K (2005). Critical period for sensory experience-dependent survival of newly generated granule cells in the adult mouse olfactory bulb. Proc Natl Acad Sci USA 102: 9697–9702.
Yoshimizu T, Chaki S (2004). Increased cell proliferation in the adult mouse hippocampus following chronic administration of group II metabotropic glutamate receptor antagonist, MGS0039. Biochem Biophys Res Commun 315: 493–496.
Zhang CL, Zou Y, He W et al (2008). A role for adult TLX-positive neural stem cells in learning and behaviour. Nature 451: 1004–1007.
Zhao X, Ueba T, Christie BR et al (2003). Mice lacking methyl-CpG binding protein 1 have deficits in adult neurogenesis and hippocampal function. Proc Natl Acad Sci USA 100: 6777–6782.
Zhao C, Teng EM, Summers RG, Jr. et al (2006). Distinct morphological stages of dentate granule neuron maturation in the adult mouse hippocampus. J Neurosci 26: 3–11.
Zhao M, Li D, Shimazu K et al (2007). Fibroblast growth factor receptor-1 is required for long-term potentiation, memory consolidation, and neurogenesis. Biol Psychiatry 62: 381–390.
Zhao C, Deng W, Gage FH (2008). Mechanisms and functional implications of adult neurogenesis. Cell 132: 645–660.

Chapter 10
Adult Neurogenesis in the Hippocampus: Lessons from Natural Populations

Jan Martin Wojtowicz

Abstract Laboratory research on adult neurogenesis is proceeding at a tremendous pace and contributing important details about the mechanism of cell development and function. A novel approach in the ultimate understanding of the functional role of young neurons and their value for species survival involves studies on natural populations of animals. Ecologically based research has already contributed to the understanding of some outstanding issues and two lessons have been learned thus far. (1) Neurogenesis plays a role in only certain types of hippocampus-dependent learning tasks. (2) The age-related decline of neurogenesis is a species-dependent adaptation related to an animal's life span. Better understanding of the adaptive role of neurogenesis will require both laboratory research on inbred rodents, and studies on diverse species adapted to their natural environment.

10.1 Introduction

The past decade has seen a great expansion of research into the mechanisms of regulation and functionality of adult-born neurons. There is now a reasonable understanding of neurogenesis as a multi-step process, with its gradual progression from mother precursor cells to a proliferating population of progenitors, and finally to differentiated and mature neurons. There are also data on dendritic and axonal extension, synaptogenesis, and basic connectivity of differentiated neurons. Finally, there is already sufficient knowledge of electrophysiological properties and afferent inputs to ascertain that the new cells are connected within the hippocampal circuitry (Wang et al., 2000; Snyder et al., 2001; Ge et al., 2005; Piatti et al., 2006; Schmidt-Hieber et al., 2004; Toni et al., 2008).

J.M. Wojtowicz (✉)
Department of Physiology, University of Toronto, Toronto, ON, Canada
e-mail: martin.wojtowicz@utoronto.ca

T. Seki et al. (eds.), *Neurogenesis in the Adult Brain I: Neurobiology*,
DOI 10.1007/978-4-431-53933-9_10, © Springer 2011

Numerous physiological and environmental factors regulate neuronal production. Common behaviours, such as running, sleeping, learning, sexual interactions or exposure to stress, influence the rate of neurogenesis. Most evidence comes from laboratory rodents; this is a problem since rodents such as rats and mice originated as species around 55 million years ago, when they diverged from other mammalian taxa. The functional adaptations of new neurons in the adult brain may have evolved differently within different species of rodents, and certainly in other non-rodent species, yet very little is known about their neurogenesis. In fact, evidence for adult neurogenesis in other mammalian species is weak and mostly qualitative. There can be no doubt that adult neurogenesis occurs in almost all mammals (Lindsey and Tropepe, 2006) and its conservation in diverse species over millions of years testifies to its evolutionary advantage. However, our understanding of its general significance is limited by short-sighted experimenters who use only a limited number of dominant experimental approaches to the exclusion of others.

One hint as to the possible role of neurogenesis is its prominence in juveniles and young adults in comparison to middle-aged and old individuals. This feature was first observed by Seki and Arai (1995) and Kuhn et al. (1996) in rats and later confirmed in mice (Kempermann et al., 2002). It is also present, although not as well documented, in monkeys (Gould et al., 1999), humans (Eriksson et al., 1998; Manganas et al., 2008) and several other species. One of the outstanding questions in neurogenesis research is the meaning of this decline with age. Is the age-related effect the result of reduced demand for new neurons or is it a pre-determined property analogous to a decline in the egg-producing capacity of the ovaries? There is also a possibility that reduced production of new neurons is not indicative of lesser function, since the proportion of new neurons that is utilized could be growing with age or at least staying the same while the functionality of other neurons in the rest of the brain declines. Finally, the evidence from laboratory rodents suggests that neurogenesis is regulated by environment, so the decline with age may be a reflection of reduced interactions of the aging animals with the environment. If the age-dependent decline of neurogenesis is closely tied to specific functions of the new neurons, we must first know what these functions are in order to understand the decline.

In 2000 a group of scientists proposed the initiation of studies on neurogenesis in natural populations, with the assertion that only through better understanding of the function of young adult-born neurons in wild animals living in their natural environment will we be able to grasp the real, i.e. adaptive role of adult neurogenesis (Boonstra et al., 2001). Two subsequent articles emphasizing the need for studies of natural populations and the potential advantages of such an approach have been published. Lindsey and Tropepe (2006) presented a survey of neurogenesis in the vertebrates. Amrein et al., (2008) contended that more studies of natural populations are needed to better understand the adaptive benefits of adult neurogenesis and gave hints as to how such studies could be performed. This article will not attempt to extend those previous efforts, but will instead address two controversial issues arising from laboratory studies that efforts,but illustrate how such specific issues can be resolved with the help of studies in natural populations.

10.2 Lesson 1: Neurogenesis Is Critical for Some Types of Memory but Dispensable for Others

Hippocampus is often associated with spatial learning and navigation. Among mammals, including primates, hippocampal lesions cause deficits in spatial learning (Driscoll and Sutherland, 2005). In humans, London cab drivers are recognized for their detailed knowledge of the London streets and the ability to locate particular sites of the city with amazing accuracy. Magnetic Resonance Imaging (MRI) on a representative group of the cab drivers showed that the posterior hippocampus was significantly larger than that of individuals within the general population (Maguire et al., 2000). The posterior part of hippocampus in primates is anatomically equivalent to the dorsal hippocampus in the four-legged species. It is this dorsal part of the hippocampus that is known to support spatial learning in laboratory rodents, in contrast to the ventral part of the hippocampus that supports anxiety-related behaviours (Moser et al., 1995; Kjelstrup et al., 2002; Bannerman et al., 2004; Pothuizen et al., 2004; Pentkowski et al., 2006). While this dorso-ventral (septo-caudal) specialization of the hippocampus holds true for the CA1 field, the relevance to the dentate gyrus is less certain. The dentate gyrus occupies about one third of the total hippocampus and most experiments involving measurements of the hippocampal size do not distinguish between the size of CA1 or CA3 vs. the dentate gyrus. Similarly, many experiments employing lesions or MRI do not distinguish between CA1 and dentate gyrus. A case can be made for CA1 being primarily responsible for spatial learning on the basis of lesions, electrophysiological and pharmacological experiments. The CA1, CA3 and dentate gyrus have evolved differently (West, 1990) and the ability to support adult neurogenesis only by the dentate gyrus is a manifestation of these differences. Thus, although dentate gyrus, CA3 and CA1 may be expected to work within the framework of the classical tri-synaptic circuit, there is also reason to postulate that direct afferent inputs from the entorhinal cortex to CA1 or CA3 may have evolved to support certain types of learning without the involvement of the dentate gyrus.

Turning now to studies of adult neurogenesis in the context of the hippocampus as a whole: the most compelling evidence for the functional role of newly-generated neurons comes from "lesion" experiments in which neurogenesis was suppressed by a known amount and the lesioned animals tested on various behavioral tasks in comparison to intact controls (Winocur et al., 2006; Wojtowicz et al., 2008; Saxe et al., 2006). For the purposes of this article we will discuss primarily the learning tasks that are dependent on the hippocampus. In this context the neurogenic subgranular zone in dentate gyrus is of high importance, being strategically located at the gateway of the hippocampus.

The perforant pathway carries over 90% of the hippocampal afferent axons originating in the entorhinal cortex and forms the first relay station of the "tri-synaptic circuit" in the dentate gyrus (Fig. 10.1). In view of the massive neuronal production in the subgranular zone and dense, forest-like dendritic branching of young neurons throughout the dentate gyrus molecular layer, it seems likely that any afferent input

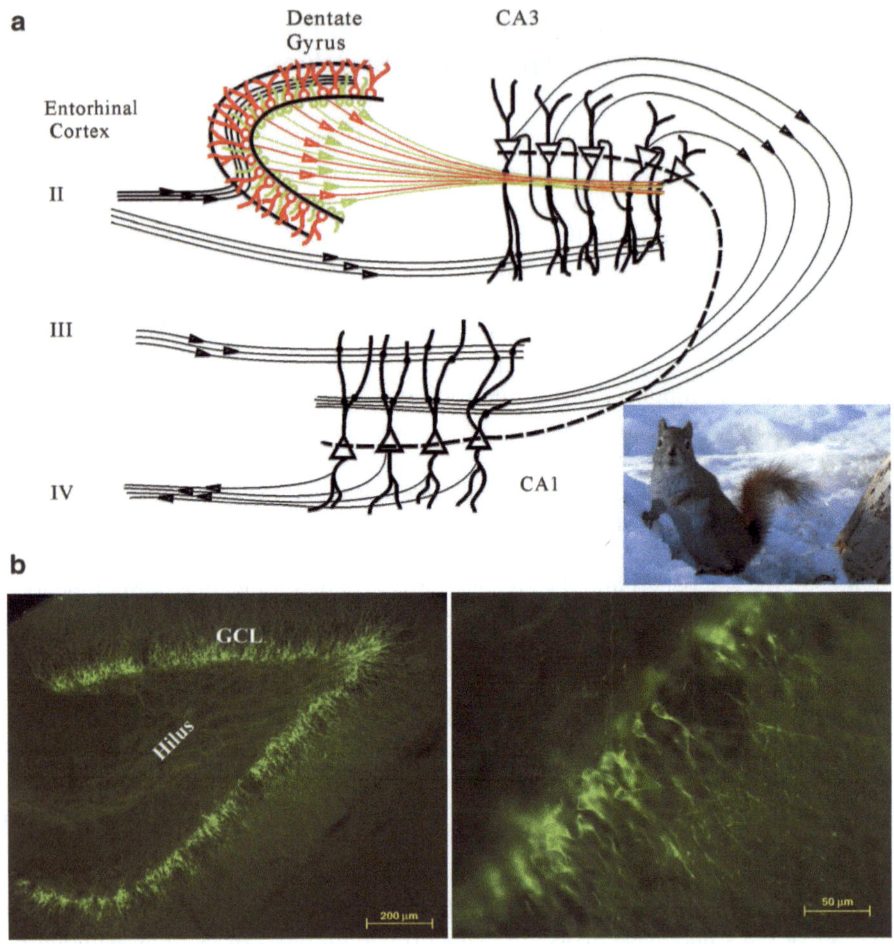

Fig. 10.1 Hippocampal circuitry showing the location of new neurons. (**a**) Diagram shows the classical tri-synaptic circuit consisting of dentate gyrus, CA3, and CA1 as well as direct pathways from Entorhinal cortex to CA3 and CA1. Dentate to CA3 mossy fiber pathway includes axons of mature (*red*) and immature (*green*) granule neurons. In contrast, the direct pathways do not include young neurons. (**b**) Images illustrating a transverse section through the dentate gyrus in a juvenile red squirrel with abundant young neurons located at the border of hilus and granule cell layer (GCL). Green stain (Doublecortin) is a marker if immature neurons. Note that Doublecortin is present in cell bodies, dendrites and mossy fibers axons of the young neurons suggesting that these neurons, while still young, are already integrated within hippocampal circuit

entering through the dentate gyrus will in some way engage these young neurons. This hypothesis is supported by evidence of enhanced plasticity of the young neurons in comparison to the pre-existing mature neurons (Abrous and Wojtowicz, 2008). The numerical prominence of young neurons is well illustrated in an image of an immunohistochemically stained section obtained from a juvenile (approximately 6 months old) red squirrel (*Tamiasciurus hudsonicus*) (Fig. 10.1). The high density of the young neurons is striking but not unique for squirrels. Juvenile rats that are approximately 1 month old have similar numbers of such neurons (see Fig. 10.3). Consequently, one might expect that any behavioural tasks involving the hippocampus should also involve the young neurons.

The evidence obtained thus far suggests large differences in sensitivities of various learning tasks to the rate of neurogenesis. These differences can be illustrated by a comparison of two prototypical hippocampal learning tasks: the spatial water maze (WM) and contextual fear conditioning (CFC). Using a beam of high-energy radiation directed to the head one can selectively and irreversibly ablate proliferating neural progenitors in the dentate gyrus, leaving mature neurons and other non-neuronal cells intact (Wojtowicz, 2006). The permanent reduction in the young neuronal population provides an opportunity to test animals with reduced yet stable levels of ongoing neurogenesis without complication of activity-induced recovery (see below). This approach can be complemented with other experiments that can enhance neurogenesis, although at this time there is no method of enhancement that can approach the selectivity and robustness of the irradiation-induced ablation.

A comparison of animals with varied degrees of neurogenesis in WM and CFC tasks showed that while the latter is extremely sensitive to changes in neurogenesis the former is not. The "readout" of CFC is the time of sitting still or "freezing" when the subject is exposed to a threatening environment (context). Less freezing time is interpreted as a weaker memory. In several studies done so far on mice and rats, a proportionality of the number of the young neurons to the performance in CFC task has been found. An 80% reduction in neurogenesis produced an 80% reduction in the freezing time (Winocur et al., 2006). In contrast, neither training performance nor the 24 h memory of the platform location was impaired in the water maze, even after a 90% reduction of neurogenesis (Snyder et al., 2005). A side by side comparison of performances on CFC and WM by the same animals showed a remarkable 69% reduction of freezing in CFC after a 69% depletion of neurogenesis, but again no effect on spatial learning (Wojtowicz et al., 2008). The proportionality of CFC is also present in individual animals that show individual variations in the densities of young neurons. This applies equally well to intact and to irradiated animals, ruling out unknown and unrelated side effects of irradiation. This correlation of memory with the number of young neurons further suggests that the difference between the two learning tasks is not simply a difference in sensitivity of one task over the other, but a qualitative, fundamental specialization of neurogenesis for certain learning tasks. These experiments raise the possibility that neurogenesis has evolved to perform specific functions within the hippocampal circuit. In relation to natural populations the CFC may be relevant to recognition of a threatening situation, possibly associated with the presence of predators. The exact proportionality of the memory to the number of available young neurons is unprecedented. In previous studies, near complete ablations of the hippocampus spared most of the learning performance in spatial tasks, suggesting redundancy in the hippocampal pathways or perhaps compensation by non-hippocampal brain regions (Moser et al., 1995). The lack of such compensation in case of CFC suggests an indispensable and fundamental requirement for neurogenesis.

The dependence of CFC on neurogenesis fits well with evidence for a role of CA3–CA1 connection via the Schaffer collaterals in CFC. Schaffer collaterals are downstream from the dentate gyrus in the tri-synaptic circuit so their blockade would be expected to have similar effect as the depletion of neurogenesis. Increasingly selective transgenic mouse models revealed that inactivation of CA3 to CA1 projection blocks the CFC memory but leaves the basic aspects of spatial learning intact (Nakashiba et al., 2008). Since all of the axons of the young dentate granule neurons project towards the CA3 their effects would be expected to be also blocked by the inactivation of the CA3 output to CA1 (see Fig. 10.1).

Spatial learning and memory are not completely immune to reductions in neurogenesis. Changes in acquisition and/or retention have been reported (Zhang et al., 2008a; Snyder et al., 2005). The most consistent defect seen in the irradiated and transgenic animals is in the lack of long-term memory (Rola et al., 2004; Imayoshi et al., 2008). There is also at least one report of enhanced learning in animals with reduced neurogenesis (Saxe et al., 2007). A possible explanation for these varied effects is that fewer young neurons make the hippocampal circuit more susceptible to interference with learning by behavioral "noise", i.e. sensory stimuli that are not directly related to the task at hand (Becker, 2005). This theory deserves further experimental scrutiny, but the interference-effect is not specific for any particular learning task (spatial or non-spatial). Thus, while the dependence of spatial learning on neurogenesis is weak and indirect, that of CFC is strong and proportional to the number of available neurons. The sharp contrast between CFC and spatial learning does not mean that only CFC is dependent on neurogenesis. Other hippocampal tasks, such as the delayed non-matching to sample (DNMTS) and trace eye-blink conditioning, are also impaired in animals with depleted neurogenesis (Shors et al., 2001, 2002; Winocur et al., 2006). Future research involving natural populations will be essential to determine which one of these laboratory tests represents a situation that may have a greater significance for an animal's survival in natural conditions. Particular adaptations are likely to be species-dependent.

Radiation-induced consequences of reduced neurogenesis were seen in adult (>3 months old) rats and mice. It is not yet known whether older animals are equally, less, or more sensitive. It is conceivable that with smaller numbers of available neurons older animals adapt to use a higher percentage of these neurons. There may also be a larger variability in older animals. For example, the well established phenomenon that older animals can be subdivided into two categories, cognitively impaired and unimpaired, has been correlated with neurogenesis, although there is no universal agreement on the nature of this correlation (Bizon et al., 2004; Drapeau et al., 2003). In one such study old rats were trained on the WM task, and their performance in terms of reduced swimming times while searching for a hidden platform was correlated with the survival of 1–2 week old neurons (Drapeau et al., 2007). This result suggests that neurogenesis can somehow be enhanced by spatial learning. The activity-dependent enhancement of neurogenesis, especially during the 1–2 week critical period, is a well established phenomenon in young and old animals, but it is difficult to understand why the type of learning that does not require neurogenesis would increase its rate. One explanation arises from a study by

Snyder et al., (2008) who found that young neurons activated during WM acquisition are predominantly found in the ventral dentate gyrus, suggesting that the activity of the young neurons may be related to the anxiety or fear-related aspect of the WM task and not to its spatial learning component.

The basic tenet of this article is that laboratory tests involving either decreases or increases in neurogenesis can be suggestive, but are unlikely to be decisive considering their artificial nature and the use of inbred animals. The CFC task, for example, involves shocking the animal with electrical current applied to the metal grid at the bottom of the cage, hardly a likely scenario in the natural environment. Ultimately, one needs a test in real-life conditions where the animals have to survive in their natural environment by learning life-saving tasks. This article argues that there is a need for studies in the natural environment in order to understand the adaptive role of neurogenesis. The relationship between spatial learning and neurogenesis is a case in point. Spatial navigation that an animal requires to find and store food and to locate appropriate nesting areas and mates is an essential survival skill, but is it dependent on neurogenesis? We have previously reported that a comparison of two species of wild living rodents with differing needs for spatial navigation did not produce good correlation with neurogenesis. Thus, grey squirrels (*Sciurus carolinensis*), typical scatter hoarders, had a relatively larger number of proliferating cells but about equal number of young neurons in comparison to chipmunks (*Tamias amoenus*). Yet, the chipmunks have much lesser spatial requirements (Barker et al., 2005). The larger ratio of proliferating to surviving cells could reflect higher turnover of new neurons, but a simpler explanation is that there may simply not be a fundamental need for neurogenesis in spatial behaviours. Another case in point is an illuminating study on several species of bats (Chiroptera) showing that a majority of bat species had no detectable neurogenesis in the hippocampus (Amrein et al., 2007) (Fig. 10.2). Considering that bats are known for their excellent spatial navigation (Holland et al., 2005) and that their hippocampus is essentially the same as those of most other mammals (Amrein et al., 2007), this striking observation appears to exclude neurogenesis as an essential property for spatial learning.

If it's not spatial learning, what other function or essential role can be assigned to neurogenesis? Further surveys of various species with particular adaptations to their environment may provide more clues. The strong dependence of CFC on neurogenesis suggests that animals that rely on freezing as a survival strategy should have high levels of neurogenesis. In contrast, species that have a high need for spatial navigation, such as most predators, should not, although specialization of the new neurons for other yet unknown functions cannot be excluded. Such a comparative approach could be complemented by the ultimate causality experiment, in which it may be possible to alter the levels of neurogenesis in species such as squirrels and test their survival skills in the natural environment. It is feasible to capture these animals, irradiate them selectively to reduce number of young neurons, and release them back to the their habitat. With the use of telemetry one could monitor their survival and ability to avoid predators, and then trap them again to verify their health, and the level of neurogenesis.

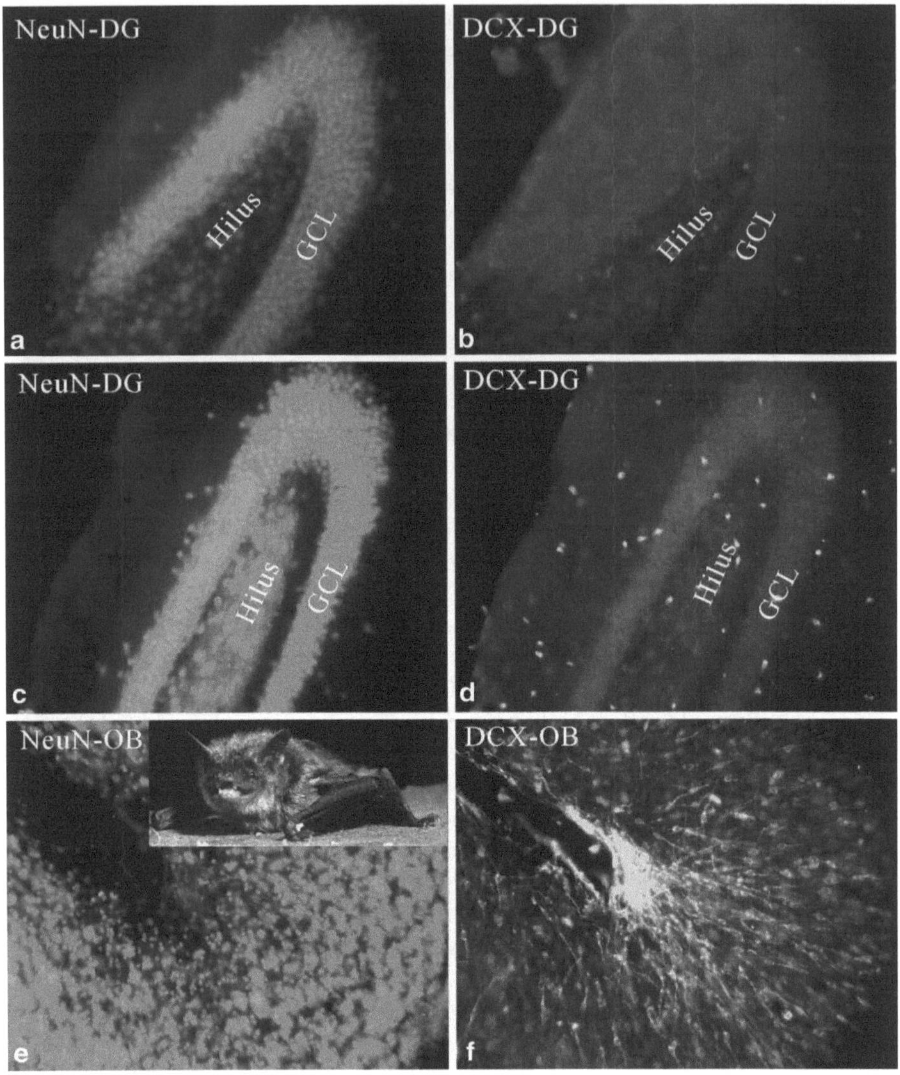

Fig. 10.2 Lack of neurogenesis in the big brown bat (*Eptesicus fuscus*) hippocampus. Representative brain sections stained for a neuronal marker (NeuN), *left column*, and immature neuronal marker (Doublecortin), *right column*. (**a**, **b**) Hippocampal section from the brain of a juvenile, 3 week old bat. (**c**, **d**) Adult (>1 year) male. Well defined granule cell layer (GCL) and Hilus can be seen at both ages. Bright puncta in the adult tissue (**d**) were probably due to endogenous fluorescence in blood vessels and are unrelated to neurogenesis. (**e**, **f**) Olfactory bulb from the same juvenile animal had strong NeuN (**e**) and doublecortin (**f**) immunoreactivity. The data were obtained in the laboratory of the author (unpublished). The bat specimens were provided by Dr. D.M. Reeder, Bucknell University, Lewisburg, PA. http://www.facstaff.bucknell.edu/dreeder

In addition to the specific learning skills that may or may not depend on neurogenesis, there is an outstanding question of why and how neurogenesis is regulated in real life conditions. Here again the laboratory tests fall short, since they are done on animals living in artificial conditions. An example is the regulation of neurogenesis by so-called "enriched environment". It has been reported that placing laboratory rats or mice in large cages equipped with running wheel and various novel objects, which the animals can explore, enhances neurogenesis several-fold (Kempermann et al., 1997). This implies that wild rats, living in infinitely complex ("enriched") surroundings, should possess comparatively larger numbers of new neurons. However, this is not the case. A study that compared wild rats with their laboratory equivalents of known ages revealed virtually the same levels of neurogenesis in both populations (Epp et al., 2008; Jennifer Barker, personal communication). This finding shows that laboratory tests can give false and misleading impression as to the importance of regulatory factors, and ultimately a real life comparison reveals the truth.

10.3 Lesson 2: Rate of Neurogenesis Is Adapted to Animal's Age

One of the areas where studies of animals in natural conditions could contribute most is the issue of how aging and neurogenesis are related. The evidence for decline in brain function with age, including the rate of neurogenesis in laboratory animals, is plentiful but is it relevant to real life conditions? Laboratory rats survive for 2–3 years but they are very different from their wild counterparts. Animals in a laboratory condition are exposed to limited environmental stimuli, surplus of water and food, and absence of predators and competitors. Additionally, the inbred strains of rats held in the laboratory are usually less aggressive and more sensitive to stress (Kunzl et al., 2003). This raises the possibility that the rapid decline of neurogenesis within 2–3 months is an artifact of the impoverished living conditions and of abnormal inbred traits that have little relevance to the real world.

Thus, the decline of neurogenesis during aging should be investigated in other species that have evolved to survive in their natural environment. It has been suggested that new neurons may be most important for learning novel, newly-encountered tasks that may be encountered in very young animals, but less important for learning in adult and aging animals that presumably had experienced all there is to know (Kempermann et al., 2004; Wiskott et al., 2006). Although this idea seems reasonable it should not be interpreted as evidence for lack of importance of neurogenesis in adult or aging animals. In fact, data from adult rats (3–5 months old) that show dramatically reduced contextual learning after irradiation suggest that the remaining young neurons in adult brain are nevertheless critically important (Wojtowicz et al., 2008).

Is the age-related decline in neurogenesis an artifact of impoverished living conditions encountered by the laboratory animals? Enticing evidence for such an argument comes from studies of van Praag and colleagues who showed that placing mice in an "enriched environment" consisting of larger cages with various objects, including running wheels, enhances neurogenesis in aging animals towards the levels encountered in the young ones (van Praag et al., 2005; Kempermann et al., 2002). If this is the case in natural conditions, what might be the levels of neurogenesis in old wild animals that presumably encounter even more enriched environment that the simulated "enriched environment" in the laboratory can provide? Surprising data emerged from a comparison of the numbers of young neurons in a population of wild mice and rats with their laboratory counterparts. Amrein and colleagues quantified the densities of cells immuno-positive for Ki-67, a marker for proliferating cells, in laboratory mice and several related species of rodents (Amrein et al., 2004). These data suggest that laboratory mice place in the middle of the range covered by the wild species and not at the low end of the scale as would be expected from their "impoverished environment". Similarly, a comparison of laboratory rats with their wild counterparts shows generally similar levels of neurogenesis and a decline of neurogenesis in both populations with a time constant of about 3–4 months, i.e. about 10–15% of the rat's average life span in the laboratory, even though the wild animals must encounter a much more complex environment (Epp et al., 2008). It is likely that the wild rat's life span is much shorter than that of the laboratory rat due to stress, predators and disease. Consequently, the decline of neurogenesis within the first year of life could be an adaptation to this short life. One study, carried out on a large population of brown rats (*Rattus norvegicus*) living in natural conditions, estimated that over 95% of the animals survive for less than 1 year. Approximately 50% of rats dies within 12 weeks of tagging but this varies with time of birth, sex etc. (Davis, 1948). Thus, although the exact chronological ages of these animals were not known, the life span of wild rats is likely to be on the order of a few months and not a few years as in the laboratory rat.

The experiments on aging in the wild are often hampered by the difficulty of not knowing the exact ages of the animals caught, but a few data sets are available. The aforementioned study on rats used the weights of the eye lenses that correlate closely with age during the first year of life (Barker et al., 2003). The older the animal, the heavier the lens. In other species surviving for several years, the yearly growth rings (annuli) in femoral bones or jaws can give a measure of age (Barker et al., 2005). One striking finding in one such study was the relatively small decline in the density of immature neurons with age in wild populations of chipmunks and squirrels. On average, the density of the young neurons in adults was about half of that seen in juveniles instead of the 90% decline observed in laboratory rats. Correspondingly, the time constant of the decay of neurogenesis was much longer in chipmunks in comparison to laboratory rats (Fig. 10.3). The oldest chipmunks that could be caught in the wild were 5 years old and the time constant of the decay of neurogenesis was equal to the real life longevity for this species, i.e. 5 years. Thus, in chipmunks the decay is much slower than in rats, taking their respective and very different life-spans into account. Similar data were obtained in eastern

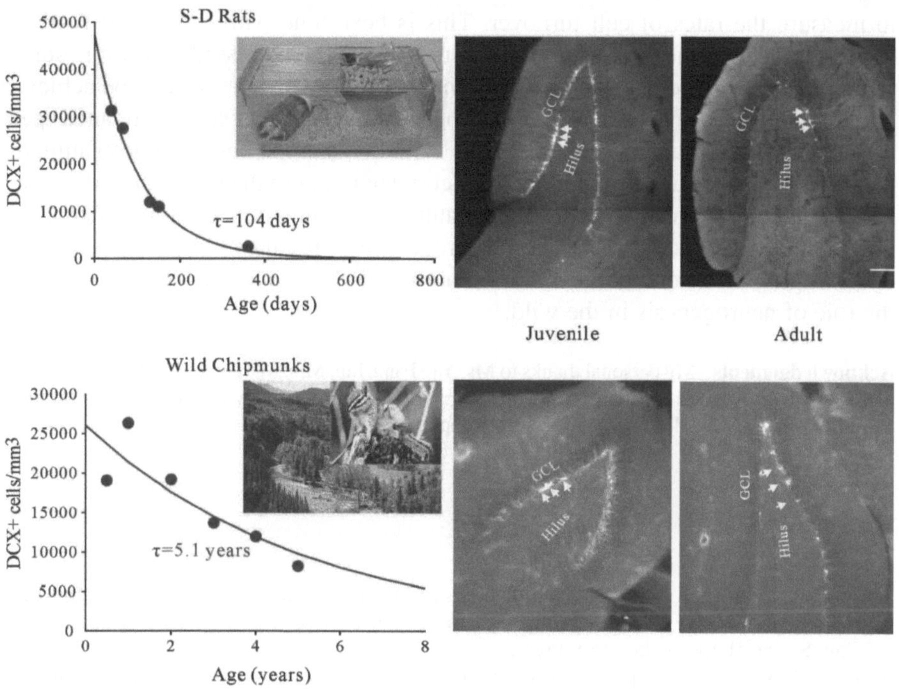

Fig. 10.3 Decline of neurogenesis with age is much slower in a wild-living species in comparison to the laboratory rat. (**a**) Decline of neurogenesis with age in the Sprague-Dawley rat. Total numbers of Doublecortin-positive (DCX+) young neurons per dentate gyrus were measured in control, untreated rats as described in (McDonald and Wojtowicz, 2005; Zhang et al., 2008b). The values were re-calculated per mm^3 of granule cell layer (GCL) and fitted to a single exponential function. The time constant of decay τ was 104 days. This is less than 15% of the average life span in laboratory conditions for this species. (**b**) Decline of neurogenesis in wild chipmunks shows a decline with a time constant of 5.1 years. This is equivalent to the maximum life span of this species in the wild. Data based on Barker et al. (2005) and Barker (2005). Note that the extrapolated, initial values of neurogenesis are comparable in the two species but the decay rates are dramatically different. This mis-match between neurogenesis and aging in the rat may be the result of artificially prolonged life under laboratory conditions

grey squirrels, i.e. the decay rate of young neurons with age was proportional to the squirrels' longevity and consequently much slower than in laboratory rats (Barker et al., 2005).

10.4 Future Directions

Further studies should take into consideration the possibility that measurements of levels of neurogenesis used thus far may not be adequately informative. Instead of measuring "snap shot" values of proliferating or young neurons one may want to measure the rates of cell turnover. This is best done with the use of BrdU in combination with the increasing repertoire of markers expressed at various stages of neurogenesis. Another quantum leap in studies of wild animals may be achieved with new, non-invasive measures of neurogenesis such as magnetic resonance spectroscopy (MRS) (Manganas et al., 2008). With this method one could trap animals, measure the magnitude of the biomarker corresponding to the rate of neurogenesis, and release the animals back to their natural environment. This type of real life longitudinal study could reveal how neurogenesis is regulated in different seasons or with aging. In combination with irradiation, MRS could be the ideal tool to study the role of neurogenesis in the wild.

Acknowledgements My personal thanks to Ms. Yao Fang Tan, Mrs. Marianne Hegström-Wojtowicz and Dr. Harold Atwood for their help with the manuscript, to Dr. Rudy Boonstra for inspiration, and to NSERC and CIHR, Canada for grant support.

References

Abrous DN, Wojtowicz JM (2008) Neurogenesis and the hippocampal memory system. In: Adult Neurogenesis (Gage FH, Kempermann G, Song H-J, eds), pp 445–462. Cold Spring Harbor: Cold Spring Harbor Laboratory Press.

Amrein I, Slomianka L, Lipp HP (2004) Granule cell number, cell death and cell proliferation in the dentate gyrus of wild-living rodents. Eur J Neurosci 20: 3342–3350.

Amrein I, Dechmann DK, Winter Y, Lipp H-P (2007) Absent or low rate of adult neurogenesis in the hippocampus of bats (chiroptera). PLoS One 2: e455.

Amrein I, Boonstra R, Lipp H-P, Wojtowicz JM (2008) Adult hippocampal neurogenesis in natural populations of mammals. In: Adult Neurogenesis (Gage FH, Kempermann G, Song H-J, eds), pp 645–659. Cold Spring Harbor: Cold Spring Harbor Laboratory Press.

Bannerman DM, Rawlins JN, McHugh SB, Deacon RM, Yee BK, Bast T, Zhang WN, Pothuizen HH, Feldon J (2004) Regional dissociations within the hippocampus–memory and anxiety. Neurosci Biobehav Rev 28: 273–283.

Barker (2005) MSc Thesis, University of Toronto.

Barker JM, Boonstra R, Schulte-Hostedde AI (2003) Age determination in yellow-pine chipmunks (*Tamias amoenus*): a comparison of eye lens masses and bone sections. Can J Zool 81: 1774–1779.

Barker J, Wojtowicz JM, Boonstra R (2005) Where is my dinner? Adult neurogenesis in free-living food-storing rodents. Genes Brain Behav 4: 89–98.

Becker S (2005) A computational principle for hippocampal learning and neurogenesis. Hippocampus 15: 722–738.

Bizon JL, Lee HJ, Gallagher M (2004) Neurogenesis in a rat model of age-related cognitive decline. Aging Cell 3: 227–234.

Boonstra R, Galea LAM, Matthews S, Wojtowicz JM (2001) Hippocampal neurogenesis in natural populations. Can J Physiol Pharmacol 79: 297–302.

Davis DE (1948) The survival of wild brown rats on a Maryland farm. Ecology 29: 437–448.

Drapeau E, Mayo W, Aurousseau C, Le Moal M, Piazza PV, Abrous DN (2003) Spatial memory performances of aged rats in the water maze predict levels of hippocampal neurogenesis. Proc Natl Acad Sci USA 100: 14385–14390.

Drapeau E, Montaron MF, Aguerre S, Abrous DN (2007) Learning-induced survival of new neurons depends on the cognitive status of aged rats. J Neurosci 27: 6037–6044.

Driscoll I, Sutherland RJ (2005) The aging hippocampus: navigating between rat and human experiments. Rev Neurosci 16: 87–121.

Epp JR, Barker JM, Galea LAM (2008) Running wild: neurogenesis in the hippocampus across the life span in wild and laboratory-bred Norway rats. Hippocampus 19: 1040–1049.

Eriksson PS, Perfilieva E, Bjork-Eriksson T, Alborn AM, Nordborg C, Peterson DA, Gage FH (1998) Neurogenesis in the adult human hippocampus. Nat Med 4: 1313–1317.

Ge S, Goh ELK, Sailor KA, Kitabatake Y, Ming G-L, Song H (2005) GABA regulates synaptic integration of newly generated neurons in the adult brain. Nature 439: 589–593.

Gould E, Reeves AJ, Fallah M, Tanapat P, Gross CG, Fuchs E (1999) Hippocampal neurogenesis in adult Old World primates. Proc Natl Acad Sci USA 96: 5263–5267.

Holland RA, Winter P, Waters DA (2005) Sensory systems and spatial memory in the fruit bat Rousettus aegyptiacus. Ethology 111: 715–525.

Imayoshi I, Sakamoto M, Ohtsuka T, Takao K, Miyakawa T, Yamaguchi M, Mori K, Ikeda T, Itohara S, Kageyama R (2008) Roles of continuous neurogenesis in the structural and functional integrity of the adult forebrain. Nat Neurosci 11: 1153–1161.

Kempermann G, Kuhn HG, Gage FH (1997) More hippocampal neurons in adult mice living in an enriched environment. Nature 386: 493–495.

Kempermann G, Gast D, Gage FG (2002) Neuroplasticity in old age: sustained fivefold induction of hippocampal neurogenesis by long term environmental enrichment. Ann Neurol 52: 135–143.

Kempermann G, Wiskott L, Gage FH (2004) Functional significance of adult neurogenesis. Curr Opin Neurobiol 14: 186–191.

Kjelstrup KG, Tuvnes FA, Steffenach HA, Murison R, Moser EI, Moser M-B (2002) Reduced fear expression after lesions of the ventral hippocampus. Proc Natl Acad Sci USA 99: 10825–10830.

Kuhn HG, Dickinson-Anson H, Gage FG (1996) Neurogenesis in the dentate gyrus of the adult rat: age-related decrease of neuronal progenitor proliferation. J Neurosci 16: 2027–2033.

Kunzl CS, Kaiser E, Meier E, Sachser N (2003) Is a wild mammal kept and reared in captivity still a wild animal? Horm Behav 43: 187–196.

Lindsey BW, Tropepe V (2006) A comparative framework for understanding the biological principles of adult neurogenesis. Prog Neurobiol 80: 281–307.

Maguire EA, Gadian DG, Johnsrude IS, Good CD, Ashburner J, Frackowiak RS, Frith CD (2000) Navigation-related structural change in the hippocampi of taxi drivers. Proc Natl Acad Sci USA 97: 4398–4403.

Manganas LN, Zhang X, Li Y, Hazel RD, Smith SD, Wagshul ME, Henn F, Benveniste H, Djuric PM, Enikolopov G, Maletic-Savatic M (2008) Magnetic resonance spectroscopy identifies neural progenitor cells in the live human brain. Science 318: 980–985.

McDonald HY, Wojtowicz JM (2005) Dynamics of neurogenesis in the dentate gyrus of adult rats. Neurosci Lett 385: 70–75.

Moser M-B, Moser EI, Forster EA, Andersen P, Morris RG (1995) Spatial learning with a minislab in the dorsal hippocampus. Proc Natl Acad Sci USA 92: 9697–9701.

Nakashiba T, Young JZ, McHugh TJ, Buhl DL, Tonegawa S (2008) Transgenic inhibition of synaptic transmission reveals role of CA3 output in hippocampal learning. Science 319: 1260–1264.

Pentkowski NS, Blanchard DC, Lever C, Litvin Y, Blanchard RJ (2006) Effects of lesions to the dorsal and ventral hippocampus on defensive behaviors in rats. Eur J Neurosci 23: 2185–2196.

Piatti VC, Esposito MS, Schinder AF (2006) The timing of neuronal development in adult hippocampal neurogenesis. The Neuroscientist 12: 463–468.

Pothuizen HH, Zhang WN, Jongen-Relo AL, Feldon J, Yee BK (2004) Dissociation of function between the dorsal and the ventral hippocampus in spatial learning abilities of the rat: a within-subject, within-task comparison of reference and working spatial memory. Eur J Neurosci 19: 705–712.

Rola R, Raber J, Rizk A, Otsuka S, VandenBerg SR, Morhardt DR, Fike JR (2004) Radiation-induced impairment of hippocampal neurogenesis is associated with cognitive deficits in young mice. Exp Neurol 188: 316–330.

Saxe MD, Battaglia F, Wang J-W, Melleret G, David DJ, Monckton JE, Garcia ADR, Sofroniev MV, Kandel ER, Santarelli L, Hen R, Drew MR (2006) Ablation of hippocampal neurogenesis impairs contextual fear conditioning and synaptic plasticity in the dentate gyrus. Proc Natl Acad Sci USA 103: 17501–17506.

Saxe MD, Malleret G, Vronskaya S, Mendez I, Garcia AD, Sofroniew MV, Kandel ER, Hen R (2007) Paradoxical influence of hippocampal neurogenesis on working memory. Proc Natl Acad Sci USA 104: 4642–4646.

Schmidt-Hieber C, Jonas P, Bischofberger J (2004) Enhanced synaptic plasticity in newly generated granule cells of the adult hippocampus. Nature 429: 184–187.

Seki T, Arai Y (1995) Age-related production of new granule cells in the adult dentate gyrus. Neuroreport 6: 2479–2482.

Shors TJ, Miesegaes G, Beylin A, Zhao M, Rydel T, Gould E (2001) Neurogenesis in the adult is involved in the formation of trace memories. Nature 410: 372–376.

Shors TJ, Townsend DA, Zhao M, Kozorovitskiy Y, Gould E (2002) Neurogenesis may relate to some but not all types of hippocampal-dependent learning. Hippocampus 12: 578–584.

Snyder JS, Kee N, Wojtowicz JM (2001) Effects of adult neurogenesis on synaptic plasticity in the rat dentate gyrus. J Neurophysiol 85: 2423–2431.

Snyder JS, Hong N, McDonald RJ, Wojtowicz JM (2005) A role for adult hippocampal neurogenesis in spatial long-term memory. Neuroscience 130: 843–852.

Snyder JS, Radik R, Wojtowicz JM, Cameron HA (2008) Anatomical gradients of neurogenesis and activity: young neurons in the ventral dentate gyrus are activated by water maze training. Hippocampus 19: 360–370.

Toni N, Laplange DA, Zhao C, Lombardi G, Ribak CE, Gage FG, Schinder AF (2008) Neurons born in the adult dentate gyrus form functional synapses with target cells. Nat Neurosci 11: 901–907.

van Praag H, Shubert T, Zhao Ch, Gage FG (2005) Exercise enhances learning and hippocampal neurogenesis in aged mice. J Neurosci 28: 8680–8685.

Wang S, Scott BW, Wojtowicz JM (2000) Heterogenous properties of dentate gyrus granule neurons in the adult rat. J Neurobiol 42: 248–257.

West MJ (1990) Stereological studies of the hippocampus: a comparison of the hippocampal subdivisions of diverse species including hedgehogs, laboratory rodents, wild mice and men. Prog Brain Res 83: 13–36.

Winocur G, Wojtowicz JM, Sekers M, Snyder JS, Wang S (2006) Inhibition of neurogenesis interferes with hippocampal-dependent memory function. Hippocampus 16: 296–304.

Wiskott L, Rasch MJ, Kempermann G (2006) A functional hypothesis for adult hippocampal neurogenesis: avoidance of catastrophic interference in the dentate gyrus. Hippocampus 16: 329–343.

Wojtowicz JM (2006) Irradiation as an experimental tool in studies of adult neurogenesis. Hippocampus 16: 261–266.

Wojtowicz JM, Askew ML, Winocur G (2008) The effects of running and inhibiting adult neurogenesis on learning and memory in rats. Eur J Neurosci 27: 1494–1502.

Zhang C-L, Zou Y, He W, Gage FG, Evans RM (2008a) A role for adult TLX-positive neural stem cells in learning and behaviour. Nature 451: 1004–1007.

Zhang W-J, Tan Y-F, Yue JT-Y, Vranic M, Wojtowicz JM (2008b) Impairment of hippocampal neurogenesis in streptozotocin-treated diabetic rats. Acta Neurol Scand 117: 205–210.

Chapter 11
Regulation of Adult Neurogenesis by Environment and Learning

Gerd Kempermann

Abstract Adult hippocampal neurogenesis is tightly linked to hippocampal function. This chapter covers the regulation of adult neurogenesis in response to either a relatively broad cognitive stimulus, an environmental enrichment, or specific learning situations. In contrast to more general behavioral stimuli, exemplified by voluntary wheel running, which induce precursor cell proliferation, such cognitive challenges promote the survival of newborn cells, presumably recruiting them for a specific function. Supposedly this function is closely related to the neuronal network, to which the new neurons are added; that is, the mossy fiber connection between the dentate gyrus, as the only location in which adult neurogenesis occurs in the hippocampus, and CA3. The key idea is that activity-dependent regulation of adult neurogenesis allows lifelong optimization of the mossy fiber connection, presumably allowing the system to efficiently respond to novelty and complexity encountered by the individual.

11.1 Introduction

"Regulation" of adult neurogenesis takes place on several conceptual levels, ranging from behavior down to the molecular control of neuronal development. Ultimately, however, all regulation must reach the molecular level. But it is the behavioral context, not these elementary molecular mechanisms, that fundamentally distinguishes adult neurogenesis from embryonic and fetal neurogenesis. In fact, adult neurogenesis shares many molecular principles with embryonic neurogenesis, although certain distinctions will exist. But what is uniquely different in adult neurogenesis compared to brain development in utero is the activity-dependent control. In embryonic neurogenesis, the activity-dependent component largely

G. Kempermann (✉)
Genomics of Regeneration, CRTD – Center for Regenerative Therapies Dresden, DFG Forschungszentrum und Excellenzcluster, Tatzberg 47 – 49, 01307 Dresden, Germany
e-mail: gerd.kempermann@crt-dresden.de

T. Seki et al. (eds.), *Neurogenesis in the Adult Brain I: Neurobiology*,
DOI 10.1007/978-4-431-53933-9_11, © Springer 2011

relates to the integration, survival, and fine-tuned integration of the neurons. In adult neurogenesis, activity also controls the precursor cells and their expansion, so that the entire process of neurogenesis is linked to how the adult individual acts in its environment. This interaction is largely subsumed under the term "learning" in the sense that learning is the persistent response of an individual to experience. This chapter covers a few concepts placing adult neurogenesis in a behavioral context by outlining how new neurons are elicited by learning.

11.2 The Experimental Concept of Enriched Environments

Originating from an anecdotal report by Donald Hebb in 1947 (Hebb, 1947), the research paradigm of environmental enrichment was established in the 1950s. It gained immense popularity in the early 1960s that reached its height in the late 1960s and 1970s (Rosenzweig and Bennett, 1996). The underlying core concept is to expose laboratory animals to an environment that is richer in stimuli of various domains than the standard housing and to assess the effect of this enrichment on the parameters of interest. Overall, consequences of this treatment are remarkable and allow the assessment of a number of ethological relevant effects under controlled conditions in the laboratory.

Despite the many strengths of the paradigm, one standard argument against this kind of research has been the notion that such "enrichment" would actually represent a severe impoverishment by feral standards. Although there is some truth to this argument, the key point is missed, first of all because the rodents used in these experiments are not feral animals either. They have been inbred for many, sometimes hundreds of, generations in a laboratory setting and have never seen wildlife conditions. Mice and rats experience the enriched environment consequently as stressing. Enrichment is still associated, however, with a large range of positive effects on almost all parameters that can be measured. "Superenrichment," that is, an enrichment to conditions more closely resembling the wild, often only increased the size of the effects but did not lead to fundamentally different results, although, for example, in males larger space allowed for larger territories and consequently reduced stress (Marashi et al., 2003). Impoverishment of laboratory animals, for example, by social isolation, in contrast had corresponding negative effects (Stranahan et al., 2006).

In general, enrichment effects are always relative to what is defined as standard. They cannot be used to gain much insight into absolute effect sizes that would be realistic under feral conditions, although for evaluating the directionality of effects and identifying the underlying principles this restriction is unlikely to matter. One needs to be aware, however, that extrapolation to a less reductionist setting that goes beyond qualitative statements will always be difficult.

In explaining the broad effects of environmental enrichment on animals, two theories prevailed: the "arousal theory" by Walsh and Cummins (Walsh and Cummins, 1975) and the "learning theory" favored by Rosenzweig and Bennett

(Rosenzweig and Bennett, 1996). Both ideas have merits, and they are not as mutually exclusive than the debates of the time might suggest. Nevertheless, in a more mechanistic (or even molecular) sense the paradigm is notoriously hard to decipher. If it is reductionist, it is so on a very different level than most other studies we do in the laboratory. The main advantage is that the manipulation is entirely noninvasive and might be the best compromise when addressing natural complexity in highly controlled laboratory research with animals.

11.3 Exposure to an Enriched Environment Stimulates Adult Hippocampal Neurogenesis

Exposure to an enriched environment increases adult hippocampal neurogenesis by exerting a survival-promoting effect on the progeny of the newborn immature neurons in the dentate gyrus (Kempermann et al., 1997b). That report in 1997 represented the first demonstration of a positive regulation of adult neurogenesis and the first functional link of adult neurogenesis to a physiological situation. The fact that experience of environmental complexity stimulated adult neurogenesis, and did so even in old age (Kempermann et al., 1998b), argued against the idea that adult neurogenesis might be a mere atavism. It seemed likely that "arousal" and "learning" here led to a relative specific response in cellular plasticity. The ensuing finding that voluntary physical activity was sufficient to increase adult neurogenesis in an acute setting (Van Praag et al., 1999) – the effect decreases again after prolonged exercise (Kronenberg et al., 2006) – suggested that the more arousing stimulus could actually be separated from the one that induced "learning." The nonspecific stimulus, as exemplified by physical activity, might set the stage on which a more specific learning stimulus, as supposedly prevalent in an enriched environment, might act. Although environmental enrichment shows primarily a recruitment effect on newborn neurons, physical activity induces the proliferation of type 2a progenitor cells, that is, cells before or shortly after neuronal determination (Kronenberg et al., 2003; Steiner et al., 2008). Sustained physical activity has additional effects in that the normally occurring age-related decrease in the population of cycling progenitor cells is reduced (Kronenberg et al., 2006). Presumably in the absence of more cognitive stimuli the maintained elevated level of precursor cell activity, however, was not translated into increased net neurogenesis. Similar preserving effects could be found in prolonged exposure to an enriched environment (Kempermann and Gage, 1999; Kempermann et al., 2002). Even when the mice were withdrawn from the enriched conditions cell proliferation remained elevated, as if in expectation of further stimuli (Kempermann and Gage, 1999). When naïve mice were exposed to environmental enrichment for the first time in older age, the relative effect was even greater than in young animals (Kempermann et al., 1998b), suggesting that the stimuli mobilized all available resources.

Neither environmental enrichment nor physical activity stimulated neurogenesis in the adult olfactory bulb (Brown et al., 1999). Exposure to olfactory complexity, however, increased neurogenesis in the olfactory bulb but was not sufficient to elicit a similar response in the hippocampus (Rochefort et al., 2002). Although the first result is plausible, the second seems surprising given the putative contribution of adult hippocampal neurogenesis to coping with the complexity and novelty of situations and the fact that the olfactory system is the primary sensory input for rodents, in which this study was done. Additional studies to solve this contradiction are required.

11.4 Genetic Influence on the Activity-Dependent Regulation of Adult Neurogenesis

Adult hippocampal neurogenesis is also influenced by a large natural variation, which in a laboratory setting can be addressed by studying strain differences (Kempermann et al., 1997a, 2006; Kempermann and Gage, 2002a,b). The genetic background also affects how the mice (and adult neurogenesis) respond to environmental enrichment. In strain 129/SvJ mice the stimulation of neurogenesis by environmental enrichment appeared much stronger than in canonical strain C57BL/6, but no side-by-side analysis of both strains was made (Kempermann et al., 1998a). The key difference was that in 129/SvJ exposure to environmental enrichment also elicited a response on the level of the proliferating precursor cells, which has never been reported for C57BL/6. This observation raised a number of interesting questions about how nature and nurture interact in the control of adult neurogenesis. The almost dichotomic situation in C57BL/6, which nicely separates effects on the precursor cell level from those on the newborn neuron depending on the type of stimulus, might be unusual. Exploiting such strain differences might be one promising strategy to decipher the genetic complexity underlying the activity-dependent regulation of adult neurogenesis.

11.5 What Benefit Does Adult Neurogenesis Confer?

129/SvJ mice that perform poorly on many behavioral tests, including those testing hippocampal learning, also showed a strong improvement in performance after having been exposed to environmental enrichment (Kempermann et al., 1998a). Such improvement was also seen in C57BL/6, but this strain performs well anyway (Kempermann et al., 1997b). Taken together, both results suggested that a presumably reciprocal link might exist between "learning" and adult neurogenesis. Learning stimulated adult neurogenesis, and increased adult neurogenesis was associated with improved learning performance. Dupret et al. were able to demonstrate this interdependency in one single experiment, in which they found that besides the active integration of the new cells, a similarly active elimination of the superfluous new neurons is required (Dupret et al., 2007).

If learning stimulates adult neurogenesis, it appears reasonable to hypothesize that the new neurons might in some way contribute to the function that induces the recruitment of the new neurons. The key question is what this function might be.

The past few years have seen a tremendous progress in research on the question of how adult neurogenesis contributes to hippocampal function (Shors et al., 2001, 2002; Snyder et al., 2005; Saxe et al., 2006, 2007; Winocur et al., 2006; Dupret et al., 2008). The new neurons contribute to the mossy fiber connection between the dentate gyrus and CA3, by itself a highly plastic structure, which constitutes a bottleneck within the network of the hippocampus. It thus seems that input to CA3 should be as lean as possible but as strong as necessary to allow the individual to cope with novelty and complexity of the environment (Kempermann, 2002). In that scenario, the new neurons would provide an effective way to prevent catastrophic interference in CA3 (Wiskott et al., 2006). Related to this idea is the concept that the new neurons contribute to temporal separation of input patterns (Aimone et al., 2006). Friedman commented on this hypothesis by pointing out that, alternatively, the dentate gyrus (and with it the new neurons) might rather play a role in determining the age of memories (Friedman, 2007). A few other computational models have been proposed (Becker, 2005; Becker and Wojtowicz, 2007), and Butz and colleagues reported the interesting idea that regulation of neurogenesis is based on a functional homeostasis (Butz et al., 2006, 2008). A "grand unifying theory" is still lacking. There can be little doubt, however, that the new neurons do contribute to hippocampal function. The reason that the "true" solution has not yet been unraveled might be a lack of appropriate behavioral tests.

11.6 Learning-Induced Regulation of Adult Neurogenesis

Initially, information on whether learning might regulate adult neurogenesis had been confusing because one of the two initial studies had found a survival-promoting effect of water-maze learning on adult neurogenesis (Gould et al., 1999) whereas the similar other study did not (Van Praag et al., 1999). It had been noted immediately that the reason for the apparently contradictory results might lie in the details of the experimental design (Greenough et al., 1999). This difference in temporal design of the studies led to the idea of a critical time window in the development of the new cells for the stimulus to elicit the neurogenic response. This idea is further discussed in the following paragraph. There were also a few confounding effects of the test itself, presumably caused by stress (Ehninger and Kempermann, 2006).

The newborn neurons show an upregulation of immediate early genes in response to a learning stimulus (Jessberger and Kempermann, 2003; Kee et al., 2007b), and the activated cells become preferentially integrated (Kee et al., 2007a). This participation of the new neurons in the neuronal network activity (so far as such integration is indicated by the appropriate upregulation of activity-dependent genes) is also found in environmental enrichment. In that situation the new neurons even showed some preference for a novel stimulus (as opposed to a familiar one) and remained associated with that stimulus (Tashiro et al., 2007). A similar observation had previously been made for adult neurogenesis in the olfactory bulb, where Magavi and colleagues discovered that the new neurons showed a preference for new odors (Magavi et al., 2005).

Nora Abrous and her colleagues have analyzed the water-maze paradigm in great detail and found that the learning stimuli elicit a characteristic sequence of events. As predicted, there was a survival-promoting effect on newborn neurons. In addition, cell death of other cells that were obviously not recruited was enhanced and, in a later phase, precursor cell proliferation was induced (Dobrossy et al., 2003; Dupret et al., 2007). The latter is consistent with the idea that the nonspecific regulation of cell proliferation is preparing the dentate gyrus for future situations of novelty and complexity, in which new neurons might be needed. This potential for preparing the hippocampus for future demands on cellular plasticity might be considered a "neurogenic reserve" (Kempermann, 2008). The "neurogenic reserve hypothesis" states that over long periods of times the functional contribution of adult neurogenesis to the hippocampus lies in providing a potential to respond to a computationally challenging situation with incorporating new neurons. The neurogenic reserve hypothesis is compatible with several theories of how individual new neurons might contribute to hippocampal function, although certain assumptions have to be made. Adult neurogenesis and its function should be cumulative rather than reflecting a turnover, and Bruel-Jungerman and colleagues found that long-term potentiation (LTP), the putative electrophysiological correlate of "learning," induced neurogenesis in the adult hippocampus in vivo (Bruel-Jungerman et al., 2006). It is not known how far environmental enrichment is associated with increased LTP in the hippocampus, and the question remained whether in this particular experimental setting LTP might not have been an equivalent of very generic neuronal activity rather than the simulation of concrete learning. Nevertheless, the observation further suggested a link between functionally relevant network activity and the regulation of adult neurogenesis.

11.7 Is There a Critical Time Window for the Effects of Environmental Enrichment on Adult Neurogenesis?

The new neurons themselves go through a period of increased synaptic plasticity. Ex vivo, the threshold to induce LTP in the new neurons was reduced compared to mature granule cells (Wang et al., 2000; Schmidt-Hieber et al., 2004). In animals with experimentally reduced adult neurogenesis, this type of LTP is abolished (Saxe et al., 2006). The period of increased synaptic plasticity constitutes a critical time window in which the recruitment of the new neurons has to take place (Ge et al., 2007). This idea is in line with other examples from developmental biology, most notably from the visual system (Hensch et al., 1998). There, the period of increased plasticity is crucially dependent on $GABA_A$ signaling (Fagiolini et al., 2004).

The early steps of adult neurogenesis also require γ-aminobutyric acid (GABA), which acts in excitatory fashion at this stage (Tozuka et al., 2005) as well as in synaptic integration (Ge et al., 2006). On the other hand, adult neurogenesis is dependent on NMDA receptor signaling (Gould et al., 1994; Cameron et al., 1995; Nacher et al., 2003; Okuyama et al., 2004), with some more or less indirect evidence suggesting that it might be the *N*-methyl-D-aspartic acid (NMDA) receptors on hilar interneurons that control the effect (Petrus et al., 2009). On the other hand, GABA and NMDA signaling affect adult neurogenesis at more than one stage, and the resulting pattern of interdependencies appears complex. For this reason, the LTP data cannot yet be directly linked to individual stages of neuronal development. It is not yet fully clear when the critical time window begins, but it appears to set in after the cells have become postmitotic. In mice this period is characterized by the transient expression of calretinin (Brandt et al., 2003), but if and how calretinin plays any functional role for the period of increased sensitivity to glutamate (and GABA) remains to be shown.

Environmental enrichment affects the late progenitor cells (type 3 cells) and postmitotic neurons (Steiner et al., 2008), which express GABA receptors (Wang et al., 2005) but probably no NMDA receptors (Nacher and McEwen, 2006; Petrus et al., 2009). At the same time, environmental enrichment strongly affects calretinin-expressing postmitotic cells (Kronenberg et al., 2003, 2006). The new neurons mature into activity-dependent responsiveness over a period of up to 7 weeks (Jessberger and Kempermann, 2003), after which they become indistinguishable from older granule cells (van Praag et al., 2002). The sensitivity of new neurons to activity elicited by environmental enrichment during this period has been directly visualized (Tashiro et al., 2007). Again, the selective and stimulus-specific integration of the new neurons is NMDA receptor dependent, and at this stage the effect appears to be direct rather than via the interneurons (Tashiro et al., 2006).

From the data on the critical time window it follows that the described separation of the effects of "activity" on adult hippocampal neurogenesis into a nonspecific and a specific phase is a convenient construct and represents a reductionist view. In reality, the two aspects intertwine.

11.8 Affective Behavior and Adult Neurogenesis

Because the hippocampus plays a role not only in learning and memory but also in affective behavior, adult hippocampal neurogenesis might also play a role in emotions (Sahay and Hen, 2007). This potential link plays a central role in the hypothesis that failing adult hippocampal neurogenesis might be involved in the pathogenesis of affective disorders such as major depression and schizophrenia (Jacobs et al., 2000; Kempermann and Kronenberg, 2003; Duman, 2004; Sahay and Hen, 2007). In addition to correlational data, one study showed by inhibiting adult neurogenesis through irradiation that antidepressants lost their anxiolytic effect, suggesting that anxiolysis (at least via serotonergic interventions) depended on adult neurogenesis (Santarelli et al., 2003). This same strong association was not detectable in another mouse strain, however, so that the functional connection might not be as straightforward as initially thought (Holick et al., 2008). Patients with affective disorders often show learning deficits, primarily in the sense of pseudodementia. Because learning is to a large degree dependent on emotional input to the hippocampus (and presumably the dentate gyrus), new neurons might also be involved in this mediation. This link has not yet been further explored. Environmental enrichment and physical activity both enhance adult neurogenesis, improve learning, and act against depression: the question is if a causal link among these three effects exists.

With respect to the regulation of adult neurogenesis by affective behavior, most data exist on stress. In numerous paradigms acute stress was found to downregulate adult neurogenesis (Cameron and Gould, 1994; Gould et al., 1997; Gould and Tanapat, 1999). In a more chronic paradigm the picture has been less clear, but overall the impression that stress is detrimental to neurogenesis holds (Heine et al., 2004, 2005). Dysregulation of the hypothalamic–pituitary–adrenal (HPA) axis has been considered as the key mechanism, but levels of glucocorticoids in serum alone cannot explain the effects. Interestingly, both physical exercise and environmental enrichment, with their robust positive effect on adult neurogenesis, also increase corticosterone levels in the animals (Van Praag et al., 1999; Kempermann et al., 2002). Interestingly, in a comparison of neurogenesis in two rat strains, Lister-Hooded and Sprague-Dawley, the difference in the response of adult neurogenesis to antidepressant treatment appeared to depend on corresponding differences in activity of the adrenal glands (Alahmed and Herbert, 2008).

In a genetic model of depression, the Flinders Sensitive Line, adult neurogenesis was found to be increased rather than decreased as one might have expected (Petersen et al., 2008). Here, an antidepressant (escitalopram) reduced cell proliferation. Presumably, this result might tell more about the complex genetics than about the interaction between affective behavior and the regulation of adult neurogenesis.

Data about "affective behavior" other than those for depression and related issues are scarce, although from the perspective of "enrichment" the influence of positive emotions on neurogenesis would be interesting. However, Mak and colleagues reported that male pheromones induced adult neurogenesis in female mice (Mak et al., 2007). The authors argue that this regulation of neurogenesis might be important for reproductive success, which from an evolutionary perspective is certainly an interesting thought.

11.9 Candidate Regulators for the Enrichment Effects on Adult Neurogenesis

To date no sound conclusions can be made whether the activity-dependent regulation of adult neurogenesis is characterized by particular factors that play no role or at least a much lesser role in controlling baseline neurogenesis. The following paragraph is not meant to cover in greater detail the topic of how adult neurogenesis is regulated on a molecular level. Only a few exemplary studies are included, in which the attempt has been made to address a particular role of these factors in activity-dependent regulation.

Historically much emphasis has been placed on the role of glucocorticoid receptor signaling, which is of particular interest in the context of aging research. Neonatal handling makes rodent pups more resistant to stress later in life (Mohammed et al., 1993). These results are consequently included in the foregoing section on "stress."

Related to the roles of affective behavior in controlling adult neurogenesis, it was found that exercise-induced regulation of cell proliferation in the adult hippocampus required beta-endorphin signaling (Koehl et al., 2008).

A few studies have addressed the question of how the exercise-induced effect is mediated. A study by Trejo and colleagues argued that insulin-like growth factor 1 (IGF-1) is necessary for the observed effects (Trejo et al., 2001). There was, however, also a strong effect of manipulating IGF-1 on baseline neurogenesis. Fabel et al. described a similar role of vascular endothelial growth factor (VEGF) (Fabel et al., 2003). Thus, obviously neither IGF-1 nor VEGF can explain the entire effect. Given the complexity of the regulation, it seems unlikely that a single gene might explain much of the net effect.

Still, similar to the findings for voluntary exercise, a few studies have found that the manipulation of individual genes affected how adult neurogenesis responded to the enrichment stimulus. For VEGF and brain-derived neurotrophic factor (BDNF), this was shown rather convincingly (Cao et al., 2004; Rossi et al., 2006), but again both studies are difficult to correlate with each other. For both factors together, a strong case could be made, but the appropriate experiment on the interaction effect is still missing. In any case, the role of both VEGF and BDNF in the control of adult neurogenesis seems so basic that the effect can hardly qualify as "specific" to the enrichment-induced role.

Along similar lines but more difficult to explain is the finding from Feng et al. that in presenilin-1 knockout mice the upregulation of adult neurogenesis in response to environmental enrichment was strongly reduced (Feng et al., 2001). On a separate note, this report is best known for its speculation that adult neurogenesis might allow the hippocampus to forget rather than to learn.

In a few other cases, exposure to an enriched environment has been included into an experiment addressing the potential role of a gene of interest in the regulation of adult neurogenesis: the lysophosphatidic receptor LPA1, for example (Matas-Rico et al., 2008). In these examples, however, manipulation of the gene affected baseline neurogenesis so that conclusions on the contribution of that factor to enrichment-induced stimulation become problematic.

Identification of factors that are effective in manipulating adult neurogenesis is only the first step in understanding "regulation."

From a systems perspective the key idea might be that neuronal activity, mediated by glutamatergic signaling and modulated by other transmitter systems, most notably serotonin, elicits the secretion of neurotrophic factors. Peripherally induced factors such as IGF-1 or hormones complement the baseline control of precursor cell activity. GABA- and glutamate-dependent mechanisms induce subsequent stages of neuronal development. Intracellularly, a series of transcription factors steers the development of the new neurons. The details of the intermediate signaling between extracellular events and transcription factor activity are still largely unknown, and so is the complex patterns of interaction between the different pathways. In the end, the precursor cells and their progeny seem to be able to integrate over a large number of signals to promote adult neurogenesis. Adult neurogenesis thereby offers the intriguing opportunity to break down a behavioral event to the molecular level, to the ensuing cellular processes, and to functional consequences that again influence behavior.

References

Aimone JB, Wiles J, Gage FH (2006) Potential role for adult neurogenesis in the encoding of time in new memories. Nat Neurosci 9:723–727.

Alahmed S, Herbert J (2008) Strain differences in proliferation of progenitor cells in the dentate gyrus of the adult rat and the response to fluoxetine are dependent on corticosterone. Neuroscience 157:677–682.

Becker S (2005) A computational principle for hippocampal learning and neurogenesis. Hippocampus 15:722–738.

Becker S, Wojtowicz JM (2007) A model of hippocampal neurogenesis in memory and mood disorders. Trends Cogn Sci 11:70–76.

Brandt MD, Jessberger S, Steiner B, Kronenberg G, Reuter K, Bick-Sander A, von der Behrens W, Kempermann G (2003) Transient calretinin expression defines early postmitotic step of neuronal differentiation in adult hippocampal neurogenesis of mice. Mol Cell Neurosci 24:603–613.

Brown ES, Rush AJ, McEwen BS (1999) Hippocampal remodeling and damage by corticosteroids: Implications for mood disorders. Neuropsychopharmacology 21:474–484.

Bruel-Jungerman E, Davis S, Rampon C, Laroche S (2006) Long-term potentiation enhances neurogenesis in the adult dentate gyrus. J Neurosci 26:5888–5893.

Butz M, Lehmann K, Dammasch IE, Teuchert-Noodt G (2006) A theoretical network model to analyse neurogenesis and synaptogenesis in the dentate gyrus. Neural Netw 19:1490–1505.

Butz M, Teuchert-Noodt G, Grafen K, van Ooyen A (2008) Inverse relationship between adult hippocampal cell proliferation and synaptic rewiring in the dentate gyrus. Hippocampus 18:879–898.

Cameron HA, Gould E (1994) Adult neurogenesis is regulated by adrenal steroids in the dentate gyrus. Neuroscience 61:203–209.

Cameron HA, McEwen BS, Gould E (1995) Regulation of adult neurogenesis by excitatory input and NMDA receptor activation in the dentate gyrus. J Neurosci 15:4687–4692.

Cao L, Jiao X, Zuzga DS, Liu Y, Fong DM, Young D, During MJ (2004) VEGF links hippocampal activity with neurogenesis, learning and memory. Nat Genet 36:827–835.

Dobrossy MD, Drapeau E, Aurousseau C, Le Moal M, Piazza PV, Abrous DN (2003) Differential effects of learning on neurogenesis: Learning increases or decreases the number of newly born cells depending on their birth date. Mol Psychiatry 8:974–982.

Duman RS (2004) Depression: A case of neuronal life and death? Biol Psychiatry 56:140–145.

Dupret D, Fabre A, Dobrossy MD, Panatier A, Rodriguez JJ, Lamarque S, Lemaire V, Oliet SH, Piazza PV, Abrous DN (2007) Spatial learning depends on both the addition and removal of new hippocampal neurons. PLoS Biol 5:e214.

Dupret D, Revest JM, Koehl M, Ichas F, De Giorgi F, Costet P, Abrous DN, Piazza PV (2008) Spatial relational memory requires hippocampal adult neurogenesis. PLoS ONE 3:e1959.
Ehninger D, Kempermann G (2006) Paradoxical effects of learning the Morris water maze on adult hippocampal neurogenesis in mice may be explained by a combination of stress and physical activity. Genes Brain Behav 5:29–39.
Fabel K, Fabel K, Tam B, Kaufer D, Baiker A, Simmons N, Kuo CJ, Palmer TD (2003) VEGF is necessary for exercise-induced adult hippocampal neurogenesis. Eur J Neurosci 18:2803–2812.
Fagiolini M, Fritschy JM, Low K, Mohler H, Rudolph U, Hensch TK (2004) Specific GABAA circuits for visual cortical plasticity. Science 303:1681–1683.
Feng R, Rampon C, Tang YP, Shrom D, Jin J, Kyin M, Sopher B, Martin GM, Kim SH, Langdon RB, Sisodia SS, Tsien JZ (2001) Deficient neurogenesis in forebrain-specific presenilin-1 knockout mice is associated with reduced clearance of hippocampal memory traces. Neuron 32:911–926.
Friedman WJ (2007) Comment on "Potential role for adult neurogenesis in the encoding of time in new memories". Hippocampus 17:503–504.
Ge S, Goh EL, Sailor KA, Kitabatake Y, Ming GL, Song H (2006) GABA regulates synaptic integration of newly generated neurons in the adult brain. Nature 439:589–593.
Ge S, Yang CH, Hsu KS, Ming GL, Song H (2007) A critical period for enhanced synaptic plasticity in newly generated neurons of the adult brain. Neuron 54:559–566.
Gould E, Tanapat P (1999) Stress and hippocampal neurogenesis. Biol Psychiatry 46:1472–1479.
Gould E, Cameron HA, McEwen BS (1994) Blockade of NMDA receptors increases cell death and birth in the developing rat dentate gyrus. J Comp Neurol 340:551–565.
Gould E, McEwen BS, Tanapat P, Galea LAM, Fuchs E (1997) Neurogenesis in the dentate gyrus of the adult tree shrew is regulated by psychosocial stress and NMDA receptor activation. J Neurosci 17:2492–2498.
Gould E, Beylin A, Tanapat P, Reeves A, Shors TJ (1999) Learning enhances adult neurogenesis in the hippocampal formation. Nat Neurosci 2:260–265.
Greenough WT, Cohen NJ, Juraska JM (1999) New neurons in old brains: Learning to survive? Nat Neurosci 2:203–205.
Hebb DO (1947) The effects of early experience on problem-solving at maturity. Am Psychol 2:306–307.
Heine VM, Maslam S, Zareno J, Joels M, Lucassen PJ (2004) Suppressed proliferation and apoptotic changes in the rat dentate gyrus after acute and chronic stress are reversible. Eur J Neurosci 19:131–144.
Heine VM, Zareno J, Maslam S, Joels M, Lucassen PJ (2005) Chronic stress in the adult dentate gyrus reduces cell proliferation near the vasculature and VEGF and FLK-1 protein expression. Eur J Neurosci 21:1304–1314.
Hensch TK, Fagiolini M, Mataga N, Stryker MP, Baekkeskov S, Kash SF (1998) Local GABA circuit control of experience-dependent plasticity in developing visual cortex. Science 282:1504–1508.
Holick KA, Lee DC, Hen R, Dulawa SC (2008) Behavioral effects of chronic fluoxetine in BALB/cj mice do not require adult hippocampal neurogenesis or the serotonin 1a receptor. Neuropsychopharmacology 33:406–417.
Jacobs BL, Praag H, Gage FH (2000) Adult brain neurogenesis and psychiatry: A novel theory of depression. Mol Psychiatry 5:262–269.
Jessberger S, Kempermann G (2003) Adult-born hippocampal neurons mature into activity-dependent responsiveness. Eur J Neurosci 18:2707–2712.
Kee N, Teixeira CM, Wang AH, Frankland PW (2007a) Preferential incorporation of adult-generated granule cells into spatial memory networks in the dentate gyrus. Nat Neurosci 10:355–362.
Kee N, Teixeira CM, Wang AH, Frankland PW (2007b) Imaging activation of adult-generated granule cells in spatial memory. Nat Protoc 2:3033–3044.
Kempermann G (2002) Why new neurons? Possible functions for adult hippocampal neurogenesis. J Neurosci 22:635–638.

Kempermann G (2008) The neurogenic reserve hypothesis: What is adult hippocampal neurogenesis good for? Trends Neurosci 31:163–169.
Kempermann G, Gage FH (1999) Experience-dependent regulation of adult hippocampal neurogenesis: Effects of long-term stimulation and stimulus withdrawal. Hippocampus 9:321–332.
Kempermann G, Gage FH (2002a) Genetic determinants of adult hippocampal neurogenesis correlate with acquisition, but not probe trial performance in the water maze task. Eur J Neurosci 16:129–136.
Kempermann G, Gage FH (2002b) Genetic influence on phenotypic differentiation in adult hippocampal neurogenesis. Brain Res Dev Brain Res 134:1–12.
Kempermann G, Kronenberg G (2003) Depressed new neurons? Adult hippocampal neurogenesis and a cellular plasticity hypothesis of major depression. Biol Psychiatry 54:499–503.
Kempermann G, Kuhn HG, Gage FH (1997a) Genetic influence on neurogenesis in the dentate gyrus of adult mice. Proc Natl Acad Sci USA 94:10409–10414.
Kempermann G, Kuhn HG, Gage FH (1997b) More hippocampal neurons in adult mice living in an enriched environment. Nature 386:493–495.
Kempermann G, Brandon EP, Gage FH (1998a) Environmental stimulation of 129/SvJ mice causes increased cell proliferation and neurogenesis in the adult dentate gyrus. Curr Biol 8:939–942.
Kempermann G, Kuhn HG, Gage FH (1998b) Experience-induced neurogenesis in the senescent dentate gyrus. J Neurosci 18:3206–3212.
Kempermann G, Gast D, Gage FH (2002) Neuroplasticity in old age: Sustained fivefold induction of hippocampal neurogenesis by long-term environmental enrichment. Ann Neurol 52:135–143.
Kempermann G, Chesler EJ, Lu L, Williams RW, Gage FH (2006) Natural variation and genetic covariance in adult hippocampal neurogenesis. Proc Natl Acad Sci USA 103:780–785.
Koehl M, Meerlo P, Gonzales D, Rontal A, Turek FW, Abrous DN (2008) Exercise-induced promotion of hippocampal cell proliferation requires beta-endorphin. FASEB J 22:2253–2262.
Kronenberg G, Reuter K, Steiner B, Brandt MD, Jessberger S, Yamaguchi M, Kempermann G (2003) Subpopulations of proliferating cells of the adult hippocampus respond differently to physiologic neurogenic stimuli. J Comp Neurol 467:455–463.
Kronenberg G, Bick-Sander A, Bunk E, Wolf C, Ehninger D, Kempermann G (2006) Physical exercise prevents age-related decline in precursor cell activity in the mouse dentate gyrus. Neurobiol Aging 27:1505–1513.
Magavi SS, Mitchell BD, Szentirmai O, Carter BS, Macklis JD (2005) Adult-born and preexisting olfactory granule neurons undergo distinct experience-dependent modifications of their olfactory responses in vivo. J Neurosci 25:10729–10739.
Mak GK, Enwere EK, Gregg C, Pakarainen T, Poutanen M, Huhtaniemi I, Weiss S (2007) Male pheromone-stimulated neurogenesis in the adult female brain: Possible role in mating behavior. Nat Neurosci 10:1003–1011.
Marashi V, Barnekow A, Ossendorf E, Sachser N (2003) Effects of different forms of environmental enrichment on behavioral, endocrinological, and immunological parameters in male mice. Horm Behav 43:281–292.
Matas-Rico E, Garcia-Diaz B, Llebrez-Zayas P, Lopez-Barroso D, Santin L, Pedraza C, Smith-Fernandez A, Fernandez-Llebrez P, Tellez T, Redondo M, Chun J, De Fonseca FR, Estivill-Torrus G (2008) Deletion of lysophosphatidic acid receptor LPA1 reduces neurogenesis in the mouse dentate gyrus. Mol Cell Neurosci 39:342–355.
Mohammed AH, Henriksson BG, Soderstrom S, Ebendal T, Olsson T, Seckl JR (1993) Environmental influences on the central nervous system and their implications for the aging rat. Behav Brain Res 57:183–191.
Nacher J, McEwen BS (2006) The role of *N*-methyl-D-asparate receptors in neurogenesis. Hippocampus 16:267–270.
Nacher J, Alonso-Llosa G, Rosell DR, McEwen BS (2003) NMDA receptor antagonist treatment increases the production of new neurons in the aged rat hippocampus. Neurobiol Aging 24:273–284.

Okuyama N, Takagi N, Kawai T, Miyake-Takagi K, Takeo S (2004) Phosphorylation of extracellular-regulating kinase in NMDA receptor antagonist-induced newly generated neurons in the adult rat dentate gyrus. J Neurochem 88:717–725.

Petersen A, Wortwein G, Gruber SH, Mathe AA (2008) Escitalopram reduces increased hippocampal cytogenesis in a genetic rat depression model. Neurosci Lett 436:305–308.

Petrus DS, Fabel K, Kronenberg G, Winter C, Steiner B, Kempermann G (2009) NMDA and benzodiazepine receptors have synergistic and antagonistic effects on precursor cells in adult hippocampal neurogenesis. Eur J Neurosci 29(2):244–252.

Rochefort C, Gheusi G, Vincent JD, Lledo PM (2002) Enriched odor exposure increases the number of newborn neurons in the adult olfactory bulb and improves odor memory. J Neurosci 22:2679–2689.

Rosenzweig MR, Bennett EL (1996) Psychobiology of plasticity: Effects of training and experience on brain and behavior. Behav Brain Res 78:57–65.

Rossi C, Angelucci A, Costantin L, Braschi C, Mazzantini M, Babbini F, Fabbri ME, Tessarollo L, Maffei L, Berardi N, Caleo M (2006) Brain-derived neurotrophic factor (BDNF) is required for the enhancement of hippocampal neurogenesis following environmental enrichment. Eur J Neurosci 24:1850–1856.

Sahay A, Hen R (2007) Adult hippocampal neurogenesis in depression. Nat Neurosci 10:1110–1115.

Santarelli L, Saxe M, Gross C, Surget A, Battaglia F, Dulawa S, Weisstaub N, Lee J, Duman R, Arancio O, Belzung C, Hen R (2003) Requirement of hippocampal neurogenesis for the behavioral effects of antidepressants. Science 301:805–809.

Saxe MD, Battaglia F, Wang JW, Malleret G, David DJ, Monckton JE, Garcia AD, Sofroniew MV, Kandel ER, Santarelli L, Hen R, Drew MR (2006) Ablation of hippocampal neurogenesis impairs contextual fear conditioning and synaptic plasticity in the dentate gyrus. Proc Natl Acad Sci USA 103:17501–17506.

Saxe MD, Malleret G, Vronskaya S, Mendez I, Garcia AD, Sofroniew MV, Kandel ER, Hen R (2007) Paradoxical influence of hippocampal neurogenesis on working memory. Proc Natl Acad Sci USA 104:4642–4646.

Schmidt-Hieber C, Jonas P, Bischofberger J (2004) Enhanced synaptic plasticity in newly generated granule cells of the adult hippocampus. Nature 429:184–187.

Shors TJ, Miesegaes G, Beylin A, Zhao M, Rydel T, Gould E (2001) Neurogenesis in the adult is involved in the formation of trace memories. Nature 410:372–376.

Shors TJ, Townsend DA, Zhao M, Kozorovitskiy Y, Gould E (2002) Neurogenesis may relate to some but not all types of hippocampal-dependent learning. Hippocampus 12:578–584.

Snyder JS, Hong NS, McDonald RJ, Wojtowicz JM (2005) A role for adult neurogenesis in spatial long-term memory. Neuroscience 130:843–852.

Steiner B, Zurborg S, Horster H, Fabel K, Kempermann G (2008) Differential 24 h responsiveness of prox1-expressing precursor cells in adult hippocampal neurogenesis to physical activity, environmental enrichment, and kainic acid-induced seizures. Neuroscience 154:521–529.

Stranahan AM, Khalil D, Gould E (2006) Social isolation delays the positive effects of running on adult neurogenesis. Nat Neurosci 9:526–533.

Tashiro A, Sandler VM, Toni N, Zhao C, Gage FH (2006) NMDA-receptor-mediated, cell-specific integration of new neurons in adult dentate gyrus. Nature 442:929–933.

Tashiro A, Makino H, Gage FH (2007) Experience-specific functional modification of the dentate gyrus through adult neurogenesis: A critical period during an immature stage. J Neurosci 27:3252–3259.

Tozuka Y, Fukuda S, Namba T, Seki T, Hisatsune T (2005) GABAergic excitation promotes neuronal differentiation in adult hippocampal progenitor cells. Neuron 47:803–815.

Trejo JL, Carro E, Torres-Aleman I (2001) Circulating insulin-like growth factor I mediates exercise-induced increases in the number of new neurons in the adult hippocampus. J Neurosci 21:1628–1634.

Van Praag H, Kempermann G, Gage FH (1999) Running increases cell proliferation and neurogenesis in the adult mouse dentate gyrus. Nat Neurosci 2:266–270.

van Praag H, Schinder AF, Christie BR, Toni N, Palmer TD, Gage FH (2002) Functional neurogenesis in the adult hippocampus. Nature 415:1030–1034.
Walsh RN, Cummins RA (1975) Mechanisms mediating the production of environmentally induced brain changes. Psychol Bull 82:986–1000.
Wang S, Scott BW, Wojtowicz JM (2000) Heterogeneous properties of dentate granule neurons in the adult rat. J Neurobiol 42:248–257.
Wang LP, Kempermann G, Kettenmann H (2005) A subpopulation of precursor cells in the mouse dentate gyrus receives synaptic GABAergic input. Mol Cell Neurosci 29:181–189.
Winocur G, Wojtowicz JM, Sekeres M, Snyder JS, Wang S (2006) Inhibition of neurogenesis interferes with hippocampus-dependent memory function. Hippocampus 16:296–304.
Wiskott L, Rasch MJ, Kempermann G (2006) A functional hypothesis for adult hippocampal neurogenesis: Avoidance of catastrophic interference in the dentate gyrus. Hippocampus 16:329–343.

Part V
Adult Neurogenesis in the Olfactory Bulb

Chapter 12
Epithelial Organization of Adult Neurogenic Germinal Niches

Zaman Mirzadeh, Young-Goo Han, José Manuel García-Verdugo, and Arturo Alvarez-Buylla

The adult mammalian brain is perhaps the most complex biological system under scientific study. Yet for all of its complexity, the brain develops from a simple epithelium. Throughout development, most neurons and macroglia are derived from embryonic neural stem cells that remain part of this neuroepithelium lining the brain ventricles. However, it is thought that neural stem cells in the adult brain, which are displaced to the subventricular zone of the lateral ventricle and the subgranular zone of the hippocampus, may have lost their epithelial properties. This chapter explores the epithelial organization of cells lining the ventricles in the adult rodent brain and reveals that a mixed neuroepithelium persists in the adult that contains neural stem cells in addition to differentiated, non-dividing ependymal cells.

12.1 Epithelia, Neuroepithelial Cells, and Radial Glia

Epithelia are sheets of tissue that line the cavities and surfaces of organs throughout the body. Epithelial sheets are cell-rich and have only a thin layer of extracellular matrix, called the basal lamina, deposited beneath the epithelial sheet. Within the sheet, epithelial cells have distinct apical and basal membranes specialized for diverse functions (Alberts et al., 1994). Characteristic protein and lipid composition

Z. Mirzadeh, Y.-G. Han, and A. Alvarez-Buylla (✉)
Department of Neurosurgery, University of California, San Francisco,
San Francisco, CA 94143, USA
and
Eli and Edythe Broad Center of Regeneration Medicine and Stem Cell Research,
University of California, San Francisco, 513 Parnassus Avenue, Room HSW-1201,
Campus Box 0525, San Francisco, CA 94143-0525, USA
e-mail: abuylla@stemcell.ucsf.edu

J.M. García-Verdugo
Laboratorio de Morfología Celular, Centro de Investigación Príncipe Felipe,
Valencia, Spain
and
CIBERNED, Spain

T. Seki et al. (eds.), *Neurogenesis in the Adult Brain I: Neurobiology*,
DOI 10.1007/978-4-431-53933-9_12, © Springer 2011

distinguish the apical and basal membranes, which are delimited by intercellular junctional complexes (Martin-Belmonte and Mostov, 2008). These complexes consist of adherens and tight junctions that recruit molecules involved in regulating apical–basal polarity of the epithelium. Although less well understood, epithelial cells also have planar polarity, which is the orientation of the cell with respect to the epithelial plane (orthogonal to the apical–basal axis) (Zallen, 2007; Simons and Mlodzik, 2008; Wu and Mlodzik, 2009). Another property of epithelia is high tensile strength, derived from cytoskeletal filaments (i.e. actin and intermediate filaments) expressed by cells within the sheet that interconnect across the epithelium at junctional complexes (Alberts et al., 1994). The expression of cytoskeletal filaments and intercellular junctional complexes, and apical–basal and planar polarity are common properties of epithelial cells, even though these cells can have different shapes and be organized into single or multiple cell layers.

The brain develops from a pseudostratified columnar epithelium. Early in development, this flat neuroepithelium (the neural plate) bends at the midline of the embryo and the lateral ridges fold over and fuse, forming an epithelial tube called the neural tube (Schoenwolf and Smith, 2000; Colas and Schoenwolf, 2001). A single layer of neuroepithelial (NE) cells forms the walls of the neural tube, each cell spanning from the lumen of the tube (prospective brain ventricle) at its apical membrane to the outer surface of the tube (prospective brain pia) at its basal membrane. The stratified appearance ('pseudostratified') of NE cells is a result of cell-cycle dependent to-and-fro movements of their nuclei, called interkinetic nuclear migration, along the apical–basal axis (Sauer, 1935; Misson et al., 1988; Noctor et al., 2001). The nucleus migrates to the basal membrane to replicate its DNA, and then migrates back to the apical luminal surface to undergo mitosis. NE cells also have other indicators of apical–basal polarity, such as the organization of their apical and basal membranes (Wodarz and Huttner, 2003; Gotz and Huttner, 2005). The apical membrane is enriched in prominin-1 (CD133) (Weigmann et al., 1997), while the basal membrane localizes integrin α_6 (Wodarz and Huttner, 2003), a transmembrane receptor that binds to the basal lamina. In addition, NE cells have adherens and tight junctions at the interface between apical and basal sides (Aaku-Saraste et al., 1996; Chenn et al., 1998). Knocking out AF-6 (also called afadin), a multi-domain scaffold protein that is a component of these junctional complexes, has been shown to disrupt the polarization of neuroepithelial cells (Ikeda et al., 1999; Zhadanov et al., 1999; Ooshio et al., 2007). Interestingly, although the apical–basal polarity of NE cells is maintained during formation of the neural tube (Schoenwolf and Smith, 2000; Colas and Schoenwolf, 2001), the composition of NE cell junctional complexes is dynamic. NE cell adherens junctions lose E-cadherin expression and increase N-cadherin expression during neural tube closure (Hatta et al., 1987; Takeichi, 1995). Occludin, a component of tight junctions, is also lost during this time, as are functional tight junctions (Aaku-Saraste et al., 1996). It has been suggested that these changes reflect a remodeling of the neuroepithelium that is necessary for neurogenesis to occur.

Neurogenesis in the mouse forebrain begins at E10-E11 (Caviness et al., 1995), resulting in the production of basally migrating neuroblasts that thicken the neural tube wall and transform it into multilayered structure (Huttner and Kosodo, 2005). In the course of this transformation, NE cells are replaced by a distinct, but related, progenitor cell type called the radial glial cell. These cells are the principal neuronal progenitors of the embryonic forebrain (Malatesta et al., 2000; Miyata et al., 2001; Noctor et al., 2001; Tamamaki et al., 2001; Anthony et al., 2004). Radial glia keep their cell body at the luminal side of the thickening neural tube in a pseudostratified layer called the ventricular zone (VZ) (The Boulder Committee, 1970). The stratified appearance of the VZ is due to interkinetic nuclear migration of radial glial cells (Misson et al., 1988; Noctor et al., 2001), a behavior very similar to that of NE cells. From their cell body in the VZ, radial glia extend a long, radially-oriented basal process that contacts the pial basal lamina (Halfter et al., 2002; Haubst et al., 2006), and a short apical process that contacts the neural tube lumen. The radial process forms additional en-passant contacts with blood vessel basal lamina as it courses through the neuronal layers towards the pia (Misson et al., 1988; Noctor et al., 2001). Radial glia, therefore, maintain the topology of NE cells (Astrom and Webster, 1991). It is thought that signals arising from both apical and basal poles of these cells are essential for proper neurogenesis [e.g. apical: (Bittman et al., 1997; Nadarajah et al., 1997; Kosodo et al., 2004; Weissman et al., 2004), e.g. basal: (Halfter et al., 2002; Haubst et al., 2006)]. In addition to maintained apical and basal contacts, radial glia share other epithelial characteristics with NE cells such as apical membrane specializations: extension of the primary cilium from the apical surface, which is associated to the basal body and centriole directly beneath the apical surface (discussed below) (Chenn et al., 1998; Tramontin et al., 2003), apical expression of prominin-1 (CD133) (Weigmann et al., 1997), and apical membrane concentration of ASIP/Par-3, Par-6, aPKC (Manabe et al., 2002; Kosodo et al., 2004; Afonso and Henrique, 2006; Imai et al., 2006; von Trotha et al., 2006; Costa et al., 2008), and Cdc42 (Cappello et al., 2006), cytoplasmic proteins that are known regulators of apical–basal polarity (Munro, 2006; Martin-Belmonte and Mostov, 2008).

Par-3, Par-6, and aPKC form a polarity complex at the apical end of NE and radial glial cells. Apically localized expression of these proteins decreases during cortical development concomitant with the decline in radial glial progenitor numbers and the increase in basally located intermediate progenitors. Intermediate progenitors are transit-amplifying cells that expand the precursor pool in the developing subventricular zone (SVZ) prior to neuronal differentiation. Interestingly, recent evidence shows that Par3 knockdown in radial glia results in early cell cycle exit and neuronal differentiation, while overexpression of Par proteins promotes symmetric proliferative divisions of radial glial progenitors (Costa et al., 2008). Consistently, deletion in radial glia of cdc42, a known activator of the par complex (Lin et al., 2000; Hutterer et al., 2004; Etienne-Manneville et al., 2005), results in loss of apical Par proteins and adherens junctions leading to early loss of radial glia and increase in intermediate progenitors (Cappello et al., 2006). Perhaps a more interesting result comes from conditional ablation of aPKC-λ, one of two isoforms of aPKC, in radial glial progenitors slightly later in development (~E15.5). Radial glia in these mutants lose apical adherens junctions and ventricular contacts yet undergo apparently normal neurogenesis (Imai et al., 2006). The milder phenotype in these mutants may be partially explained by the later inactivation of the polarity complex. However, one tantalizing possibility suggested by these studies altogether is that while apical–basal polarity is essential for the function of radial glia as primary progenitors, maintenance of this polarity does not absolutely require apical adherens junctions or ventricular contact.

As in NE cells, junctional complexes including adherens junctions (Mollgard et al., 1987; Chenn et al., 1998; Kosodo et al., 2004) delimit the apical and basolateral membranes of radial glia. Surprisingly, the integrity of radial glial adherens junctions was recently shown to require numb, a molecule well studied for its roles in neuronal cell fate determination (Kuo et al., 2006; Rasin et al., 2007). Loss of Numb and Numb-like in radial glia disrupts proper basolateral insertion of cadherins and results in abnormal adherens junctions and polarity (Rasin et al., 2007).

Planar polarity is an epithelial property that has been studied in NE cells early in neural tube development. Before the neural tube closes, NE cells undergo mediolateral intercalation within the plane of the neural plate (Schoenwolf and Alvarez, 1989). Net cell movement is towards the midline. This polarized rearrangement of cells, called convergent extension, simultaneously narrows and elongates the neural plate, and brings together the elevating lateral folds that will fuse to form the neural tube (Copp et al., 2003; Wallingford, 2006). Failure of this planar polarized movement of NE cells results in open neural tube defects (Greene et al., 1998; Wallingford et al., 2000; Wallingford and Harland, 2002; Murdoch et al., 2003).

In contrast to the early neuroepithelium, planar polarized behavior of radial glial cells has not been demonstrated. However, recent evidence shows that the large majority of radial glial mitoses in the cortical VZ occur with a vertical cleavage plane (perpendicular to the ventricular surface) (Noctor et al., 2008), suggesting that the mitotic spindle is oriented within the epithelial plane of the VZ. Although it has been previously suggested that such cleavage planes produce symmetric divisions, it is now understood that vertical or near vertical cleavages are capable of producing asymmetric divisions (Huttner and Brand, 1997; Huttner and Kosodo, 2005). This may occur through asymmetric inheritance of the apical membrane (Kosodo et al., 2004) or a fate-determining molecule, such as the EGF receptor (Sun et al., 2005). The prevalence of vertical cleavage planes indicates that asymmetric distribution of these factors would occur with an orientation orthogonal to the apical–basal axis. It will be interesting to determine whether radial glia in the VZ have planar polarity that is predictive of this asymmetric distribution.

In sum, NE and radial glial cells are epithelial cells with shared apical–basal polarity, evident in the organization of their apical and basolateral membranes separated by junctional complexes. Other epithelial characteristics shared by NE cells and radial glia include expression of the intermediate filament protein nestin (Hockfield and McKay, 1985; Frederiksen and McKay, 1988) and the intermediate filament associated protein RC2 (Misson et al., 1988). However, despite the shared epithelial characteristics, radial glia are distinct from NE cells. Specifically, radial glia acquire several astroglial properties: glycogen granules (Gadisseux and Evrard, 1985), and the expression of the intermediate filament protein vimentin (Pixley and De Vellis, 1984), the astrocyte-specific glutamate transporter GLAST (Shibata et al., 1997), brain lipid binding protein (BLBP) (Feng and Heintz, 1995), tenascin-C (Peretto et al., 2005), and in primates (Levitt and Rakic, 1980) and some murine radial glia (Rickmann et al., 1987), the intermediate filament protein and astrocyte marker GFAP (Glial Fibrillary Acidic Protein). It has been suggested that these molecular changes are related to the function of radial glia as guides for neuronal migration (Rakic, 1972; Hatten, 1999; Malatesta et al., 2003). Radial glia and NE cells are therefore distinct cell types, but these embryonic neural stem cells are both epithelial cells that remain part of an epithelium lining the brain ventricles throughout development.

12.2 Historical Perspective on Adult Neurogenesis

Shortly after birth in mammals, radial glia transform into parenchymal astrocytes (Schmechel and Rakic, 1979; Voigt, 1989) and ependymal cells (discussed below) (Spassky et al., 2005) and little remains of their unique morphology and behavior. The VZ is replaced by a simple ependymal epithelium. This stark transformation of the brain ventricular walls in part fueled the notion that neurogenesis was a prenatal or early postnatal phenomenon. In addition, primate studies, where adult neurogenesis may be poorly represented (Rakic, 1985), supported the notion that the adult brain was too complex for the useful integration of new neurons. However, studies in the 1960s started to challenge this notion. A series of papers by Joseph Altman and colleagues suggested that specific regions in the adult rodent brain did generate new neurons, which could be labeled by [^{3}H]-thymidine (Altman, 1962, 1963, 1967, 1969; Altman and Das, 1965). A decade later, electron microscopy studies of [^{3}H]-thymidine labeled neurons added more support to the controversial claim of adult neurogenesis (Kaplan and Hinds, 1977). Finally, experiments in songbirds would provide definitive evidence that neurons generated in the adult did functionally integrate into existing circuits. Goldman and Nottebohm showed that new neurons generated in the adult canary brain could be labeled with [^{3}H]-thymidine (Goldman and Nottebohm, 1983). Paton and Nottebohm showed that new neurons, labeled with [^{3}H]-thymidine, could be impaled with microelectrodes and (1) exhibited action potentials, (2) responded to auditory stimuli, and (3) could be filled with peroxidase to reveal typical dendritic and axonal morphology (Paton and Nottebohm, 1984; Nottebohm, 2004). Several studies have since shown that the adult rodent brain contains two major proliferative regions, the lateral wall of the lateral ventricle (LV) and the dentate gyrus of the hippocampus (Kaplan and Bell, 1983; Reynolds and Weiss, 1992; Lois and Alvarez-Buylla, 1993; Alvarez-Buylla and Garcia-Verdugo, 2002; van Praag et al., 2002; Zhao et al., 2008). With adult neurogenesis becoming a more widely accepted phenomenon, the search was on for the adult neural stem cells in vivo, the primary progenitors from which new neurons in the adult brain are derived. To date, it has been shown that in adult songbirds (Alvarez-Buylla et al., 1990), lizards (Garcia-Verdugo et al., 2002), turtles (Russo et al., 2004), and fish (Zupanc, 2006), neurogenic radial glia are retained as adult primary progenitors for the continual generation of new neurons and glial cells. However, in mammals, the adult neural stem cells appear to have an astroglial phenotype (discussed below) (Doetsch et al., 1999b; Seri et al., 2001).

12.3 Identification of the Adult Mammalian Neural Stem Cell

In contrast to neural stem cells (neuroepithelial cells and radial glia) of the embryonic mammalian brain, neural stem cells in the adult are not considered to be part of an epithelium. Adult neural stem cells are displaced from the ventricular lining to either the subventricular zone (SVZ) of the LV (Alvarez-Buylla and Garcia-Verdugo, 2002) or the subgranular zone (SGZ) of the hippocampal dentate gyrus (Zhao et al., 2008). Here we will review the epithelial properties, such as apical–basal polarity, of adult neural stem cells in the SVZ. A similar analysis has not been conducted for the progenitors in the adult hippocampal SGZ, although these cells have been studied extensively (Altman and Bayer, 1990; Eves et al., 1992; Palmer et al., 2000; Pleasure et al., 2000; Alvarez-Buylla et al., 2002; Seaberg and van der Kooy, 2002; Seri and Alvarez-Buylla, 2002; Filippov et al., 2003; Fukuda et al., 2003; Seri et al., 2004; Zhao et al., 2008).

The first evidence that the walls of the adult lateral ventricle retained neural stem cells came from in vitro studies in which dissociated cells were cultured in the presence of EGF and FGF (Reynolds and Weiss, 1992; Morshead et al., 1994). Under these conditions, some cells proliferated to form free-floating balls of cells called neurospheres. Neurospheres could be passaged and differentiated into neurons, astrocytes, and oligodendrocytes, thereby exhibiting self-renewal and multipotency. These criteria, which are used to define stem cells in other systems, have been used to define the adult NSC in vitro (Tropepe et al., 1999; Geschwind et al., 2001; Capela and Temple, 2002). However, the walls of the LV, the lateral wall in particular, contain at least four different cell types identified by morphology, ultrastructure, and molecular markers (Fig. 12.1) (Doetsch et al., 1997). For this reason, and the concern that culturing cells may alter their potential (Doetsch et al., 2002; Gabay et al., 2003), determining the identity of the adult neural stem cell required in vivo analysis.

The walls of the adult LV are lined by a simple epithelium of multiciliated ependymal (Type E) cells. Although these cells have been suggested to be the stem cells (Johansson et al., 1999), further analysis has shown that they do not divide in the adult brain (Spassky et al., 2005) and do not act as progenitors in vitro (Chiasson et al., 1999; Doetsch et al., 1999b). Beneath the ependymal lining lies the subependymal or subventricular zone, occupied by three additional cell types (Types A, B, and C cells), described by immunohistochemistry and electron microscopy (Doetsch et al., 1997). Type A cells are neuroblasts that have a migratory morphology and express the markers typical of young neurons: Tuj1, Dcx, and PSA-NCAM. These cells migrate in chains, found throughout the lateral wall of the LV (Doetsch and Alvarez-Buylla, 1996), that coalesce in the anterior SVZ to form the rostral migratory stream (RMS) to the olfactory bulb (OB) (Luskin, 1993; Lois and Alvarez-Buylla, 1994). Once they have reached their final destination in the OB, the neuroblasts mature into different subtypes of OB interneurons (Petreanu and Alvarez-Buylla, 2002; Carleton et al., 2003; Hack et al., 2005; Kohwi et al., 2005, 2007; Kelsch et al., 2007; Merkle et al., 2007). Chains of migrating A cells are ensheathed by glial tubes formed by astrocytic processes (Jankovski and Sotelo,

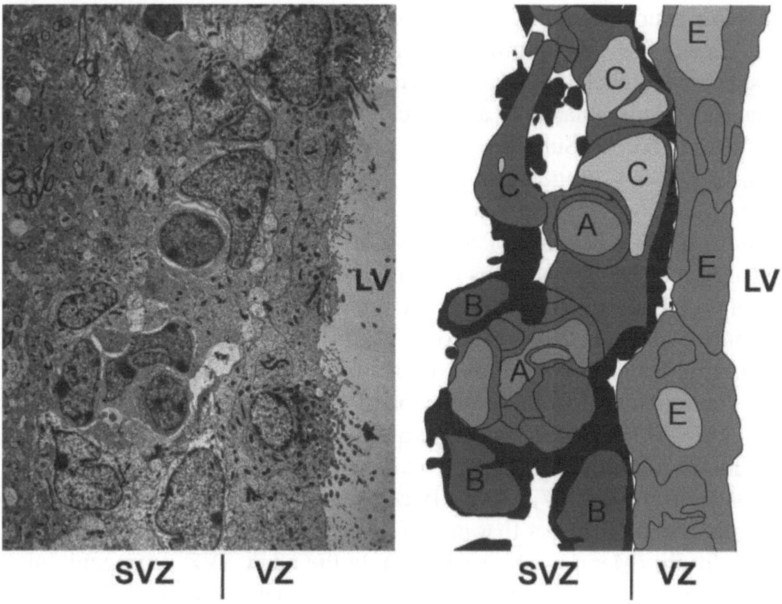

Fig. 12.1 Characterization of SVZ cell types by serial ultrathin section reconstruction at the electron microscope. The ultramicrograph to the left illustrates the complexity of the SVZ. While this initial work allowed the identification of the different cell types, it did not show any obvious epithelial organization of the progenitors. Micrograph montages of the SVZ (*left*) were studied and cell types were identified and traced (for original color coding see: Doetsch et al., 1997). On the *right*, cell types were named A–E: Ependymal (E) cells (*light gray*) line the lateral ventricle. Beneath the layer of E cells, three cell types (B, C, A) are located in the SVZ. Type B cells (*B*) are adult neural stem cells that generate migratory neuroblasts (A) through an intermediate progenitor (C cell). Type B cells form glial tubes that ensheath chains of migrating A cells destined for the olfactory bulb

1996; Lois et al., 1996; Peretto et al., 1997), some of which belong to Type B cells. The astrocytic properties of B cells include GFAP expression, abundant intermediate filaments, specialized contact with blood vessels, and glycogen granules. B cells can be further subtyped into type B1 and B2 cells. While both of these subtypes have the typical ultrastructure of astrocytes, B1 cells have lighter cytoplasm, a larger cell body, and less condensed chromatin than B2 cells. In addition, B1 cells are located adjacent to E cells with processes that form a lamina separating E cells from A cells; B2 cells are located deeper in the parenchyma at the SVZ-striatal interface. Type C cells are similar in size to B cells but have a simpler morphology, contain deeply invaginated nuclei, no intermediate filaments, and express the markers Mash1 (Parras et al., 2004) and Dlx2 (Doetsch et al., 2002). A subpopulation of C cells also expresses Olig2 (Menn et al., 2006). Of these three cell types, B and C cells can act as stem cells in vitro and give rise to large colonies of A cells (Lim and Alvarez-Buylla, 1999; Doetsch et al., 2002), whereas A cells themselves have limited self-renewal (Lim and Alvarez-Buylla, 1999).

The adult neural stem cell was identified in vivo by its ability to generate the other SVZ cell types. First, intracranial infusion of cytosine arabinoside (Ara-C), a nucleoside analog chemotherapeutic drug, was used to deplete the SVZ of rapidly dividing cells. This ablated C and A cells but not ependymal cells or the more slowly-dividing B cells. Surprisingly, this revealed that Type B cells (SVZ astrocytes) function as primary progenitors in the regeneration of the SVZ; they divide to generate C cells, which in turn divide to generate A cells (Doetsch et al., 1999b). In addition, specific labeling of SVZ astrocytes by an avian retrovirus resulted in the generation of labeled neurons that migrated to and integrated in the OB, again through a C cell intermediate. Much like intermediate progenitors of the embryonic SVZ (Haubensak et al., 2004; Noctor et al., 2004, 2007), it appears that C cells are a transit-amplifying cell type that expands the progenitor pool that ultimately produces neuroblasts. More recent work in transgenic mice, tracing the fate of (Garcia et al., 2004), or ablating (Imura et al., 2003), GFAP+ SVZ astrocytes, has confirmed the finding that Type B cells are the primary progenitors for adult neurogenesis from the SVZ. Type B cells were also recently shown to produce oligodendrocyte progenitors that migrate to, and generate oligodendrocytes in, the corpus callosum (Menn et al., 2006). While the adult SVZ neurogenic niche was initially described as residing along the anterior lateral wall of the LV (Luskin, 1993), recent studies have shown that the neurogenic niche is far more extensive, including the RMS (Bonfanti et al., 1997; Pencea et al., 2001; Gritti et al., 2002; Liu and Martin, 2003; Hack et al., 2005; Merkle et al., 2007) and more posterior regions of the lateral wall (Merkle et al., 2007), and also the dorsal or pallial wall of the LV (Kohwi et al., 2007; Merkle et al., 2007; Ventura and Goldman, 2007), the anterior medial wall of the LV (Doetsch et al., 1999b), and the subcallosal zone (Seri et al., 2006). Unlike in the embryo, neural stem cells in these regions occupy a subventricular or subependymal position, displaced from the ventricle by postnatal development of ependymal cells.

12.4 The Ependymal Epithelium

A bona fide epithelium, with both apical–basal and planar polarity, does remain lining the ventricles in the adult brain. This epithelium consists of a single layer of multi-ciliated ependymal cells, which are postmitotic (Spassky et al., 2005). The ependyma plays an important role as a protective barrier at the brain–cerebrospinal fluid (CSF) interface (Del Bigio, 1995). Ependymal cells have been conceived as a filtration system, transporting ions, small molecules, and fluid between the brain and CSF (Kuchler et al., 1994; Bruni, 1998). In addition, they have the enzymatic and structural characteristics necessary for scavenging and detoxifying various substances in the CSF (Del Bigio, 1995). Monoamine oxidase enzymes, which metabolize biogenic amines, are expressed in ependymal cells and suggest that the ependyma functions as an amine-barrier system (Williams et al., 1979; Saura et al., 1994). Structurally, ependymal cells are specialized for the propulsion of CSF, which helps redistribute and dilute local concentrations of toxins and metabolites

(Del Bigio, 1995). In mice, ependymal cells extend 30–70 motile (9+2) cilia from their apical surface into the ventricles (Bruni, 1998). Through coordinated, planar-polarized beating of these cilia across the epithelium, ependymal cells generate directed cerebrospinal fluid (CSF) flow in the ventricles (Worthington and Cathcart, 1963; Ibanez-Tallon et al., 2004; Sawamoto et al., 2006) and help maintain CSF homeostasis (Cathcart and Worthington, 1964).

Although functional assays that measure the orientation of fluid flow across the epithelium (Ibanez-Tallon et al., 2004; Sawamoto et al., 2006) have for a long time demonstrated the planar polarity of ependymal cells, until recently there were no anatomical correlates within these cells indicative of their planar polarity. In other multiciliated epithelia, such as the lining of the quail oviduct (Boisvieux-Ulrich et al., 1985), the mammalian respiratory tract (Satir and Dirksen, 1985), or the *Xenopus* larval skin (Mitchell et al., 2007), the basal foot, an accessory structure of the basal body (discussed below), points in the direction of the ciliary beat. Recently, en face ultrastructural analysis of the ependymal epithelium showed that the basal foot of ependymal cell basal bodies is also aligned with the direction of the ciliary beat and CSF flow (Mirzadeh et al., 2010). Additionally, this study showed that ependymal cells have an indicator of planar polarity not described in other multiciliated epithelia. Ependymal cell basal bodies are clustered in a well circumscribed patch comprising ~20% apical surface. Within neighboring ependymal cells, the basal body patch is displaced to the same "downstream" side of the cell with respect to CSF flow. Altogether, ependymal cells therefore have two types of planar polarity: rotational, based on the direction of the basal foot as it projects from the basal body; and translational, based on the position of the basal body patch on the apical surface (Mirzadeh et al., 2010).

Recent studies have shown that both ependymal cells (Spassky et al., 2005) and SVZ Type B cells (Merkle et al., 2004) are derived from radial glia (discussed below), providing evidence of continuity between the embryonic and adult germinal niches. Furthermore, there is evidence that the epithelial organization of ependymal cells, including their apical–basal polarity and their planar polarity, is inherited from their radial glial progenitors. In radial glia undergoing differentiation into ependymal cells, ultrastructural analyses reveal the presence of deuterosomes, electron-dense spheres of material in the supranuclear or apical compartment of the cell, used to build multiple basal bodies (Spassky et al., 2005; Dawe et al., 2007). The nascent basal bodies then migrate to the apical surface of the formerly uni-ciliated (9 + 0) radial glial cell, where they nucleate multiple (9+2) cilia (Jose Manuel García-Verdugo, personal communication). Remarkably, the planar polarity of ependymal cells also appears to be inherited from their radial glial progenitors. Analogous to the translational polarity of the basal body patch observed in ependymal cells, the single basal body at the base of the primary cilium in radial glia is displaced to the "downstream" side of the apical surface in a manner predictive of prospective ependymal polarity (Mirzadeh et al., 2010). In conditional mutants lacking radial glial primary cilia, subsequent translational polarity of the ependymal basal body patch is disrupted. Interestingly, in a slightly different conditional mutant where cilia are ablated after radial glial transformation into ependymal cells (radial glia retain primary cilia but ependymal cells lack motile cilia), translational polarity is relatively well preserved but rotational polarity is disrupted. Altogether, these studies reveal that while ependymal cell translational polarity is inherited from their radial glial progenitors and is regulated by primary cilia, rotational polarity is determined by ependymal motile cilia (Mirzadeh et al., 2010).

Evidence from other multiciliated epithelia, such as the Xenopus larval skin (Twitty, 1928; Tung and Tung, 1940) or the quail oviduct (Boisvieux-Ulrich and

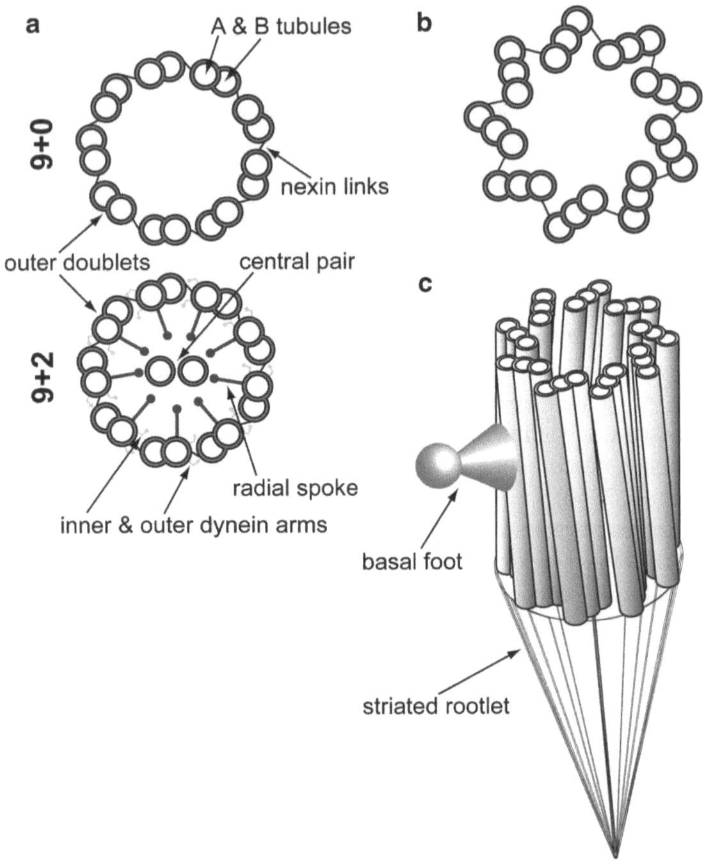

Fig. 12.2 Axonemal structure of cilia and basal bodies. (**a**) Cross-section of (9+2) and (9+0) axonemes. (9+2) axonemes have a ring of 9 doublet microtubules (A and B tubules) surrounding 2 central singlet microtubules. (9+0) axonemes lack the central pair of microtubules. (9+2) axonemes also have associated molecular motors, the inner and outer dynein arms, which are required for ciliary motility and radial spokes, which regulate ciliary bending. (**b**) Cross-section of a basal body, showing a ring of 9 triplet microtubules, which act as a template for building the ciliary axoneme. (**c**) Three-dimensional illustration of the basal body showing two of its accessory structures: the basal foot and striated rootlet. The basal foot bestows rotational asymmetry to the basal body. In several multiciliated epithelia, the basal foot points in the direction of the ciliary beat and is a structural basis of the planar polarity in these epithelial cells

Sandoz, 1991), also suggest that the establishment of planar polarity precedes multiciliated cell differentiation. Interestingly, a recent study of hair cells in the mammalian cochlea suggests that the primary cilium may relay planar polarity signals from the cell membrane to the cytoskeleton (Jones et al., 2008). Determining how radial glial planar polarity is transmitted via the primary cilium or its basal body to ependymal progeny is a provocative and unresolved question.

12.5 Cilia and Basal Bodies

Cilia are hair-like cell surface projections that contain a specialized microtubule cytoskeleton (axoneme) bounded by cell membrane. Despite their continuum with the cell body, cilia are specialized organelles with distinct sets of proteins and diverse functions. Defective primary cilia are associated with various human diseases including primary ciliary dyskinesia, polycystic kidney disease, Bardet-Biedl syndrome, retinal degeneration and neural tube defects (Bisgrove and Yost, 2006; Marshall, 2008; Gerdes et al., 2009).

Cilia are generally divided into two different types, (9 + 2) and (9 + 0), based on their axonemal structure (Satir and Christensen, 2007). (9 + 2) axonemes consist of a ring of 9 doublet microtubules surrounding a central pair of singlet microtubules, whereas (9 + 0) axonemes lack the central pair (Fig. 12.2a). In multicellular organisms, these two types of cilia are found in different tissue types and have different basic functions. (9 + 2) cilia are found, up to 200–300 per cell, on epithelial cells lining the respiratory airway, the oviduct, and the ependymal lining of the brain ventricles, where they beat to sweep mucus or fluid along the epithelium. A single (9 + 2) cilium propels the sperm cell. The motility of (9 + 2) cilia is generated by the axonemal dyneins, molecular motors associated with the axoneme (Gibbons and Rowe, 1965). These motors form the inner and outer dynein arms (Fig. 12.2a), which extend from one doublet microtubule to the next around the ring, acting to slide one doublet with respect to the other (Satir, 1965; Summers and Gibbons, 1971; Satir and Christensen, 2007). This is the basic motion underlying ciliary motility, called doublet sliding (Lindemann, 2004). (9 + 0) cilia include sensory cilia of invertebrate sensory neurons and vertebrate photoreceptors (outer segment of photoreceptor is a modified cilium), and primary cilia, which are found one per cell on almost all vertebrate cell types (Singla and Reiter, 2006). These cilia, with the exception of nodal (9 + 0) cilia found in the embryonic node (Nonaka et al., 1998), do not have the associated molecular motors required for doublet sliding and are largely thought to be immotile (Satir and Christensen, 2007).

The ciliary axoneme is constructed from a template provided by the basal body, which has a ring of nine triplet microtubules (Fig. 12.2b) (Anderson, 1972; Bettencourt-Dias and Glover, 2007; Dawe et al., 2007). The basal body is a modified centriole, an important component of the mitotic spindle apparatus (Rieder et al., 2001; Bornens, 2002). During interphase, the centriole acquires several accessory structures and docks at the plasma membrane, where it nucleates the axoneme (Dawe et al., 2007).

Accessory structures of the basal body include the striated rootlet, which emanates from the proximal end of the basal body and extends longitudinally into the cytoplasm, and the basal foot, a single laterally projecting cone-shaped structure (Fig. 12.2c) (Anderson, 1972). The basal foot bestows rotational asymmetry to the basal body. In several multiciliated epithelia, the basal foot points in the direction of the ciliary beat (Boisvieux-Ulrich et al., 1985; Satir and Dirksen, 1985; Mitchell et al., 2007) and is a structural basis of the planar polarity in these epithelial cells. Structural specializations are also found at the distal end of the basal body where it meets the ciliary axoneme. These include the terminal plate and transition fibers (Anderson, 1972; Bettencourt-Dias and Glover, 2007). It has been suggested that these structures form a molecular sieve that restricts the entry of proteins into the cilium.

Cilia lack protein synthesis machinery. Construction of cilia therefore requires a specialized intracellular trafficking system called intraflagellar transport (IFT) to supply ciliary components (Rosenbaum and Witman, 2002). IFT shuttles ciliary components between the cell body and the tip of the cilium in large complexes, called IFT particles, carried by microtubule-based motors. This transport occurs in both the anterograde and retrograde directions (Marshall and Rosenbaum, 2001; Qin et al., 2004). Anterograde transport, driven by the heterotrimeric motor kinesin-2 (Kozminski et al., 1995), moves axonemal components from the cell body to the distal tip of the cilium, where the axoneme is assembled. Retrograde transport, driven by an isoform of cytoplasmic dynein called dynein 2 (or dynein 1b) (Pazour et al., 1998), returns IFT particles to the cell body. An understanding of IFT and its role in ciliogenesis has been instrumental to recent functional studies of primary (9+0) cilia. Mouse mutants for IFT motors (KIF3A, a subunit of kinesin-2, and DYNC2H1, a subunit of cytoplasmic dynein) and IFT particle subunits (IFT52, IFT57, IFT88, IFT139, and IFT172) have revealed that primary (9+0) cilia are essential for mammalian development (Marszalek et al., 1999; Takeda et al., 1999; Huangfu et al., 2003; Haycraft et al., 2005; Huangfu and Anderson, 2005; Liu et al., 2005; May et al., 2005; Houde et al., 2006; Tran et al., 2008).

Recent studies in the past decade have revealed that the primary cilium plays a central role in several developmental signaling pathways, including hedgehog and Wnt signaling (Singla and Reiter, 2006; Eggenschwiler and Anderson, 2007). Studies from mutant mice defective for primary cilia have shown that hedgehog signaling requires the primary cilium for both activation and repression of the pathway (Huangfu et al., 2003; Haycraft et al., 2005; Huangfu and Anderson, 2005; Liu et al., 2005; May et al., 2005; Han et al., 2008, 2009; Spassky et al., 2008; Wong et al., 2009). In the absence of hedgehog ligand, Patched1, the receptor for hedgehog, localizes to cilia and inhibits signaling by preventing Smoothened, a downstream activator, from entering cilia (Rohatgi et al., 2007). Hedgehog binding to Patched1 allows Smoothened to enter cilia. This ciliary localization of Smoothened is essential for it to signal; a mutant Smoothened unable to enter cilia fails to activate the downstream pathway (Corbit et al., 2005). Primary cilia are also required for repression of the hedgehog pathway. In the absence of hedgehog ligand, Gli3, a downstream transcription factor in the hedgehog pathway, is constitutively processed to its repressor form, which represses hedgehog target genes. The production of Gli3 repressor depends on the presence of cilia; without cilia, Gli3 repressors fail to form, resulting in the de-repression of hedgehog target genes (Huangfu et al., 2003; Haycraft et al., 2005; Huangfu and Anderson, 2005; Liu et al., 2005; May et al., 2005; Han et al., 2009; Wong et al., 2009). Importantly, recent studies have showed that transit neural progenitors require primary cilia to receive hedgehog signaling necessary for their expansion in the cerebellum (Chizhikov et al., 2007; Spassky et al., 2008) and for the formation of adult neural stem cells in the dentate gyrus of the hippocampus (Han et al., 2008). Wnt signaling, which diverges into at least two pathways – the canonical (β-catenin dependent) pathway and the non-canonical (planar cell polarity or PCP) pathway (Veeman et al., 2003; Logan and Nusse, 2004) – also appears to be regulated by the primary cilium. These two pathways diverge at the level of the Wnt pathway protein Dishevelled (Dsh); cytoplasmic Dsh functions in the canonical pathway while membrane-bound Dsh functions in the non-canonical pathway (Axelrod et al., 1998). A recent study showed that Inversin, which localizes to primary cilia, acts as a switch from the canonical to the non-canonical pathway by targeting cytoplasmic Dsh for degradation (Simons et al., 2005). Several additional studies, in the kidney (Lin et al., 2003), the pancreas (Cano et al., 2004), and in mouse embryos and embryonic stem cells (Corbit et al., 2008), have supported the conclusion that the primary cilium normally suppresses canonical Wnt signaling. Disrupting ciliogenesis in these systems results in hyper-activity of the canonical Wnt pathway. In addition, disrupting the primary cilium leads to PCP defects, such as misalignment of cochlear stereociliary bundles and an open neural tube due to failed convergent extension (Ross et al., 2005; Jones et al., 2008). Notably, recent studies also suggest that Wnt signaling requires basal body function, but not the primary cilium (Huang and Schier, 2009; Ocbina et al., 2009). Interestingly, disrupting the PCP pathway in Xenopus results not only in planar polarity defects, but also defects in ciliogenesis and consequently, Hedgehog signaling (Park et al., 2006). The primary cilium has therefore been likened to an antenna, where several intercellular signaling pathways intersect (Singla and Reiter, 2006; Eggenschwiler and Anderson, 2007).

12.6 SVZ Type B Cells Are Derived from Radial Glia

Radial glia, the embryonic neural stem cells, transform into astrocytes (Schmechel and Rakic, 1979; Voigt, 1989) and ependymal cells (Spassky et al., 2005) shortly after birth. In the lateral wall of the LV, some of these cells (striatal radial glia) persist for approximately 1 week postnatally. They can be identified by their radial morphology, ultrastructure, and their expression of RC2 (Merkle et al., 2004). In addition, these neonatal radial glia exhibit the typical properties of interkinetic nuclear migration (Tramontin et al., 2003). In contrast, there are no SVZ astrocytes observed in the lateral wall of the LV at birth. Over the course of the first week, RC2 expression

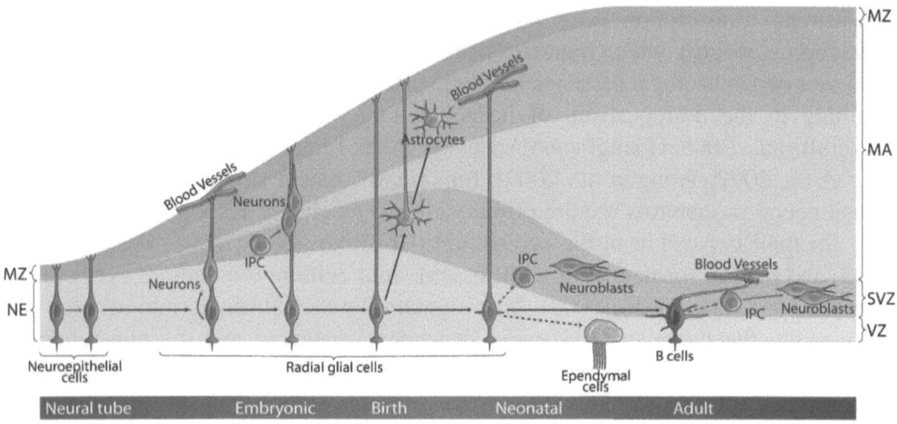

Fig. 12.3 The neural stem cell lineage from the embryonic neural tube to the adult brain. Neuroepithelial cells of the neural tube, radial glia of the developing brain, and type B cells of the adult SVZ share common epithelial characteristics including apical–basal polarity, apical contact with the ventricle, and basal contact with pial basal lamina and/or blood vessels. *Illustration by Kenneth Xavier Probst of Xavier Studio, adapted with permission from the Annual Review of Neuroscience*

and radial glial morphology gradually disappear and are replaced by the expression of GFAP and S100B, markers of SVZ astrocytes and ependymal cells, respectively (Merkle et al., 2004; Spassky et al., 2005). Although the transformation of radial glia into parenchymal astrocytes had been shown (Schmechel and Rakic, 1979; Voigt, 1989), definitive evidence was lacking that radial glia also generate SVZ astrocytes, the Type B cells that function as adult neural stem cells. To provide this evidence, Merkle et al. (2004) traced the lineage of neonatal striatal radial glia in Z/EG mice, a Cre reporter transgenic line. Z/EG mice carry a constitutively-expressed transgene encoding lacZ and a STOP cassette, flanked by loxP sites, followed by enhanced GFP (eGFP). Cre-mediated excision replaces the expression of lacZ with eGFP, providing a permanent labeling of cells and their progeny. To specifically label radial glial cell bodies in the striatal VZ, adenovirus expressing Cre was injected into the ventrolateral striatum of neonatal Z/EG mice. This resulted in infection of local striatal neurons and radial glial processes that passed through the injection site. Because adenovirus is efficiently retrogradely transported in the radial process to the radial glial cell body, this resulted in permanent labeling of radial glia in the lateral wall of the lateral ventricle. Importantly, radial glia were the only cell types in the lateral wall labeled by this method, confirmed by electron microscopy at 2 days after injection. At longer survivals, labeled radial glia had produced labeled SVZ Type B cells, which were shown to behave like neural stem cells in vitro and generate C cells, A cells and olfactory bulb interneurons in vivo. Control experiments demonstrated that viral diffusion was limited to a small area around the injection site and that infective viral activity disappeared in less than an hour after infection, confirming that Type B cells were not labeled directly. In addition, transplantation experiments confirmed that labeled radial glia, not labeled striatal cells at the injection site, were the progenitors of the Type B cells. These experiments provided a unified view on the lineage of neural stem cells as belonging to a radial-astroglial lineage (Fig. 12.3).

12.7 Evidence for Apical–Basal Polarity of Type B cells

The finding of a continuous lineage of radial-astroglial neural stem cells for the embryonic and adult brain suggests that SVZ Type B cells, like ependymal cells, may inherit apical–basal polarity and epithelial organization from their radial glial progenitors. There are several additional clues that SVZ Type B cells may be organized into a specialized epithelium. One clue comes from early experiments aimed at identifying the neural stem cells in the adult SVZ. Following Ara-C treatment, it was observed that a few Type B cells (B1 subtype) made apical contact with the ventricle (Doetsch et al., 1999a). This contact occurred through an apical extension interposed between ependymal cells. Often, a primary cilium was observed extending from the apical surface into the ventricular lumen. Type B cells in contact with the ventricle were rarely observed in untreated animals. It was inferred from these experiments that ventricular contact may be an indication of an "activated" Type B cell, stimulated to regenerate the SVZ. In addition to Ara-C treatment, other methods of SVZ stimulation, including intraventricular infusions of epidermal growth factor (EGF) (Doetsch et al., 2002) or activation of Ephrin B signaling (Conover et al., 2000), also result in increased numbers of Type B cells making apical contact with the ventricle. This apical contact, including the extension of a primary cilium into the ventricle, is reminiscent of the apical ventricular contact of NE and radial glial cells in the embryo and suggests that at least some Type B cells have the potential to form apical specializations. However, it was not known whether this was an induced, transient organization or a stable property of Type B cells. In addition, since all these studies relied on EM to identify the apical structures, it was difficult to quantify or provide a comprehensive view on the relationship of Type B cells touching the ventricle to other cells within this region.

In the embryonic brain, the apical membrane domain of NE and radial glial cells is segregated from the basolateral membrane by adherens and tight junctions located just beneath the apical ventricular surface (Aaku-Saraste et al., 1996; Chenn et al., 1998; Kosodo et al., 2004). The basal compartment of these cells is readily identified by the radial fiber and its basal contacts with both cortical blood vessels and pial basal lamina (Alvarez-Buylla et al., 1987; Misson et al., 1988; Noctor et al., 2001; Halfter et al., 2002; Haubst et al., 2006). It has been shown that soluble factors secreted by endothelial cells promote stem cell self-renewal and enhance neuronal production of both embryonic VZ and adult SVZ neural stem cells in vitro (Shen et al., 2004). Type B cells of the SVZ may therefore be regulated by factors in and around blood vessels similar to their radial glial progenitors (Jiang et al., 1998; Shingo et al., 2003; Jackson et al., 2006; Kerever et al., 2007). Recent evidence shows that an extensive network of blood vessel basal lamina infiltrates the adult

SVZ neurogenic niche and makes contact with various SVZ cell types and astrocytic processes (Mercier et al., 2002; Alvarez-Buylla and Lim, 2004). This suggested that Type B cells may also have basal membrane domains in contact with basal lamina. In this context, it is intriguing that vascular development in the embryonic brain begins at the surface and progresses inward towards the ventricles (Mancuso et al., 2008). Could this outside-in pattern mean that embryonic radial glia and adult Type B cells have a topologically equivalent interaction with blood vessel basal lamina? Interestingly, SVZ astrocytes in the ventral part of the adult LV have a long GFAP+ basal process that extends deep into the parenchyma (Sundholm-Peters et al., 2004), reminiscent of the radial fiber of radial glia. Radial glial morphology can also be re-induced in some adult SVZ astrocytes of the pallial wall of the LV by re-expression of ErbB2, one of the EGF receptors (Ghashghaei et al., 2007). Finally, recent evidence from en-face imaging of the adult SVZ provides a new anatomical perspective of this germinal niche, demonstrating that Type B1 cells have epithelial organization including both apical contact with the ventricle and basal contact with blood vessels (Mirzadeh et al., 2008; Shen et al., 2008; Tavazoie et al., 2008).

12.8 Epithelial Organization of Type B cells

En-face imaging reveals that in addition to ependymal cells, type B1 cells contact the lateral ventricles in the adult brain. Type B1 cells have an apical surface in contact with the ventricle and exhibit apical specializations including intercellular junc-

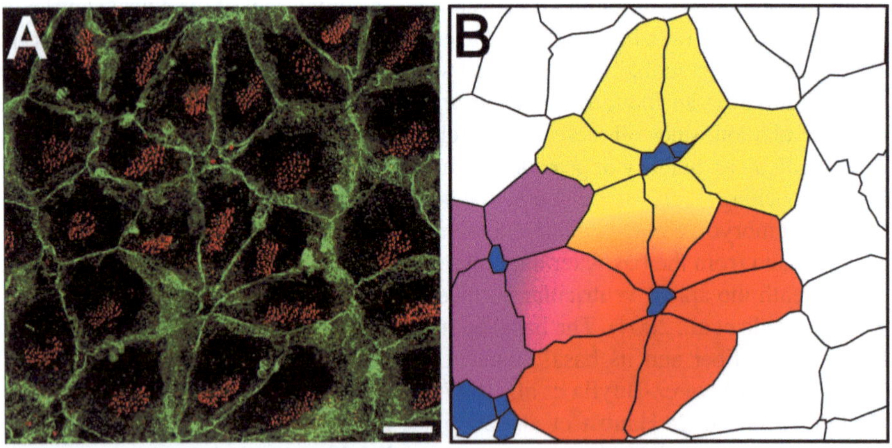

Fig. 12.4 Pinwheel architecture of the ventricular surface in the mixed neuroepithelium of the adult brain. (**a**) Confocal image of the surface of the lateral wall of the lateral ventricle stained for γ-tubulin (*red*) and β-catenin (*green*) reveals pinwheels with centrally located B cell apical surfaces surrounded by ependymal cells. *Scale bar* 10 μm. (**b**) Tracing of the β-catenin+ membranes and color coding of B cells (*blue*) and ependymal cells (*yellow*, *red*, *magenta*) to illustrate three pinwheels. Some ependymal cells occupy positions in more than one pinwheel

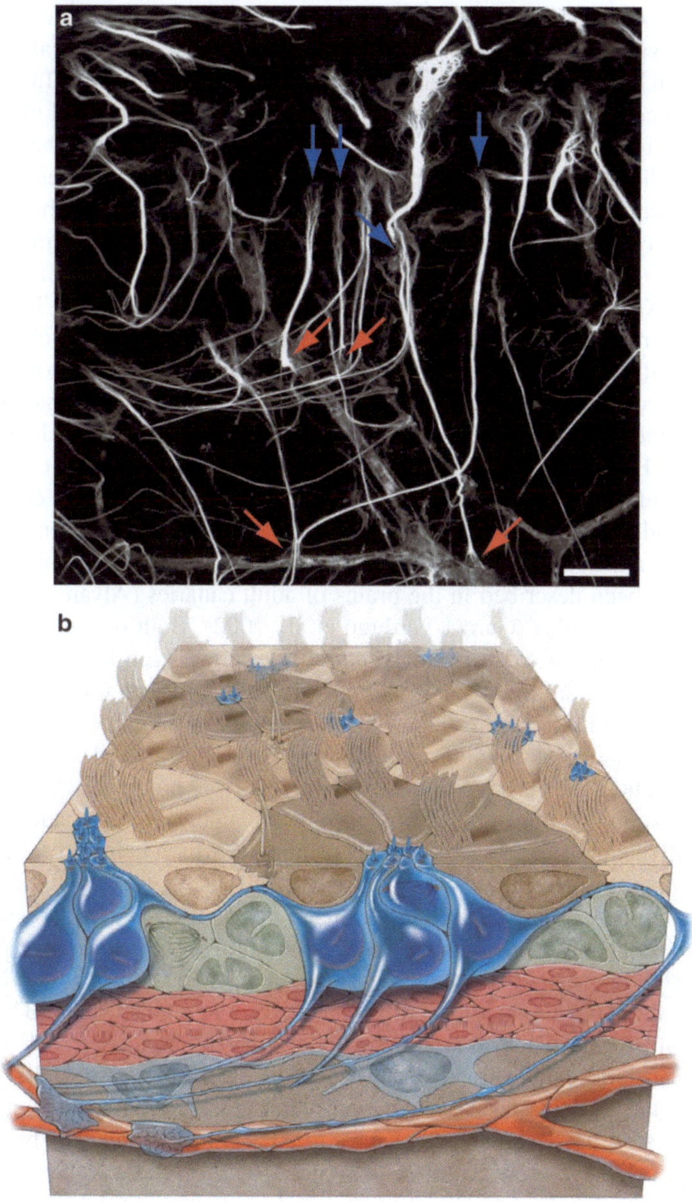

Fig. 12.5 The adult VZ neurogenic niche. (**a**) Maximum projection of a confocal stack taken of a lateral wall wholemount stained for GFAP. *Blue arrows* indicate the ventricle-contacting apical process and *red arrows* indicate the blood vessel-contacting basal process. *Scale bar* 10 μm. (**b**) Three-dimensional model of the adult VZ neurogenic niche illustrating B cells (*blue*), C cells (*green*), and A cells (*red*). B cells have a long basal process that terminates on blood vessels (*orange*) and an apical ending at the ventricle surface. Note the pinwheel organization (*light* and *dark brown*) composed of ependymal cells encircling B apical surfaces. *Illustration by Kenneth Xavier Probst of Xavier Studio, adapted with permission from Cell Stem Cell*

tional complexes and a primary cilium (Mirzadeh et al., 2008). En-face imaging also reveals that type B1 cells contain a previously unappreciated long basal process that terminates on blood vessels (Mirzadeh et al., 2008; Shen et al., 2008; Tavazoie et al., 2008); these cells are therefore morphologically similar to their radial glial progenitors. In fact, type B1 cells may be considered modified radial glia that maintain a "neuroepithelium" in the adult brain. A view emerges of the neural stem cell lineage from early embryonic development through adulthood (neuroepithelial cells to radial glia to type B cells), as consisting of cell types exhibiting a progressive transition towards more glial characteristics while maintaining epithelial properties (Fig. 12.3) (Kriegstein and Alvarez-Buylla, 2009). That is, although Type B1 cells in the adult brain are astrocyte-like cells, they share many epithelial characteristics with neuroepithelial cells of the neural tube. However, unlike the embryonic neuroepithelium that contains only neural stem cells (neuroepithelial cells or radial glia), the adult neuroepithelium is highly modified and contains ependymal cells in addition to B1 cells. Ependymal cells occupy the majority of the surface of this mixed neuroepithelium and squeeze the apical cytoplasm of neighboring B1 cells, pushing their cell bodies away from the lateral ventricle toward the subventricular zone (Mirzadeh et al., 2008). A mixed neuroepithelium of type B cells and ependymal cells has also been described in the brains of adult canaries (Alvarez-Buylla et al., 1998) and adult reptiles (Garcia-Verdugo et al., 2002). Both of these studies were done by serial section electron microscopy. Interestingly, en-face imaging of the mixed neuroepithelium in the adult mammalian brain reveals a previously unappreciated center-surround organization, where the apical surfaces of B1 cells are surrounded by ependymal cells forming pinwheels (Fig. 12.4). The pinwheels are present throughout the neurogenic niche, but absent from non-neurogenic ventricular walls, suggesting that this organization plays a fundamental role for adult neurogenesis. Ependymal cells express high levels of noggin, which helps create a niche for type B1 cells to function as neuronal progenitors (Lim et al., 2000; Peretto et al., 2004). Ependymal cells also form specialized apical tight and adherens junctions with B1 cells. The pinwheels may provide a permissive architecture for interaction and regulation between differentiated ependymal cells and neurogenic B1 cells.

As indicated above, in addition to apical ventricular contact, Type B1 cells have a long basal process that penetrates the underlying brain parenchyma (Fig. 12.5). This process terminates directly on blood vessels forming specialized endfeet (Mirzadeh et al., 2008). Basal contact with blood vessels and the surrounding basal lamina indicates that Type B1 cells have a true basal membrane. The basal process of some B1 cells also forms en-passant endings, thereby contacting multiple blood vessels. Similar endings are observed in radial glial fibers in the developing neocortex (Noctor et al., 2001) and radial cell fibers in song birds (Alvarez-Buylla et al., 1987). Interestingly, the basal processes of B1 cells do not always terminate on nearby blood vessels within the SVZ. Instead, they terminate on blood vessels, often at a distance, located along the trajectory of the long basal process. Together with evidence that Type B1 cells are derived from radial glia, this suggests that the basal processes are remnants of radial glial fibers, which have retracted away from the pial surface but have maintained contact with the invading vasculature.

As discussed above, recent evidence indicates that contact with blood vessels may be essential for neural stem cell function in both the embryonic and adult brain (Palmer et al., 2000; Alvarez-Buylla and Lim, 2004; Cao et al., 2004; Shen et al., 2004, 2008; Tavazoie et al., 2008). New anatomical analysis has revealed that the vascular network in the SVZ has a leaky blood–brain barrier, with areas along the blood vessels that lack endothelial tight junctions, pericytes and astrocytic endfeet (Tavazoie et al., 2008). This study found that transit amplifying C cells were often located at these sites in direct contact with endothelial cells. Interestingly, results from another recent study suggest that SVZ astrocytes, like NE cells, express α6β1 integrin, a laminin receptor, which they use to bind to blood vessel basal lamina (Shen et al., 2008). Intraventricular infusions of a blocking antibody against this receptor lead to an increase in the distance between SVZ astrocytes and blood vessels and a statistically significant increase in proliferation in the SVZ. Maintenance of basal contact with blood vessels may therefore allow B1 cells access to regulatory cues from the vascular microenvironment.

Furthermore, radial glia of the developing neocortex (Rakic, 1972; Noctor et al., 2001), radial astrocytes in the hippocampus (Seri et al., 2004), and radial cells in song birds (Alvarez-Buylla and Nottebohm, 1988) use the long basal process to maintain contact with neuronal progeny during early stages of differentiation and migration. This contact may be important for guiding the migration and maturation of nascent progeny. Something similar may take place in the adult SVZ. Many of the basal processes of B1 cells are oriented in the same direction as neuroblast migration, especially in the dense dorsal corridor of the SVZ (Sawamoto et al., 2006; Mirzadeh et al., 2008; Shen et al., 2008; Tavazoie et al., 2008). In this region, chains of migrating neuroblasts appear to run along the basal process of B1 cells. B1 cells' basal processes may contribute to their initial directional migration and differentiation. Recent work suggests that chemorepellants in cerebrospinal fluid and its directional flow also contribute to directional migration (Sawamoto et al., 2006). B1 cells may bridge these two compartments: they may detect chemorepellants through their apical ventricular ending and transfer this information to the SVZ along their basal processes that run parallel to chains of migrating neuroblasts. Therefore the organization of B1 cell's basal extensions could play important roles in both the regulation of stem cell proliferation by maintaining specialized contacts with blood vessels and the interaction of these stem cells with their progeny.

12.9 Conclusion

Understanding the epithelial organization of Type B1 cells and their topological relationship with the ventricle (apically) and blood vessels (basally) provides an anatomical model where molecules can be placed. Several intercellular signaling pathways are known to regulate the germinal activities of type B cells, including FGF (Kuhn et al., 1997), PDGF (Jackson et al., 2006), Shh (Machold et al., 2003; Palma et al., 2005), Wnt (Yu et al., 2006; Adachi et al., 2007), Noggin/BMP

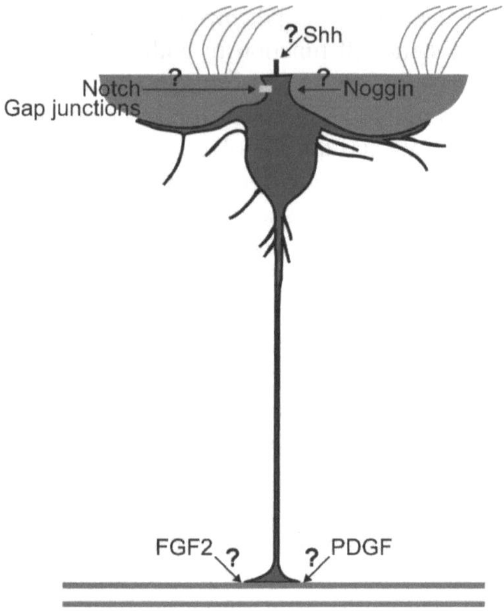

Fig. 12.6 Hypothetical model of signaling molecules impinging upon B1 cells in distinct apical and basal compartments. B cells are in a position to integrate signals derived apically from ependymal cells and the ventricular surface with signals derived basally from blood vessels and the basal lamina. Integration of signals from these distinct compartments may be essential to the germinal activity of B cells

(Lim et al., 2000), and Ephrin signaling (Conover et al., 2000). In addition, type B1 cells are regulated by signaling pathways dependent on cell–cell contact, such as Notch signaling (Nyfeler et al., 2005; Givogri et al., 2006), and are likely also regulated by intercellular junctional complexes, such as gap junctions. These diverse regulatory mechanisms may impinge upon type B1 cells with spatially specific organization. Conversely, the elaborate morphology of type B1 cells may allow these cells to bridge between different regulatory microenvironments, such as the ventricular surface and the vasculature. In addition, the ability of type B1 cells to respond to specific signaling pathways may depend on localization of receptors to specific micro-domains within the cell. For example, the ability of these cells to respond to Shh may depend on the localization of Smoothened, the Shh co-receptor, to the primary cilium (Huangfu et al., 2003; Corbit et al., 2005). Noggin, which is highly expressed by ependymal cells and creates a neurogenic niche for type B1 cells (Lim et al., 2000), may act at the level of the apical process within the pinwheel organization. Other signaling pathways may operate at the level of the basal process: a recent study showed that blood vessel basal lamina in the SVZ concentrates FGF2 (Kerever et al., 2007), which promotes the neurogenic activity of type B cells. A hypothetical model is provided in Fig. 12.6. Altogether, this indicates that the epithelial organization of type B1 cells may allow them to establish topological relationships with essential apical and basal regulatory microenvironments.

Determining the functional significance of the pinwheel organization may yield insights into the role of B1 cell apical specializations such as the primary cilium and its associated basal body. These structures may be essential for progenitor function in the adult ventricular zone. Interestingly, a recent study of asymmetric radial glial cell division in the developing cortex suggested that inheritance of the mother centriole, which transforms into the basal body during interphase, determined which daughter cell retained apical contact and progenitor identity (Wang et al., 2009). A similar mechanism may be operating in the adult ventricular zone, to maintain the population of neurogenic ventricle-contacting type B1 cells.

Here we have reviewed the epithelial organization of the neuroepithelial-radial glial-type B1 cell lineage of primary progenitors in the developing and adult brain. It will be interesting to investigate how these epithelial properties are essential to the common function of these cells as progenitors and the key developmental transitions that make these cells distinct.

Acknowledgements This work was supported by the National Institutes of Health (NIH) grants HD-32116 and NS28478, the Sandler Family Supporting Foundation, and the John G. Bowes Research Fund. A.A.-B. holds the Heather and Melanie Muss Endowed Chair. We thank Kenneth Xavier Probst of Xavier Studio for assisting with the design and preparation of Figs. 12.3 and 12.5.

References

Aaku-Saraste E, Hellwig A, Huttner WB (1996) Loss of occludin and functional tight junctions, but not ZO-1, during neural tube closure – remodeling of the neuroepithelium prior to neurogenesis. Dev Biol 180:664–679.

Adachi K, Mirzadeh Z, Sakaguchi M, Yamashita T, Nikolcheva T, Gotoh Y, Peltz G, Gong L, Kawase T, Alvarez-Buylla A, Okano H, Sawamoto K (2007) Beta-catenin signaling promotes proliferation of progenitor cells in the adult mouse subventricular zone. Stem Cells 25:2827–2836.

Afonso C, Henrique D (2006) PAR3 acts as a molecular organizer to define the apical domain of chick neuroepithelial cells. J Cell Sci 119:4293–4304.

Alberts B, Bray D, Lewis J, Raff M, Roberts K, Watson JD (1994) Molecular Biology of the Cell, 3rd Edition. New York: Garland Publishing, Inc.

Altman J (1962) Are new neurons formed in the brains of adult mammals? Science 135:1127–1128.

Altman J (1963) Autoradiographic investigation of cell proliferation in the brains of rats and cats. Anat Rec 145:573–591.

Altman J (1967) Postnatal growth and differentiation of the mammalian brain, with implications for a morphological theory of memory. In: The Neurosciences: First Study Program (Quarton GC, Melnechuck T, Schmitt FO, eds), pp 763–743. New York: Rockefeller University Press.

Altman J (1969) Autoradiographic and histological studies of postnatal neurogenesis. IV. Cell proliferation and migration in the anterior forebrain, with special reference to persisting neurogenesis in the olfactory bulb. J Comp Neurol 137:433–458.

Altman J, Bayer SA (1990) Mosaic organization of the hippocampal neuroepithelium and the multiple germinal sources of dentate granule cells. J Comp Neurol 301:325–342.

Altman J, Das GD (1965) Autoradiographic and histological evidence of postnatal hippocampal neurogenesis in rats. J Comp Neurol 124:319–336.

Alvarez-Buylla A, Garcia-Verdugo JM (2002) Neurogenesis in adult subventricular zone. J Neurosci 22:629–634.

Alvarez-Buylla A, Lim DA (2004) For the long run: maintaining germinal niches in the adult brain. Neuron 41:683–686.
Alvarez-Buylla A, Nottebohm F (1988) Migration of young neurons in adult avian brain. Nature 335:353–354.
Alvarez-Buylla A, Buskirk DR, Nottebohm F (1987) Monoclonal antibody reveals radial glia in adult avian brain. J Comp Neurol 264:159–170.
Alvarez-Buylla A, Theelen M, Nottebohm F (1990) Proliferation "hot spots" in adult avian ventricular zone reveal radial cell division. Neuron 5:101–109.
Alvarez-Buylla A, García-Verdugo JM, Mateo A, Merchant-Larios H (1998) Primary neural precursors and intermitotic nuclear migration in the ventricular zone of adult canaries. J Neurosci 18:1020–1037.
Alvarez-Buylla A, Seri B, Doetsch F (2002) Identification of neural stem cells in the adult vertebrate brain. Brain Res Bull 57:751–758.
Anderson RG (1972) The three-dimensional structure of the basal body from the rhesus monkey oviduct. J Cell Biol 54:246–265.
Anthony TE, Klein C, Fishell G, Heintz N (2004) Radial glia serve as neuronal progenitors in all regions of the central nervous system. Neuron 41:881–890.
Astrom KE, Webster HD (1991) The early development of the neopallial wall and area choroidea in fetal rats. A light and electron microscopic study. Adv Anat Embryol Cell Biol 123:1–76.
Axelrod JD, Miller JR, Shulman JM, Moon RT, Perrimon N (1998) Differential recruitment of Dishevelled provides signaling specificity in the planar cell polarity and Wingless signaling pathways. Genes Dev 12:2610–2622.
Bettencourt-Dias M, Glover DM (2007) Centrosome biogenesis and function: centrosomics brings new understanding. Nat Rev Mol Cell Biol 8:451–463.
Bisgrove BW, Yost HJ (2006) The roles of cilia in developmental disorders and disease. Development 133:4131–4143.
Bittman K, Owens DF, Kriegstein AR, LoTurco JJ (1997) Cell coupling and uncoupling in the ventricular zone of developing neocortex. J Neurosci 17:7037–7044.
Boisvieux-Ulrich E, Sandoz D (1991) Determination of ciliary polarity precedes differentiation in the epithelial cells of quail oviduct. Biol Cell 72:3–14.
Boisvieux-Ulrich E, Laine MC, Sandoz D (1985) The orientation of ciliary basal bodies in quail oviduct is related to the ciliary beating cycle commencement. Biol Cell 55:147–150.
Bonfanti L, Peretto P, Merighi A, Fasolo A (1997) Newly-generated cells from the rostral migratory stream in the accessory olfactory bulb of the adult rat. Neuroscience 81:489–502.
Bornens M (2002) Centrosome composition and microtubule anchoring mechanisms. Curr Opin Cell Biol 14:25–34.
Bruni JE (1998) Ependymal development, proliferation, and functions: a review. Microsc Res Tech 41:2–13.
Cano DA, Murcia NS, Pazour GJ, Hebrok M (2004) Orpk mouse model of polycystic kidney disease reveals essential role of primary cilia in pancreatic tissue organization. Development 131:3457–3467.
Cao L, Jiao X, Zuzga DS, Liu Y, Fong DM (2004) VEGF links hippocampal activity with neurogenesis, learning and memory. Nat Genet 36:827.
Capela A, Temple S (2002) LeX/ssea-1 is expressed by adult mouse CNS stem cells, identifying them as nonependymal. Neuron 35:865–875.
Cappello S, Attardo A, Wu X, Iwasato T, Itohara S, Wilsch-Brauninger M, Eilken HM, Rieger MA, Schroeder TT, Huttner WB, Brakebusch C, Gotz M (2006) The Rho-GTPase cdc42 regulates neural progenitor fate at the apical surface. Nat Neurosci 9:1099–1107.
Carleton A, Petreanu LT, Lansford R, Alvarez-Buylla A, Lledo PM (2003) Becoming a new neuron in the adult olfactory bulb. Nat Neurosci 6:507–518.
Cathcart RS, 3rd, Worthington WC, Jr. (1964) Ciliary movement in the rat cerebral ventricles: clearing action and directions of currents. J Neuropathol Exp Neurol 23:609–618.
Caviness VS, Jr., Takahashi T, Nowakowski RS (1995) Numbers, time and neocortical neuronogenesis: a general developmental and evolutionary model. Trends Neurosci 18:379–383.

Chenn A, Zhang YA, Chang BT, McConnell SK (1998) Intrinsic polarity of mammalian neuroepithelial cells. Mol Cell Neurosci 11:183–193.

Chiasson BJ, Tropepe V, Morshead CM, Van der Kooy D (1999) Adult mammalian forebrain ependymal and subependymal cells demonstrate proliferative potential, but only subependymal cells have neural stem cell characteristics. J Neurosci 19:4462–4471.

Chizhikov VV, Davenport J, Zhang Q, Shih EK, Cabello OA, Fuchs JL, Yoder BK, Millen KJ (2007) Cilia proteins control cerebellar morphogenesis by promoting expansion of the granule progenitor pool. J Neurosci 27:9780–9789.

Colas JF, Schoenwolf GC (2001) Towards a cellular and molecular understanding of neurulation. Dev Dyn 221:117–145.

Conover JC, Doetsch F, Garcia-Verdugo JM, Gale NW, Yancopoulos GD, Alvarez-Buylla A (2000) Disruption of Eph/ephrin signaling affects migration and proliferation in the adult subventricular zone. Nat Neurosci 3:1091–1097.

Copp AJ, Greene ND, Murdoch JN (2003) Dishevelled: linking convergent extension with neural tube closure. Trends Neurosci 26:453–455.

Corbit KC, Aanstad P, Singla V, Norman AR, Stainier DY, Reiter JF (2005) Vertebrate smoothened functions at the primary cilium. Nature 437:1018–1021.

Corbit KC, Shyer AE, Dowdle WE, Gaulden J, Singla V, Reiter JF (2008) Kif3a constrains beta-catenin-dependent Wnt signalling through dual ciliary and non-ciliary mechanisms. Nat Cell Biol 10:70–76.

Costa MR, Wen G, Lepier A, Schroeder T, Gotz M (2008) Par-complex proteins promote proliferative progenitor divisions in the developing mouse cerebral cortex. Development 135:11–22.

Dawe HR, Farr H, Gull K (2007) Centriole/basal body morphogenesis and migration during ciliogenesis in animal cells. J Cell Sci 120:7–15.

Del Bigio MR (1995) The ependyma: a protective barrier between brain and cerebrospinal fluid. Glia 14:1–13.

Doetsch F, Alvarez-Buylla A (1996) Network of tangential pathways for neuronal migration in adult mammalian brain. Proc Natl Acad Sci USA 93:14895–14900.

Doetsch F, Garcia-Verdugo JM, Alvarez-Buylla A (1997) Cellular composition and three-dimensional organization of the subventricular germinal zone in the adult mammalian brain. J Neurosci 17:5046–5061.

Doetsch F, Garcia-Verdugo JM, Alvarez-Buylla A (1999a) Regeneration of a germinal layer in the adult mammalian brain. Proc Natl Acad Sci USA 96:11619–11624.

Doetsch F, Caille I, Lim DA, García-Verdugo JM, Alvarez-Buylla A (1999b) Subventricular zone astrocytes are neural stem cells in the adult mammalian brain. Cell 97:1–20.

Doetsch F, Petreanu L, Caille I, Garcia-Verdugo JM, Alvarez-Buylla A (2002) EGF converts transit-amplifying neurogenic precursors in the adult brain into multipotent stem cells. Neuron 36:1021–1034.

Eggenschwiler JT, Anderson KV (2007) Cilia and developmental signaling. Annu Rev Cell Dev Biol 23:345–373.

Etienne-Manneville S, Manneville JB, Nicholls S, Ferenczi MA, Hall A (2005) Cdc42 and Par6-PKCzeta regulate the spatially localized association of Dlg1 and APC to control cell polarization. J Cell Biol 170:895–901.

Eves EM, Tucker MS, Roback JD, Downen M, Rosner MR, Wainer BH (1992) Immortal rat hippocampal cell lines exhibit neuronal and glial lineages and neurotrophin gene expression. Proc Natl Acad Sci USA 89:4373–4377.

Feng L, Heintz N (1995) Differentiating neurons activate transcription of the brain lipid-binding protein gene in radial glia through a novel regulatory element. Development 121:1719–1730.

Filippov V, Kronenberg G, Pivneva T, Reuter K, Steiner B, Wang LP, Yamaguchi M, Kettenmann H, Kempermann G (2003) Subpopulation of nestin-expressing progenitor cells in the adult murine hippocampus shows electrophysiological and morphological characteristics of astrocytes. Mol Cell Neurosci 23:373–382.

Frederiksen K, McKay RDG (1988) Proliferation and differentiation of rat neuroepithelial precursor cells in vivo. J Neurosci 8:1144–1151.

Fukuda S, Kato F, Tozuka Y, Yamaguchi M, Miyamoto Y, Hisatsune T (2003) Two distinct subpopulations of nestin-positive cells in adult mouse dentate gyrus. J Neurosci 23:9357–9366.
Gabay L, Lowell S, Rubin LL, Anderson DJ (2003) Deregulation of dorsoventral patterning by FGF confers trilineage differentiation capacity on CNS stem cells in vitro. Neuron 40:485–499.
Gadisseux JF, Evrard P (1985) Glial neuronal relationship in the developing central nervous system. Dev Neurosci 7:12–32.
Garcia AD, Doan NB, Imura T, Bush TG, Sofroniew MV (2004) GFAP-expressing progenitors are the principal source of constitutive neurogenesis in adult mouse forebrain. Nat Neurosci 7:1233–1241.
Garcia-Verdugo JM, Ferron S, Flames N, Collado L, Desfilis E, Font E (2002) The proliferative ventricular zone in adult vertebrates: a comparative study using reptiles, birds, and mammals. Brain Res Bull 57:765–775.
Gerdes JM, Davis EE, Katsanis N (2009) The vertebrate primary cilium in development, homeostasis, and disease. Cell 137:32–45.
Geschwind DH, Ou J, Easterday MC, Dougherty JD, Jackson RL, Chen Z, Antoine H, Terskikh A, Weissman IL, Nelson SF, Kornblum HI (2001) A genetic analysis of neural progenitor differentiation. Neuron 29:325–339.
Ghashghaei HT, Weimer JM, Schmid RS, Yokota Y, McCarthy KD, Popko B, Anton ES (2007) Reinduction of ErbB2 in astrocytes promotes radial glial progenitor identity in adult cerebral cortex. Genes Dev 21:3258–3271.
Gibbons IR, Rowe AJ (1965) Dynein: a protein with adenosine triphosphatase activity from cilia. Science 149:424–426.
Givogri MI, de Planell M, Galbiati F, Superchi D, Gritti A, Vescovi A, de Vellis J, Bongarzone ER (2006) Notch signaling in astrocytes and neuroblasts of the adult subventricular zone in health and after cortical injury. Dev Neurosci 28:81–91.
Goldman SA, Nottebohm F (1983) Neuronal production, migration, and differentiation in a vocal control nucleus of the adult female canary brain. Proc Natl Acad Sci USA 80:2390–2394.
Gotz M, Huttner WB (2005) The cell biology of neurogenesis. Nat Rev Mol Cell Biol 6:777–788.
Greene ND, Gerrelli D, Van Straaten HW, Copp AJ (1998) Abnormalities of floor plate, notochord and somite differentiation in the loop-tail (Lp) mouse: a model of severe neural tube defects. Mech Dev 73:59–72.
Gritti A, Bonfanti L, Doetsch F, Caille I, Alvarez-Buylla A, Lim DA, Galli R, Verdugo JM, Herrera DG, Vescovi AL (2002) Multipotent neural stem cells reside into the rostral extension and olfactory bulb of adult rodents. J Neurosci 22:437–445.
Hack MA, Saghatelyan A, de Chevigny A, Pfeifer A, Ashery-Padan R, Lledo PM, Gotz M (2005) Neuronal fate determinants of adult olfactory bulb neurogenesis. Nat Neurosci 8:865–872.
Halfter W, Dong S, Yip YP, Willem M, Mayer U (2002) A critical function of the pial basement membrane in cortical histogenesis. J Neurosci 22:6029–6040.
Han YG, Spassky N, Romaguera-Ros M, Garcia-Verdugo JM, Aguilar A, Schneider-Maunoury S, Alvarez-Buylla A (2008) Hedgehog signaling and primary cilia are required for the formation of adult neural stem cells. Nat Neurosci 11:277–284.
Han YG, Kim HJ, Dlugosz AA, Ellison DW, Gilbertson RJ, Alvarez-Buylla A (2009) Dual and opposing roles of primary cilia in medulloblastoma development. Nat Med 15:1062–1065.
Hatta K, Takagi S, Fujisawa H, Takeichi M (1987) Spatial and temporal expression pattern of N-cadherin cell adhesion molecules correlated with morphogenetic processes of chicken embryos. Dev Biol 120:215–227.
Hatten ME (1999) Central nervous system neuronal migration. [Review] [124 refs]. Annu Rev Neurosci 22:511–539.
Haubensak W, Attardo A, Denk W, Huttner WB (2004) Neurons arise in the basal neuroepithelium of the early mammalian telencephalon: a major site of neurogenesis. Proc Natl Acad Sci USA 101:3196–3201.
Haubst N, Georges-Labouesse E, De Arcangelis A, Mayer U, Gotz M (2006) Basement membrane attachment is dispensable for radial glial cell fate and for proliferation, but affects positioning of neuronal subtypes. Development 133:3245–3254.

Haycraft CJ, Banizs B, Aydin-Son Y, Zhang Q, Michaud EJ, Yoder BK (2005) Gli2 and Gli3 localize to cilia and require the intraflagellar transport protein polaris for processing and function. PLoS Genet 1:e53.

Hockfield S, McKay RDG (1985) Identification of major cell classes in the developing mammalian nervous system. J Neurosci 5, No 12:3310–3328.

Houde C, Dickinson RJ, Houtzager VM, Cullum R, Montpetit R, Metzler M, Simpson EM, Roy S, Hayden MR, Hoodless PA, Nicholson DW (2006) Hippi is essential for node cilia assembly and Sonic hedgehog signaling. Dev Biol 300:523–533.

Huang P, Schier AF (2009) Dampened Hedgehog signaling but normal Wnt signaling in zebrafish without cilia. Development 136:3089–3098.

Huangfu D, Anderson KV (2005) Cilia and Hedgehog responsiveness in the mouse. Proc Natl Acad Sci USA 102:11325–11330.

Huangfu D, Liu A, Rakeman AS, Murcia NS, Niswander L, Anderson KV (2003) Hedgehog signalling in the mouse requires intraflagellar transport proteins. Nature 426:83–87.

Hutterer A, Betschinger J, Petronczki M, Knoblich JA (2004) Sequential roles of Cdc42, Par-6, aPKC, and Lgl in the establishment of epithelial polarity during Drosophila embryogenesis. Dev Cell 6:845–854.

Huttner WB, Brand M (1997) Asymmetric division and polarity of neuroepithelial cells. Curr Opin Neurobiol 7:29–39.

Huttner WB, Kosodo Y (2005) Symmetric versus asymmetric cell division during neurogenesis in the developing vertebrate central nervous system. Curr Opin Cell Biol 17:648–657.

Ibanez-Tallon I, Pagenstecher A, Fliegauf M, Olbrich H, Kispert A, Ketelsen UP, North A, Heintz N, Omran H (2004) Dysfunction of axonemal dynein heavy chain Mdnah5 inhibits ependymal flow and reveals a novel mechanism for hydrocephalus formation. Hum Mol Genet 13:2133–2141.

Ikeda W, Nakanishi H, Miyoshi J, Mandai K, Ishizaki H, Tanaka M, Togawa A, Takahashi K, Nishioka H, Yoshida H, Mizoguchi A, Nishikawa S, Takai Y (1999) Afadin: a key molecule essential for structural organization of cell-cell junctions of polarized epithelia during embryogenesis. J Cell Biol 146:1117–1132.

Imai F, Hirai S, Akimoto K, Koyama H, Miyata T, Ogawa M, Noguchi S, Sasaoka T, Noda T, Ohno S (2006) Inactivation of aPKClambda results in the loss of adherens junctions in neuroepithelial cells without affecting neurogenesis in mouse neocortex. Development 133:1735–1744.

Imura T, Kornblum HI, Sofroniew MV (2003) The predominant neural stem cell isolated from postnatal and adult forebrain but not early embryonic forebrain expresses GFAP. J Neurosci 23:2824–2832.

Jackson EL, Garcia-Verdugo JM, Gil-Perotin S, Roy M, Quinones-Hinojosa A, Vandenberg S, Alvarez-Buylla A (2006) PDGFRalpha-positive B cells are neural stem cells in the adult SVZ that form glioma-like growths in response to increased PDGF signaling. Neuron 51:187–199.

Jankovski A, Sotelo C (1996) Subventricular zone-olfactory bulb migratory pathway in the adult mouse: cellular composition and specificity as determined by heterochronic and heterotopic transplantation. J Comp Neurol 371:376–396.

Jiang J, McMurtry J, Niedzwiecki D, Goldman SA (1998) Insulin-like growth factor-1 is a radial cell-associated neurotrophin that promotes neuronal recruitment from the adult songbird ependyma/subependyma. J Neurobiol 36:1–15.

Johansson CB, Momma S, Clarke DL, Risling M, Lendahl U, Frisén J (1999) Identification of a neural stem cell in the adult mammalian central nervous system. Cell 96:25–34.

Jones C, Roper VC, Foucher I, Qian D, Banizs B, Petit C, Yoder BK, Chen P (2008) Ciliary proteins link basal body polarization to planar cell polarity regulation. Nat Genet 40:69–77.

Kaplan MS, Bell DH (1983) Neuronal proliferation in the 9-month-old rodent-radioautographic study of granule cells in the hippocampus. Exp Brain Res 52:1.

Kaplan MS, Hinds JW (1977) Neurogenesis in the adult rat: electron microscopic analysis of light radioautographs. Science 197:1092–1094.

Kelsch W, Mosley CP, Lin CW, Lois C (2007) Distinct mammalian precursors are committed to generate neurons with defined dendritic projection patterns. PLoS Biol 5:e300.

Kerever A, Schnack J, Vellinga D, Ichikawa N, Moon C, Arikawa-Hirasawa E, Efird JT, Mercier F (2007) Novel extracellular matrix structures in the neural stem cell niche capture the neurogenic factor fibroblast growth factor 2 from the extracellular milieu. Stem Cells 25:2146–2157.
Kohwi M, Osumi N, Rubenstein JL, Alvarez-Buylla A (2005) Pax6 is required for making specific subpopulations of granule and periglomerular neurons in the olfactory bulb. J Neurosci 25:6997–7003.
Kohwi M, Petryniak MA, Long JE, Ekker M, Obata K, Yanagawa Y, Rubenstein JL, Alvarez-Buylla A (2007) A subpopulation of olfactory bulb GABAergic interneurons is derived from Emx1- and Dlx5/6-expressing progenitors. J Neurosci 27:6878–6891.
Kosodo Y, Roper K, Haubensak W, Marzesco AM, Corbeil D, Huttner WB (2004) Asymmetric distribution of the apical plasma membrane during neurogenic divisions of mammalian neuroepithelial cells. Embo J 23:2314–2324.
Kozminski KG, Beech PL, Rosenbaum JL (1995) The Chlamydomonas kinesin-like protein FLA10 is involved in motility associated with the flagellar membrane. J Cell Biol 131:1517–1527.
Kriegstein A, Alvarez-Buylla A (2009) The glial nature of embryonic and adult neural stem cells. Annu Rev Neurosci 32:149–184.
Kuchler S, Graff MN, Gobaille S, Vincendon G, Roche AC, Delaunoy JP, Monsigny M, Zanetta JP (1994) Mannose dependent tightening of the rat ependymal cell barrier. In vivo and in vitro study using neoglycoproteins. Neurochem Int 24:43–55.
Kuhn HG, Winkler J, Kempermann G, Thal LJ, Gage FH (1997) Epidermal growth factor and fibroblast growth factor-2 have different effects on neural progenitors in the adult rat brain. J Neurosci 17:5820–5829.
Kuo CT, Mirzadeh Z, Soriano-Navarro M, Rasin M, Wang D, Shen J, Sestan N, Garcia-Verdugo J, Alvarez-Buylla A, Jan LY, Jan YN (2006) Postnatal deletion of Numb/Numblike reveals repair and remodeling capacity in the subventricular neurogenic niche. Cell 127:1253–1264.
Levitt P, Rakic P (1980) Immunoperoxidase localization of glial fibrillary acidic protein in radial glial cells and astrocytes of the developing rhesus monkey brain. J Comp Neurol 193:815–840.
Lim DA, Alvarez-Buylla A (1999) Interaction between astrocytes and adult subventricular zone precursors stimulates neurogenesis. Proc Natl Acad Sci USA 96:7526–7531.
Lim DA, Tramontin AD, Trevejo JM, Herrera DG, Garcia-Verdugo JM, Alvarez-Buylla A (2000) Noggin antagonizes BMP signaling to create a niche for adult neurogenesis. Neuron 28:713–726.
Lin D, Edwards AS, Fawcett JP, Mbamalu G, Scott JD, Pawson T (2000) A mammalian PAR-3-PAR-6 complex implicated in Cdc42/Rac1 and aPKC signalling and cell polarity. Nat Cell Biol 2:540–547.
Lin F, Hiesberger T, Cordes K, Sinclair AM, Goldstein LS, Somlo S, Igarashi P (2003) Kidney-specific inactivation of the KIF3A subunit of kinesin-II inhibits renal ciliogenesis and produces polycystic kidney disease. Proc Natl Acad Sci USA 100:5286–5291.
Lindemann CB (2004) Testing the geometric clutch hypothesis. Biol Cell 96:681–690.
Liu Z, Martin LJ (2003) Olfactory bulb core is a rich source of neural progenitor and stem cells in adult rodent and human. J Comp Neurol 459:368–391.
Liu A, Wang B, Niswander LA (2005) Mouse intraflagellar transport proteins regulate both the activator and repressor functions of Gli transcription factors. Development 132:3103–3111.
Logan CY, Nusse R (2004) The Wnt signaling pathway in development and disease. Annu Rev Cell Dev Biol 20:781–810.
Lois C, Alvarez-Buylla A (1993) Proliferating subventricular zone cells in the adult mammalian forebrain can differentiate into neurons and glia. Proc Natl Acad Sci USA 90:2074–2077.
Lois C, Alvarez-Buylla A (1994) Long-distance neuronal migration in the adult mammalian brain. Science 264:1145–1148.
Lois C, Garcia-Verdugo JM, Alvarez-Buylla A (1996) Chain migration of neuronal precursors. Science 271:978–981.
Luskin MB (1993) Restricted proliferation and migration of postnatally generated neurons derived from the forebrain subventricular zone. Neuron 11:173–189.
Machold R, Hayashi S, Rutlin M, Muzumdar MD, Nery S, Corbin JG, Gritli-Linde A, Dellovade T, Porter JA, Rubin LL, Dudek H, McMahon AP, Fishell G (2003) Sonic hedgehog is required for progenitor cell maintenance in telencephalic stem cell niches. Neuron 39:937–950.

Malatesta P, Hartfuss E, Gotz M (2000) Isolation of radial glial cells by fluorescent-activated cell sorting reveals a neuronal lineage. Development 127:5253–5263.
Malatesta P, Hack MA, Hartfuss E, Kettenmann H, Klinkert W, Kirchhoff F, Gotz M (2003) Neuronal or glial progeny: regional differences in radial glia fate. Neuron 37:751–764.
Manabe N, Hirai S, Imai F, Nakanishi H, Takai Y, Ohno S (2002) Association of ASIP/mPAR-3 with adherens junctions of mouse neuroepithelial cells. Dev Dyn 225:61–69.
Mancuso MR, Kuhnert F, Kuo CJ (2008) Developmental angiogenesis of the central nervous system. Lymphat Res Biol 6:173–180.
Marshall WF (2008) The cell biological basis of ciliary disease. J Cell Biol 180:17–21.
Marshall WF, Rosenbaum JL (2001) Intraflagellar transport balances continuous turnover of outer doublet microtubules: implications for flagellar length control. J Cell Biol 155:405–414.
Marszalek JR, Ruiz-Lozano P, Roberts E, Chien KR, Goldstein LS (1999) Situs inversus and embryonic ciliary morphogenesis defects in mouse mutants lacking the KIF3A subunit of kinesin-II. Proc Natl Acad Sci USA 96:5043–5048.
Martin-Belmonte F, Mostov K (2008) Regulation of cell polarity during epithelial morphogenesis. Curr Opin Cell Biol 20:227–234.
May SR, Ashique AM, Karlen M, Wang B, Shen Y, Zarbalis K, Reiter J, Ericson J, Peterson AS (2005) Loss of the retrograde motor for IFT disrupts localization of Smo to cilia and prevents the expression of both activator and repressor functions of Gli. Dev Biol 287:378–389.
Menn B, Garcia-Verdugo JM, Yaschine C, Gonzalez-Perez O, Rowitch D, Alvarez-Buylla A (2006) Origin of oligodendrocytes in the subventricular zone of the adult brain. J Neurosci 26:7907–7918.
Mercier F, Kitasako JT, Hatton GI (2002) Anatomy of the brain neurogenic zones revisited: fractones and the fibroblast/macrophage network. J Comp Neurol 451:170–188.
Merkle FT, Tramontin AD, Garcia-Verdugo JM, Alvarez-Buylla A (2004) Radial glia give rise to adult neural stem cells in the subventricular zone. Proc Natl Acad Sci USA 101:17528–17532.
Merkle FT, Mirzadeh Z, Alvarez-Buylla A (2007) Mosaic organization of neural stem cells in the adult brain. Science 317:381–384.
Mirzadeh Z, Han YG, Soriano-Navarro M, Garcia-Verdugo JM, Alvarez-Buylla A (2010) Cilia organize ependymal planar polarity. J Neurosci 30:2600–2610.
Mirzadeh Z, Merkle FT, Soriano-Navarro M, Garcia-Verdugo JM, Alvarez-Buylla A (2008) Neural stem cells confer unique pinwheel architecture to the ventricular surface in neurogenic regions of the adult brain. Cell Stem Cell 3:265–278.
Misson JP, Edwards MA, Yamamoto M, Caviness VS, Jr.(1988) Mitotic cycling of radial glial cells of the fetal murine cerebral wall: a combined autoradiographic and immunohistochemical study. Dev Brain Res 38:183–190.
Mitchell B, Jacobs R, Li J, Chien S, Kintner C (2007) A positive feedback mechanism governs the polarity and motion of motile cilia. Nature 447:97–101.
Miyata T, Kawaguchi A, Okano H, Ogawa M (2001) Asymmetric inheritance of radial glial fibers by cortical neurons. Neuron 31:727–741.
Mollgard K, Balslev Y, Lauritzen B, Saunders NR (1987) Cell junctions and membrane specializations in the ventricular zone (germinal matrix) of the developing sheep brain: a CSF-brain barrier. J Neurocytol 16:433–444.
Morshead CM, Reynolds BA, Craig CG, McBurney MW, Staines WA, Morassutti D, Weiss S, van der Kooy D (1994) Neural stem cells in the adult mammalian forebrain: a relatively quiescent subpopulation of subependymal cells. Neuron 13:1071–1082.
Munro EM (2006) PAR proteins and the cytoskeleton: a marriage of equals. Curr Opin Cell Biol 18:86–94.
Murdoch JN, Henderson DJ, Doudney K, Gaston-Massuet C, Phillips HM, Paternotte C, Arkell R, Stanier P, Copp AJ (2003) Disruption of scribble (Scrb1) causes severe neural tube defects in the circletail mouse. Hum Mol Genet 12:87–98.
Nadarajah B, Jones AM, Evans WH, Parnavelas JG (1997) Differential expression of connexins during neocortical development and neuronal circuit formation. J Neurosci 17:3096–3111.

Noctor SC, Flint AC, Weissman TA, Dammerman RS, Kriegstein AR (2001) Neurons derived from radial glial cells establish radial units in neocortex. Nature 409:714–720.

Noctor SC, Martinez-Cerdeno V, Ivic L, Kriegstein AR (2004) Cortical neurons arise in symmetric and asymmetric division zones and migrate through specific phases. Nat Neurosci 7:136–144.

Noctor SC, Martinez-Cerdeno V, Kriegstein AR (2007) Contribution of intermediate progenitor cells to cortical histogenesis. Arch Neurol 64:639–642.

Noctor SC, Martinez-Cerdeno V, Kriegstein AR (2008) Distinct behaviors of neural stem and progenitor cells underlie cortical neurogenesis. J Comp Neurol 508:28–44.

Nonaka S, Tanaka Y, Okada Y, Takeda S, Harada A, Kanai Y, Kido M, Hirokawa N (1998) Randomization of left-right asymmetry due to loss of nodal cilia generating leftward flow of extraembryonic fluid in mice lacking KIF3B motor protein. Cell 95:829–837.

Nottebohm F (2004) The road we travelled: discovery, choreography, and significance of brain replaceable neurons. Ann NY Acad Sci 1016:628.

Nyfeler Y, Kirch RD, Mantei N, Leone DP, Radtke F, Suter U, Taylor V (2005) Jagged1 signals in the postnatal subventricular zone are required for neural stem cell self-renewal. Embo J 24:3504–3515.

Ocbina PJ, Tuson M, Anderson KV (2009) Primary cilia are not required for normal canonical Wnt signaling in the mouse embryo. PLoS One 4:e6839.

Ooshio T, Fujita N, Yamada A, Sato T, Kitagawa Y, Okamoto R, Nakata S, Miki A, Irie K, Takai Y (2007) Cooperative roles of Par-3 and afadin in the formation of adherens and tight junctions. J Cell Sci 120:2352–2365.

Palma V, Lim DA, Dahmane N, Sanchez P, Brionne TC, Herzberg CD, Gitton Y, Carleton A, Alvarez-Buylla A, Ruiz i Altaba A (2005) Sonic hedgehog controls stem cell behavior in the postnatal and adult brain. Development 132:335–344.

Palmer TD, Willhoite AR, Gage FH (2000) Vascular niche for adult hippocampal neurogenesis. J Comp Neurol 425:479–494.

Park TJ, Haigo SL, Wallingford JB (2006) Ciliogenesis defects in embryos lacking inturned or fuzzy function are associated with failure of planar cell polarity and Hedgehog signaling. Nat Genet 38:303–311.

Parras CM, Galli R, Britz O, Soares S, Galichet C, Battiste J, Johnson JE, Nakafuku M, Vescovi A, Guillemot F (2004) Mash1 specifies neurons and oligodendrocytes in the postnatal brain. Embo J 23:4495–4505.

Paton JA, Nottebohm F (1984) Neurons generated in the adult brain are recruited into functional circuits. Science 225:1046–1048.

Pazour GJ, Wilkerson CG, Witman GB (1998) A dynein light chain is essential for the retrograde particle movement of intraflagellar transport (IFT). J Cell Biol 141:979–992.

Pencea V, Bingaman KD, Freedman LJ, Luskin MB (2001) Neurogenesis in the subventricular zone and rostral migratory stream of the neonatal and adult primate forebrain. Exp Neurol 172:1–16.

Peretto P, Merighi A, Fasolo A, Bonfanti L (1997) Glial tubes in the rostral migratory stream of the adult rat. Brain Res Bull 42:9–21.

Peretto P, Dati C, De Marchis S, Kim HH, Ukhanova M, Fasolo A, Margolis FL (2004) Expression of the secreted factors noggin and bone morphogenetic proteins in the subependymal layer and olfactory bulb of the adult mouse brain. Neuroscience 128:685–696.

Peretto P, Giachino C, Aimar P, Fasolo A, Bonfanti L (2005) Chain formation and glial tube assembly in the shift from neonatal to adult subventricular zone of the rodent forebrain. J Comp Neurol 487:407–427.

Petreanu L, Alvarez-Buylla A (2002) Maturation and death of adult-born olfactory bulb granule neurons: role of olfaction. J Neurosci 22:6106–6113.

Pixley SKR, De Vellis J (1984) Transition between immature radial glia and mature astrocytes studied with a monoclonal antibody to vimentin. Dev Brain Res 15:201–209.

Pleasure SJ, Collins AE, Lowenstein DH (2000) Unique expression patterns of cell fate molecules delineate sequential stages of dentate gyrus development. J Neurosci 20:6095–6105.

Qin H, Diener DR, Geimer S, Cole DG, Rosenbaum JL (2004) Intraflagellar transport (IFT) cargo: IFT transports flagellar precursors to the tip and turnover products to the cell body. J Cell Biol 164:255–266.

Rakic P (1972) Mode of cell migration to the superficial layers of fetal monkey neocortex. J Comp Neurol 145:61–84.

Rakic P (1985) Limits of neurogenesis in primates. Science 227:1054–1056.

Rasin MR, Gazula VR, Breunig JJ, Kwan KY, Johnson MB, Liu-Chen S, Li HS, Jan LY, Jan YN, Rakic P, Sestan N (2007) Numb and Numbl are required for maintenance of cadherin-based adhesion and polarity of neural progenitors. Nat Neurosci 10:819–827.

Reynolds B, Weiss S (1992) Generation of neurons and astrocytes from isolated cells of the adult mammalian central nervous system. Science 255:1707–1710.

Rickmann M, Amaral DG, Cowan WM (1987) Organization of radial glial cells during the development of the rat dentate gyrus. J Comp Neurol 264:449–479.

Rieder CL, Faruki S, Khodjakov A (2001) The centrosome in vertebrates: more than a microtubule-organizing center. Trends Cell Biol 11:413–419.

Rohatgi R, Milenkovic L, Scott MP (2007) Patched1 regulates hedgehog signaling at the primary cilium. Science 317:372–376.

Rosenbaum JL, Witman GB (2002) Intraflagellar transport. Nat Rev Mol Cell Biol 3:813–825.

Ross AJ et al. (2005) Disruption of Bardet-Biedl syndrome ciliary proteins perturbs planar cell polarity in vertebrates. Nat Genet 37:1135–1140.

Russo RE, Fernandez A, Reali C, Radmilovich M, Trujillo-Cenoz O (2004) Functional and molecular clues reveal precursor-like cells and immature neurones in the turtle spinal cord. J Physiol 560:831–838.

Satir P (1965) Studies on cilia. II. Examination of the distal region of the ciliary shaft and the role of the filaments in motility. J Cell Biol 26:805–834.

Satir P, Christensen ST (2007) Overview of structure and function of mammalian cilia. Annu Rev Physiol 69:377–400.

Satir P, Dirksen ER (1985) Function-structure correlations in cilia from mammalian respiratory tract. In: Handbook of Physiology – The Respiratory System (Fishman AP, Cherniak NS, Widdicombe JG, Gieger SR, eds), pp 473–494. Bethesda: American Physiological Society.

Sauer FC (1935) Mitosis in the neural tube. J Comp Neurol 62:377–405.

Saura J, Richards JG, Mahy N (1994) Differential age-related changes of MAO-A and MAO-B in mouse brain and peripheral organs. Neurobiol Aging 15:399–408.

Sawamoto K, Wichterle H, Gonzalez-Perez O, Cholfin JA, Yamada M, Spassky N, Murcia NS, Garcia-Verdugo JM, Marin O, Rubenstein JL, Tessier-Lavigne M, Okano H, Alvarez-Buylla A (2006) New neurons follow the flow of cerebrospinal fluid in the adult brain. Science 311:629–632.

Schmechel DE, Rakic P (1979) A Golgi study of radial glia cells in developing monkey telencephalon: morphogenesis and transformation into astrocytes. Anat Embryol 156:115–152.

Schoenwolf GC, Alvarez IS (1989) Roles of neuroepithelial cell rearrangement and division in shaping of the avian neural plate. Development 106:427–439.

Schoenwolf GC, Smith JL (2000) Mechanisms of neurulation. Methods Mol Biol 136:125–134.

Seaberg RM, van der Kooy D (2002) Adult rodent neurogenic regions: the ventricular subependyma contains neural stem cells, but the dentate gyrus contains restricted progenitors. J Neurosci 22:1784–1793.

Seri B, Alvarez-Buylla A (2002) Neural stem cells and the regulation of neurogenesis in the adult hippocampus. Clin Neurosci Res 2:11–16.

Seri B, Garcia-Verdugo JM, McEwen BS, Alvarez-Buylla A (2001) Astrocytes give rise to new neurons in the adult mammalian hippocampus. J Neurosci 21:7153–7160.

Seri B, Garcia-Verdugo JM, Collado-Morente L, McEwen BS, Alvarez-Buylla A (2004) Cell types, lineage, and architecture of the germinal zone in the adult dentate gyrus. J Comp Neurol 478:359–378.

Seri B, Herrera DG, Gritti A, Ferron S, Collado L, Vescovi A, Garcia-Verdugo JM, Alvarez-Buylla A (2006) Composition and organization of the SCZ: a large germinal layer

containing neural stem cells in the adult mammalian brain. Cereb Cortex 16 Suppl 1:i103–111.
Shen Q, Goderie SK, Jin L, Karanth N, Sun Y, Abramova N, Vincent P, Pumiglia K, Temple S (2004) Endothelial cells stimulate self-renewal and expand neurogenesis of neural stem cells. Science 304:1338–1340.
Shen Q, Wang Y, Kokovay E, Lin G, Chuang SM, Goderie SK, Roysam B, Temple S (2008) Adult SVZ stem cells lie in a vascular niche: a quantitative analysis of niche cell-cell interactions. Cell Stem Cell 3:289–300.
Shibata T, Yamada K, Watanabe M, Ikenaka K, Wada K, Tanaka K, Inoue Y (1997) Glutamate transporter GLAST is expressed in the radial glia-astrocyte lineage of developing mouse spinal cord. J Neurosci 17:9212–9219.
Shingo T, Gregg C, Enwere E, Fujikawa H, Hassam R, Geary C, Cross JC, Weiss S (2003) Pregnancy-stimulated neurogenesis in the adult female forebrain mediated by prolactin. Science 299:117–120.
Simons M, Mlodzik M (2008) Planar cell polarity signaling: from fly development to human disease. Annu Rev Genet 42:517–540.
Simons M, Gloy J, Ganner A, Bullerkotte A, Bashkurov M, Kronig C, Schermer B, Benzing T, Cabello OA, Jenny A, Mlodzik M, Polok B, Driever W, Obara T, Walz G (2005) Inversin, the gene product mutated in nephronophthisis type II, functions as a molecular switch between Wnt signaling pathways. Nat Genet 37:537–543.
Singla V, Reiter JF (2006) The primary cilium as the cell's antenna: signaling at a sensory organelle. Science 313:629–633.
Spassky N, Merkle FT, Flames N, Tramontin AD, Garcia-Verdugo JM, Alvarez-Buylla A (2005) Adult ependymal cells are postmitotic and are derived from radial glial cells during embryogenesis. J Neurosci 25:10–18.
Spassky N, Han YG, Aguilar A, Strehl L, Besse L, Laclef C, Ros MR, Garcia-Verdugo JM, Alvarez-Buylla A (2008) Primary cilia are required for cerebellar development and Shh-dependent expansion of progenitor pool. Dev Biol 317:246–259.
Summers KE, Gibbons IR (1971) Adenosine triphosphate-induced sliding of tubules in trypsin-treated flagella of sea urchin sperm. Proc Natl Acad Sci USA 68:3092–3096.
Sun Y, Goderie SK, Temple S (2005) Asymmetric distribution of EGFR receptor during mitosis generates diverse CNS progenitor cells. Neuron 45:873–886.
Sundholm-Peters NL, Yang HK, Goings GE, Walker AS, Szele FG (2004) Radial glia-like cells at the base of the lateral ventricles in adult mice. J Neurocytol 33:153–164.
Takeda S, Yonekawa Y, Tanaka Y, Okada Y, Nonaka S, Hirokawa N (1999) Left-right asymmetry and kinesin superfamily protein KIF3A: new insights in determination of laterality and mesoderm induction by kif3A-/- mice analysis. J Cell Biol 145:825–836.
Takeichi M (1995) Morphogenetic roles of classic cadherins. Curr Opin Cell Biol 7:619–627.
Tamamaki N, Nakamura K, Okamoto K, Kaneko T (2001) Radial glia is a progenitor of neocortical neurons in the developing cerebral cortex. Neurosci Res 41:51–60.
Tavazoie M, Van der Veken L, Silva-Vargas V, Louissaint M, Colonna L, Zaidi B, Garcia-Verdugo JM, Doetsch F (2008) A specialized vascular niche for adult neural stem cells. Cell Stem Cell 3:279–288.
Tramontin AD, Garcia-Verdugo JM, Lim DA, Alvarez-Buylla A (2003) Postnatal development of radial glia and the ventricular zone (VZ): a continuum of the neural stem cell compartment. Cereb Cortex 13:580–587.
Tran PV, Haycraft CJ, Besschetnova TY, Turbe-Doan A, Stottmann RW, Herron BJ, Chesebro AL, Qiu H, Scherz PJ, Shah JV, Yoder BK, Beier DR (2008) THM1 negatively modulates mouse sonic hedgehog signal transduction and affects retrograde intraflagellar transport in cilia. Nat Genet 40:403–410.
Tropepe V, Sibilia M, Ciruna BG, Rossant J, Wagner EF, van der Kooy D (1999) Distinct neural stem cells proliferate in response to EGF and FGF in the developing mouse telencephalon. Dev Biol 208:166–188.

The Boulder Committee B (1970) Embryonic vertebrate central nervous system: revised terminology. Anat Rec 166:257–262.

Tung T-C, Tung Y-FY (1940) Experimental studies on the determination of polarity of ciliary action of anuran embryos. Arch Biol (Liege) 51:203–218.

Twitty VC (1928) Experimental studies on the ciliary action of amphibian embryos. J Exp Zool 50:310–344.

van Praag H, Schinder AF, Christie BR, Toni N, Palmer TD, Gage FH (2002) Functional neurogenesis in the adult hippocampus. Nature 415:1030–1034.

Veeman MT, Axelrod JD, Moon RT (2003) A second canon. Functions and mechanisms of beta-catenin-independent Wnt signaling. Dev Cell 5:367–377.

Ventura RE, Goldman JE (2007) Dorsal radial glia generate olfactory bulb interneurons in the postnatal murine brain. J Neurosci 27:4297–4302.

Voigt T (1989) Development of glial cells in the cerebral wall of ferrets: direct tracing of their transformation from radial glia into astrocytes. J Comp Neurol 289:74–88.

von Trotha JW, Campos-Ortega JA, Reugels AM (2006) Apical localization of ASIP/PAR-3:EGFP in zebrafish neuroepithelial cells involves the oligomerization domain CR1, the PDZ domains, and the C-terminal portion of the protein. Dev Dyn 235:967–977.

Wallingford JB (2006) Planar cell polarity, ciliogenesis and neural tube defects. Hum Mol Genet 15 Spec No 2:R227–234.

Wallingford JB, Harland RM (2002) Neural tube closure requires Dishevelled-dependent convergent extension of the midline. Development 129:5815–5825.

Wallingford JB, Rowning BA, Vogeli KM, Rothbacher U, Fraser SE, Harland RM (2000) Dishevelled controls cell polarity during Xenopus gastrulation. Nature 405:81–85.

Wang X, Tsai JW, Imai JH, Lian WN, Vallee RB, Shi SH (2009) Asymmetric centrosome inheritance maintains neural progenitors in the neocortex. Nature 461:947–955.

Weigmann A, Corbeil D, Hellwig A, Huttner WB (1997) Prominin, a novel microvilli-specific polytopic membrane protein of the apical surface of epithelial cells, is targeted to plasmalemmal protrusions of non-epithelial cells. Proc Natl Acad Sci USA 94:12425–12430.

Weissman TA, Riquelme PA, Ivic L, Flint AC, Kriegstein AR (2004) Calcium waves propagate through radial glial cells and modulate proliferation in the developing neocortex. Neuron 43:647–661.

Williams D, Gascoigne JE, Street M, Williams ED (1979) Histochemical characterization of monoamine oxidase in ependyma of rat hypothalamus. Histochem J 11:83–95.

Wodarz A, Huttner WB (2003) Asymmetric cell division during neurogenesis in Drosophila and vertebrates. Mech Dev 120:1297–1309.

Wong SY, Seol AD, So PL, Ermilov AN, Bichakjian CK, Epstein EH, Jr., Dlugosz AA, Reiter JF (2009) Primary cilia can both mediate and suppress Hedgehog pathway-dependent tumorigenesis. Nat Med 15:1055–1061.

Worthington WC, Jr., Cathcart RS, 3rd (1963) Ependymal cilia: distribution and activity in the adult human brain. Science 139:221–222.

Wu J, Mlodzik M (2009) A quest for the mechanism regulating global planar cell polarity of tissues. Trends Cell Biol 19:295–305.

Yu JM, Kim JH, Song GS, Jung JS (2006) Increase in proliferation and differentiation of neural progenitor cells isolated from postnatal and adult mice brain by Wnt-3a and Wnt-5a. Mol Cell Biochem 288:17–28.

Zallen JA (2007) Planar polarity and tissue morphogenesis. Cell 129:1051–1063.

Zhadanov AB, Provance DW, Jr., Speer CA, Coffin JD, Goss D, Blixt JA, Reichert CM, Mercer JA (1999) Absence of the tight junctional protein AF-6 disrupts epithelial cell-cell junctions and cell polarity during mouse development. Curr Biol 9:880–888.

Zhao C, Deng W, Gage FH (2008) Mechanisms and functional implications of adult neurogenesis. Cell 132:645–660.

Zupanc GK (2006) Neurogenesis and neuronal regeneration in the adult fish brain. J Comp Physiol A Neuroethol Sens Neural Behav Physiol 192:649–670.

Chapter 13
Neurogenesis in the Adult Rabbit: From Olfactory System to Cerebellum

Giovanna Ponti, Federico Luzzati, Paolo Peretto, and Luca Bonfanti

Abstract Adult neurogenesis is a feature that is phylogenetically highly conserved, yet most of the current knowledge in this field is based on laboratory rodents. In addition to the differences among vertebrates, remarkable peculiarities are also emerging in the location and structural and functional features of neurogenic systems in mammals. Although the orders Lagomorpha and Rodentia are quite similar, unexpected features of structural plasticity and neurogenesis have been found to occur in the rabbit central nervous system. The structural and temporal characteristics of protracted and persistent neurogenesis in rabbits are herewith addressed. Attention is focused on the subventricular zone cytoarchitecture and extensions, as well as on processes of persistent neurogenesis that have been recently demonstrated to occur in the striatum and in the cerebellum. The existence of neural cell progenitors persisting within the rabbit brain parenchyma (so-called nonneurogenic in rodents) are discussed.

13.1 Introduction

The New Zealand white rabbit (*Orictolagus cuniculus*) belongs to the lagomorphs (order Lagomorpha), forming a distinct group compared with rodents (order Rodentia), within the monophyletic group of Glires (Douzery and Huchon 2004).

G. Ponti and L. Bonfanti (✉)
Department of Veterinary Morphophysiology, University of Turin, Turin, Italy
e-mail: luca.bonfanti@unito.it

F. Luzzati and P. Peretto
Department of Animal and Human Biology, University of Turin, Turin, Italy

P. Peretto and L. Bonfanti
National Institute of Neuroscience-Italy (INN), Turin, Italy

G. Ponti, F. Luzzati, P. Peretto, and L. Bonfanti
Istituto scientifico Fondazione Cavalieri Ottolenghi, Turin, Italy

T. Seki et al. (eds.), *Neurogenesis in the Adult Brain I: Neurobiology*,
DOI 10.1007/978-4-431-53933-9_13, © Springer 2011

Rabbits, in addition to being domestic, are also laboratory animals, yet in contrast with the deep knowledge of mouse and rat neuroanatomy only a few studies have addressed the same issue in lagomorphs systematically, making it difficult to set up detailed comparative considerations.

During recent years, by focusing on the anatomical structure and neural progenitor activity in postnatal germinative regions of the rabbit brain, we have described the localization, organization, and expansion of germinal matrices, showing relevant anatomical and functional differences between lagomorphs and rodents (see Bonfanti and Ponti 2008; Luzzati et al. 2007). In addition to differences concerning the "classic" neurogenic sites, such as the subventricular zone (SVZ) and the subgranular zone (SGZ; Gage 2000), these studies led us to the discovery that rabbit neurogenesis is more widespread, being also detectable in other regions of the central nervous system (CNS), commonly referred to in rodents as "nonneurogenic" parenchyma.

Some peculiar features concerning rabbit brain anatomy should be considered with respect to rodents: the occurrence of more prominent cerebral ventricular extensions and more expanded temporal lobes. Differences in the overall mammalian brain anatomy can be related to a different shape and extension of the cerebral ventricles, which in turn correlates well with that of the embryonic germinal matrices, thus potentially reflecting changes of configuration and size of the adult brain (McFarland et al. 1969; Chenn and Walsh 2002; Kriegstein et al. 2006; Fig. 13.1a). The persistence of the *olfactory ventricle* during postnatal life in rabbits, such as in other mammals (dogs, cats, cows, sheep, guinea pig, humans; McFarland et al. 1969; Leonhardt 1972; Rodriguez-Perez et al. 2003; Curtis et al. 2007), does not seem to be related to brain size because the olfactory ventricle is replaced by a solid chord of tissue in most rodents and primates (McFarland et al. 1969; Leonhardt 1972). Consequently, the tissue forming the rostral migratory stream (RMS) in these latter species is replaced in rabbits by a SVZ rostral extension ensheathing the dorsal part of the olfactory ventricle (Fig. 13.1b,c). Another extension of the lateral ventricular system called the *inferior horn* and protruding ventrally within the temporal lobe, absent in rodents but present in rabbits and primates, correlates well with the expanded temporal lobes of these species (McFarland et al. 1969; Leonhardt 1972; Fig. 13.1a). This trend is confirmed by the occurrence of a *posterior horn* in species with massive development of the occipital lobes and high degree of visual specialization, maximally evident in primates and humans.

In the first part of this chapter, the overall anatomy and cytoarchitecture of the rabbit SVZ are compared with current well-established knowledge in rodents. Most interestingly, in the second part of the chapter we describe the occurrence of spontaneous neurogenic activity in the mature rabbit striatum (Luzzati et al. 2006) and cerebellum (Ponti et al. 2006a, 2008), two CNS regions that are considered nonneurogenic under physiological condition in mammals. In particular, the local origin of newly born neural cells from parenchymal (striatal and cerebellar) progenitors is discussed.

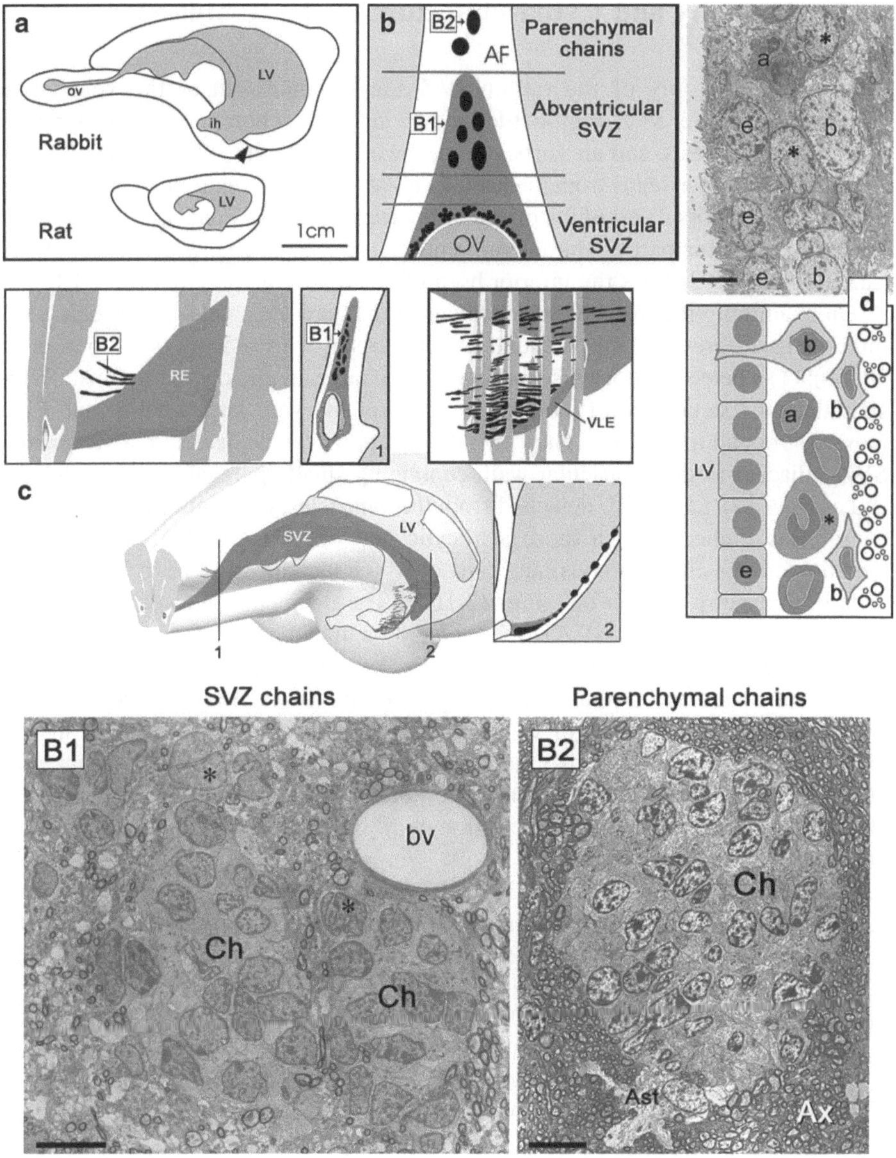

Fig. 13.1 Organization of the rabbit subventricular zone (SVZ) and its extensions. (**a**) The lateral ventricle (*LV*) shows an open olfactory ventricle (*ov*) and an inferior horn (*ih*), the latter in correspondence with an enlarged temporal lobe (*arrowhead*). (**b**) Large chains of neuroblasts in the *abventricular SVZ* and in the anterior forceps (*AF*) of the corpus callosum. SVZ area in *dark gray*. (**c**) Parenchymal chains (*black*) leave the SVZ at the level of the rostral extension (*RE*) and the ventral lateral extension (*VLE*). (**B1, B2**) Ultrastructure of the large SVZ and parenchymal chains (*Ch*): *bv* blood vessel, *Ast* astrocyte, *Ax* axons, *asterisk* type C-like cell. (**d**) Ultrastructure of the stem cell niche at the LV level. *e* Ependyma, *a* type A cells, *b* type B cells (astrocytes); these latter form a ribbon, rather than glial tubes

13.2 Anatomy and Cytoarchitecture of the Rabbit SVZ

The overall anatomy of the rabbit SVZ adapts somewhat to the ventricular conformation. The SVZ region (defined as an area wherein cell proliferation, chains of neuroblasts, and an astrocytic meshwork do overlap) (Luzzati et al. 2003; Ponti et al. 2006b) extends from the lateral wall of the lateral ventricle to the olfactory bulb, lining the dorsal part of the olfactory ventricle (rostral extension; Fig. 13.1b,c). From the posterior part of the lateral ventricle, a *ventral lateral extension* (VLE; Luzzati et al. 2003) lines the inferior horn of the ventricular cavity (Fig. 13.1c).

On examining the internal arrangement of the New Zealand white rabbit SVZ, three main differences can be found in comparison with rodents (see Ponti et al. 2006b): (1) a heterogeneous distribution of the chain and glia compartments; (2) the simultaneous occurrence of isolated neuroblasts and very large chains; and (3) a looser astrocytic meshwork. Thus, two distinct parts can be detected: a *ventricular* SVZ, adjacent to the ventricular wall, containing small aggregates of neuroblasts, and an *abventricular* SVZ, detached from the ventricles, containing large chains of neuroblasts immersed in a loose astrocytic network (Fig. 13.1b). About 10–20 medium to large chains are visible, some of which include up to 15 to 20 nucleated cells in transverse section (see Fig. 13.1B1). The distinction between rabbit ventricular and abventricular SVZ is reinforced by a thick band of tissue, poor in cells and enriched in nerve fibers and glial processes, reminiscent of the *hypocellular gap* of the human SVZ (Quinones-Hinojosa et al. 2006). Another feature shared with humans is the absence of well-organized glial tubes, replaced by an incomplete astrocyte row (Fig. 13.1d), quite similar to the *astrocyte ribbon* (Sanai et al. 2004). Yet, in contrast to the human SVZ, most of the rabbit neuroblasts form large chains located laterally to both the astrocyte ribbon and hypocellular tissue (Ponti et al. 2006b; Fig. 13.1b,d).

The finding of a similar pattern in the bovine SVZ (Rodriguez-Perez et al. 2003) suggests that in nonrodent mammals the glial meshwork could be less tightly packed than in rodents and the SVZ differently compartmentalized, which could be a shared feature in brains characterized by larger size. However, a striking difference between rabbit and human SVZ consists of large chains in the rabbit abventricular SVZ, in contrast with the absence of evident chains in humans (Sanai et al. 2004; Quinones-Hinojosa et al. 2006; Curtis et al. 2007).

Although a certain degree of heterogeneity in the arrangement of SVZ compartments has been described in rodents, involving medial/lateral asymmetry and even rat/mouse species differences (Peretto et al. 2005), the rabbit appears substantially different as to the kind of contact between the newly generated cell precursors and their substrates. Instead of an almost complete glial ensheathment (the *glial tubes* of rodents; Peretto et al. 1997), chains of the rabbit abventricular SVZ are in contact with three types of substrates – astrocytic glia, blood vessels, and axons – the latter representing the highest percentage (Ponti et al. 2006b; Fig. 13.1B1). Also, the postnatal assembly of the rabbit SVZ differs. Indeed, although SVZ chains are not present in rodents at birth, appearing later around the third postnatal week

(Peretto et al. 2005), chain formation is observed very early in rabbits, starting from postnatal day 10 (Ponti et al. 2006b; Fig. 13.2a). Such a difference is strengthened by the different temporal shift in postnatal development, which is earlier in rodents (puberty occurs in the first to second postnatal months) than in rabbits (around the fourth month).

The ultrastructural analysis of the periventricular SVZ, where the stem cell niche is expected to occur (see Chap. 12), revealed the same cell types described in rodents, although with a different spatial organization (Fig. 13.1d). Again, the absence of glial tubes in the rabbit ventricular SVZ, along with the occurrence of astrocytes forming a row parallel to the ependyma, makes this structure somehow more similar to the human stem cell niche, at least under the morphological profile (Sanai et al. 2004; Quinones-Hinojosa et al. 2006; Bonfanti and Ponti 2008).

Thus, both chain and glia compartments can differ as to their architecture and mutual relationships in relatively closely related mammalian orders, namely, lagomorphs and rodents. In parallel, common features can be shared by relatively distant species, such as rabbits and humans.

Quantitative and qualitative studies carried out after bromodeoxyuridine (BrdU) systemic administration and cell tracer green (CTG) intraventricular injection (see Luzzati et al. 2003, 2006) indicate that the different structural organization of the rabbit SVZ seems not to affect cell migration toward the olfactory bulb. Neuroblasts generated at the SVZ periventricular level appear to reach the core of the olfactory bulb in 7–10 days. At 1 month survival time, several newly generated cells have radially migrated in the olfactory bulb granular and glomerular layers, whereby they have differentiated into NeuN-positive mature neurons. It is worth noting that at this survival time several newly generated neurons are still located in periventricular position (Ponti and Luzzati, unpublished results). Accordingly, this pattern of migration is placed between the fast cell displacement described in rodents (about 2 weeks to reach the final destination into the olfactory bulb layers; Lois and Alvarez-Buylla 1994) and the far more slow pattern observed in primates (about 3 months; Kornack and Rakic 2001).

13.2.1 SVZ Extensions Within the Rabbit Brain Parenchyma

Besides its peculiar cytoarchitecture, the rabbit SVZ is characterized by the occurrence of extensions entering the surrounding mature parenchyma (Figs. 13.1b,c and 13.2a,c). These *parenchymal chains* were identified after serial reconstruction of brain tissue sections immunostained for the polysialylated form of the neural cell adhesion molecule (PSA-NCAM). Two groups of chain-like aggregates can be observed: (1) anterior chains, leaving the SVZ rostral extension through the anterior forceps of the corpus callosum, beneath the frontal cortex, and (2) posterior chains, apparently leaving the ventral–lateral extension, along the external capsule (Luzzati et al. 2003).

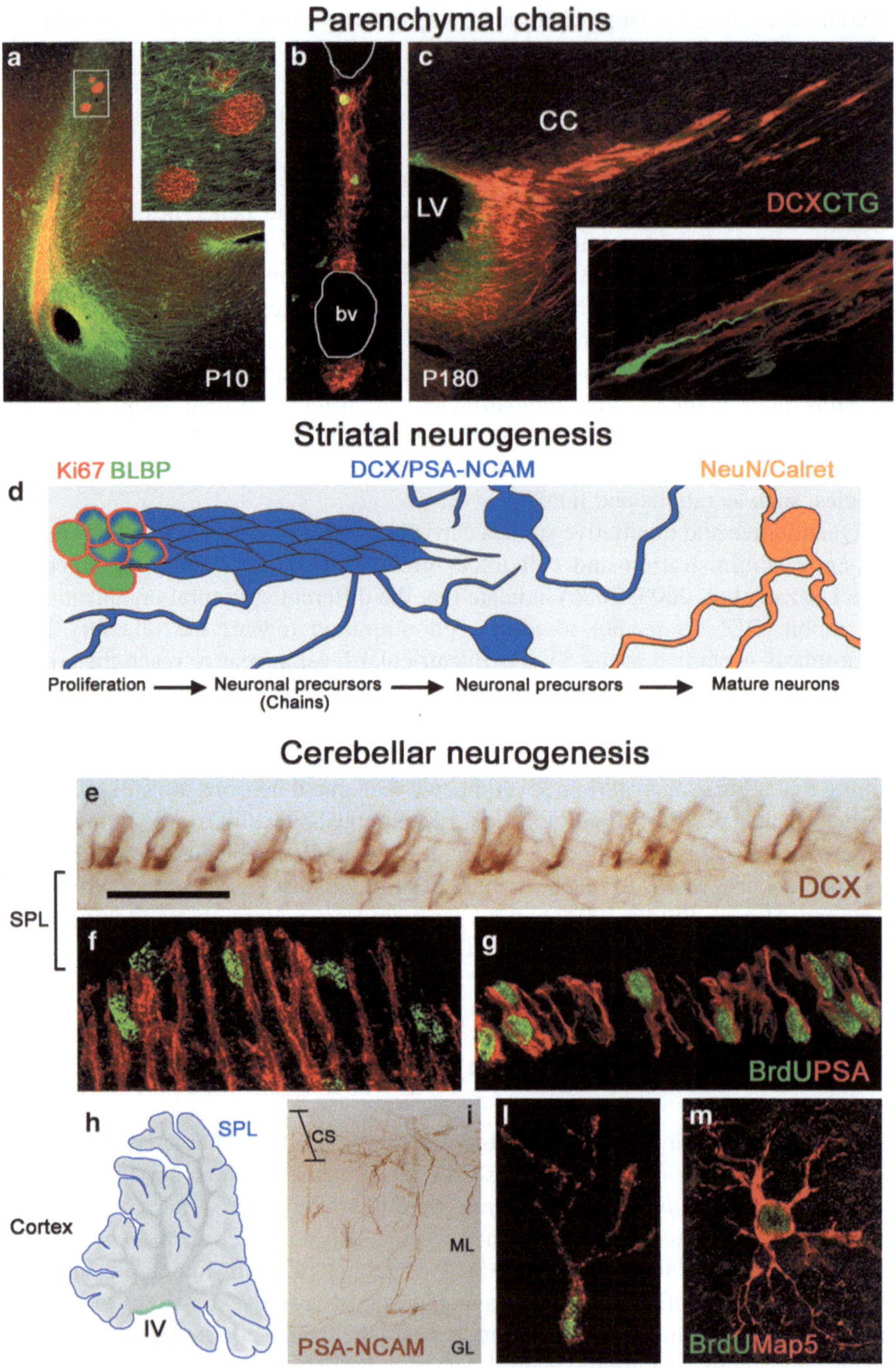
Parenchymal chains
a
b
c
CC
LV
DCXCTG
bv
P10
P180
Striatal neurogenesis
Ki67 BLBP
DCX/PSA-NCAM
NeuN/Calret
d
Proliferation
Neuronal precursors (Chains)
Neuronal precursors
Mature neurons
Cerebellar neurogenesis
e
DCX
SPL
f
g
BrdUPSA
h
SPL
Cortex
IV
CS
i
ML
GL
PSA-NCAM
l
m
BrdUMap5

The real nature of migrating chains has been confirmed by ultrastructural serial reconstruction, BrdU treatment analyzed at different postinjection survival times, and confocal analysis after intraventricu lar injection of tracers (Luzzati et al. 2003, 2006; Ponti et al. 2006b; Figs. 13.1c, B2 and 13.2a–c). Immunoreactivity for PSA-NCAM and doublecortin on these tangentially oriented bulks of cells along with ultrastructural features displayed by their cellular components are both features of "chain migration" (Lois et al. 1996). Cell migration from the SVZ within the anterior chains has been confirmed by intraventricular injections of CTG (Luzzati et al. 2006). No mitoses or BrdU+ proliferating cells are detectable within the parenchymal chains, yet the occurrence of newly generated cells can be observed 5–10 days after BrdU incorporation (Fig. 13.2b). By contrast, the intraventricular injections of tracers failed to stain any posterior chains cells up to 9 days of survival time. These data indicate that anterior chains contain migratory cells that are generated within the SVZ, whereas posterior chains could be made up of locally generated elements characterized by inconstant relationships with the SVZ or scarce migratory capacity.

Similar to abventricular SVZ chains, parenchymal chains are very large, abundantly surpassing the average size of chains found in rodents (usually limited to 3–5 cells in cross section). The compact morphology of anterior chains (Fig. 13.1B2) is consistently different from the more irregular, laminar arrangement of the posterior ones, made up of noncontinuous clusters and rows of cells (Ponti et al. 2006b). In contrast to SVZ chains of rodents, the rabbit parenchymal chains are glia independent, being directly in contact with axons (anterior chains; Fig. 13.1B2) or at the interface between white and gray matter (posterior chains). Anterior chains show continuity with large chains of the abventricular SVZ, in some cases following blood vessels in their shift from the SVZ to white matter (Ponti et al. 2006b).

The occurrence of rabbit anterior parenchymal chains is mainly limited to postnatal/young ages and is highly variable in different individuals. Interestingly, newly generated neurons have been proposed to exist in the frontal neocortex of adult monkeys and rats (Gould et al. 1999, 2001), and migrating neuroblasts have been observed in the corpus callosum of these species. These data open the possibility that rabbit anterior parenchymal chains could represent a migratory pathway providing new neurons in cortical or callosal regions of mammals in young individuals. Nevertheless, in spite of the presence of some PSA-NCAM+ cells with the

Fig. 13.2 Anterior parenchymal chains in the corpus callosum (*CC*) are already present postnatally (postnatal day 10, *P10*; **a**) and still detectable at 6 months of age (P180; **c**). They contain migrating neuroblasts as shown by intraventricular injection of bromodeoxyuridine (BrdU; **b**) and cell tracer-green (*CTG*; **c**). (**a**) *Red* doublecortin (*DCX*); *green* vimentin; (**b**) *red* polysialylated neural cell adhesion molecule (PSA-NCAM); *green* BrdU (5 days post injection). These chains are frequently associated with blood vessels (*bv*). *LV* lateral ventricle. (**d**) Neurogenic sequence in the adult rabbit caudate nucleus. *BLBP* brain lipid-binding protein. (**e–g**) Subpial layer (*SPL*) on the cerebellar surface of the peripuberal rabbit. Chains of neuroblasts form after cell division: *green*, *BrdU* after 2 h (**f**) and 5 days (**g**) post injection; *red*, *PSA*. (**h**) The SPL is present on the whole subpial surface (*blue*) but not in the roof of the fourth ventricle (*IV*, *green*). After the fifth month of life, newly generated cells are detectable in the whole cerebellar cortex even in the absence of SPL (*cs* cortical surface). Neuronal-like (*L*) and glial-like cell precursors (*M*) occur in both the peripuberal and adult rabbit typical bipolar-shaped morphology of migrating elements, radially oriented in the cortical/subcortical areas adjacent to parenchymal chains, no newly generated neurons are detectable in the cortex, apart from rare newly born cells occasionally detected in the amygdala and posterior putamen of rabbits 1 month after BrdU treatment (Luzzati et al. 2003). Although it is very unlikely that such alternative routes could provide functional cell addition or renewal in the brain parenchyma, their existence in peripuberal rabbits indicates how SVZ structural plasticity could be heterogeneous in different mammals, appearing particularly protracted in young lagomorphs.

13.3 Local Neurogenesis Independent from Persistent Germinal Layers

13.3.1 Adult Striatal Neurogenesis

Another intriguing feature of the adult rabbit brain is the occurrence of newly formed neurons in the caudate nucleus (Luzzati et al. 2006). This region represents the most mediodorsal part of the corpus striatum, and it is generally considered nonneurogenic during postnatal life under physiological conditions.

By using BrdU injections coupled to confocal analyses performed at different survival times from the treatment, the genesis of new neurons can be demonstrated in the caudate nucleus of sexually mature and adult animals. Stereological time-course studies indicate that BrdU/NeuN-positive cells are detectable at long survival times, although after 2 months only 0.7% of surviving BrdU+ cells actually correspond to mature neurons (Luzzati et al. 2006). Phenotypic analyses showed that about one-sixth of these cells differentiate into the calretinin striatal interneurons, a number that is extremely low but nonnegligible if referred to the whole striatal calretinin interneuronal population (see also Luzzati et al. 2007). This finding suggests that adult striatal neurogenesis is partly devoted to replacing and/or adding cells within a small subpopulation of interneurons.

By contrast, the immunocytochemical analysis of early neuronal markers (i.e., PSA-NCAM and doublecortin) shows that the rabbit caudate nucleus harbors a significant presence of newly generated neuroblasts. Interestingly, at 65 days, the NeuN+/BrdU+ neurons represent less than 10% of the mean number of the doublecortin+/BrdU+ cells found at 5 days survival, thus indicating the occurrence of a strong selection. Newly generated neuroblasts are frequently associated to form longitudinally arranged, compact structures that have been referred to as *striatal chains*, being reminiscent of the SVZ and RMS chains (Fig. 13.2d). However, intraventricular injections of a cell tracer (cell tracker green, CTG) failed to specifically label these neuroblasts, and three-dimensional reconstruction analyses of the striatal chains revealed no connections with the SVZ (Luzzati et al. 2006). Thus, striatal and SVZ chains are two separate populations.

These data suggest that rabbit striatal neurogenesis does originate from the activity of local neuronal progenitors, rather than from SVZ extensions, an hypothesis that was confirmed by in vitro approaches, such as the culture of caudate nucleus tissue explants (Luzzati et al. 2006). Consistent with these results, the analysis of cycling cells in the rabbit striatum revealed the existence of proliferating clusters associated to the striatal chains. The phenotypic analysis indicated that most of these elements express the astroglial marker BLBP (brain lipid-binding protein; Anthony et al. 2004), a protein abundant in radial glial cells serving as progenitors for the adult SVZ and SGZ neurogenesis. Moreover, nearly 60% of proliferating elements of the same clusters also express the early neuronal marker doublecortin, suggesting a clonal relationship between BLBP- and doublecortin-expressing cells (Fig. 13.2d). Thus, the mature parenchyma of the rabbit striatum would contain glial cells acting as primary neuronal progenitors associated to transit amplifying elements.

On the whole, these experiments indicate a striking ability of the rabbit striatum in maintaining adult neurogenesis under physiological condition. Although in rodents and in some primates only small numbers of neuroblasts/new neurons have been described in the mature striatum under physiological conditions (Dayer et al. 2005; Bédard et al. 2002), neurogenesis in this region can be experimentally stimulated (Fallon et al. 2000; Arvidsson et al. 2002). Following these treatments, also the striatum of rodents show neuroblasts organized both as individual cells or as chains, followed by strong cell selection and poor integration of new neurons (Arvidsson et al. 2002).

These observations support the general vision of the mature mammalian striatum as an unfavorable environment for the survival rather than the genesis of newly formed neuroblasts. In most cases, whether the occurrence of local multipotent progenitors into the mature brain parenchyma can be caused by either translocation of embryonic pools of progenitor cells or migration of transit-amplifying elements from the persistent neurogenic zones remains to be determined. Accordingly, the origin, identity, and multipotentiality of local parenchymal progenitors of the rabbit CNS remain as yet unresolved, but the rabbit could be a suitable model to address these issues.

13.3.2 Neurogenesis in the Postnatal and Adult Cerebellum

13.3.2.1 Cerebellar Neurogenesis in Mammals

Striking neurogenesis has been described to occur throughout life in the cerebellum of some nonmammalian vertebrates, such as fish (Zupanc 2006). By contrast, the mammalian cerebellum is known as one of the most static structures in the CNS, its plasticity being limited to synaptic rearrangement of preexisting circuits (Ito 2006). In contrast to the forebrain, the cerebellum is considered incapable of any spontaneous cell genesis after the exhaustion of the external germinal layer (EGL).

The EGL is a secondary actively proliferating zone made up of cell precursors that originate from the germinal trigone following tangential subpial displacement, then leading to protracted genesis of the granule cell population in the postnatal developing cerebellar cortex (Altman and Bayer 1997). In all species studied, this transitory germinal zone progressively reduces its thickness as the granule cell precursors migrate deep into the cortex to form the inner granule layer (Fig. 13.3a,b). In altricial mammals, which are not able to walk and run soon after birth, the EGL is active in the pre- and postnatal period (Sanchez-Villagra and Sultan 2001). Later, it disappears at specific ages (reviewed in Bonfanti and Ponti 2008; Fig. 13.3a). In parallel, the genesis of cerebellar interneurons originating from the neuroepithelium and migrating through the white matter is completed before the end of granule cell genesis (see Leto et al. 2008). This population of interneuron precursors of neuroepithelial origin specifically expresses the transcription factor Pax2 (Maricich and Herrup 1999; Weisheit et al. 2006), whereas Pax6 is expressed by EGL-derived granule cell precursors (Engelkamp et al. 1999; Yamasaki et al. 2001).

After EGL exhaustion, the parallel fibers of the molecular layer come in direct contact with the pial surface, with the interposition of the glia limitans (Bergmann glial end feet; Altman and Bayer 1997; Fig. 13.3b′). No sign of cell proliferation or cells expressing early differentiation markers such as PSA-NCAM and doublecortin can be detected in the cerebellar cortex after this stage (P21 in mice; Bonfanti et al. 1992; Ponti et al. 2006a). By contrast, recent work carried out in the New Zealand white rabbit described the existence of a secondary germinal matrix persisting beyond puberty in the subpial position (Ponti et al. 2006a). This germinal matrix has been called the *subpial layer* (SPL; Ponti et al. 2006a) because it undergoes postnatal structural modifications similar to those described in the forebrain SVZ.

13.3.2.2 The Subpial Layer

Cell proliferation is detectable on the rabbit cerebellar surface up to the fifth month of age, thus extending beyond puberty (Ponti et al. 2006a). Analyses of the rabbit SPL at different postnatal/peripuberal stages by combination of different techniques (i.e., detection of Ki67 and BrdU at 2-h survival, electron microscopy, and detection of BrdU at different survival times in association with antigens of cell specification/differentiation) reveal that proliferating elements are in subpial position forming a single, noncontinuous layer not affecting the meninges (Figs. 13.2 and 13.3).

The SPL contains *tangential chains* of neuroblasts that express molecules strictly associated to adult newly generated neuronal precursors and required for their migration, such as PSA-NCAM and doublecortin (reviewed in Bonfanti 2006; see Fig. 13.2e–g). The real nature of the chains of these cell aggregates was confirmed by electron microscopy serial reconstruction of cerebellar coronal and parasagittal sections (Ponti et al. 2006a). Small subpial cells (about 3–4 μm in diameter) sharing the same cytology of neuroblasts (type A cells) in the SVZ of

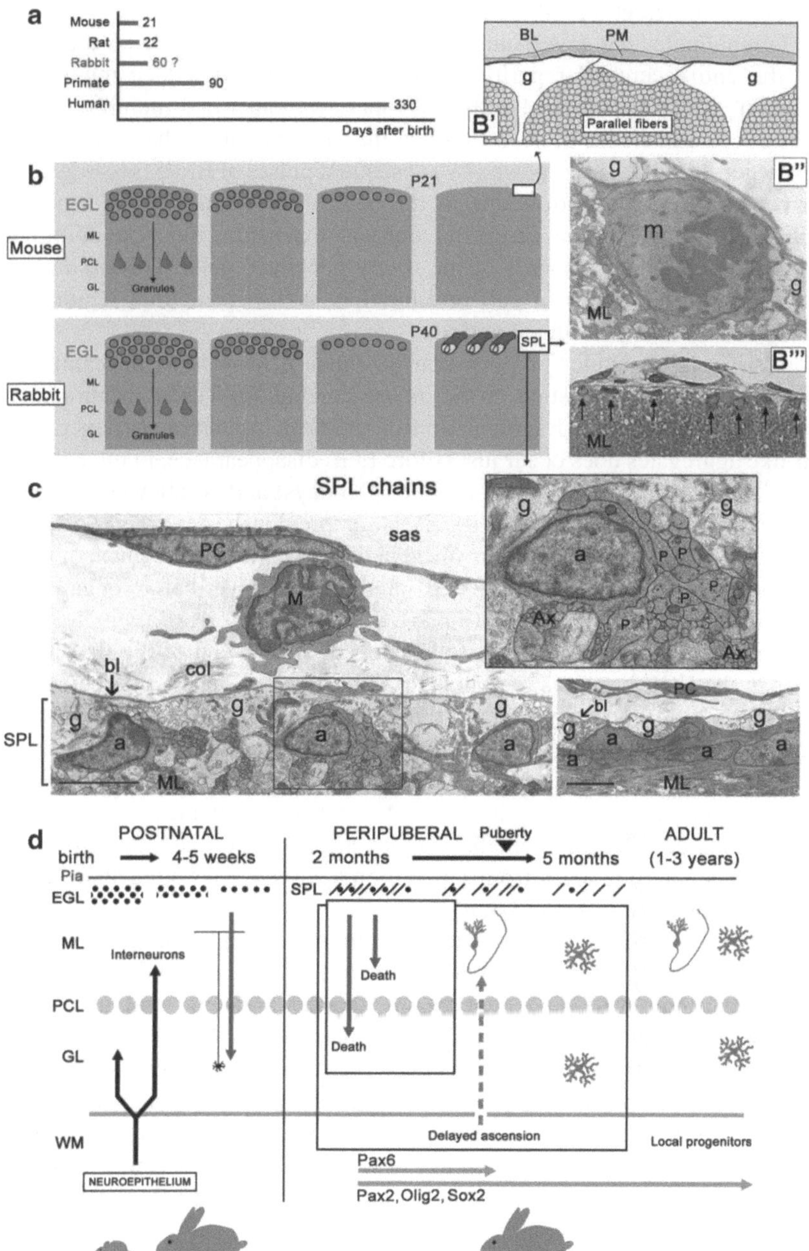

Fig. 13.3 Rabbit cerebellar neurogenesis. Cerebellar neurogenesis in mammals is thought to cease at specific early postnatal (prepuberal) ages (**a**); this coincides with the exhaustion of the external germinal layer (*EGL*, **b**), leaving a direct contact between the molecular layer and the pia mater (*PM*) (**B′**). *BL* basal lamina, *g* glia. In rabbits, around P40 the EGL is replaced by the SPL, a proliferative layer characterized by chains of neuroblasts (**B″**, **B‴**, **c**). (**d**) Dynamics of postnatal, peripuberal, and adult neurogenesis in rabbits (see text)

both mice (Doetsch et al. 1997) and rabbits (Ponti et al. 2006b) are present at the basal lamina/molecular layer interface (Fig. 13.3c). Thousands of these chains cover the entire cerebellar perimeter, regularly arranged with a medial–lateral orientation, and spaced about 10–20 μm from one another in 3-month-old animals. Dividing cells and chains of neuroblasts in the SPL are part of the same population of newly generated cells, as showed by subsequent pulses of BrdU (see Fig. 13.2f,g). Both cell proliferation and chains in the SPL show homogeneous topographic distribution throughout the cerebellar cortical extension, thus excluding that a restricted site could be a source of the newly generated cells. This assumption is confirmed by the absence of cell proliferation in the ependymal regions of the fourth ventricle at any postnatal periods.

To clarify the dynamics of cerebellar postnatal neurogenesis in mammals, the last phases of EGL exhaustion in the mouse cerebellum were reexamined (Ponti et al. 2006a). Although a fragmentation of the EGL premigratory layer in small chain-like aggregates does occur just before EGL disappearance, in mice the occurrence of these structures is limited in time (4–5 days) and is followed by complete disappearance after postnatal day 21 (P21), as previously described (Altman and Bayer 1997). SPL chains could be involved in the tangential displacement of neuronal precursors, as shown in the mouse EGL premigratory layer before engagement in radial migration (Komuro et al. 2001).

The transition from rabbit EGL to SPL occurs at the end of the first month of life, when the EGL becomes a monolayer. After P40, no distinction between proliferative and premigratory layers is detectable, and a clear pattern of chain fragmentation is maintained up to 4–5 months of age, with a progressive increase in the distance among chains and decrease in their number. From the second month of age up to puberty (fourth month), the neurogenic process in the SPL is remarkable. Then, dividing cells on the cerebellar surface disappear between the fifth and sixth months. Thus, the SPL spans from 1 to 5 months of age, persisting shortly beyond puberty (summarized in Fig. 13.3c).

The rabbit SPL shares features with the forebrain SVZ. It originates from structural modification of a secondary germinal layer, and it generates neuronal precursors that assemble into tangential chains sharing antigenic and ultrastructural features with SVZ chains (Lois et al. 1996; Doetsch et al. 1997; Peretto et al. 1997; reviewed in Bonfanti 2006), although smaller and far more numerous. On the other hand, the transitory character of the rabbit SPL highly differs from persisting brain neurogenic areas, suggesting that it contains a population of neuronal progenitors rather than stem cells. This assumption is supported by the absence of astrocytic cell bodies within the SPL, an essential element in the SVZ and hippocampal neural stem cell niches (Doetsch et al. 1997; Seri et al. 2004).

13.3.2.3 Newly Formed Cells in the Cerebellar Cortex

Besides the protracted proliferative activity of the SPL, new cells continue to be generated in the whole cerebellar cortex of peripuberal rabbits. A remnant of this process is still detectable in the adult, in the absence of the SPL.

Peripuberal Rabbits

In spite of a progressive reduction in cell genesis within the SPL, and in sharp contrast with our common knowledge concerning the mammalian cerebellum, a remarkable genesis of cells is detectable in the peripuberal rabbit cerebellar cortex (Figs. 13.2 and 13.3). Systematically administered BrdU detected at different postinjection survival times (from 5 to 15 days) reveals that a substantial amount of these cells is still alive 2 weeks after their birth (Ponti et al. 2006a) and, to a lesser extent, after 2 months (Ponti et al. 2008; Fig. 13.3d). Their morphology can be visualized by double labelings with antibodies raised against markers of structural plasticity implicated in dynamic cellular events of developmental and adult neurogenesis, such as PSA-NCAM, doublecortin, and the cytoskeletal protein Map5 (Brown et al. 2003; Bonfanti 2006). The newly generated cortical cells fall into three main morphological types: *bipolar*, *polarized neuronal-like*, and *multipolar* (see Fig. 13.2l,m), homogeneously distributed in all dorsal/ventral and medial/lateral lamellae. PSA-NCAM and doublecortin are detectable in bipolar and polarized cells but not in multipolar cells. The latter are immunoreactive for the microtubule-associated protein Map5 (Map1B; Wu et al. 2001; Fig. 13.2m).

The persistence of neurogenic processes in the SPL raises the question as to how long is the genesis of granule cells in rabbits prolonged. A subpopulation of PSA-NCAM+/doublecortin+ bipolar cortical cells express the markers for cerebellar granule cell precursors Pax6 (Yamasaki et al. 2001) and NeuN (Weyer and Schilling 2003). However, Pax6 and NeuN in the rabbit cerebellum disappear starting from the third postnatal month (Ponti et al. 2008; Fig. 13.3d), thus indicating that the descent of granule cell precursors is exhausted before puberty, as in rodents.

The immunoreactivity for the microtubule-binding protein doublecortin (Brown et al. 2003) in virtually all PSA-NCAM+ cells confirms their neuronal nature. Bipolar cells show the typical morphology of migrating elements. Polarized neuronal-like cells are restricted to the molecular layer. They have a piriform cell body (15–20 μm long), a long, thin axonal-like process with a U-shaped direction remaining within the molecular layer, and one to three thicker, dendritic-like processes on the opposite side of the cell (Fig. 13.2i,l). BrdU/PSA-NCAM double staining carried out at 5, 10, and 15 days postinjection survival times support the existence of a neurogenic sequence leading to a young neuronal cell type in the molecular layer during a 2-week temporal window. These neuronal cells also express the transcription factor *Pax2*, a nuclear marker for GABAergic cerebellar interneurons of neuroepithelial origin (Maricich and Herrup 1999; Weisheit et al. 2006). In addition, most of them show a remarkable staining for the neurotransmitter γ-aminobutyric acid (GABA), also present in inhibitory interneurons. By considering the coexpression of all molecules, it can be concluded that a subset of GABAergic *interneurons of neuroepithelial origin* are generated within the peripuberal rabbit cerebellar cortex (Ponti et al. 2008).

The Map5+ multipolar cells are characterized by a spheric cell body, about 8–10 μm in diameter, and a stellate, highly ramified morphology. Using BrdU/Map5 double staining, they also are shown to be newly generated (Ponti et al. 2008;

Fig. 13.2m). In contrast with the neuronal cells described above, the multipolar cells are more homogeneously distributed in all cerebellar layers and are not immunoreactive for PSA-NCAM, doublecortin, or glial fibrillary acidic protein (GFAP). They are consistently double stained for both the transcription factors Sox2 and Olig2 (Zhou and Anderson 2002; Takebayashi et al. 2002) but not for Pax2. Thus, the PSA-NCAM+/doublecortin+ cells and the Map5+ cells actually represent two distinct populations, at least under the antigenic expression profile.

Based on their morphology, multipolar cells could belong to the population referred in rodents to as *synantocytes* (elements that form multiple contact with neurons, astrocytes, and oligodendrocytes; Butt et al. 2005) or as *polydendrocytes* (related to oligodendrocytes but with more processes and functions; Nishiyama 2007). Because the expression of Map5 remains at relatively high level in neurogenic/structurally plastic regions of the adult brain (Viereck et al. 1989; Nothias et al. 1996), in the cerebellum of lagomorphs anti-Map5 antibodies seem to act as a marker for a reservoir of dividing or young progenitor cells related to the synantocyte/polydendrocyte lineage. However, because of the lack of antibodies capable of localizing in the rabbit tissue the general marker for synantocytes/polydendrocytes in rodents (Ng2 antigen), we do not know if the rabbit Map5+ cells represent a subpopulation of Ng2+ progenitors.

An analysis carried out in postnatal mice (Ponti et al. 2006a) confirmed that no cell proliferation and no PSA-NCAM+ or doublecortin+ cells can be detected in the whole cerebellar cortex after P21, thus supporting peripuberal cerebellar neurogenesis as specific to rabbits.

Adult Rabbits

Although in adults the cerebellar surface is completely devoid of SPL, newly generated cells are detectable in the cortex using cell proliferation markers. The occurrence of BrdU+ nuclei (after 15 days survival) associated with PSA-NCAM+/doublecortin+ neuronal-shaped cells and Map5+ multipolar cells confirms that both cell populations continue to be generated in fully adult animals up to 3 years of age (Ponti et al. 2008). Similarly to what is observed in peripuberal stages, the neuronal-shaped PSA-NCAM+/doublecortin+ cells coexpress the transcription factor Pax2 and are immunoreactive for GABA. Accordingly, the Map5+ stellate-shaped cells coexpress Sox2 and Olig2. As expected, no Pax6+ cells could be detected in the adult cerebellar cortex.

Some differences concern the absence of PSA-NCAM+/doublecortin+ cells within the white matter of adult cerebella and the very rare occurrence of bipolar-shaped cells in the cortex. Counting carried out on adult animals indicates that a progressive reduction in the number of newly generated cerebellar progenitor cells does occur (1/10 with respect to the 6-month old rabbit, and 1/100 with respect to the 3-month-old). About one-half of these cells are neuronal progenitors in the molecular layer.

13.4 Concluding Remarks

Persistent neurogenesis is a phylogenetically highly conserved feature in the animal world (Lindsey and Tropepe 2006). The genesis of new neurons is rather widespread in the CNS of invertebrates and nonmammalian vertebrates such as fish, reptiles, and birds (Garcia-Verdugo et al. 2002; Zupanc 2006), whereas in mammals it is more restricted to specific brain areas (Gage 2000; Alvarez-Buylla and Garcìa-Verdugo 2002; Bonfanti and Ponti 2008). Recent comparative studies have started to show the existence of important differences even among mammals: these involve the topographic location of neurogenesis, the internal arrangement of adult neurogenic sites, and the mode/direction of cell migration (reviewed in Bonfanti and Ponti 2008; Gould 2007; see Part III of this book).

The use of rabbits as a model to address issues of comparative neurogenesis has revealed intriguing examples of structural plasticity. Differences in the internal arrangement of SVZ cellular compartments seem to be associated with the extension and shape of the ventricular cavities. A peculiar feature in rabbits concerns the occurrence of chains of neuroblasts in regions of the mature brain parenchyma other than the SVZ-RMS pathway. These regions include the corpus callosum, the frontal cortex, and the amygdala (parenchymal chains), the caudate nucleus (striatal chains), and the cerebellar surface (SPL chains). A common feature of all these "extra-SVZ" chains is their relative glia independence. Parenchymal chains in the forebrain and SPL chains in the cerebellum are temporally restricted to young, growing animals, whereas those in the striatum persist throughout life. Forebrain parenchymal chains are nonproliferative extensions of a neurogenic site, whereas SPL chains are substantially independent from the protracted neurogenic activity detectable within the cerebellar cortex. In this context, only the striatal chains are part of a quantitatively low but complete neurogenic sequence.

The most peculiar feature of rabbits seems to be related to the occurrence of local neural progenitors in the mature striatum and cerebellum, namely, in the so-called nonneurogenic parenchyma of rodents. It is now well established that cell proliferation in the adult mammalian CNS is not restricted to the germinative regions of hippocampus and olfactory bulb. Yet, in rodents it is generally thought that cell proliferation in nonneurogenic regions gives rise primarily to glial cells. Neuronal progenitors could naturally exist in several mature regions including septum, striatum, cerebral cortex, and spinal cord, but their potentialities would be extrinsecated only in vitro, under experimental conditions (Palmer et al. 1999; Kondo and Raff 2000). A still open question concerns the origin and specification of stem cells and local parenchymal progenitors (Temple 2001; Wexler and Palmer 2002). Future studies should be addressed to establish whether adult local parenchymal progenitors represent displaced periventricular embryonic progenitors and whether their specification can be regulated by external factors. This knowledge is crucial in the perspective of designing new efficient strategies for mobilizing endogenous progenitors to replace dying neurons in CNS diseases.

The overall features emerging from the rabbit model indicate that in this mammalian species the CNS is permissive to newly born cells at different levels, including cell migration through the brain parenchyma from “classic” neurogenic sites, along with the possibility of cell migration and differentiation from local cell progenitors capable of dividing in situ, both in brain and in the cerebellum. In this perspective, rabbits could offer the opportunity of studying in vivo mechanisms to efficiently mobilizing endogenous progenitors in wide areas of the CNS.

A possible functional explanation for differences in neurogenic potential between rodents and lagomorphs could reside in the pronounced variation in lifespan among these animal species. Distinct patterns of growth, maturation, and senescence may have an impact on the extent of plasticity and neurogenesis throughout life (Lindsey and Tropepe 2006; Berardi et al. 2000). This theory opens the intriguing possibility that a delayed genesis of neuronal and glial progenitor cells could exist in other mammalian species characterized by slow growth and long lifespan, including humans. In this context, the rabbit brain and cerebellum could be a suitable model for studying CNS permissiveness in mammals.

Acknowledgments This work was supported by Compagnia di San Paolo (Progetto NEUROTRANSPLANT-2004.2019).

References

Altman J, Bayer SA (1997) Development of the cerebellar system. CRC Press, Boca Raton, USA

Alvarez-Buylla A, Garcìa-Verdugo JM (2002) Neurogenesis in adult subventricular zone. J Neurosci 22:629–634

Anthony TE, Klein C, Fishell G, Heintz N (2004) Radial glia serve as neuronal progenitors in all regions of the central nervous system. Neuron 41:881–890

Arvidsson A, Collin T, Kirik D et al (2002) Neuronal replacement from endogenous precursors in the adult brain after stroke. Nat Med 8:963–970

Bédard A, Cossette M, Lévesque M, Parent A (2002) Proliferating cells can differentiate into neurons in the striatum of normal adult monkey. Neurosci Lett 328:213–216

Berardi N, Pizzorusso T, Maffei L (2000) Critical periods during sensory development. Curr Opin Neurobiol 10:138–145

Bonfanti L (2006) PSA-NCAM in mammalian structural plasticity and neurogenesis. Prog Neurobiol 80:129–164

Bonfanti L, Ponti G (2008) Adult neurogenesis and the New Zealand white rabbit. Vet J 175:310–331

Bonfanti L, Olive S, Poulain DA, Theodosis, DT (1992) Mapping of the distribution of polysialylated neural cell adhesion molecule throughout the central nervous system of the adult rat: an immunohistochemical study. Neuroscience 49:419–436

Brown JP, Couillard-Després S, Cooper-Kuhn CM et al (2003) Transient expression of doublecortin during adult neurogenesis. J Comp Neurol 467:1–10

Butt AM, Hamilton N, Hubbard P, Pugh M, Ibrahim M (2005) Synantocytes: the fifth element. J Anat 207:695–706

Chenn A, Walsh CA (2002) Regulation of cerebral cortical size by control of cell cycle exit in neural precursors. Science 297:365–369

Curtis MA, Kam M, Nannmark U et al (2007) Human neuroblasts migrate to the olfactory bulb via a lateral ventricular extension. Science 315:1243–1249

Dayer AG, Cleaver KM, Abouantoun T, Cameron HA (2005) New GABAergic interneurons in the adult neocortex and striatum are generated from different precursors. J Cell Biol 168, 415–427

Doetsch F, Garcìa-Verdugo JM, Alvarez-Buylla A (1997) Cellular composition and three-dimensional organization of the subventricular germinal zone in the adult mammalian brain. J Neurosci 17:5046–5041

Douzery EJ, Huchon D (2004) Rabbits, if anything, are likely Glires. Mol Phylogenet Evol 33:922–935

Engelkamp D, Rashbass P, Seawright A, van Heyningen V (1999) Role of Pax6 in development of the cerebellar system. Development 126:3585–3596

Fallon J, Reid S, Kinyamu R et al (2000) In vivo induction of massive proliferation, directed migration, and differentiation of neural cells in the adult mammalian brain. Proc Natl Acad Sci USA 97:14686–14691

Gage FH (2000) Mammalian neural stem cells. Science 287:1433–1438

Garcia-Verdugo JM, Ferron S et al (2002) The proliferative ventricular zone in adult vertebrates: a comparative study using reptiles, birds, and mammals. Brain Res Bull 57:765–775

Gould E (2007) How widespread is adult neurogenesis in mammals? Nat Rev Neurosci 8:481–488

Gould E, Reeves AJ, Graziano MS, Gross CG (1999)Neurogenesis in the neocortex of adult primates. Science 286:548–552

Gould E, Vail N, Wagers M, Gross CG (2001) Adult-generated hippocampal and neocortical neurons in macaques have a transient existence. Proc Natl Acad Sci USA 98:10910–10917

Ito M (2006) Cerebellar circuitry as a neuronal machine. Prog Neurobiol 78:272–303

Komuro H, Yacubova E, Yacubova E, Rakic P (2001) Mode and tempo of tangential cell migration in the cerebellar external granule layer. J Neurosci 21:527–540

Kondo T, Raff M (2000) Oligodendrocyte precursor cells reprogrammed to become multipotential CNS stem cells. Science 289:1754–1757

Kornack DR, Rakic P (2001) The generation, migration, and differentiation of olfactory neurons in the adult primate brain. Proc Natl Acad Sci USA 98:4752–4757

Kriegstein A, Noctor S, Martinez-Cerdeno V (2006) Patterns of neural stem and progenitor cell division may underlie evolutionary cortical expansion. Nat Rev Neurosci 7:883–890

Leonhardt H (1972) Topographic distribution of subependymal basement labyrinths of the ventricular system of brain in rabbit. Z Zellforsch Mikrosk Anat 127:392–406

Leto K, Bartolini A, Rossi F (2008) Neurogenesis in the cerebellum of rodents. In: Bonfanti L, ed. Postnatal and Adult Neurogenesis. Research Signpost, Trivandrum, India pp. 63–81

Lindsey BW, Tropepe V (2006) A comparative framework for understanding the biological principles of adult neurogenesis. Prog Neurobiol 80:281–307

Lois C, Alvarez-Buylla A (1994) Long-distance neuronal migration in the adult mammalian brain. Science 264:1145–1148

Lois C, Garcìa-Verdugo J, Alvarez-Buylla A (1996) Chain migration of neuronal precursors. Science 271:978–981

Luzzati F, Peretto P, Aimar P et al (2003) Glia-independent chains of neuroblasts through the subcortical parenchyma of the adult rabbit brain. Proc Natl Acad Sci USA 100:13036–13041

Luzzati F, De Marchis S, Fasolo A, Peretto P (2006) Neurogenesis in the caudate nucleus of the adult rabbit. J Neurosci 26:609–621

Luzzati F, De Marchis S, Fasolo A, Peretto P (2007) Adult neurogenesis and local neuronal progenitors in the striatum. Neurodegen Dis 4:322–327

Maricich SM, Herrup K (1999) Pax-2 expression defines a subset of GABAergic interneurons and their precursors in the developing murine cerebellum. J Neurobiol 41:281–294

McFarland WL, Morgane PJ, Jacobs MS (1969) Ventricular system of the brain of the dolphin, *Tursiops truncatus*, with comparative anatomical observations and relations to brain specializations. J Comp Neurol 135:275–368

Nishiyama A (2007) Polydendrocytes: NG2 cells with many roles in development and repair of the CNS. Neuroscientist 13:62–76

Nothias F, Fischer I, Murray M et al (1996) Expression of a phosphorylated isoform of MAP1B is maintained in adult central nervous system areas that retain capacity for structural plasticity. J Comp Neurol 368:317–334

Palmer TD, Markakis EA, Willhoite AR, Safar F, Gage, FH (1999) Fibroblast growth factor-2 activates a latent neurogenic program in neural stem cells from diverse regions of the adult CNS. J Neurosci 19:8487–8497

Peretto P, Merighi A, Fasolo A, Bonfanti L (1997) Glial tubes in the rostral migratory stream of the adult rat. Brain Res Bull 42:9–21

Peretto P, Giachino C, Aimar P et al (2005) Chain formation and glial tube assembly in the shift from neonatal to adult subventricular zone of the rodent forebrain. J Comp Neurol 487:407–427

Ponti G, Peretto P, Bonfanti L (2006a) A subpial, transitory germinal zone forms chains of neuronal precursors in the rabbit cerebellum. Dev Biol 294:168–180

Ponti G, Aimar P, Bonfanti L (2006b) Cellular composition and cytoarchitecture of the rabbit subventricular zone (SVZ) and its extensions in the forebrain. J Comp Neurol 498:491–507

Ponti G, Peretto P, Bonfanti L (2008) Genesis of neuronal and glial progenitors in the cerebellar cortex of peripuberal and adult rabbits. PLoS ONE 3:e2366

Quinones-Hinojosa A, Sanai N, Soriano-Navarro M et al (2006) Cellular composition and cytoarchitecture of the adult human subventricular zone: a niche of neural stem cells. J Comp Neurol 494:415–434

Rodriguez-Perez LM, Perez-Martin M, Jimenez, AJ et al (2003) Immunocytochemical characterization of the wall of the bovine ventricle. Cell Tissue Res 314:325–335

Sanai N, Tramontin AD, Quinones-Hinojosa A et al (2004) Unique astrocyte ribbon in adult human brain contains neural stem cells but lacks chain migration. Nature 427:740–744

Sanchez-Villagra MR, Sultan F (2001) The cerebellum at birth in therian mammals, with special reference to rodents. Brain Behav Evol 59:101–113

Seri B, Garcìa-Verdugo JM, Collaudo-Morente L et al (2004) Cell types, lineage, and architecture of the germinal zone in the adult dentate gyrus. J Comp Neurol 478:359–378

Takebayashi H, Nabeshima Y, Yoshida, S et al (2002) The basic helix-loop-helix factor olig2 is essential for the development of motoneuron and oligodendrocyte lineages. Curr Biol 12:1157–1163

Temple S (2001) The development of neural stem cells. Nature 414:112–117

Viereck C, Tucker RP, Matus A (1989) The adult rat olfactory system expresses microtubule-associated proteins found in the developing brain. J Neurosci 9:3547–3557

Weisheit G, Gliem M, Endl E et al (2006) Postnatal development of the murine cerebellar cortex: formation and early dispersal of basket, stellate and Golgi neurons. Eur J Neurosci 24:466–478

Wexler E, Palmer T (2002) Where, oh where, have my stem cells gone? Trends Neurosci 25:225–227

Weyer A, Schilling K (2003) Developmental and cell type-specific expression of the neuronal marker NeuN in the murine cerebellum. J Neurosci Res 73:400–409

Wu HY, Dawson MR, Reynolds R, Hardy RJ (2001) Expression of QKI proteins and MAP1B identifies actively myelinating oligodendrocytes in adult rat brain. Mol Cell Neurosci 17:292–302

Yamasaki T, Kawaji K, Ono K et al (2001) Pax6 regulates granule cell polarization during parallel fiber formation in the developing cerebellum. Development 128:3133–3144

Zhou Q, Anderson DJ (2002) The bHLH transcription factors OLIG2 and OLIG1 couple neuronal and glial subtype specification. Cell 109:61–73

Zupanc GK (2006) Neurogenesis and neuronal regeneration in the adult fish brain. J Comp Physiol A 192:649–670

Chapter 14
Neuronal Migration in the Adult Brain

Masato Sawada, Shi-hui Huang, Yuki Hirota, Naoko Kaneko, and Kazunobu Sawamoto

Abstract Neuronal migration is an important process in brain development and homeostasis. It is not only a phenomenon of embryogenesis: it also occurs in the adult brain, following adult neurogenesis. In fact, throughout life, numerous new neurons generated in the adult subventricular zone take the long journey to the olfactory bulb. The adult-born neurons form chain-like aggregates and migrate inside glial tubes. In this chapter, we summarize current knowledge on the mechanisms of neuronal migration occurring in the adult brain of various animals under physiological and pathological conditions.

14.1 Introduction

New neurons generated in the adult mouse subventricular zone (SVZ) migrate over a long distance (about 8 mm) at a high speed (more than 100 μm/h), toward the olfactory bulb (OB) (Luskin, 1993; Lois and Alvarez-Buylla, 1994; Alvarez-Buylla and Garcia-Verdugo, 2002). After their arrival at the core of the OB, they migrate radially to their final destination, and each differentiates into one of two major types of interneurons in the granule and glomerular layers (Luskin, 1993). Studies have established that neuronal migration from the SVZ to the OB is conserved from rodents to primates (Kornack and Rakic, 2001; Pencea et al., 2001). Moreover, considerable evidence suggests that some newly generated neurons in the SVZ migrate toward damaged areas and differentiate into mature local neurons in injured brain tissue, indicating that the adult brain has the potential to regenerate damaged neuronal circuitry (Arvidsson et al., 2002; Yamashita et al., 2006). In this chapter,

M. Sawada, S-H. Huang, Y. Hirota, N. Kaneko, and K. Sawamoto (✉)
Department of Developmental and Regenerative Biology, Institute of Molecular Medicine, Nagoya City University Graduate School of Medical Sciences, 1 Kawasumi, Mizuho-cho, Mizuho-ku, Nagoya 467-8601, Japan
e-mail: sawamoto@med.nagoya-cu.ac.jp

T. Seki et al. (eds.), *Neurogenesis in the Adult Brain I: Neurobiology*,
DOI 10.1007/978-4-431-53933-9_14, © Springer 2011

we describe the mechanisms of neuronal migration from the SVZ toward the OB in the adult rodent brain. We also compare these mechanisms with neuronal migratory mechanisms in the injured brain and in primates.

14.2 Neuronal Migration from the Subventricular Zone to the Olfactory Bulb in the Adult Rodent Brain

14.2.1 Chain Migration and Cell–Cell Interactions

In rodents, new neurons generated in the adult SVZ migrate a long distance at a high speed toward the OB (Fig. 14.1). The migrating new neurons (neuroblasts) form elongated cell aggregates called "chains," in which each cell migrates along the others in the chain (Lois et al., 1996). In contrast to the radial glia-dependent neuronal migration in the developing cerebral cortex, the movement of neuroblasts in chains in the adult SVZ does not require axons or glial fibers (Wichterle et al., 1997). The neuroblasts migrate rostrally, forming an extensive network of chains covering the lateral wall of the lateral ventricles. They then enter a restricted pathway called the rostral migratory stream (RMS), which leads to the OB (Doetsch and Alvarez-Buylla, 1996). In the RMS, the neuronal chains are surrounded by tubes composed of astrocytic processes (Lois et al., 1996). The chains of neuroblasts and surrounding glial tubes are distinct characteristics of the adult RMS. The studies described in the following sections suggest that strictly regulated interactions between the neuroblasts and glia contribute to the fast and directional migration of the neuroblasts.

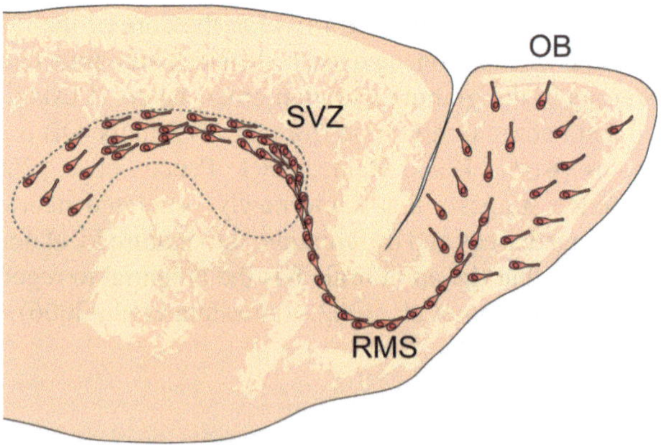

Fig. 14.1 Neuronal migration from the subventricular zone (*SVZ*) to the olfactory bulb (*OB*) in the adult rodent brain. Newly generated neuroblasts (*red*) migrate rostrally, forming an extensive network of chains that cover the lateral wall of the lateral ventricles and then enter the rostral migratory stream (*RMS*), which leads to the OB

14.2.1.1 Neuroblast Interactions

Chain formation is a unique characteristic of neuroblasts generated in the adult SVZ. Neuroblasts bunch up together into chains and migrate along their neighbors efficiently, rather than being tangled with each other, suggesting that neuroblast interactions with other factors are important for chain formation. Several extracellular ligands and their transmembrane receptors have been implicated in regulating the interactions between neuroblasts (Fig. 14.2). The binding of ApoER2 (apolipoprotein E receptor 2) and its major ligand Reelin is a well-known signal for precise cell positioning in the developing cerebral cortex. In the adult mouse, ApoER2 and its downstream adaptor protein Dab1 (disabled-1) are expressed in the RMS, where they are involved in chain formation independent of Reelin (Andrade et al., 2007).

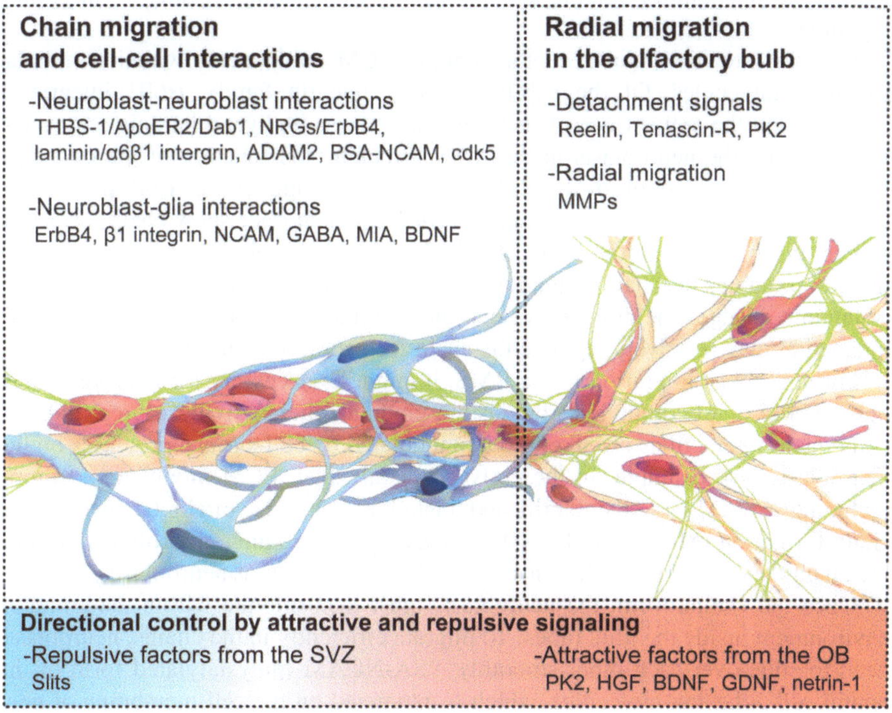

Fig. 14.2 Chain migration in the adult RMS and radial migration in the OB. The aggregation of migrating neuroblasts (*red*) into chains, and the formation of the surrounding glial tube (*blue*), are distinct features of neuronal migration in the RMS. Cell–cell interactions, including those between neuroblasts and between neuroblasts and glia, contribute to the formation of these structures. Chain-forming neuroblasts migrate in close association with blood vessels (*orange*) inside the glial tube, where extracellular matrix (ECM) components (*green*) accumulate around the chains. In addition, guidance cues from the SVZ and OB regulate the fast and directional migration of neuroblasts toward the OB. After arriving at the core of the OB, the neuroblasts disperse and begin to migrate radially as individual cells. Some extracellular proteins may act as signals for detachment from the chains, and their activities seem to be regulated by sensory input

Instead, the glycoprotein THBS-1 (thrombospondin-1) has been identified as a novel ligand for ApoER2 in the adult RMS in the promotion of neuroblast chain formation (Blake et al., 2008).

Receptor tyrosine kinases also contribute to chain formation (Conover et al., 2000; Anton et al., 2004). The tyrosine kinase ErbB4 is predominantly expressed on the cell membrane of neuroblasts. Mice lacking ErbB4 show a disrupted RMS and misrouted neuroblasts (Anton et al., 2004). NRG1 (neuregulin-1) and NRG2, ligands of ErbB4, are expressed in the RMS and appear to be involved in controlling interactions between neuroblasts during chain migration.

In addition to these soluble ligands and their receptors, extracellular matrix (ECM)-mediated signaling also appears to regulate chain formation. Several ECM components, including Tenascin-C and laminins, accumulate around the chains of neuroblasts inside the glial tubes. ECM components in the RMS are known to act not only as anchors for soluble factors but also as signaling ligands for migrating neuroblasts.

Integrins are the major receptors linking ECM components with intracellular signaling molecules. Of the α,β-heterodimer integrin family, α6β1 integrin is expressed on the cell surface of neuroblasts, whereas its ligand laminin is concentrated around the neuroblasts in the RMS. Blocking the function of the α6 or β1 subunit causes disruption of the RMS (Emsley and Hagg, 2003). Both β1 integrin conditional knockout mice and laminin α2/α4-deficient mice show chain formation defects in vivo (Belvindrah et al., 2007). Moreover, the binding of laminin to β1 integrin promotes the chain formation of neuroblasts in vitro, and an ectopic infusion of laminin changes the direction of neuronal migration. Thus, the integrin–laminin interaction may positively regulate the chain formation of neuroblasts in the RMS.

For neuroblast migration in the RMS, it might be necessary for neuroblasts not only to bind ECM components but also to modulate their activities. Integrin–ECM component interactions are likely to be modulated by cell-surface proteases on the neuroblasts. For example, in the early postnatal stage, ECM modulatory factors, including ADAM2 (a disintegrin and metalloprotease-2), might be involved in chain formation (Murase et al., 2008). Thus, the extracellular environment around the chains may have multiple roles in the chain formation of neuroblasts.

Once the chains form, the neuroblasts must then migrate within the complex environment inside the glial tubes. To migrate efficiently in the chains, neuroblasts need to slide over each other smoothly. PSA-NCAM (polysialylated form of the neural cell adhesion molecule), which is expressed on the cell membrane of neuroblasts, may be involved in this process. Neither the genetic deletion of the *NCAM* gene (Cremer et al., 1994) nor the elimination of the PSA moiety by enzymatic digestion (Ono et al., 1994) interferes with chain formation, but both manipulations cause neuroblasts to accumulate in the RMS and result in a smaller OB, suggesting that the PSA moiety is important for smooth cell migration, rather than for chain formation itself (Chazal et al., 2000). It is possible that the negatively charged PSA moiety decreases the affinity between the NCAMs on neighboring neuroblasts, thereby increasing the mobility of cells migrating in the chains (Gascon et al., 2010).

Extracellular stimuli mediated by soluble ligands and ECM components regulate cytoskeletal rearrangements, which are likely to be involved in the control of cell movement during chain migration (see Sect. 3). In the embryonic brain, cdk5 (cyclin-dependent kinase 5) plays an essential role in neuronal migration by regulating the cytoskeleton. We found that a conditional knockout of cdk5 causes defects in neuroblast chain formation, migration speed, and direction in the postnatal RMS (Hirota et al., 2007). Thus, chain formation and migration seem to depend on coordinated neuroblast interactions mediated by a variety of extracellular signals, some of which regulate the cytoskeleton.

14.2.1.2 Neuroblast–Glia Interactions

The glial tube structure is a separate and distinct feature of the adult RMS. SVZ astrocytes, some of which act as neural stem cells (see Chap. 12 of this book), form tunnels surrounding the chains of neuroblasts in the RMS (Fig. 14.2). These astrocytes appear to separate the migrating chain-forming neuroblasts physically from the brain parenchyma.

In spite of its unique structure, the exact role of the glial tube is largely unknown. Mice with chain migration defects also show altered glial tube organization (Chazal et al., 2000; Anton et al., 2004; Belvindrah et al., 2007), suggesting that interactions between the neuroblasts and the tube-forming astrocytes may be required for both chain migration and glial tube formation. On the other hand, morphological studies of nonrodent mammals such as the rabbit (see Chap. 13 of this book) have suggested that the glial tubes are not essential for neuroblast migration in the adult brain.

The glial tube seems to provide a special microenvironment that enables neuroblasts to migrate efficiently. Migrating neuroblasts in the RMS release GABA (γ-aminobutyric acid), which activates the $GABA_AR$ ($GABA_A$ receptor) in a paracrine manner and negatively regulates neuroblast migration speed. RMS astrocytes express GAT4, a high-affinity GABA transporter, which takes up excess GABA and contributes to the tight regulation of neuronal migration (Bolteus and Bordey, 2004). RMS astrocytes also release a novel attractive activity (migration-inducing activity, MIA) for neuroblasts, although its identity remains unknown (Mason et al., 2001).

In addition to interactions between neuroblasts and glia, interactions among neuroblasts, glia, and vascular endothelial cells in the adult RMS were recently observed. The neuroblasts in chains are closely associated with blood vessels that parallel the RMS (Fig. 14.2). The processes of the RMS astrocytes, which ensheath blood vessels as well as the neuroblasts, capture BDNF (brain-derived neurotrophic factor) produced by the vascular endothelial cells and induce the blood vessel-associated migration in the adult RMS (Snapyan et al., 2009).

The defective cell migration phenotypes reported in various mutants with impaired chain formation and/or glial tubes suggest that neuroblast–neuroblast and neuroblast–glia interactions contribute to neuronal migration in the adult brain. The continued identification of molecules that regulate the formation of chains and glial tubes will help elucidate the complex mechanisms underlying cell migration in the adult rodent brain.

14.2.2 Directional Control by Attractive and Repulsive Signaling

For neuroblasts generated in the SVZ, the journey to the OB is a long one. Even though the migratory chains provide a directional structure, the chains must follow the correct route. There is increasing evidence that the highly directional migration of neuroblasts toward the OB is controlled by a combination of attractive and repulsive signaling molecules (see Fig. 14.2).

In the SVZ, Slit proteins appear to act as repulsive factors for neuroblasts. During neural development, Slits are reported to bind their receptors, the Robos (roundabouts), to regulate the cytoskeleton to control axonal guidance (Brose et al., 1999; Li et al., 1999). In the adult brain, Slits secreted from the septum and choroid plexus (Hu and Rutishauser, 1996; Hu, 1999; Wu et al., 1999), located at the caudal region of the lateral ventricles, have been implicated in the regulation of neuronal migration. Neuroblasts generated in the SVZ express Robos and can be repelled by Slits (Hu, 1999; Wu et al., 1999). However, how the Slit proteins from the caudal lateral ventricles might reach neuroblasts migrating in the SVZ tissue is unclear.

We found that the migration of neuroblasts parallels the direction of cerebrospinal fluid (CSF) flow and the beating of ependymal cilia. Slit protein injected into the ventricle forms a concentration gradient in the SVZ in the direction of ciliary beating and CSF flow (Sawamoto et al., 2006). These results suggest that the concentration gradient of Slit protein formed by the CSF flow is involved in the directional control of neuronal migration. Interestingly, the neuroblasts express Slit1 as well as its receptors Robo2/3 in the RMS, indicating that Slit1 may act cell-autonomously in the RMS (Nguyen-Ba-Charvet et al., 2004). Recently, we reported that neuroblasts use Slit1 to interact with Robo-expressing astrocytes in their path for rapid migration (Kaneko et al., 2010).

A gradient of CSF-derived repulsive molecules such as Slit could explain the neuroblast migration in the SVZ along the ventricular wall. However, it is unlikely to be effective for neuroblasts migrating in the RMS toward the OB. One possible mechanism for guiding neuroblasts to their destinations is a chemotactic attraction to the OB. One study showing that neuroblasts can still migrate rostrally after the OB is eliminated by surgical aspiration seemed to indicate that attractive factors in the OB are not essential (Kirschenbaum et al., 1999). However, subsequent quantitative analyses indicated that removing the OB significantly reduces the number of neuroblasts migrating toward it (Liu and Rao, 2003). In addition, the role of the OB in the directional migration of neuroblasts is supported by co-culture experiments showing that the OB glomerular layer contains at least one chemoattractant for neuroblasts (Liu and Rao, 2003).

In fact, more recently, diffusible proteins expressed in the adult OB have been reported to act as attractants for neuroblasts. A secreted ligand, PK2 (prokineticin 2), is expressed in the granule and glomerular layers in the OB (Ng et al., 2005), whereas its receptor Prokr2 is expressed in neuroblasts in the RMS (Puverel et al., 2009). PK2 shows an attractive activity for neuroblasts in vitro. In addition, in PK2 knockout mice and Prokr2 mutant mice neuroblasts accumulate in the rostral RMS, resulting in a smaller than normal OB. Other putative attractants in the adult OB that have receptors expressed by neuroblasts include HGF (hepatocyte growth factor)/Met, BDNF/TrkB, and GDNF (glial cell line-derived neurotrophic factor)/GFRα1 (Paratcha et al., 2006; Chiaramello et al., 2007; Garzotto et al., 2008). Netrin-1/DCC signaling is also involved in the attraction of neuroblasts toward the OB. However, netrin-1 expression is reduced by postnatal day 4 (Murase and Horwitz, 2002), suggesting that it might be involved in neuronal migration only in the early postnatal stage. Taken together, the long journey of migrating neuroblasts toward the core of the OB seems to be guided by a combination of mechanisms mediated by repulsive factors from the SVZ and attractive factors from the OB.

14.2.3 Radial Migration in the Olfactory Bulb

Once neuroblasts arrive at the core of the OB, their chain migration halts. They migrate radially as individual cells and differentiate into interneurons in the granule and glomerular layers. This process, the transition from chain (tangential) to radial migration, is required for the final differentiation of the new neurons (see Fig. 14.2), and several molecules involved in this step have been reported.

In the adult OB, the extracellular protein Reelin is predominantly expressed by mitral cells, whereas its receptor, ApoER2, and downstream adaptor protein, Dab1, are expressed in neuroblasts migrating in the RMS (Hack et al., 2002). In vitro, Reelin induces the dispersion of neuroblast chains, without affecting the distance or direction of neuroblast migration. In the Reelin mutant mouse *reeler*, the neuroblasts form thick cell aggregates in the rostral RMS, and moreover the neuroblasts remain in their chains even in the granule layer of the OB. Individual radially migrating cells are rarely seen in this mutant, suggesting that Reelin acts as a detachment signal for the chain-forming neuroblasts.

Another ECM glycoprotein, Tenascin-R, also disperses the chain-forming neuroblasts and initiates their radial migration (Saghatelyan et al., 2004). In the adult OB, Tenascin-R expression is restricted to the granule layer, and it is absent from the SVZ and the RMS. Tenascin-R-deficient mice show accumulated neuroblasts in the core of the OB that remain in chains in the granule layer, indicating Tenascin-R is crucial for initiating radial migration.

Importantly, the Tenascin-R expression level is regulated by sensory input (Saghatelyan et al., 2004). Unilateral sensory deprivation by closing the naris results in a subtle but significant decrease in Tenascin-R mRNA and protein, and this decreased expression can be completely rescued by reopening the naris. Reelin immunoreactivity is also reduced after the main olfactory epithelium is lesioned by zinc sulfate treatment, which induces sensory deprivation (Okuyama-Yamamoto et al., 2005). Thus, detachment signals and radial migration in the OB seem to be regulated in a sensory activity-dependent manner. PK2/Prokr2, an attractant activity for neuroblasts in the RMS, also functions as a detachment signal in the OB (Ng et al., 2005); however, it remains unclear whether PK2 expression is regulated by sensory input.

Because the adult OB contains highly packed neuronal circuitry, once the neuroblasts disperse from the RMS, they probably need to alter their microenvironment to permit radial migration by remodeling surrounding components such as the ECM. A recent study suggests that MMPs (matrix metalloproteases) are involved in the radial migration in the OB. MMP inhibitors significantly reduce the speed of radially migrating neuroblasts in slice culture time-lapse imaging (Bovetti et al., 2007a).

In the RMS, neuroblasts migrate by sliding along each other, using their neighbors as a scaffold. During their radial migration, the neuroblasts move as individual cells without forming chains. One possible candidate for the scaffold used by individual neuroblasts after their dispersion from the chains is blood vessels. Radially migrating neuroblasts in the granule layer, but not in the glomerular layer, are associated with blood vessels, suggesting that blood vessel-associated migration might be an initial mechanism after detachment from the chains (Bovetti et al., 2007b). Another candidate scaffold, particularly for the periglomerular cells, which migrate further into the region close to the OB surface, is the dendrites of mitral/tufted cells, although no such role has yet been reported for them. Because the interneuron subtype that each neuroblast will adopt in the OB is determined when it is generated in the SVZ (Merkle et al., 2007), neuroblasts with different fates may use distinct modes of radial migration in the OB. Thus, the mechanisms of radial migration in the OB remain largely obscure.

14.3 Cytoskeletal Regulation

In contrast to migratory morphology of other cell types such as fibroblasts and keratinocytes, typical neuroblasts in the adult RMS have a bipolar shape with a relatively long leading process and shorter trailing process. Migrating neuroblasts move in a saltatory manner by repeating three steps: (1) extension of the leading process, (2) movement of the centrosome, and (3) nuclear movement and retraction of the rear of the cell (Wichterle et al., 1997; Schaar and McConnell, 2005; Higginbotham et al., 2006). In the first step, the neuroblast initiates movement by extending its leading process, which points to its destination along the neighboring cells. This stage is followed by movement of the centrosome and then of the nucleus (often called nuclear–centrosome (N–C) coupling). Similar cell behavior has been reported in the migration of various types of cells in the developing brain, although some neurons, such as cerebellar granule cells, show irregular movement of the centrosome (Umeshima et al., 2007). Cytoskeletal components such as F-actin or microtubules and their regulatory proteins seem to be important for the dynamic cytoskeletal reorganization at each step of neuroblast migration. Here we summarize the molecular basis of the three steps in neuroblast migration in the adult brain (Fig. 14.3).

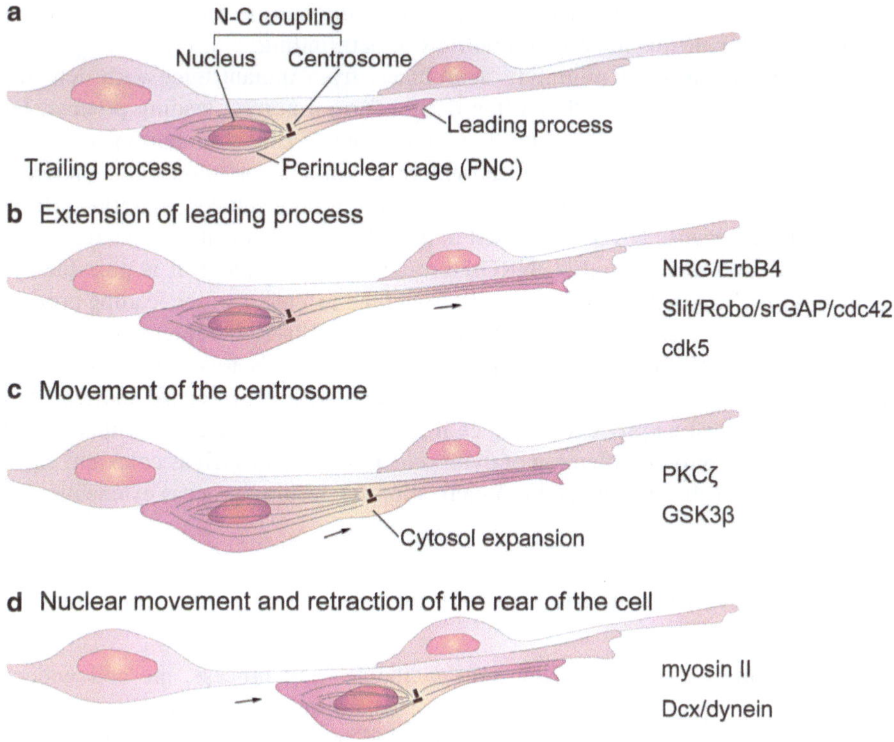

Fig. 14.3 Cytoskeletal regulation of chain-forming neuroblasts. The neuroblasts in chains have a relatively long leading process and short trailing process. The nucleus of a migrating neuroblast is linked with the centrosome by the perinuclear cage (*PNC*); this process is called N–C coupling (**a**). When a neuroblast starts to migrate, it extends its leading process toward the direction of movement along its neighbors. The receptors for directional cues as well as cytoskeletal regulators are involved in this process (**b**). Next, the cell widens, expanding the cytosol anterior to the nucleus, and the centrosome moves forward into it. Centrosomal positioning is predominantly regulated by several cell polarity-related factors (**c**). The subsequent movement of the nucleus can be mediated by both actomyosin-derived force and PNC-related factors (**d**). The chain-forming neuroblasts in the RMS migrate in a saltatory manner by repeating these three steps (**b–d**)

14.3.1 Extension of the Leading Process

When neuroblasts start to migrate, they first form a long leading process in the direction of movement, which is one of the distinct features of neuronal migration. The direction and length of the leading process appear to depend on interactions of cell-surface receptors with extracellular stimuli, including ECM components and various growth factors. Of these guidance molecules, attractive cues usually activate the Rho family small GTPases Rac and cdc42, leading to membrane extension, whereas repellent cues induce retraction of the rear of the cell via the activation of Rho and suppression of Rac and/or cdc42 (Guan and Rao, 2003).

Current evidence suggests that the extension of the neuronal leading process has molecular mechanisms in common with axon pathfinding.

Close examination of migrating neuroblasts from mutant mice with defective chain formation shows that they often do not form a normal leading process. For example, neuroblasts in cdk5 conditional knockout or ErbB4 mutant mice exhibit, respectively, a shorter than normal leading process or a misoriented leading process: both phenotypes lead to irregularly directed migration (Anton et al., 2004; Hirota et al., 2007). Because NRGs expressed in the RMS are chemoattractants for neuroblasts and cdk5 is activated by NRGs through ErbB receptors, it is possible that ErbB4 is involved in the response to NRGs and the extension of the leading process. In addition, a repulsive signal generated by the Slit–Robo interaction recruits srGAP (Slit-Robo GTPase-activating protein) and leads to cdc42 inactivation, possibly inducing reorganization of the actin cytoskeleton to cause shortening or turning of the leading process (Wong et al., 2001). Thus, these extracellular stimuli could converge into modification of cytoskeletal regulators such as cdk5 and cdc42 and in turn alter the cytoskeletal reorganization for leading process extension.

14.3.2 Movement of the Centrosome

After neuroblasts form new adhesions at the leading process, the cell widens, expanding the cytosolic region anterior to the nucleus. Then, the centrosome, a cytosolic organelle that includes two centrioles and integrates the peripheral and perinuclear microtubule networks, moves forward into the expanded region. Several reports have shown that the positioning of the centrosome is regulated by highly conserved cell polarity-related factors, such as a complex containing Par3/6 (partition-defective 3/6), aPKC (atypical protein kinase C), and microtubules (Etienne-Manneville and Hall, 2003; Solecki et al., 2004).

The cell polarity-related factors PKCζ and GSK3β (glycogen synthase kinase-3β) play key roles in centrosome reorientation in the migration of various neuronal cells, including SVZ neuroblasts (Higginbotham et al., 2006). Specific inhibitors of PKCζ and GSK3β disrupt the centrosomal reorientation. Moreover, a GSK3β inhibitor causes impaired extension of the leading process in response to Slit activity, suggesting that the mechanisms for extension of the leading process are integrated with those for centrosome movement.

14.3.3 Nuclear Movement and Retraction of the Rear of the Cell

Microtubules extending from the centrosome form a perinuclear cage (PNC) around the nucleus. The PNC may contribute to N–C coupling, which appears to be important for nuclear movement. Two models have been proposed for nuclear movement. One involves microtubule–dynein interaction-mediated nuclear movement, in which dyneins on the PNC move toward the centrosome along bundles of microtubules in an ATP-dependent manner, which generates force for the nuclear movement (Tsai and Gleeson, 2005). The second involves F-actin-myosin (actomyosin)-mediated nuclear movement. In this case, activation of the Rho family small GTPase Rho leads to the phosphorylation of myosin II in the rear of the cell, and then an actomyosin interaction-derived force pushes the nucleus toward the centrosome in the expanded cytosolic region (Bellion et al., 2005).

In the SVZ neuroblasts, the major driving force for nuclear movement appears to be actomyosin-mediated signals (Schaar and McConnell, 2005). Nocodazole, a microtubule-depolymerizing agent, does not completely disrupt the nuclear movement or the retraction of the rear of the cell, whereas the myosin II-specific inhibitor Blebbistatin rapidly inhibits them, indicating that myosin II activation at the rear of the cell has a prominent role in retraction. Dcx (doublecortin), expressed in migrating neuroblasts, is also involved in this process. It is reported that Dcx is localized to the PNC with dynein and contributes to N–C coupling (Tanaka et al., 2004). Dcx mutant mice show a nuclear movement defect with no effect on centrosome positioning (Koizumi et al., 2006). These studies suggest that the microtubule-dynein and actomyosin systems might not be completely independent; instead, they probably share common mechanisms or molecular partners in this process.

The precise molecular mechanisms regulating SVZ neuroblast migration are still poorly understood. However, recent studies suggest that the SVZ neuroblasts share some mechanisms of migration with other cell types. Because neuroblasts migrating in chains use their neighbors as a scaffold, it is possible that homotypic cell adhesion-mediated signals play a critical role in their cytoskeletal regulation. Further studies are still needed to unravel the adult-specific cytoskeletal regulations of neuronal migration in the adult brain.

14.4 Neuronal Migration in Primates

Most of the studies on adult neuronal migration have been performed in rodents. However, similarities in the cytoarchitecture of the SVZ and RMS, such as the chain formation of neuroblasts and glial tubes, are observed in nonrodent mammals, including primates. The significance of these structures and why they are conserved are important issues that are still not entirely clear.

Macaque monkeys (*Macaca mulatta* and *Macaca fascicularis*), nonhuman Old World primates, are reported to possess an RMS structure similar to that of rodents (Kornack and Rakic, 2001; Pencea et al., 2001; Gil-Perotin et al., 2009). As in rodents, the macaque SVZ contains newly generated neuroblasts that display a bipolar shape, a typical morphology for active migration, and that express the immature neuronal markers PSA-NCAM and Tuj1 (neuron-specific class III β-tubulin). These neuroblasts aggregate and form chains ensheathed by GFAP (glial fibrillary acidic protein)-positive astrocytic processes in the core of the olfactory tract, indicating the existence of a typical RMS structure. In *Macaca fascicularis*, the chains of neuroblasts are not continuous in the RMS; rather, they seem to disperse into smaller clusters of three cells or fewer as they approach the OB (Gil-Perotin et al., 2009). Newly generated neuroblasts labeled with bromodeoxyuridine (BrdU) reach the OB by 97 days, and some of these cells express NeuN, a mature neuronal marker (Kornack and Rakic, 2001). These features of neuronal migration persist from an early postnatal stage (P2) to adulthood (8 years) (Pencea et al., 2001). Thus, new OB interneurons generated in the SVZ migrate in chains through the RMS to the OB in adult macaques.

The adult human SVZ also appears to contain neural stem cells that generate new neurons, but the mechanisms of their migration remain controversial. In contrast to macaque monkeys, PSA-NCAM- or Tuj1-positive neuroblasts were only rarely observed in two studies on the adult human SVZ–olfactory trigone connection, which is enriched in astrocytic processes and corresponds to the RMS, and they did not appear to form chains (Sanai et al., 2004; Quinones-Hinojosa et al., 2006). On the other hand, another group reported that this region contains migratory neuroblasts and has a typical RMS structure, similar to that of rodents and macaques (Curtis et al., 2007; Kam et al., 2009). Thus, neural stem cells reside in the SVZ, and their progeny may migrate in chains or as individual cells toward the OB in the adult human. Interestingly, in the adult common marmoset (a New World primate), in which the SVZ has an organization similar to that of human rather than macaque monkeys, neuroblasts do not exhibit typical chain migration in the RMS (Sawamoto et al., 2011).

The finding of a rodent-like RMS in nonhuman primates indicates that it is an important structure in many animal species. The difference in RMS structure between humans and nonhumans might reflect a species difference in the dependency on olfaction. Indeed, most rodents and nonhuman primates depend strongly on olfaction, and the size of their OB relative to their brain is greater than in humans. Humans depend much more on the visual system and therefore may not need to use chain migration. Nevertheless, at least in many nonhuman species, chain migration appears to have been evolutionally conserved as a mechanism for efficiently transporting a large number of neurons into the OB of the adult brain.

14.5 Neuronal Migration Toward Injured Brain Tissue

There are many reports that adult neurogenesis in the SVZ is affected by various brain injuries, including damage caused by ischemia and neurodegenerative diseases. Among many brain injury models, an ischemic stroke model induced by middle cerebral artery occlusion (MCAO) is widely used for studying the ectopic neurogenesis that occurs after injury. MCAO causes infarction in the striatum and adjacent parietal cortex (Fig. 14.4). After MCAO, neuronal precursors in the SVZ proliferate, and a significant number of neuroblasts then appear in the vicinity of the damaged area in the striatum; some differentiate into mature neurons (Arvidsson et al., 2002; Parent et al., 2002). To determine the origin of these neuroblasts, we used viral infection-mediated cell-lineage tracing methods, which showed that the neuroblasts were derived from GFAP-positive astrocytes in the SVZ that migrated toward the damaged striatum (Yamashita et al., 2006).

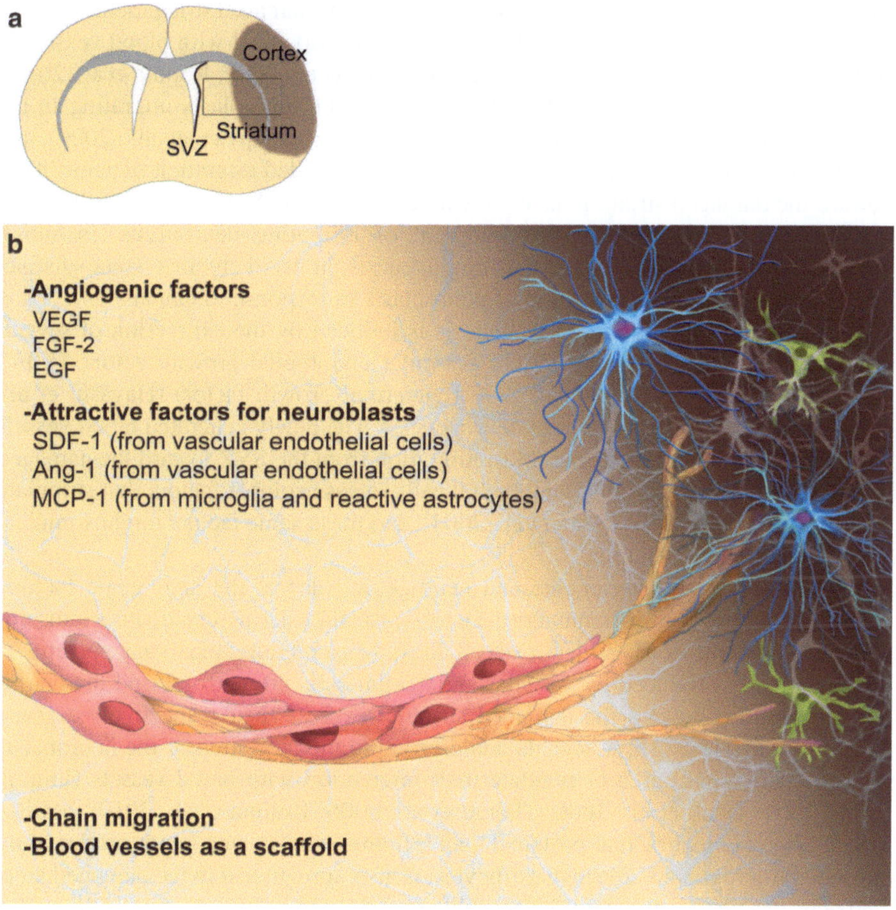

Fig. 14.4 Neuronal migration toward injured brain tissue. The ischemic stroke model induced by middle cerebral artery occlusion (MCAO) causes infarction in the striatum and adjacent parietal cortex (*brown*) (**a**). (**b**) After ischemic damage, inflammatory cells such as reactive astrocytes (*blue*) and microglia (*green*) accumulate in the vicinity of the damaged area and may release inflammatory cytokines. Angiogenic factors are also upregulated in the vicinity of the damaged area and subsequently induce angiogenesis. Subsequently, some of the newly generated neuroblasts (*red*) in the SVZ change direction and migrate toward the infarct area. They frequently form chains and are closely associated with blood vessels (*orange*) and astrocytic processes, similar to neuroblast migration in the RMS. Angiogenic factors released from vascular endothelial cells as well as inflammatory cytokines from reactive astrocytes and microglia may recruit neuroblasts to the area of infarction

This finding leads to questions about how neuroblasts migrate to the damaged area, and whether the mechanisms of neuronal migration differ in physiological versus pathological conditions. In fact, in rodents, some of the morphological features of neuroblasts migrating in the RMS are also seen in those migrating toward a damaged area (see Fig. 14.4). At first, they frequently form chain-like cell aggregates (Parent et al., 2002; Yamashita et al., 2006; Thored et al., 2007). These chains are largely oriented in the mediolateral direction toward the damaged striatum. In addition, most of the chain-forming neuroblasts are closely associated with blood vessels or astrocytic processes (Yamashita et al., 2006; Zhang et al., 2009; Kojima et al., 2010). Although the interaction between blood vessels and neuroblasts migrating in the RMS appears to be mediated by BDNF signaling (Snapyan et al., 2009), the molecular mechanisms regulating the blood vessel-guided migration of neuroblasts toward the damaged striatum have not yet been reported. As ischemic insult causes various changes in the microenvironment of migrating neuroblasts, including inflammation, reactive gliosis, and angiogenesis, it is likely that some distinct mechanisms induce the migration of neuroblasts in response to damage.

In the ischemic striatum, angiogenesis is induced by the expression of several angiogenic factors, including VEGF (vascular endothelial growth factor), FGF-2 (fibroblast growth factor-2), and EGF (epidermal growth factor) (Hayashi et al., 2003), and angiogenesis precedes neurogenesis. The newly generated blood vessels may act as the scaffold for neuroblast migration from the SVZ toward the damaged striatum. Moreover, vascular endothelial cells release soluble factors that stimulate neurogenesis in the SVZ (Shen et al., 2004), and these same factors might stimulate neurogenesis after ischemia.

In addition to the angiogenic and neurogenic roles of the angiogenic factors, some angiogenic and inflammatory factors probably act as chemoattractants for migrating neuroblasts. SDF-1 (stromal cell-derived factor-1) and Ang-1 (angiopoietin-1) are abundantly produced by vascular endothelial cells after stroke, and the neuroblasts migrating toward the damaged striatum express their respective receptors, CXCR-4 and Tie-2. These signals could promote the migration of the neuroblasts into the peri-infarct area or regulate their interaction with blood vessels (Imitola et al., 2004; Ohab et al., 2006; Thored et al., 2006; Kojima et al., 2010). MCP-1 (monocyte chemoattractant protein-1), an inflammation-related factor produced by activated microglia and reactive astrocytes, is also upregulated in the damaged area, and acts as a chemoattractant for neuroblasts that express its receptor, CCR-2 (Yan et al., 2007). Therefore, these angiogenic and inflammation-related factors induced by ischemia could guide neuroblasts toward the damaged area.

In contrast to the neuroblasts in the RMS, which is a permissive environment for migration in the adult brain, those in the damaged striatum need to move through developed neuronal circuitry to reach the damaged area. To migrate through the striatum to the damaged area after MCAO, neuroblasts utilize MMPs (Lee et al., 2006), most likely because they must remodel the ECM to enable their movement toward the damaged area.

Thus, neuroblasts migrating toward the damaged area form "chains" that are closely associated with blood vessels, similar to those in the RMS under physiological conditions. In addition, recent studies have revealed several unique mechanisms for neuroblast migration in the damaged brain. Further studies of the regeneration process in the adult damaged brain may reveal new insights into the interactions between migrating neuroblasts and their microenvironments and clarify the mechanisms underlying all types of neuronal migration in the adult brain.

14.6 Conclusion

In contrast to the embryonic brain, the adult brain contains highly packed neuronal circuitry and numerous glia, an environment that is traditionally viewed as nonpermissive for neuronal migration. However, neuroblasts generated in the SVZ migrate a long distance toward the OB at great speed, about four times faster than the radial glia-dependent neuronal migration that occurs in the developing cerebral cortex. Two major distinguishing features of neuronal migration in the RMS, neuroblast chain formation and glial tubes, may be important for this efficient migration in the adult brain. Furthermore, these features are conserved at least from rodents to nonhuman primates, although their existence in humans remains controversial. This mode of neuronal migration also seems to be used in the adult injured brain as a mechanism for supplying new neurons to the damaged area, at least in rodents.

At the single-cell level, the regulatory mechanisms of cytoskeletal reorganization seem common to the neuronal migration of embryonic and adult neuroblasts. However, the neuroblasts generated in the adult brain apparently use distinct directional cues, and require special scaffolds to move through the relatively disadvantageous environment. Thus, regulatory mechanisms that are common from the embryonic to the adult stage and those that are adult specific both contribute to neuronal migration in the adult brain. Further study should help clarify the detailed regulation as well as the functions of neuronal migration in the adult brain.

References

Alvarez-Buylla, A., and Garcia-Verdugo, J.M. (2002). Neurogenesis in adult subventricular zone. J Neurosci *22*, 629–634.

Andrade, N., Komnenovic, V., Blake, S.M., Jossin, Y., Howell, B., Goffinet, A., Schneider, W.J., and Nimpf, J. (2007). ApoER2/VLDL receptor and Dab1 in the rostral migratory stream function in postnatal neuronal migration independently of Reelin. Proc Natl Acad Sci USA *104*, 8508–8513.

Anton, E.S., Ghashghaei, H.T., Weber, J.L., McCann, C., Fischer, T.M., Cheung, I.D., Gassmann, M., Messing, A., Klein, R., Schwab, M.H., *et al.* (2004). Receptor tyrosine kinase ErbB4 modulates neuroblast migration and placement in the adult forebrain. Nat Neurosci *7*, 1319–1328.

Arvidsson, A., Collin, T., Kirik, D., Kokaia, Z., and Lindvall, O. (2002). Neuronal replacement from endogenous precursors in the adult brain after stroke. Nat Med *8*, 963–970.

Bellion, A., Baudoin, J.P., Alvarez, C., Bornens, M., and Metin, C. (2005). Nucleokinesis in tangentially migrating neurons comprises two alternating phases: forward migration of the Golgi/centrosome associated with centrosome splitting and myosin contraction at the rear. J Neurosci *25*, 5691–5699.

Belvindrah, R., Hankel, S., Walker, J., Patton, B.L., and Muller, U. (2007). Beta1 integrins control the formation of cell chains in the adult rostral migratory stream. J Neurosci *27*, 2704–2717.

Blake, S.M., Strasser, V., Andrade, N., Duit, S., Hofbauer, R., Schneider, W.J., and Nimpf, J. (2008). Thrombospondin-1 binds to ApoER2 and VLDL receptor and functions in postnatal neuronal migration. EMBO J *27*, 3069–3080.

Bolteus, A.J., and Bordey, A. (2004). GABA release and uptake regulate neuronal precursor migration in the postnatal subventricular zone. J Neurosci *24*, 7623–7631.

Bovetti, S., Bovolin, P., Perroteau, I., and Puche, A.C. (2007a). Subventricular zone-derived neuroblast migration to the olfactory bulb is modulated by matrix remodelling. Eur J Neurosci *25*, 2021–2033.

Bovetti, S., Hsieh, Y.C., Bovolin, P., Perroteau, I., Kazunori, T., and Puche, A.C. (2007b). Blood vessels form a scaffold for neuroblast migration in the adult olfactory bulb. J Neurosci *27*, 5976–5980.

Brose, K., Bland, K.S., Wang, K.H., Arnott, D., Henzel, W., Goodman, C.S., Tessier-Lavigne, M., and Kidd, T. (1999). Slit proteins bind Robo receptors and have an evolutionarily conserved role in repulsive axon guidance. Cell *96*, 795–806.

Chazal, G., Durbec, P., Jankovski, A., Rougon, G., and Cremer, H. (2000). Consequences of neural cell adhesion molecule deficiency on cell migration in the rostral migratory stream of the mouse. J Neurosci *20*, 1446–1457.

Chiaramello, S., Dalmasso, G., Bezin, L., Marcel, D., Jourdan, F., Peretto, P., Fasolo, A., and De Marchis, S. (2007). BDNF/ TrkB interaction regulates migration of SVZ precursor cells via PI3-K and MAP-K signalling pathways. Eur J Neurosci *26*, 1780–1790.

Conover, J.C., Doetsch, F., Garcia-Verdugo, J.M., Gale, N.W., Yancopoulos, G.D., and Alvarez-Buylla, A. (2000). Disruption of Eph/ephrin signaling affects migration and proliferation in the adult subventricular zone. Nat Neurosci *3*, 1091–1097.

Cremer, H., Lange, R., Christoph, A., Plomann, M., Vopper, G., Roes, J., Brown, R., Baldwin, S., Kraemer, P., Scheff, S., *et al.* (1994). Inactivation of the N-CAM gene in mice results in size reduction of the olfactory bulb and deficits in spatial learning. Nature *367*, 455–459.

Curtis, M.A., Kam, M., Nannmark, U., Anderson, M.F., Axell, M.Z., Wikkelso, C., Holtas, S., van Roon-Mom, W.M., Bjork-Eriksson, T., Nordborg, C., *et al.* (2007). Human neuroblasts migrate to the olfactory bulb via a lateral ventricular extension. Science *315*, 1243–1249.

Doetsch, F., and Alvarez-Buylla, A. (1996). Network of tangential pathways for neuronal migration in adult mammalian brain. Proc Natl Acad Sci USA *93*, 14895–14900.

Emsley, J.G., and Hagg, T. (2003). alpha6beta1 integrin directs migration of neuronal precursors in adult mouse forebrain. Exp Neurol *183*, 273–285.

Etienne-Manneville, S., and Hall, A. (2003). Cdc42 regulates GSK-3beta and adenomatous polyposis coli to control cell polarity. Nature *421*, 753–756.

Garzotto, D., Giacobini, P., Crepaldi, T., Fasolo, A., and De Marchis, S. (2008). Hepatocyte growth factor regulates migration of olfactory interneuron precursors in the rostral migratory stream through Met-Grb2 coupling. J Neurosci *28*, 5901–5909.

Gascon E, Vutskits L, Kiss JZ (2010) The role of PSA-NCAM in adult neurogenesis. Adv Exp Med Biol *663*, 127–136.

Gil-Perotin, S., Duran-Moreno, M., Belzunegui, S., Luquin, M.R., and Garcia-Verdugo, J.M. (2009). Ultrastructure of the subventricular zone in *Macaca fascicularis* and evidence of a mouse-like migratory stream. J Comp Neurol *514*, 533–554.

Guan, K.L., and Rao, Y. (2003). Signalling mechanisms mediating neuronal responses to guidance cues. Nat Rev Neurosci *4*, 941–956.

Hack, I., Bancila, M., Loulier, K., Carroll, P., and Cremer, H. (2002). Reelin is a detachment signal in tangential chain-migration during postnatal neurogenesis. Nat Neurosci *5*, 939–945.

Hayashi, T., Noshita, N., Sugawara, T., and Chan, P.H. (2003). Temporal profile of angiogenesis and expression of related genes in the brain after ischemia. J Cereb Blood Flow Metab *23*, 166–180.

Higginbotham, H., Tanaka, T., Brinkman, B.C., and Gleeson, J.G. (2006). GSK3beta and PKCzeta function in centrosome localization and process stabilization during Slit-mediated neuronal repolarization. Mol Cell Neurosci *32*, 118–132.

Hirota, Y., Ohshima, T., Kaneko, N., Ikeda, M., Iwasato, T., Kulkarni, A.B., Mikoshiba, K., Okano, H., and Sawamoto, K. (2007). Cyclin-dependent kinase 5 is required for control of neuroblast migration in the postnatal subventricular zone. J Neurosci *27*, 12829–12838.

Hu, H. (1999). Chemorepulsion of neuronal migration by Slit2 in the developing mammalian forebrain. Neuron *23*, 703–711.

Hu, H., and Rutishauser, U. (1996). A septum-derived chemorepulsive factor for migrating olfactory interneuron precursors. Neuron *16*, 933–940.

Imitola, J., Raddassi, K., Park, K.I., Mueller, F.J., Nieto, M., Teng, Y.D., Frenkel, D., Li, J., Sidman, R.L., Walsh, C.A., *et al.* (2004). Directed migration of neural stem cells to sites of CNS injury by the stromal cell-derived factor 1alpha/CXC chemokine receptor 4 pathway. Proc Natl Acad Sci USA *101*, 18117–18122.

Kam, M., Curtis, M.A., McGlashan, S.R., Connor, B., Nannmark, U., and Faull, R.L. (2009). The cellular composition and morphological organization of the rostral migratory stream in the adult human brain. J Chem Neuroanat *37*, 196–205.

Kaneko, N., Marín, O., Koike, M., Hirota, Y., Uchiyama, Y., Wu, J.Y., Lu, Q., Tessier-Lavigne, M., Alvarez-Buylla, A., Okano, H., *et al.* (2010). New neurons clear the path of astrocytic processes for their rapid migration in the adult brain. Neuron *67*, 213–223.

Kirschenbaum, B., Doetsch, F., Lois, C., and Alvarez-Buylla, A. (1999). Adult subventricular zone neuronal precursors continue to proliferate and migrate in the absence of the olfactory bulb. J Neurosci *19*, 2171–2180.

Koizumi, H., Higginbotham, H., Poon, T., Tanaka, T., Brinkman, B.C., and Gleeson, J.G. (2006). Doublecortin maintains bipolar shape and nuclear translocation during migration in the adult forebrain. Nat Neurosci *9*, 779–786.

Kojima, T., Hirota, Y., Ema, M., Takahashi, S., Miyoshi, I., Okano, H., and Sawamoto, K. (2010). Subventricular zone-derived neural progenitor cells migrate along a blood vessel scaffold toward the post-stroke striatum. Stem Cells *28*, 545–554.

Kornack, D.R., and Rakic, P. (2001). The generation, migration, and differentiation of olfactory neurons in the adult primate brain. Proc Natl Acad Sci USA *98*, 4752–4757.

Lee, S.R., Kim, H.Y., Rogowska, J., Zhao, B.Q., Bhide, P., Parent, J.M., and Lo, E.H. (2006). Involvement of matrix metalloproteinase in neuroblast cell migration from the subventricular zone after stroke. J Neurosci *26*, 3491–3495.

Li, H.S., Chen, J.H., Wu, W., Fagaly, T., Zhou, L., Yuan, W., Dupuis, S., Jiang, Z.H., Nash, W., Gick, C., *et al.* (1999). Vertebrate slit, a secreted ligand for the transmembrane protein roundabout, is a repellent for olfactory bulb axons. Cell *96*, 807–818.

Liu, G., and Rao, Y. (2003). Neuronal migration from the forebrain to the olfactory bulb requires a new attractant persistent in the olfactory bulb. J Neurosci *23*, 6651–6659.

Lois, C., and Alvarez-Buylla, A. (1994). Long-distance neuronal migration in the adult mammalian brain. Science *264*, 1145–1148.

Lois, C., Garcia-Verdugo, J.M., and Alvarez-Buylla, A. (1996). Chain migration of neuronal precursors. Science *271*, 978–981.

Luskin, M.B. (1993). Restricted proliferation and migration of postnatally generated neurons derived from the forebrain subventricular zone. Neuron *11*, 173–189.

Mason, H.A., Ito, S., and Corfas, G. (2001). Extracellular signals that regulate the tangential migration of olfactory bulb neuronal precursors: inducers, inhibitors, and repellents. J Neurosci *21*, 7654–7663.

Merkle, F.T., Mirzadeh, Z., and Alvarez-Buylla, A. (2007). Mosaic organization of neural stem cells in the adult brain. Science *317*, 381–384.

Murase, S., and Horwitz, A.F. (2002). Deleted in colorectal carcinoma and differentially expressed integrins mediate the directional migration of neural precursors in the rostral migratory stream. J Neurosci *22*, 3568–3579.

Murase, S., Cho, C., White, J.M., and Horwitz, A.F. (2008). ADAM2 promotes migration of neuroblasts in the rostral migratory stream to the olfactory bulb. Eur J Neurosci *27*, 1585–1595.

Ng, K.L., Li, J.D., Cheng, M.Y., Leslie, F.M., Lee, A.G., and Zhou, Q.Y. (2005). Dependence of olfactory bulb neurogenesis on prokineticin 2 signaling. Science *308*, 1923–1927.

Nguyen-Ba-Charvet, K.T., Picard-Riera, N., Tessier-Lavigne, M., Baron-Van Evercooren, A., Sotelo, C., and Chedotal, A. (2004). Multiple roles for slits in the control of cell migration in the rostral migratory stream. J Neurosci *24*, 1497–1506.

Ohab, J.J., Fleming, S., Blesch, A., and Carmichael, S.T. (2006). A neurovascular niche for neurogenesis after stroke. J Neurosci *26*, 13007–13016.

Okuyama-Yamamoto, A., Yamamoto, T., Miki, A., and Terashima, T. (2005). Changes in reelin expression in the mouse olfactory bulb after chemical lesion to the olfactory epithelium. Eur J Neurosci *21*, 2586–2592.

Ono, K., Tomasiewicz, H., Magnuson, T., and Rutishauser, U. (1994). N-CAM mutation inhibits tangential neuronal migration and is phenocopied by enzymatic removal of polysialic acid. Neuron *13*, 595–609.

Paratcha, G., Ibanez, C.F., and Ledda, F. (2006). GDNF is a chemoattractant factor for neuronal precursor cells in the rostral migratory stream. Mol Cell Neurosci *31*, 505–514.

Parent, J.M., Vexler, Z.S., Gong, C., Derugin, N., and Ferriero, D.M. (2002). Rat forebrain neurogenesis and striatal neuron replacement after focal stroke. Ann Neurol *52*, 802–813.

Pencea, V., Bingaman, K.D., Freedman, L.J., and Luskin, M.B. (2001). Neurogenesis in the subventricular zone and rostral migratory stream of the neonatal and adult primate forebrain. Exp Neurol *172*, 1–16.

Puverel, S., Nakatani, H., Parras, C., and Soussi-Yanicostas, N. (2009). Prokineticin receptor 2 expression identifies migrating neuroblasts and their subventricular zone transient-amplifying progenitors in adult mice. J Comp Neurol *512*, 232–242.

Quinones-Hinojosa, A., Sanai, N., Soriano-Navarro, M., Gonzalez-Perez, O., Mirzadeh, Z., Gil-Perotin, S., Romero-Rodriguez, R., Berger, M.S., Garcia-Verdugo, J.M., and Alvarez-Buylla, A. (2006). Cellular composition and cytoarchitecture of the adult human subventricular zone: a niche of neural stem cells. J Comp Neurol *494*, 415–434.

Saghatelyan, A., de Chevigny, A., Schachner, M., and Lledo, P.M. (2004). Tenascin-R mediates activity-dependent recruitment of neuroblasts in the adult mouse forebrain. Nat Neurosci *7*, 347–356.

Sanai, N., Tramontin, A.D., Quinones-Hinojosa, A., Barbaro, N.M., Gupta, N., Kunwar, S., Lawton, M.T., McDermott, M.W., Parsa, A.T., Manuel-Garcia Verdugo, J., *et al.* (2004). Unique astrocyte ribbon in adult human brain contains neural stem cells but lacks chain migration. Nature *427*, 740–744.

Sawamoto, K., Hirota, Y., Alfaro-Cervello, C., Soriano-Navarro, M., He, X., Hayakawa-Yano, Y., Yamada, M., Hikishima, K., Tabata, H., Iwanami, A., *et al.* (2011). Cellular composition and organization of the subventricular zone and rostral migratory stream in the adult and neonatal common marmoset brain. J Comp Neurol *519*, 690–713.

Sawamoto, K., Wichterle, H., Gonzalez-Perez, O., Cholfin, J.A., Yamada, M., Spassky, N., Murcia, N.S., Garcia-Verdugo, J.M., Marin, O., Rubenstein, J.L., *et al.* (2006). New neurons follow the flow of cerebrospinal fluid in the adult brain. Science *311*, 629–632.

Schaar, B.T., and McConnell, S.K. (2005). Cytoskeletal coordination during neuronal migration. Proc Natl Acad Sci USA *102*, 13652–13657.

Shen, Q., Goderie, S.K., Jin, L., Karanth, N., Sun, Y., Abramova, N., Vincent, P., Pumiglia, K., and Temple, S. (2004). Endothelial cells stimulate self-renewal and expand neurogenesis of neural stem cells. Science *304*, 1338–1340.

Snapyan, M., Lemasson, M., Brill, M.S., Blais, M., Massouh, M., Ninkovic, J., Gravel, C., Berthod, F., Gotz, M., Barker, P.A., *et al.* (2009). Vasculature guides migrating neuronal precursors in the adult mammalian forebrain via brain-derived neurotrophic factor signaling. J Neurosci *29*, 4172–4188.

Solecki, D.J., Model, L., Gaetz, J., Kapoor, T.M., and Hatten, M.E. (2004). Par6alpha signaling controls glial-guided neuronal migration. Nat Neurosci *7*, 1195–1203.

Tanaka, T., Serneo, F.F., Higgins, C., Gambello, M.J., Wynshaw-Boris, A., and Gleeson, J.G. (2004). Lis1 and doublecortin function with dynein to mediate coupling of the nucleus to the centrosome in neuronal migration. J Cell Biol *165*, 709–721.

Thored, P., Arvidsson, A., Cacci, E., Ahlenius, H., Kallur, T., Darsalia, V., Ekdahl, C.T., Kokaia, Z., and Lindvall, O. (2006). Persistent production of neurons from adult brain stem cells during recovery after stroke. Stem Cells *24*, 739–747.

Thored, P., Wood, J., Arvidsson, A., Cammenga, J., Kokaia, Z., and Lindvall, O. (2007). Long-term neuroblast migration along blood vessels in an area with transient angiogenesis and increased vascularization after stroke. Stroke *38*, 3032–3039.

Tsai, L.H., and Gleeson, J.G. (2005). Nucleokinesis in neuronal migration. Neuron *46*, 383–388.

Umeshima, H., Hirano, T., and Kengaku, M. (2007). Microtubule-based nuclear movement occurs independently of centrosome positioning in migrating neurons. Proc Natl Acad Sci USA *104*, 16182–16187.

Wichterle, H., Garcia-Verdugo, J.M., and Alvarez-Buylla, A. (1997). Direct evidence for homotypic, glia-independent neuronal migration. Neuron *18*, 779–791.

Wong, K., Ren, X.R., Huang, Y.Z., Xie, Y., Liu, G., Saito, H., Tang, H., Wen, L., Brady-Kalnay, S.M., Mei, L., *et al.* (2001). Signal transduction in neuronal migration: roles of GTPase activating proteins and the small GTPase Cdc42 in the Slit-Robo pathway. Cell *107*, 209–221.

Wu, W., Wong, K., Chen, J., Jiang, Z., Dupuis, S., Wu, J.Y., and Rao, Y. (1999). Directional guidance of neuronal migration in the olfactory system by the protein Slit. Nature *400*, 331–336.

Yamashita, T., Ninomiya, M., Hernandez Acosta, P., Garcia-Verdugo, J.M., Sunabori, T., Sakaguchi, M., Adachi, K., Kojima, T., Hirota, Y., Kawase, T., *et al.* (2006). Subventricular zone-derived neuroblasts migrate and differentiate into mature neurons in the post-stroke adult striatum. J Neurosci *26*, 6627–6636.

Yan, Y.P., Sailor, K.A., Lang, B.T., Park, S.W., Vemuganti, R., and Dempsey, R.J. (2007). Monocyte chemoattractant protein-1 plays a critical role in neuroblast migration after focal cerebral ischemia. J Cereb Blood Flow Metab *27*, 1213–1224.

Zhang, R.L., Chopp, M., Gregg, S.R., Toh, Y., Roberts, C., Letourneau, Y., Buller, B., Jia, L., S, P.N.D., and Zhang, Z.G. (2009). Patterns and dynamics of subventricular zone neuroblast migration in the ischemic striatum of the adult mouse. J Cereb Blood Flow Metab *29*, 1240–1250.

Chapter 15
Development and Survival of Adult-Born Olfactory Neurons

Masahiro Yamaguchi

Abstract In rodents, the olfactory bulb is located at the rostral end of the brain, where it forms part of the telencephalon. The olfactory bulb first develops in the midembryonic period as an evagination from the rostral tip of the telencephalon (Hinds, 1968b). Mitral and tufted cells, the bulbar excitatory projection neurons, are produced during the mid- to late embryonic period (Hinds, 1968a; Bayer, 1983), whereas granule cells and periglomerular cells, the bulbar local interneurons, are generated extensively during the late embryonic and early neonatal periods, with production continuing in adulthood. This adult neurogenesis of bulbar interneurons may be responsible for the high plasticity that characterizes information processing in the olfactory bulb. Animals are required to adapt to environments with various odors in daily life. Bulbar interneurons generated in adulthood may increase the plasticity of olfaction such that animals can choose appropriate behavioral responses in circumstances with changing odors. Because not all new neurons are integrated into preexisting circuits, adult neurogenesis is in one sense wasteful; in the olfactory bulb also, only half of new interneurons are integrated into the bulbar circuitry while the other half are eliminated (Petreanu and Alvarez-Buylla, 2002; Winner et al., 2002). Adult neurogenesis therefore resembles embryonic development, in that neurons are initially prepared in excess, many of which are subsequently eliminated. This selection process during development is crucial to ensuring that the neuronal circuitry has been appropriately tuned to provide proper information processing (Buss et al., 2006). A similar selection process occurs in the olfactory bulb of adult mammals, where it allows the bulbar circuit to select the appropriate type and number of neurons from among excess pools of new neurons in accordance with the particular requirements of the odor situation in which the animal finds itself. The mechanism by which adult-born neurons are selected, in other words, their fate of survival or death, is therefore key to understanding how adult-born bulbar interneurons contribute to the functional plasticity of the bulbar circuitry.

M. Yamaguchi (✉)
Department of Physiology, Graduate School of Medicine,
The University of Tokyo, Tokyo 113-0033, Japan
e-mail: yamaguti@m.u-tokyo.ac.jp

T. Seki et al. (eds.), *Neurogenesis in the Adult Brain I: Neurobiology*,
DOI 10.1007/978-4-431-53933-9_15, © Springer 2011

In this section, we first address how the survival or death of adult-born neurons is controlled. We then compare the development and survival of adult- and neonate-born neurons to investigate how the integrity of the bulbar circuit is maintained and remodeled by adult-born neurons. Although two major types of interneurons are found, periglomerular cells and granule cells, the major population of adult-born interneurons are granule cells. This chapter therefore focuses mainly on the properties of granule cells.

15.1 Part I: Survival and Death of Adult-Born Interneurons

The olfactory bulb is considered a unique region in the brain, in that inhibitory interneurons outnumber excitatory projection neurons. The total number of granule cells in the adult rat olfactory bulb is calculated to be around ten million, which is two orders of magnitude larger than that of mitral/tufted cells (Kaplan et al., 1985). The number of adult-born bulbar interneurons is also large. In rodents, at least several tens of thousands of neurons enter the olfactory bulb each day (Alvarez-Buylla et al., 2001; Winner et al., 2002; Lledo et al., 2006), corresponding to roughly 1% of the total granule cell number. Thus, a simple calculation suggests that adult-born interneurons will outnumber preexisting interneurons within 100 days. This numerical profile suggests a critical role for interneurons, including adult-born interneurons, in the processing of odor information in the olfactory bulb.

Because the selection process through which adult-born interneurons pass necessarily involves extensive neuronal elimination (Buss et al., 2006), the number of apoptotic interneurons in the adult olfactory system is accordingly large. Apoptotic cells are found in the subventricular zone, rostral migratory stream, and olfactory bulb (Biebl et al., 2000; Mandairon et al., 2003), indicating that selection occurs throughout the neurogenic routes. When the absolute number of apoptotic cells was compared among different neurogenic areas, close to 80% were found in the olfactory bulb (Biebl et al., 2000), indicating that the selection of adult-born interneurons occurs mostly after they enter the olfactory bulb. Among newly arrived bulbar interneurons, nearly half are integrated into the bulbar circuitry and survive over a period of months whereas the rest are eliminated within several weeks (Petreanu and Alvarez-Buylla, 2002; Winner et al., 2002; Yamaguchi and Mori, 2005). Regulation of the survival and death of adult-born interneurons is crucial to appropriate structural and functional improvements in the bulbar circuitry. The following sections introduce several examples of how the survival and death of adult-born interneurons in the olfactory bulb are regulated.

15.1.1 Olfactory Sensory Experience

Olfaction plays a key role in many areas of animal behavior, including feeding, mating, avoidance of predators, and territorial marking (Doty, 1986). Animals experience various odors in daily life. Considering the importance of adult-born interneurons in adapting to the changing odorous environment, their survival and death is likely under the control of olfactory sensory experience, and many studies have shown that deprivation of olfactory sensory input increases the number of apoptotic cells in the olfactory bulb (Corotto et al., 1994; Najbauer and Leon, 1995; Fiske and Brunjes, 2001a,b). It has been postulated that apoptotic cells in the olfactory bulb represent the cell death of adult-born interneurons and that olfactory sensory deprivation decreases the survival of adult-born interneurons. In this regard, survival of adult-born granule cells was decreased in anosmic mice lacking a cyclic nucleotide-gated channel in olfactory sensory neurons (Petreanu and Alvarez-Buylla, 2002) and was increased in mice repeatedly exposed to novel odors (odor-enriched environment) (Rochefort et al., 2002). These findings clearly show that olfactory sensory experience promotes the survival of adult-born interneurons whereas sensory deprivation decreases it. Modulation of olfactory sensory experience did not significantly influence the proliferation of progenitors in the subventricular zone (Petreanu and Alvarez-Buylla, 2002; Rochefort et al., 2002). Thus, olfactory sensory experience regulates the final number of adult-born interneurons integrated into the bulbar circuitry mainly by controlling the survival and death of these cells within the olfactory bulb.

In general, sensory experience-dependent plastic change in the central nervous system is subject to the influence of time windows. One well-known example is the plasticity of ocular dominance (Hensch, 2005): deprivation of visual input from one eye shifts the response property of binocular zone neurons in the visual cortex preferentially toward the nondeprived eye input. This ocular dominance plasticity occurs during a specific period after birth, called the critical period. The dependence of the fate choice of adult-born bulbar interneurons between survival and death on sensory experience raises the possibility that it may have a critical time window, and that this window may depend on the age of the individual adult-born neurons themselves rather than on the age of the animals. In anosmic mice, the number of adult-born granule cells at day 14 after generation was comparable to that of wild-type mice, whereas that at day 45 was significantly decreased, suggesting that the influence of sensory deprivation on survival becomes conspicuous at later than 14 days after generation (Petreanu and Alvarez-Buylla, 2002). When wild-type mice were deprived of olfactory sensory input at various periods by naris cauterization, deprivation at 14–28 days after generation remarkably reduced the survival of new interneurons, whereas that before or after this period showed no significant effect (Yamaguchi and Mori, 2005). Many of the apoptotic granule cells were aged 14–28 days (Fig. 15.1). These observations indicate that the period of 14–28 days after generation is the critical period for bulbar interneurons when survival and death are strongly regulated by olfactory sensory experience.

What mechanisms determine the survival and death of adult-born interneurons during the critical period? Morphological analysis of adult-born interneurons by retroviral labeling showed the extensive development of synaptic structures on the dendrites of new granule cells on days 14–28 after generation (Petreanu and Alvarez-Buylla, 2002). Functional analysis by patch-clamp recording showed that new granule cells begin to receive glutamatergic synaptic input at around day 14, and

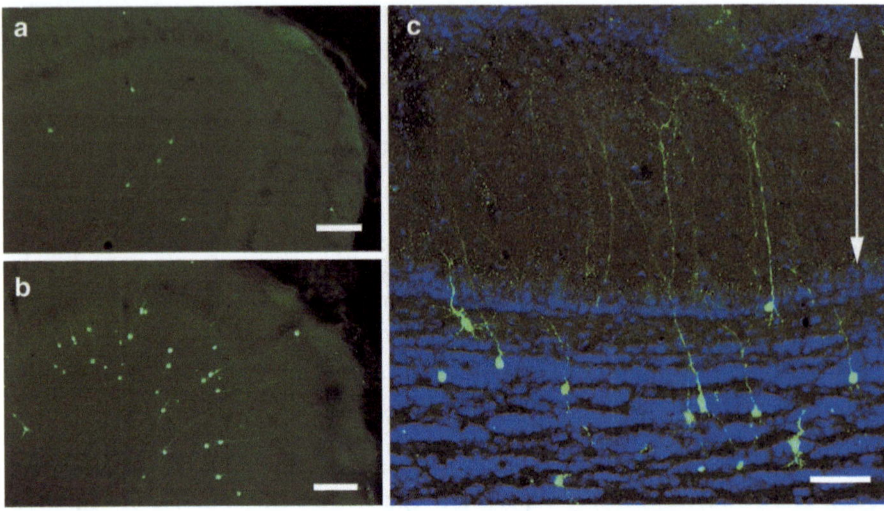

Fig. 15.1 Apoptotic granule cells in the olfactory bulb. Apoptotic granule cells can be seen by immunostaining with anti-activated caspase-3 antibody (*green*). (**a**) Apoptotic granule cells in the control olfactory bulb. (**b**) Increased number of apoptotic granule cells in the sensory-deprived olfactory bulb. (**c**) Many apoptotic granule cells in the sensory-deprived olfactory bulb have apical dendrites that elongate into external plexiform layer (*arrow*). *Blue*, DAPI (4′,6″-diamino-2-phenylindole·2HCl) staining. *Bars* (**a**,**b**) 100 μm; (**c**) 20 μm. (Modified from Yamaguchi and Mori 2005. Copyright 2005, National Academy of Sciences, USA.)

that these connections continue to develop before plateauing at a similar degree of development to that in preexisting mature granule cells by around day 28 (Carleton et al., 2003). The critical period for the survival and death decision of adult-born granule cells roughly matches the period when they develop glutamatergic synaptic connections with preexisting neurons, and glutamatergic input onto new bulbar interneurons is thus a plausible candidate for the control of the fate of these cells during the critical period (Fig. 15.2). Mature granule cells in the olfactory bulb receive glutamatergic input from three origins (Shepherd et al., 2004): in the external plexiform layer, they make dendrodendritic reciprocal synapses with secondary dendrites of mitral/tufted cells and receive glutamatergic input from these structures; in the granule cell layer, they make axodendritic glutamatergic synapses with centrifugal axons of olfactory cortical neurons; and granule cell dendrites in the granule cell layer also make axodendritic glutamatergic synapses with axon collaterals of mitral/tufted cells. The time course of synapse formation from these different origins onto adult-born interneurons was recently clarified (Whitman and Greer, 2007; Kelsch et al., 2008). Adult-born granule cells first form synapses in the granule cell layer at around day 14, whereas synapse formation in the external plexiform layer is delayed, becoming apparent around day 21. By day 28, all synaptic structures are indistinguishable from those of preexisting mature granule cells. The development of glutamatergic synaptic input onto adult-born granule cells is thus both spatially and temporally regulated, and development occurs mostly during the period critical

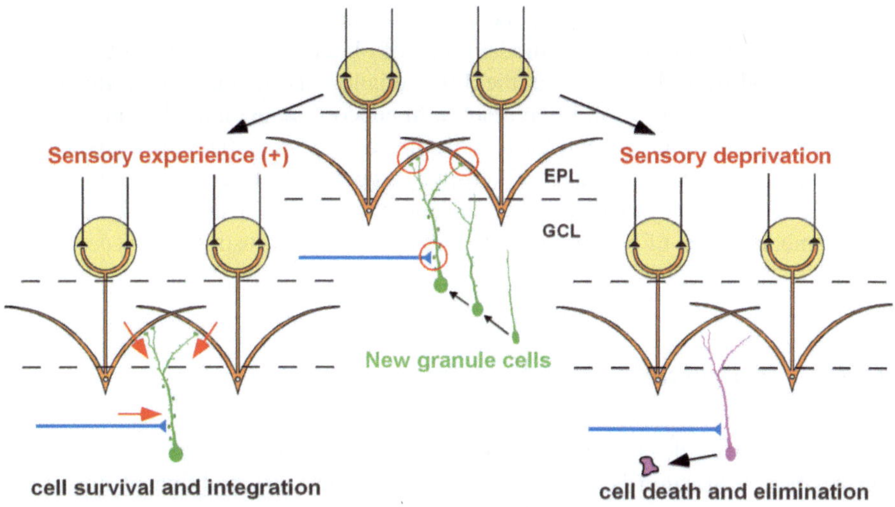

Fig. 15.2 Fate decision of adult-born granule cells between survival and death. Survival and death of adult-born bulbar interneurons are regulated by olfactory sensory experience. *Middle:* The critical period for the sensory experience-dependent fate decision corresponds to the period of synaptogenesis. Adult-born granule cells around the critical period form synapses in the external plexiform layer (*EPL*) with secondary dendrites of mitral/tufted cells and in the granule cell layer (*GCL*) with several origins including centrifugal axons of olfactory cortical neurons and axon collaterals of mitral/tufted cells. *Left:* Synaptic input through these structures may support the survival of adult-born granule cells during the critical period. Then, the new granule cells are integrated in the bulbar circuitry. *Right:* Deprivation of synaptic input enhances cell death to the adult-born granule cells during the critical period. Those cells are eliminated from the bulbar circuitry

to the sensory experience-dependent survival and death decision. An understanding of the individual and combined roles of these glutamatergic synaptic inputs from different origins may provide important clues to determining the fate decision mechanism of adult-born interneurons.

Overall, these findings suggest that olfactory sensory experience promotes the survival of adult-born interneurons. This concept does not always hold true, however: although the survival of new interneurons in adult mice increased when they were subject to an olfactory discrimination task on day 21–28 after generation, supporting the survival-promoting effect of olfactory sensory experience (Alonso et al., 2006), survival conversely decreased when a similar but not identical task was conducted on day 30–40 after generation (Mandairon et al., 2006a). To preclude differences in experimental conditions other than the period of task application, the same odor task was applied to adult mice at various periods after bromodeoxyuridine (BrdU) labeling of adult-born interneurons (Mouret et al., 2008). Testing during days 11–30 after cell generation increased their survival whereas that during days 30–38 decreased it; testing before or after these periods showed no significant influence. These findings strengthen the view that the survival of adult-born interneurons is subject to a period of sensitivity to olfactory sensory experience, but further show that survival can be conversely decreased when exposure occurs outside the so-called critical period. This bimodal effect of olfactory sensory experience might be understood from differences in the differentiation and maturation of glutamatergic input on interneurons of different ages. In neocortical pyramidal neurons, deprivation of sensory input prevents the loss of spines, suggesting that sensory experience plays a positive role in the loss of synaptic structures (Zuo et al., 2005). Another example of the sensory experience-dependent refinement of neuronal circuitry might be the enhancement of bulbar neuronal loss by olfactory sensory experience.

15.1.2 Neuromodulatory Systems

The sense of smell does not depend solely on peripheral odor input from the external environment. Olfaction is also influenced by the state or condition of the animal itself, such as its attention to and motivation toward odor detection (Doty, 1986). These states are at least partly mediated by neuromodulatory systems. The olfactory bulb is targeted by several subcortical neuromodulatory systems (Shipley and Ennis, 1996), namely, cholinergic input from the horizontal limb of the diagonal band, noradrenergic input from the locus coeruleus, and serotonergic input from the raphe nuclei. Because most of the neuromodulatory fibers terminate on granule and periglomerular cells (Shepherd et al., 2004), these neuromodulatory inputs are also candidates as direct regulators of the survival and death of adult-born interneurons.

Cholinergic fibers innervate all layers of the olfactory bulb, from the granule cell layer to the glomerular layer (Kasa et al., 1995). Adult-born interneurons also receive these neuromodulatory inputs during their maturation. Acetylcholine receptors are expressed in adult-born interneurons, and cholinergic axon terminals make contact with these interneurons (Kaneko et al., 2006; Whitman and Greer, 2007). Modulation of cholinergic input has been shown to influence the survival and death of adult-born interneurons in a variety of ways. For example, ablation of cholinergic neurons of the diagonal band by toxin infusion decreased the survival of adult-born bulbar interneurons (Cooper-Kuhn et al., 2004), whereas enhancement of the cholinergic system by systemic administration of an acetylcholinesterase inhibitor increased the survival of adult-born interneurons (Kaneko et al., 2006). On the other hand, selective ablation of nicotinic acetylcholine beta-2 receptor enhanced granule cell survival (Mechawar et al., 2004). These various effects of the cholinergic system may depend on the expression of different types of acetylcholine receptors on adult-born interneurons or may reflect the indirect as well as direct influence of cholinergic systems on the survival of bulbar interneurons.

With regard to the adrenergic and noradrenergic systems in the olfactory bulb, noradrenergic fibers from the locus coeruleus are distributed predominantly in the granule cell layer (McLean et al., 1989; McLean and Shipley, 1991), whereas adrenergic receptors (beta-1 and beta-2) are expressed widely in all layers (Woo and Leon, 1995). Application of a noradrenergic agonist to cultured bulbar neurons activated their Erk/MAPK pathway, which was assumed to enhance the survival of young neurons (Miwa and Storm, 2005), while that of a noradrenergic antagonist blocked the odor enrichment-induced enhancement of survival of adult-born bulbar interneurons (Veyrac et al., 2009). The importance of noradrenergic signaling in odor memory formation was first documented in neonatal rats (Sullivan et al., 1989). Given the extensive development of noradrenergic connections during the postnatal period (McLean and Shipley, 1991), noradrenergic input may regulate both olfactory memory and neuronal survival in the adult animals. In fact, application of a noradrenergic receptor antagonist attenuated the formation of short-term olfactory memory in adult mice (Veyrac et al., 2009).

The contribution of neuromodulation to olfactory memory has been well documented in the mating behavior of rodents. Pregnant female mice experience abortion when exposed to pheromonal signals from an unfamiliar male but not when exposed to those of the mated male (Brennan et al., 1990). This type of olfactory memory (blockade of abortive response to a particular pheromonal signal) requires the combination of peripheral pheromonal inputs of mating males and centrifugal noradrenergic inputs to the accessory olfactory bulb during mating behavior (Rosser and Keverne, 1985). This example indicates that the neuromodulatory system works in concert with peripheral sensory input, possibly under the mediation of glutamatergic synaptic input. From this, we assume that neuromodulatory inputs on adult-born bulbar interneurons regulate their survival in combination with their glutamatergic inputs, which convey odor information about the external environment. Such integration of peripheral glutamatergic and centrifugal neuromodulatory inputs in adult-born interneurons might be the key regulatory mechanism of their survival and death. Application of an olfactory discrimination task more effectively promotes the survival of adult-born interneurons than the simple delivery of novel odors (Alonso et al., 2006). Neuromodulatory input might underlie the importance of behavioral context to the olfactory experience-dependent survival and death of adult-born interneurons.

15.1.3 Long-Term Survival of Adult-Born Interneurons

After passing through the critical survival or death decision period, adult-born interneurons become relatively stable and tend to exist for an extended period in the preexisting bulbar circuitry. Regulation of this long-term survival is not well understood, however. Some observations show that new interneurons selected for survival during the sensitive period continue to survive for more than a year, for example, while others show a gradual decrease in number, with half eliminated within a year (Winner et al., 2002; Petreanu and Alvarez-Buylla, 2002). Survival may differ between species (rats and mice), as well as by housing conditions and the animal age at cell labeling. Of note, the survival-promoting effect of olfactory sensory experience is not necessarily long lasting. Olfactory sensory experience for 3 weeks after BrdU labeling enhanced the survival of adult-born granule cells, but withdrawal of the experience after that period reduced survival to a level similar to that of mice without the experience (Rochefort and Lledo, 2005). This result indicates that the survival of adult-born interneurons does not become refractory to odor stimulation after the critical period and suggests that their long-term survival may also be regulated by sensory experience after the critical period.

15.2 Part II: Preexisting and New Interneurons

15.2.1 Overall Contribution of Adult-Born Interneurons in the Bulbar Circuitry

Adult-born interneurons integrate into the preexisting bulbar circuitry, which contains projection neurons generated in the embryonic period and interneurons generated during the embryonic and younger postnatal periods (Hinds, 1968a,b; Bayer, 1983; Lemasson et al., 2005). Although the number of new bulbar neurons declines with age, neurogenesis continues even in 2-year-old mice (Enwere et al., 2004). To what extent do adult-born interneurons contribute to the overall constitution of the bulbar circuitry in the adult? Volumetric studies in rodents suggest that the olfactory bulb continues to grow during adulthood. The volume of the granule cell layer of the mouse olfactory bulb increases about 1.2-fold from age 6 to 24 months (Mirich et al., 2002); the absolute number of granule cells in the rat olfactory bulb increases twofold from age 3 to 24 months (Kaplan et al., 1985). These observations suggest that adult-born neurons are continually added to the bulbar circuit and contribute to the increase in granule cell number and olfactory bulb size. Because pulse labeling of adult-born interneurons with BrdU analogues or retroviral infection identifies only a small fraction of adult-born interneurons, however, measuring the proportion of adult-born among all interneurons in the adult bulbar circuitry has been difficult. To overcome this, a recent genetic technique has allowed the labeling of a much larger proportion of adult-born interneurons (Lagace et al., 2007; Ninkovic et al., 2007; Imayoshi et al., 2008). A Cre-mediated recombination system was modified so that recombination can be triggered in adult progenitors by the application of an exogenous ligand (tamoxifen). Tamoxifen application irreversibly induces recombination of the genome of the adult progenitors, which allows the continual expression of marker proteins in the progenitors and their descendent cells. Results for the proportion of labeled adult-born interneurons among all interneurons differ among studies, but all have shown that adult-born interneurons gradually accumulate and increase in number for at least several months after the initiation of gene recombination. In one study, labeled adult-born granule cells accounted for nearly 20% of the total number of granule cells, a proportion that remained stable between 4 and 9 months. This finding raises the possibility that adult-born granule cells may constitute a specific cohort of cells that are repeatedly replaced by much younger adult-born granule cells (Ninkovic et al., 2007). In another study, the proportion of labeled adult-born granule cells continued to increase up to age 12 months to 60–70% of total granule cells, raising the possibility that these cells continue to accumulate and finally account for the majority of all granule cells in the adult olfactory bulb (Imayoshi et al., 2008). The actual proportion of adult-born granule cells remains unclear, but the two models just described, repeated replacement and continual addition, are not necessarily mutually exclusive; rather, these two modes of cell integration may work together, to differing extents under different conditions. It is intriguing to postulate the regulatory mechanisms that balance these two modes of integration of adult-born interneurons.

15.2.2 Comparison of Neonate- and Adult-Born Interneurons

Understanding the role of adult-born interneurons in the structure and function of the adult bulbar circuitry requires clarification of whether adult neurogenesis is distinct from embryonic and neonatal neurogenesis. Results to date have identified distinctions in not only the site of cell proliferation but also in the migration speed of progenitors and the layer distribution of new interneurons in the olfactory bulb (Lemasson et al., 2005). The survival rate of new granule cells in the olfactory bulb is also different. When neonate-born granule cells were labeled with BrdU, the number of surviving cells at day 22 and day 60 did not significantly differ, whereas the number of labeled adult-born granule cells at day 60 was decreased to nearly half that at day 22 (Lemasson et al., 2005). Neonatal-born interneurons may be heavily utilized in the construction of the functional bulbar circuitry, whereas adult-born interneurons might be utilized on demand when required for plastic modulation of preexisting bulbar circuitry.

A second question is the long-term survival of embryonic or neonate-born interneurons. The markedly high number of adult-born interneurons, representing as much as 1% of total interneurons per day, suggests that preexisting interneurons, including those neonate born, may be gradually eliminated and replaced by adult-born interneurons. In the in vivo time-lapse imaging study, periglomerular cells in the mouse olfactory bulb showed the disappearance of old cells and the appearance of new ones in the same observation field, with a turnover rate of approximately 3% per month (Mizrahi et al., 2006). Although the cellular age of the lost neurons was unknown, this study was the first to show the replacement of preexisting bulbar interneurons with new bulbar interneurons in vivo. An interesting question is then whether the loss of preexisting bulbar interneurons depends on the arrival of new bulbar interneurons: in other words, do new interneurons drive out preexisting interneurons, or do preexisting interneurons disappear by themselves without the arrival of new interneurons? In genetically manipulated mice, the majority of adult-born bulbar interneurons were ablated by tamoxifen-induced toxin expression (Imayoshi et al., 2008). In intact mice, the number of BrdU-labeled neonate-born granule cells gradually decreased during the adult period, indicating that neonate-born preexisting granule cells are gradually eliminated. In mice in which adult-born interneurons were ablated, the number of BrdU-labeled neonate-born granule cells decreased at a similar rate to that in intact animals. This result suggests that the elimination of preexisting neonate-born granule cells can occur independently from the arrival of adult-born new granule cells. Preexisting bulbar interneurons may make "vacant rooms" in which the adult-born interneurons can integrate, and thereby potentiate the plastic properties of the adult bulbar circuitry by replacing old, stubborn interneurons with new, flexible interneurons.

Whatever the proportion of adult-born interneurons, it appears that not all preexisting interneurons are replaced in the adult. BrdU-labeled neonate-born interneurons can be found in large numbers in the adult olfactory bulb (Lemasson et al., 2005; Imayoshi et al., 2008). What kinds of cells are eliminated and what kinds survive? Understanding this may reveal how some odor memories are kept stable throughout life whereas others are forgotten or renewed during adulthood. Both periglomerular and granule cells consist of distinct subtypes, classified by molecular expression, layer distribution of the somata, dendritic arborization, and targeting domain on principal neurons (Kosaka et al., 1995; Imamura et al., 2006; Parrish-Aungst et al., 2007; Lledo et al., 2008; Naritsuka et al., 2009). Interestingly, recent analyses have revealed that the subtypes generated in neonates and adults are distinct (De Marchis et al., 2007; Merkle et al., 2007; Ninkovic et al., 2007; Batista-Brito et al., 2008). For example, neonate-born granule cells tend to be located in the superficial portion of the granule cell layer and adult-born granule cells in the deep portion (Lemasson et al., 2005; Imayoshi et al., 2008). Viral labeling experiments show that neonate-born granule cells extend dendrites into the superficial portion of the external plexiform layer and adult-born granule cells into the deep portion of the layer (Kelsch et al., 2007). These observations are consistent with the overall tendency of the dendritic arborization of granule cells whose somata are positioned in different depths in the granule cell layer (Mori, 1987). Control of the survival and death of individual subtypes of bulbar interneurons is an important point that remains to be addressed. Olfactory sensory deprivation by naris cauterization reduced the survival of adult-born granule cells in the deep portion of the granule cell layer but did not significantly affect that of cells in the superficial portion (Mandairon et al., 2006b), suggesting that sensory experience differently influences the survival of different interneuron subtypes.

15.3 Summary

Olfaction plays a key role in many areas of animal behavior. The selection of appropriate behavioral responses in circumstances with various odors requires that the olfactory system possess high plasticity. Adult-born bulbar interneurons may greatly contribute to the plasticity of odor information processing in adult animals. The selection of appropriate interneurons from among excess pools of new neurons is considered a key mechanism in the remodeling of bulbar circuitry in accordance with the requirements of the odorous situation. Many studies indicate the importance of olfactory sensory experience and the behavioral state of an animal to the survival or death decision of adult-born interneurons. Plausible candidate mediators of this "external" and "internal" information are glutamatergic synaptic input and neuromodulatory input. Adult-born interneurons integrate into preexisting bulbar circuitry. Recent studies have revealed the differences in and relationships between neonatal and adult neurogenesis. A clear understanding of this would help clarify the functional significance of adult neurogenesis in the plasticity of olfaction in adult animals.

References

Alonso, M., Viollet, C., Gabellec, M.M., Meas-Yedid, V., Olivo-Marin, J.C. & Lledo, P.M. (2006) Olfactory discrimination learning increases the survival of adult-born neurons in the olfactory bulb. Journal of Neuroscience, 26, 10508–10513.

Alvarez-Buylla, A., Garcia-Verdugo, J. & Tramontin A.D. (2001) A unified hypothesis on the lineage of neural stem cells. Nature Reviews in Neuroscience, 2, 287–293.

Batista-Brito, R., Close, J., Machold, R. & Fishell, G. (2008) The distinct temporal origins of olfactory bulb interneuron subtypes. Journal of Neuroscience, 28, 3966–3975.

Bayer, S.A. (1983) 3H-thymidine-radiographic studies of neurogenesis in the rat olfactory bulb. Experimental Brain Research, 50, 329–340.

Biebl, M., Cooper, C.M., Winkler, J. & Kuhn, H.G. (2000) Analysis of neurogenesis and programmed cell death reveals a self-renewing capacity in the adult rat brain. Neuroscience Letters, 291, 17–20.

Brennan, P., Kaba, H. & Keverne, E.B. (1990) Olfactory recognition: a simple memory system. Science, 250, 1123–1126.

Buss, R.R., Sun, W. & Oppenheim, R.W. (2006) Adaptive roles of programmed cell death during nervous system development. Annual Review of Neuroscience, 29, 1–35.

Carleton, A., Petreanu, L.T., Lansford, R., Alvarez-Buylla, A. & Lledo, P.M. (2003) Becoming a new neuron in the adult olfactory bulb. Nature Neuroscience, 6, 507–518.

Cooper-Kuhn, C.M., Winkler, J. & Kuhn, H.G. (2004) Decreased neurogenesis after cholinergic forebrain lesion in the adult rat. Journal of Neuroscience Research, 77, 155–165.

Corotto, F., Henegar, J. & Maruniak, J. (1994) Odor deprivation leads to reduced neurogenesis and reduced neuronal survival in the olfactory bulb of the adult mouse. Neuroscience, 61, 739–744.

De Marchis, S., Bovetti, S., Carletti, B., Hsieh, Y.C., Garzotto, D., Peretto, P., Fasolo, A., Puche, A.C. & Rossi, F. (2007) Generation of distinct types of periglomerular olfactory bulb interneurons during development and in adult mice: implication for intrinsic properties of the subventricular zone progenitor population. Journal of Neuroscience, 27, 657–664.

Doty, R.L. (1986) Odor-guided behavior in mammals. Experientia. 42, 257–271.

Enwere, E., Shingo, T., Gregg, C., Fujikawa, H., Ohta, S. & Weiss, S. (2004) Aging results in reduced epidermal growth factor receptor signaling, diminished olfactory neurogenesis, and deficits in fine olfactory discrimination. Journal of Neuroscience, 24, 8354–8365.

Fiske, B. & Brunjes, P. (2001a) Cell death in the developing and sensory-deprived rat olfactory bulb. Journal of Comparative Neurology, 431, 311–319.

Fiske, B. & Brunjes, P. (2001b) NMDA receptor regulation of cell death in the rat olfactory bulb. Journal of Neurobiology, 47, 223–232.

Hensch, T.K. (2005) Critical period plasticity in local cortical circuits. Nature Reviews in Neuroscience, 6, 877–888.

Hinds, J.W. (1968a) Autoradiographic study of histogenesis in the mouse olfactory bulb. I. Time of origin of neurons and neuroglia. Journal of Comparative Neurology, 134, 287–304.

Hinds, J.W. (1968b) Autoradiographic study of histogenesis in the mouse olfactory bulb. II. Cell proliferation and migration. Journal of Comparative Neurology, 134, 305–322.

Imamura, F., Nagao, H., Naritsuka, H., Murata, Y., Taniguchi, H. & Mori, K. (2006) A leucine-rich repeat membrane protein, 5T4, is expressed by a subtype of granule cells with dendritic arbors in specific strata of the mouse olfactory bulb. Journal of Comparative Neurology, 495, 754–768.

Imayoshi, I., Sakamoto, M., Ohtsuka, T., Takao, K., Miyakawa, T., Yamaguchi, M., Mori, K., Ikeda, T., Itohara, S. & Kageyama, R. (2008) Roles of continuous neurogenesis in the structural and functional integrity of the adult forebrain. Nature Neuroscience, 11, 1153–1161.

Kaneko, N., Okano, H. & Sawamoto, K. (2006) Role of cholinergic system in regulating survival of newborn neurons in the adult mouse dentate gyrus and olfactory bulb. Genes to Cells. 11, 1145–1159.

Kaplan, M.S., McNelly, N.A. & Hinds, J.W. (1985) Population dynamics of adult-formed granule neurons of the rat olfactory bulb. Journal of Comparative Neurology, 239, 117–125.

Kasa, P., Hlavati, I., Dobo, E., Wolff, A., Joo, F. & Wolff, J.R. (1995) Synaptic and non-synaptic cholinergic innervation of the various types of neurons in the main olfactory bulb of adult rat: immunocytochemistry of choline acetyltransferase. Neuroscience, 67, 667–677.

Kelsch, W., Mosley, C.P., Lin, C.W. & Lois, C. (2007) Distinct mammalian precursors are committed to generate neurons with defined dendritic projection patterns. PLoS Biology, 5, 2501–2512.

Kelsch, W., Lin, C.W. & Lois, C. (2008) Sequential development of synapses in dendritic domains during adult neurogenesis. Proceedings of the National Academy of Sciences of the USA, 105, 16803–16808.

Kosaka, K., Aika, Y., Toida, K., Heizmann, C.W., Huinziker, W., Jacobowitz, D.M., Nagatsu, I., Streit, P., Visser, T.J. & Kosaka, T. (1995) Chemically defined neuron groups and their subpopulations in the glomerular layer of the rat main olfactory bulb. Neuroscience Research, 23, 73–88.

Lagace, D.C., Whitman, M.C., Nooman, M.A., Ables, J.L., DeCarolis, N.A., Arguella, A.A., Donovan, M.H., Fischer, S.J., Farnbauch, L.A., Beech, R.D., DiLeone, R.J., Greer, C.A., Mandyam, C.D. & Eisch, A.J. (2007) Dynamic contribution of nestin-expressing stem cells to adult neurogenesis. Journal of Neuroscience, 27, 12623–12629.

Lemasson, M., Saghatelyan, A., Olivo-Marin, J.C. & Lledo, P.M. (2005) Neonatal and adult neurogenesis provide two distinct populations of newborn neurons to the mouse olfactory bulb. Journal of Neuroscience, 25, 6816–6825.

Lledo, P.M., Alonso, M. & Grubb, M.S. (2006) Adult neurogenesis and functional plasticity in neuronal circuits. Nature Reviews in Neuroscience. 7, 179–193.

Lledo, P.M., Merkle, F.T. & Alvarez-Buylla, A. (2008) Origin and function of olfactory bulb interneuron diversity. Trends in Neuroscience, 31, 392–400.

Mandairon, N., Jourdan, F. & Didier, A. (2003) Deprivation of sensory inputs to the olfactory bulb upregulates cell death and proliferation in the subventricular zone of adult mice. Neuroscience, 119, 507–516.

Mandairon, N., Sacquet, J., Garcia, S., Ravel, N., Jourdan, F. & Didier, A. (2006a) Neurogenic correlates of an olfactory discrimination task in the adult olfactory bulb. European Journal of Neuroscience, 24, 3578–3588.

Mandairon, N., Sacquet, J., Jourdan, F. & Didier, A. (2006b) Long-term fate and distribution of newborn cells in the adult mouse olfactory bulb: influences of olfactory deprivation. Neuroscience, 141, 443–451.

McLean, J.H. & Shipley, M.T. (1991) Postnatal development of the noradrenergic projection from locus coeruleus to the olfactory bulb in the rat. Journal of Comparative Neurology, 304, 467–477.

McLean, J.H., Shipley, M.T., Nickell, W.T., Aston-Jones, G. & Reyher, C.K.H. (1989) Chemoanatomical organization of the noradrenergic input from locus coeruleus to the olfactory bulb of the adult rat. Journal of Comparative Neurology, 285, 339–349.

Mechawar, N., Saghatelyan, A., Grailhe, R., Scoriels, L., Gheusi, G., Gabellec, M.M., Lledo, P.M. & Changeux, J.P. (2004) Nicotinic receptors regulate the survival of newborn neurons in the adult olfactory bulb. Proceedings of the National Academy of Sciences of the USA, 101, 9822–9826.

Merkle, F.T., Mirzadeh, Z. & Alvarez-Buylla, A. (2007) Mosaic organization of neural stem cells in the adult brain. Science, 317, 381–384.

Mirich, J.M., Williams, N.C., Berlau, D.J. & Brunjes, P.C. (2002) Comparative study of aging in the mouse olfactory bulb. Journal of Comparative Neurology, 454, 361–372.

Miwa, N. & Storm, D.R. (2005) Odorant-induced activation of extracellular signal-regulated kinase/mitogen-activated protein kinase in the olfactory bulb promotes survival of newly formed granule cells. Journal of Neuroscience, 25, 5404–5412.

Mizrahi, A., Lu, J., Irving, R., Feng, G & Katz, L.C. (2006) *In vivo* imaging of juxtaglomerular neuron turnover in the mouse olfactory bulb. Proceedings of the National Academy of Sciences of the USA, 103, 1912–1917.

Mori, K. (1987) Membrane and synaptic properties of identified neurons in the olfactory bulb. Progress in Neurobiology, 29, 275–320.

Mouret, A., Gheusi, G., Gabellec, M.M., de Chaumont, F., Olivo-Martin, J.C. & Lledo, P.M. (2008) Learning and survival of newly generated neurons: when time matters. Journal of Neuroscience, 28, 11511–11516.

Najbauer, J. & Leon, M. (1995) Olfactory experience modulates apoptosis in the developing olfactory bulb. Brain Research, 674, 245–251.

Naritsuka, H., Sakai, K., Hashikawa, T., Mori, K. & Yamaguchi, M. (2009) Perisomatic-targeting granule cells in the mouse olfactory bulb. Journal of Comparative Neurology, 515, 409–426.

Ninkovic, J., Mori, T. & Gotz, M. (2007) Distinct modes of neuron addition in adult mouse neurogenesis. Journal of Neuroscience, 27, 10906–10911.

Parrish-Aungst, S., Shipley, M.T., Erdelyi, F., Szabo, G. & Puche, A.C. (2007) Quantitative analysis of neuronal diversity in the mouse olfactory bulb. Journal of Comparative Neurology, 501, 825–836.

Petreanu, L. & Alvarez-Buylla, A. (2002) Maturation and death of adult-born olfactory bulb granule neurons: role of olfaction. Journal of Neuroscience, 22, 6106–6113.

Rochefort, C. & Lledo, P.M. (2005) Short-term survival of newborn neurons in the adult olfactory bulb after exposure to a complex odor environment. European Journal of Neuroscience, 22, 2863–2870.

Rochefort, C., Gheusi, G., Vincent, J. & Lledo, P.M. (2002) Enriched odor exposure increases the number of newborn neurons in the adult olfactory bulb and improves odor memory. Journal of Neuroscience, 22, 2679–2689.

Rosser, A.E. & Keverne, E.B. (1985) The importance of central noradrenergic neurons in the formation of an olfactory memory in the prevention of pregnancy block. Neuroscience, 15, 1141–1147.

Shepherd, G.M., Chen, W. & Greer, C. (2004) Olfactory bulb. In Shepherd, G. (ed) The synaptic organization of the brain, fifth Edition. Oxford University Press, New York, pp. 165–216.

Shipley, M.T. & Ennis, M. (1996) Functional organization of olfactory system. Journal of Neurobiology, 30, 123–176.

Sullivan, R.M., Wilson, D.A. & Leon, M. (1989) Norepinephrine and learning-induced plasticity in infant rat olfactory system. Journal of Neuroscience, 9, 3998–4006.

Veyrac, A., Sacquet, J., Nguyen, V., Marien, M., Jourdan, F. & Didier, A. (2009) Novelty determines the effects of olfactory enrichment on memory and neurogenesis through noradrenergic mechanisms. Neuropsychopharmacology, 34, 786–795.

Whitman, M.C. & Greer, C.A. (2007) Synaptic integration of adult-generated olfactory bulb granule cells: basal axodendritic centrifugal input precedes apical dendrodendritic local circuits. Journal of Neuroscience, 27, 9951–9961.

Winner, B., Cooper-Kuhn, C.M., Aigner, R., Winkler, J. & Kuhn, H.G. (2002) Long-term survival and cell death of newly generated neurons in the adult rat olfactory bulb. European Journal of Neuroscience, 16, 1681–1689.

Woo, C.C. & Leon, M. (1995) Distribution and development of beta-adrenergic receptors in the rat olfactory bulb. Journal of Comparative Neurology, 352, 1–10.

Yamaguchi, M. & Mori, K. (2005) Critical period for sensory experience-dependent survival of newly generated granule cells in the adult mouse olfactory bulb. Proceedings of the National Academy of Sciences of the USA, 102, 9697–9702.

Zuo, Y., Yang, G., Kwon, E. & Gan, W.B. (2005) Long-term sensory deprivation prevents dendritic spine loss in primary somatosensory cortex. Nature, 436, 261–265.

Chapter 16
Wiring New Neurons with Old Circuits

Pierre-Marie Lledo

Abstract The brain derives much of its function from its ability to adapt to tasks on a wide range of time scales, from milliseconds to days, weeks, and months. This adaptability at multiple time scales is found across all brain areas, from cognitive areas all the way to the peripheral areas, in which sensory information is encoded so as to facilitate the subsequent extraction of relevant information. The adaptability of the brain is achieved through multiple morphological and physiological changes occurring at all levels: from molecules, to spines, and to dendrites and axons. It is clear today that brain plasticity also operates at the level of entire cells. In at least two areas of the adult brain, new neurons are constitutively generated throughout life and form an integral part of the networks. Because the capacity of the adult brain to rewire itself depends on external influences, ongoing neuronal production represents also a plastic mechanism by which brain performance can be optimized according to the prevailing environment. However, it remains unclear when and why brain performance has to be optimized.

This chapter focuses on the functional issues linked to neurogenesis in the olfactory system. After outlining the processes of adult neurogenesis in the olfactory system and discussing their regulation by various factors, I consider how existing circuits can continue to work in the face of constant arrivals and departures of cells and explore the possible functional role of adult-born neurons in the host microcircuit. Concentrating exclusively on mammalian systems, I demonstrate throughout this chapter that adult neurogenesis is a plastic mechanism by which brain performance can be optimized to cope with changing environment or internal state.

P.-M. Lledo (✉)
Laboratory for Perception and Memory, Department of Neuroscience,
Institut Pasteur, 75015 Paris, France
and
CNRS, URA2182, 75015 Paris, France
e-mail: pmlledo@pasteur.fr

T. Seki et al. (eds.), *Neurogenesis in the Adult Brain I: Neurobiology*,
DOI 10.1007/978-4-431-53933-9_16, © Springer 2011

16.1 Introduction

Olfaction is a vital sense for most species. Animals recognize and discriminate between a large number of chemical signals in their environment that profoundly influence their behavior and provide them with essential information for survival (Eisthen 1997). Remarkably, the sense of smell has been conserved throughout evolution, and its initial processing stage is very similar across vertebrates (Hildebrand and Shepherd 1997; Ache and Young 2005). The first central station of the olfactory system, the so-called olfactory bulb, is endowed with the complex connectivity required for processing the large number of dimensions of olfactory space, making it possible to process the information entering the olfactory system effectively (Shepherd *et al.* 2004). This highly complex system has to meet a number of challenges, including changes in the olfactory world or substantial changes in its significance with time. Such shifts could be seasonal, or related to pregnancy and giving birth, for instance. However, odor representation must be sufficiently stable for higher brain regions to recognize relevant odorants even if the encoding of the odor has changed: the smell of a rose, say, should be recognized as that of a rose, be it spring or fall.

The continuous production of new olfactory neurons raises a series of important questions about the possible contribution of adult neurogenesis to the functioning of mature circuits. If adult neurogenesis is beneficial, what advantages does it confer? Are the newly generated neurons special in any way other than having been formed late? How are these neurons integrated into existing networks? How could they possibly contribute to brain function?

I address these questions here, using the adult olfactory bulb as a model system. I discuss how adult neurogenesis may provide qualitatively new aspects of plasticity beyond those afforded by synaptic changes. For the sake of conciseness, this chapter focuses on the olfactory bulb of mammals, primarily those of rodents.

16.2 Synaptic Organization of the Mammalian Olfactory Bulb

The mammalian olfactory epithelium recognizes more than a thousand airborne volatile molecules known as odorant compounds (or odorants) (Buck and Axel 1991). This sensory organ is connected to the olfactory bulb, the first central station for processing olfactory information (Shipley and Ennis 1996; Shepherd *et al.* 2004). Odor detection begins with the binding of odorant molecules to their cognate receptors, expressed by olfactory sensory neurons (OSNs) (Young *et al.* 2002). Individual OSNs project in a "receptor–topic" arrangement, such that an individual glomerulus receives sensory input from only one type of receptor (Fig. 16.1). Each glomerular structure results from the convergence of 5 to 40,000 axon terminals, leading to the establishment of synapses with dendrites of mitral and tufted cells (i.e., the projection neurons of the olfactory bulb) and of various classes of local interneurons. As each group of glomerulus-specific output neurons is odorant receptor specific, they form a defined morphological network (Willhite *et al.* 2006)

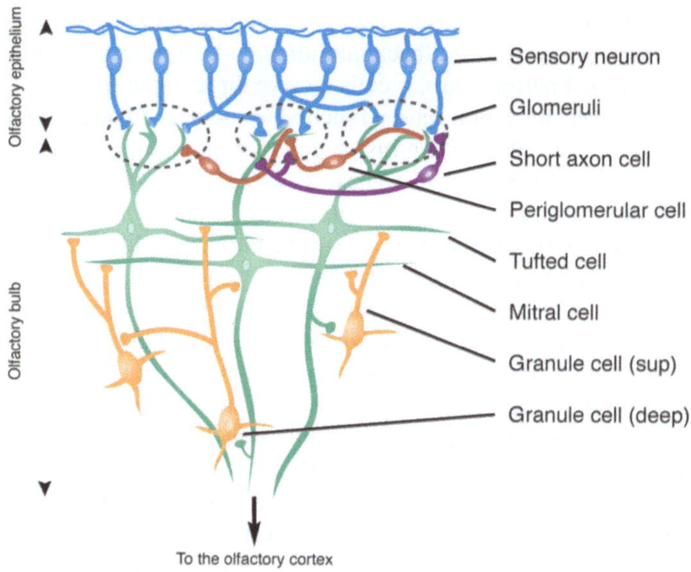

Fig. 16.1 Basic circuitry of the main olfactory bulb in rodent. Axons of sensory neurons in the main olfactory epithelium project to the olfactory bulb (OB). Olfactory sensory neurons (in *blue*) that express the same odorant receptor gene project their axons to either of two glomeruli in the OB. Their axons converge on specific glomeruli, where they synapse with the dendrite of local interneurons (periglomerular cell in *brown*) and short axon cell (in *purple*) and second-order neurons (mitral and tufted cells in green). Short axon cells contact both apical and lateral dendrites of mitral cells. The lateral dendrites of mitral cells contact the apical dendrites of deep granule cells whereas those from tufted cells contact superficial (*sup*) granule cells. Note that the second-order neuron projections of the OB target the primary olfactory cortex, which includes the anterior olfactory nucleus, the piriform cortex, the olfactory tubercle, the lateral part of the cortical amygdala, and the entorhinal cortex

somewhat analogous to ocular-dominance columns in the visual cortex or barrels in the somatosensory cortex (Shepherd *et al.* 2004; Shepherd 2006).

As an individual odorant may activate multiple olfactory receptors, glomerular input maps are fragmented and highly distributed (Malnic *et al.* 1999). Thus, information about odorant identity is initially transmitted through a combination of the activation of glomeruli and the relative levels of activation of different glomeruli (Wachowiak and Shipley 2006). Once activated, glomeruli may elicit a precisely timed action potential in a group of projection neurons. Thus, odor encoding in the olfactory bulb results from the complex and dynamic activation of sensory afferents, which activate large numbers of glomeruli (*spatial dimension*) before the processing and refining of the signal (*temporal dimension*) within the olfactory bulb network (Mori *et al.* 2006) via intense interactions between projection neurons and different categories of local interneurons (Wilson and Mainen 2006).

Local olfactory bulb interneurons comprise periglomerular cells, short axon cells, and granule cells (see Fig. 16.1). Periglomerular cells form three-way synapses with mitral and tufted cell apical dendrites and with the presynaptic termini of OSNs (Kosaka *et al.* 1998). They are GABAergic and/or dopaminergic and are usually considered to mediate feedback inhibition (GABAergic and/or dopaminergic interneurons). Type I periglomerular cells receive direct input from the OSN axons and therefore regulate projection neuron function via both feed-back and feed-forward circuits (Kosaka *et al.* 1997). In contrast, type II periglomerular cells do not receive direct input from the OSN axons. These cells receive synaptic input only via dendrodendritic synapses with projection neurons (Kosaka *et al.* 1997). Remarkably, many types of periglomerular cells are generated throughout life, but a large proportion of these cells are type I cells. Thus, the main goal of adult periglomerular cell neurogenesis may be to regulate the synaptic activity of primary afferents.

Short axon cells receive excitation from external tufted cells and, in turn, make contact with other short axon cells and external tufted cells in other glomeruli (Aungst *et al.* 2003), potentially leading to the formation of an extensive lateral network across the bulb. The cell bodies of mitral and deep and middle tufted cells lie in deeper layers of the olfactory bulb. Mitral cells project laterally across the bulb within the external plexiform layer, in which they interact with the dendrites of GABAergic granule cells via reciprocal synapses. This connectivity is believed to contribute to the establishment of different inhibitory processes within the bulb and to stimulus-evoked oscillations.

Mitral and tufted cell basal dendrites receive synaptic contacts from granule cells, which constitute the most numerous populations of bulbar interneurons (Schoppa and Urban 2003), accounting for about 90% of the total bulbar interneuron population. In contrast to periglomerular cells, granule cells have no axon and therefore make contacts exclusively through dendrites and cell bodies (Schoppa and Urban 2003; Shepherd *et al.* 2007). All granule cells are GABAergic, and these cells mediate various types of inhibition over a wide range of spatial scales (Lledo and Lagier 2006). In short, the diversity of inhibition facilitates the spread of neuronal activity throughout the olfactory bulb network, allowing sensory input patterns to be shaped and projection neuron activity to be synchronized. If this dense circuit of inhibitory interneurons is so crucial, why does it have to be continually replaced?

16.3 Sensory Processing Requires the Correct Excitatory–Inhibitory Balance

Excitatory and inhibitory connections between glomeruli and between projection neurons and interneurons have been proposed as mechanisms for contrast enhancement and gain control. Rodents have about 5×10^4 mitral cells and 5×10^6 granule cells (Shepherd *et al.* 2004). Thus, the ratio of excitatory neurons to local interneurons in the olfactory bulb (1:100) is much higher than in other parts of the brain. For comparison, this ratio is only about 5:1 in the neocortex. This balance between excitation and inhibition in the olfactory bulb is conserved among mammals and is thought to play a key role in the temporal aspects of olfactory processing (Wilson and Mainen 2006), although the key features of this temporal dimension remain unclear.

Within the respiratory slow cycle, different patterns can be observed on a fine timescale. At the end of each inhalation, a fast oscillation (~40–100 Hz, also known as a gamma rhythm) is initiated. This odor-evoked oscillation was first described by Adrian (1942) and has been studied in detail at both the physiological and computational levels (Cleland and Linster 2005). Most researchers agree that olfactory bulb gamma oscillations arise from reciprocal dendrodendritic interactions between mitral and granule cells in the external plexiform layer, in a negative feedback circuit. The firing times of olfactory bulb mitral cells are probabilistically related to gamma oscillation at population level. Gamma oscillation is often referred to as a source of synchrony between individual projection neurons, but it represents, more precisely, the synchrony between individual neurons and the emergent local field potential.

Activity in the gamma-frequency band has been associated with odor discrimination circuitry in many species. Walter J. Freeman was the first to demonstrate the existence of a common gamma band waveform of the EEG over the olfactory bulb (Freeman and Schneider 1982). The spatial amplitude patterns of this waveform are the best indicators of an odor, and reliable patterns are produced only when meaning (positive or negative reinforcement) is associated with an odorant. Gamma band oscillatory population synchrony is a specific mechanism associated with the discrimination of more difficult or highly overlapping odors in rodents. The disruption of these oscillations leads to a selective decrease in the discrimination of highly overlapping odorants (fine discrimination). The greater olfactory bulb gamma power in $GABA_A$ receptor-knockout mice leads to a selective increase in fine odor discrimination, with no effect on coarse discrimination. Rats devote much more power to gamma oscillation during fine odor discrimination than during coarse discrimination in a two-alternative choice task, consistent with enhancement of the temporal precision of mitral cell firing patterns. Overall, the mixture of ordered but highly complex maps of olfactory input to the olfactory bulb and oscillatory dynamic states contribute to odor perception. The extent to which input patterns overlap predicts odorant similarity and the difficulty of discrimination, allowing animals to adjust their olfactory system dynamics for the interpretation of an odorant response. I now discuss the functional consequences of challenging the interneuron population in the early stages of olfactory information processing.

16.4 Neurogenesis in the Adult Olfactory Bulb

Postdevelopmental neurogenesis has been conserved throughout evolution, from crustaceans (Beltz and Sandeman 2003) and insects (Scotto-Lomassese *et al.* 2003) to higher vertebrates, including birds (Goldman and Nottebohm 1983; Alvarez-Buylla 1990), rodents (Altman and Das 1965; Hinds 1968; Lois and Alvarez-Buylla 1994), primates (Gould *et al.* 1999; Kornack and Rakic 1999), and humans (Eriksson *et al.* 1998). Despite this conservation of adult neurogenesis throughout evolution, fundamental differences in this process are observed between species. For instance, lower vertebrates can regenerate entire regions of the brain whereas other animals cannot.

It seems indeed that the extent of adult neurogenesis decreases with increasing brain complexity (Kempermann 2006). In mammals, adult neurogenesis under normal conditions is confined to the subgranular zone of the hippocampal dentate gyrus and the subventricular zone (SVZ), which contributes interneurons to the olfactory bulb (Rakic 2002; Alvarez-Buylla and Garcia-Verdugo 2002). There is some evidence to suggest that constitutive neurogenesis occurs in other regions of the adult brain, but methodological issues (e.g., Kornack and Rakic 2001; Koketsu *et al.* 2003; Frielingsdorf *et al.* 2004) make it impossible to add these regions to the list of zones in which adult neurogenesis is known to occur, at least for the time being.

16.4.1 Producing New Neurons

The astrocytes lining the border between the striatum and the lateral ventricle act as slowly dividing neural stem cells, generating a progeny of neuroblast precursors (Doetsch *et al.* 1999; Laywell *et al.* 2000). Once generated in the germinal niche (see the glossary in Box 16.1), these neuroblasts proceed toward the olfactory bulb along an intricate pathway called the rostral migratory stream (RMS), which may be as much as 5 mm long in rats. More than 30,000 neuroblasts leave the rodent SVZ for the RMS each day (Alvarez-Buylla *et al.* 2001). Rather than being guided by radial glia, as in the developing brain, adult-born neuroblasts migrate tangentially in chains through tubular structures formed by specialized astrocytes. Time-lapse video imaging can be used to analyze this migration directly and has shown that nuclear translocation always occurs in the direction of the leading process extension, often leaving a thin trailing process behind the cell body. Remarkably, cells move in both directions, with occasional migration back toward the SVZ observed. Several studies have reported migration to be rapid in explants (100 μm/h) and ex vivo (50 μm/h). However, it remains unclear how neuroblasts migrate in vivo, as it is not currently possible to monitor migration appropriately from the adult neurogenic zone, through the migratory pathway, or at the integration site. These areas are too deep for conventional fluorescence microscopy. A method based on integrated fiber confocal fluorescence microscopy has therefore been developed, facilitating minimally invasive, long-term recording several millimeters into anesthetized animals. This novel technique has made it possible, for the first time, to visualize migrating newly formed neurons in living animals (Davenne *et al.* 2005). A large number of SVZ progenitor cells are labeled, and the migration of neuroblasts can then be traced over several hours. The speed of migration in vivo is not constant. Instead, there seem to be spurts of migration interspersed with periods of inactivity.

After reaching the olfactory bulb and detaching from the chains, the new cells migrate radially, giving rise to two main types of neuron, granule cells and periglomerular cells, located in different layers (see Fig. 16.1). The two cell types make only local contacts in the bulb, directly or indirectly modulating the processing of sensory information by the projection neurons.

Box 16.1 Neurogenic Stem Cell Niches of the Adult Mammalian Brain

Neural Stem/Precursor Cells. Adult central nervous system (CNS) stem cells have several cardinal features, including an unlimited capacity for self-renewal, the ability to proliferate indefinitely in presence of mitogens, and multipotency for the various neuroectodermal lineages of the CNS. The multipotent progenitors of the adult brain are proliferative cells with a limited capacity for self-renewal that can differentiate into at least two different cell lineages. In contrast, lineage-specific precursors (or progenitors) are cells restricted to only one particularly lineage: neuronal, astroglial, or oligodendroglial. CNS stem cells and all types of precursor/progenitor may be considered, in general terms, to be neural precursor cells (NPCs).

The Germinal Niche. NPC self-renewal and differentiation are regulated by a specialized microenvironment, conventionally referred to as the germinal niche, in which these cells reside. Both environmental cues and intrinsic genetic programs are required to maintain stem cell properties and to direct, or regulate, stem cell proliferation and differentiation within niches (Doetsch 2003).

Niche Cytoarchitecture. The subventricular zone (SVZ) of the lateral ventricle wall and the subgranular zone (SGZ) of the dentate gyrus of the hippocampus are germinal regions of the brain in which neural stem cells reside and support neurogenesis and gliogenesis throughout adult life. Cells with the structural and molecular characteristics of astrocytes are responsible for neurogenesis (true stem cells or type B cells) in the SVZ and SGZ (Doetsch *et al.* 1999; Laywell *et al.* 2000; Garcia *et al.* 2004). The SVZ astrocytes, which produce glial fibrillary acidic protein (GFAP), are in intimate contact with all other SVZ cell types, including the rapidly dividing transit-amplifying (type C) cells and the lineage-committed (postmitotic) migratory neuroblasts (type A cells). The differentiation of the cell lineage passes from type B through type C to type A cells, with type B cells believed to be the self-renewing primary precursors (Doetsch *et al.* 1999). In the SGZ, GFAP-expressing astrocytes function as stem (type B) cells, undergoing self-renewal, proliferation, and differentiation into transit-amplifying (type D) cells, which then differentiate into lineage-committed migratory granule neurons (type G cells) (Seri *et al.* 2001). The maintenance and differentiation of neural stem cells in brain neurogenic areas depend on these cells being in physical contact with the basal lamina, which acts as a scaffold and sequesters and/or modulates cytokines and growth factors derived from local cells (such as fibroblasts, macrophages, and pericytes). Type B cells in the SVZ are in close contact (interdigitated) with both the basal lamina and the blood vessels. In the SGZ, bursts of endothelial cell division are spatially and temporally related to clusters of neurogenesis.

16.4.2 Integrating Newly Generated Interneurons

By transducing newly generated cells in the adult SVZ with replication-incompetent retroviruses, resulting in the green fluorescent protein (GFP) labeling of these cells (Box 16.2), researchers have been able to characterize the morphological and electrophysiological properties of newly generated neurons during their migration and differentiation (Petreanu and Alvarez-Buylla 2002; Belluzzi *et al.* 2003; Carleton *et al.* 2003; Whitman and Greer 2007). The morphological complexity of the newly generated cells increases within a few weeks of their generation. Granule cells form more elaborate dendrites, extending into the external plexiform layer of the bulb and becoming fully morphologically mature within 2 weeks of their generation (Petreanu and Alvarez-Buylla 2002; Whitman and Greer 2007). By contrast, periglomerular cells take about 4 weeks to develop their full dendritic and axonal morphology (Belluzzi *et al.* 2003). During this maturation period, the immature bulbar interneurons express transient marker proteins, including TUJ1 and TUC4 (Winner *et al.* 2002).

The sequential formation of GABAergic and glutamatergic synapses is crucial for the construction of stereotypic neural networks during brain development. Studies of neurogenesis in the adult olfactory bulb have revealed major differences in glutamatergic synaptogenesis between newly generated periglomerular cells and granule cells. As periglomerular cells mature, the ratio of 2-amino-3-(5-methyl-3-oxo-1,2-oxazol-4-yl) propanoic acid to *N*-methyl-D-aspartic acid (the AMPA:NMDA ratio) at OSN synapses increases, and the contribution of NR2B subunit-containing receptors to NMDA responses (a hallmark of juvenile NMDA synapses) decreases (Grubb *et al.* 2008). Surprisingly, however, these postsynaptic transformations are not mirrored by presynaptic changes. Once a periglomerular cell receives inputs from the OSN, the presynaptic function at the first olfactory connection remains remarkably constant. This finding suggests that newly generated periglomerular cells preferentially make initial contacts with preexisting functioning boutons (i.e., either already involved in a synapse or formed before the postsynaptic site).

As for periglomerular cells, the GABAergic and glutamatergic synapses involving new granule cells are established sequentially, with the GABAergic synapses established first. However, by contrast to the sequential acquisition of synapses in newly generated periglomerular cells, NMDA-mediated synaptic events occur in new interneurons that are not yet excitable but which have started to receive γ-aminobutyric acid A ($GABA_A$) and AMPA inputs (Carleton *et al.* 2003). In periglomerular cells, the maturation of voltage-dependent sodium current and, consequently, the capacity of the newly generated cells to fire action potentials precedes the formation of synaptic contacts (Belluzzi *et al.* 2003; Grubb *et al.* 2008). It remains unclear why the constraints on the granule cell network integration should differ so greatly from those in periglomerular cells. However, it should be borne in mind that new periglomerular cells integrate into microcircuits characterized by the ongoing replacement of primary afferents whereas new granule cells receive synaptic inputs from the dendrites of projection neurons known to be stable over extended periods of time (Mizrahi and Katz 2003). The rules governing the incorporation of adult-generated olfactory bulb interneurons thus differ according to the neuronal subtype (granule cells are axon-less interneurons whereas periglomerular cells have axons), and neuronal microcircuit as well.

Whatever the mechanism involved, this difference indicates that the functional integration of adult-born neurons follows a cell type-specific, rather than stereotyped, pattern. The delayed maturation of granule cell excitability may help to prevent disruption of the existing circuitry during integration of the new cells. Further studies are required to determine whether the newly generated neurons form functional synapses with their downstream target neurons and release appropriate neurotransmitters, providing an unequivocal demonstration of the integration of these cells into adult networks.

Maturing neurons integrate into circuits by forming functional synapses. In the adult olfactory bulb, synapses in maturing cells have been identified by staining for presynaptic markers (Belluzzi *et al.* 2003; Magavi *et al.* 2005) or electron microscopy (Whitman and Greer 2007). This approach has highlighted the precocious formation of synapses between the basal compartments of newly generated granule cells and centrifugal and mitral cell axon collaterals (Whitman and Greer 2007). Such early connections may be crucial for the differentiation and stabilization of adult-generated olfactory bulb interneurons.

Dendritic spine density changes over time in adult-born granule cells: it increases rapidly from 14 to 28 days after generation of the cell, remains steady until 42 days, and then decreases after 56 days (Whitman and Greer 2007). This decrease in spine density may result from pruning in cells with synaptic connections lasting more than 42 days, or the preferential death of cells with a high spine density, resulting in a population with an overall lower density of spines. Periglomerular cell spines mature more slowly than those of granule cells. Periglomerular cells have elaborate dendritic arbors at day 14, but immature filopodium-like processes do not appear until day 28, with mature pedunculate spines absent until day 42 (Whitman and Greer 2007). Spine density depends not only on cell age, but also on the degree of neuronal activity in the host circuit. Dendritic spine density decreases to 50% control levels when a new granule cell reaches a silenced olfactory bulb network, whereas it is unaffected in preexisting granule cells (Saghatelyan *et al.* 2005). This activity-dependent regulation is consistent with the selective stabilization theory, according to which an initial phase of proliferation, involving the production of large numbers of new neurons, is followed by massive cell elimination through apoptosis (Oppenheim 1991). Based on this model, synapse pruning should help to stabilize a particular set of microcircuits, favoring the survival of this set of contacts over the many others, and thereby sculpting the precise network required for a given brain function (Changeux and Danchin 1976).

Box 16.2 Viral Tools for Labeling Subsets of Cells Generated in Adults

Studies of the processes of maturation, migration, and integration of newly generated neurons in the adult brain can be simplified by the reliable labeling and visualization of particular types of cell. In recent years, the use of viral vectors encoding green fluorescent protein (GFP) has revolutionized the investigation of adult neurogenesis. In contrast to many of the methods commonly used for fluorescent cell labeling (e.g., the use of lipophilic carbocyanin vital dyes such as DiI or DiO), the GFP label remains within the cell and its progeny so long as the transgene encoding GFP remains present and there are appropriate transcriptional elements to transgene expression. Thus, internalization, cell division, and cell growth do not decrease GFP fluorescence, whereas they would decrease the intensity of chemical dyes. In addition, GFP is a large protein that is not transferred from cell to cell, a critical property for the unequivocal tracing of cell lineages.

However, for imaging cells in intact tissues, GFP should ideally be produced in a large subset of cells, but not in every cell. Many viral vectors are now available for gene transfer into neuronal cells (see, for instance, Washbourne and McAllister 2002), offering a wide range of options in terms of infection specificity and efficiency and the delivery and regulation of particular transgenes (Davidson and Breakefield 2003).

Many viral vectors, based on pseudorabies, herpesvirus, adenovirus, and retrovirus backbones, have been used to introduce reporter genes, such as GFP into cells in culture or within intact animals. For example, Carlén *et al.* (2002) used a pseudorabies virus producing GFP to cross the synapse between connected neurons. They injected this virus into the piriform cortex of adult mice exposed to bromodeoxyuridine (BrdU) in drinking water. BrdU and viral colabeling of the periglomerular and granule cells showed that new neurons reached the olfactory bulb. Furthermore, the triple labeling of newly generated neurons showed that some BrdU–NeuN colabeled cells were also positive for c-Fos induction in adult mice exposed to odorants. Such results clearly demonstrate that adult-born olfactory neurons are functionally integrated into the adult circuitry.

Differing from other viral systems, which preferentially infect quiescent cells, Moloney murine leukemia virus (MoMuLV)-derived recombinant retroviral vectors selectively infect dividing cells, integrating permanently into their genome, making it possible to visualize neuronal development from early progenitor cell stages. In this case, cycling neuroblasts can be labeled and their progeny followed for extended periods of time by injecting Moloney-based retroviral vectors encoding GFP into the adult SVZ. These vectors have made it possible to study the successive steps by which a neuroblast undergoes morphological maturation to become a mature neuron (Petreanu and Alvarez-Buylla 2002). Following on from this work, the consecutive electrophysiological stages through which the

(continued)

Box 16.2 (continued)

neuroblast passes before becoming a fully mature neuron have been described, using similar GFP-encoding retroviral vectors. The proportion of different cell types generated from transduced SVZ cells differs from that generated by transduced RMS cells (Hack *et al.* 2005). The construction of Moloney-based retroviral vectors carrying instructive genes (Hack *et al.* 2005; Kohwi *et al.* 2005; Marshall *et al.* 2005) makes it possible to alter the phenotypic fate of progenitor cells both predictably and efficiently, an intervention with potentially important clinical applications. As recombinant retroviral vectors target only those cells involved in postnatal neurogenesis, retroviral transduction could provide specific cell types for reconstruction of the brain after brain injury.

As for the Moloney-based retroviral vector approach, oncoretroviral vectors have been injected into the SVZ to transduce dividing cells (Falk *et al.* 2002; Rogelius *et al.* 2005), but not neural stem cells, which are moderately quiescent (Morshead *et al.* 1994). Both types of retroviral vector transduce only rapidly dividing precursors and neuroblasts, the so-called A- and C-type cells (Garcia-Verdugo *et al.* 1998). In contrast, lentiviral vectors, which transduce both dividing and nondividing cells, preferentially transduce neural stem cells (B-type cells) when injected into the SVZ because of the higher density of these cells in this forebrain region (Rogelius *et al.* 2005).

As MoMuLV-derived vectors cannot deliver genes to nondividing cells, these retroviral particles have limited applications in the adult CNS. Lentiviruses are currently the most promising retroviruses for in vivo applications as they are of low toxicity and can provide long-term reporter gene expression. Lentiviruses containing specific promoters have been used in vivo for the specific labeling of neurons or glial cells (Jakobsson *et al.* 2003), and lentiviral systems have also been developed for RNAi (Rubinson *et al.* 2003; Tiscornia *et al.* 2003). Lentiviruses can also infect a wide range of species, and the lentiviral infection of embryos results in germline transmission of the integrated virus, providing a method for the stable expression of reporter genes in species other than mice (Lois *et al.* 2002; Rubinson *et al.* 2003; Tiscornia *et al.* 2003). These developments suggest that lentiviruses may be routinely used both to label neurons and to silence genes in vivo.

Thus, the use of viral vectors to deliver genes to the nervous system is very promising, both for basic research and for therapeutic applications. Several landmark applications of in vivo neuronal labeling have been reported in recent years. However, the continual development of new labeling strategies and advances in imaging techniques suggest that this is just the beginning. In the near future, the genetic manipulation of sparsely labeled neurons should become possible. In addition, improved GFP-based reporters for neuronal physiology studies are likely to find a greater number of applications in vivo. It may even become possible to use viral vectors for RNAi-based genetic screens for neuronal function in species in which traditional forward genetics has not previously proved feasible.

16.5 Adult Neurogenesis: A Neural Basis for Experience-Induced Plasticity

Bulbar interneurons generated in adults may play an important role in olfactory learning and memory, possibly by connecting new combinations of projection neurons that increase in relevance during the life of the animal (Lledo *et al.* 2005). This feature may result from the unique properties of the newcomers, which are endowed with juvenile functions. During the development of the neuronal circuitry, environment-induced activity shapes the establishment of appropriate patterns in the neuronal circuitry. This experience-induced activity is particularly influential during a sensitive period of development – the critical period – during which maturation of the GABAergic network is thought to be one of the key factors contributing to functional remodeling (Hensch 2005). Thus, in contrast to other sensory systems, the olfactory bulb interneurons display constant neurogenesis, raising the possibility of an endless critical period for olfaction.

A previous study on the functional response of olfactory granule neurons generated in adult animals showed that, unlike preexisting neurons, newly generated granule cells preferentially respond to new odors after their synaptic integration, consistent with the view that these neurons have unique properties and make a specific contribution to the olfactory circuitry (Magavi *et al.* 2005). The greater sensitivity of adult-born cells to new odors seems to emerge after a period of familiarization with these odors, indicating that the greater responsiveness of the newly formed neurons is both stimulus specific and experience dependent. Thus, the bulbar interneurons generated in adult animals seem to have unusual properties that make a unique contribution to the olfactory system. It has been suggested that bulbar, and particularly, granule cells neurogenesis in adults is involved in learning and memory, but this phenomenon should also be considered in other contexts. In addition to contributing to endogenous repair, adult-born interneurons may be involved in inducing, facilitating, or maintaining olfactory-mediated behavior, such as food selection, mate selection, or caring for young (Shingo *et al.* 2003). Studies of the potential role of newly formed neurons in these processes may provide insight into the functional significance of adult neurogenesis.

Active neurogenesis clearly occurs in the adult olfactory bulb, but only a small percentage of the newly formed neurons survive. These surviving interneurons integrate into bulbar circuits, whereas the others are destroyed. Quantitative studies have demonstrated that almost half of all newly generated interneurons are eliminated 1–2 months after their creation (Petreanu and Alvarez-Buylla 2002; Winner *et al.* 2002). The other 50%, which survive this initial period of cell elimination, generally survive for at least another 19 months (Winner *et al.* 2002). Consistent with these findings, a previous study based on nostril occlusion at various time points suggested that the survival of newly generated olfactory bulb granule cells was affected by sensory experience, but only when the cells concerned were between 14 and 28 days old (Yamaguchi and Mori 2005). It would therefore be of interest to determine the regulatory mechanisms controlling the survival of newly generated neurons. Several studies have suggested that long-lasting, odorant-dependent activity in the olfactory bulb may control the survival of the new neurons and improve olfactory learning (Petreanu and Alvarez-Buylla 2002; Rochefort *et al.* 2002; Mandairon *et al.* 2003). Mitogen-activated protein (MAP) kinase activation may contribute to the odorant-induced increase in the survival of adult-born bulbar interneurons (Miwa and Storm 2005).

16.6 Contribution of Adult-Born Neurons to Sensory Information Processing

A myriad of factors regulate the production, migration, and development of newly generated neurons in adults and determine whether these neurons die or survive. However, once a newly generated cell has reached its target, matured into a neuron, and survived, what does it do then? Why is it there? What is its function? How long does it survive?

16.6.1 A Role in Sensory Information Processing

In theory, the new neurons may contribute to adult brain functions through actions at the cellular, network, and systems levels. At the cellular level, the function of these new cells is quite obvious: they are just neurons. However, they may be restricted to a subcategory of preexisting interneurons [in the olfactory bulb, hydroxylase-expressing (TH), calbindin (CalB)-, calretinin (CalR)-, and parvalbumin (ParV)-positive cells]. The newcomers may also differ from the preexisting neurons in terms of target identity. Previous studies of the maturation of newly generated cells have suggested that their cellular function may differ from that of the existing cells, at least transiently. Generally, the newly generated neurons are likely to be more excitable and more adaptable than the older, established neurons. In many ways, they act as do the neurons of the embryonic brain, readily making new synapses with other neurons and changing the strength of these connections following their integration into mature circuits. Young granule cells in the adult olfactory bulb, for example, display much greater synaptic plasticity than older granule cells (Nissant *et al.* 2009). In addition, newly generated granule and periglomerular cells have active membrane properties markedly different from those of the established neurons surrounding them (Carleton *et al.* 2003; Belluzzi *et al.* 2003), and display greater plasticity in response to sensory deprivation (Saghatelyan *et al.* 2005).

There is now clear evidence that olfactory interneurons produced postnatally in the olfactory bulb are physiologically integrated (Carlén *et al.* 2002; Belluzzi *et al.* 2003; Carleton *et al.* 2003). However, the functional significance of the newly generated interneurons in the olfactory bulb remains unclear. Nevertheless, growing interest in adult neurogenesis has led to advances in our understanding of the molecular, cellular, and physiological properties and location of newly generated neurons. Evidence is accumulating that newborn neurons have properties lacked by their established counterparts. These unique properties provide a fundamental framework for elucidating the functional role of the neurons generated in adult animals. Newly generated neurons in the olfactory bulb have different characteristics at early and late postnatal stages, suggesting that postnatal neurogenesis does not give rise to a uniform population of bulbar interneurons. The newly generated interneurons are positioned deeper in the granule cell layer in late postnatal than in early postnatal periods (Lemasson *et al.* 2005). As no difference in apoptosis across the granule cell layer is observed between stages, this gradient may reflect differences in the recruitment of newly generated interneurons, which suggests that the cells generated in the early postnatal period and in adulthood respond differently to common signals or respond to different molecules during the life of the animal. The neurons generated later in the postnatal period also migrate to the olfactory bulb more rapidly than those generated early in the postnatal period (Lemasson *et al.* 2005). The glial tube surrounding the migrating neuroblasts is difficult to detect before the third week of postnatal development (Peretto *et al.* 1997, 1999). The observed difference in migration speed may therefore be the result of molecules being released from the glial processes of astrocytes along the RMS of the adult, facilitating neuroblast migration. The newly generated interneurons of the mature olfactory bulb also differ from those established during the development of the olfactory bulb in having a much lower survival rate. It has been suggested that interneurons generated early in life, with a high rate of survival, play a critical role in determining the control of precocious olfactory processes over lifelong behavior (Lemasson *et al.* 2005), whereas interneurons generated during adulthood, which have a much lower rate of survival, fine tune olfaction (Gheusi *et al.* 2000; Rochefort *et al.* 2002; Enwere *et al.* 2004). Investigations of the causes of these differences in the survival potential of neurons produced at different ages would help to broaden our understanding of these functions.

At the network level, the newly generated neurons may contribute to the properties driving the concerted activity of groups of cells. In the olfactory bulb, the oscillations and synchrony of neuronal activity are major network phenomena serving as potential targets for adult neurogenesis. Indeed, the olfactory bulb receives and processes signals from OSNs and transmits them to higher processing centers. The mapping of sensory inputs has revealed that each odorant generates a reproducible spatial pattern of restricted activation in the glomerular layer of the olfactory bulb. However, within the bulb, these confined patterns are transformed into widely distributed patterns by lateral interactions between relay neurons and local interneurons. Thus, odor information processing requires the spatial patterning of both sensory inputs and synaptic interactions mediated by local interneurons. Odor representation is thus highly dynamic and temporally orchestrated by local interneurons (Lledo and Lagier 2006). Adult neurogenesis may lead to persistent or transient enhancement of the lateral inhibition mediated by GABAergic interneurons of the olfactory bulb network, facilitating odorant discrimination and increasing the chances of survival of newly generated interneurons. The continuous recruitment of local interneurons throughout life is therefore a key means of continually adapting transposition from the spatial to the temporal dimension, to optimize olfactory information processing.

Finally, in terms of functional plasticity, neurogenesis at the network level should be considered as a potential permissive or instructive mechanism. It is still unknown whether the new cells have predetermined functions, enabling them to instruct further changes in the networks they join, or whether they simply facilitate changes to occur in these circuits. The perceptual and mnemonic functions of the olfactory bulb have understandably led to investigations of the importance of adult neurogenesis in these processes. The gap between cell generation and cognitive processes is large, but we are beginning to understand how the plasticity afforded by adult neurogenesis might contribute to shaping the behavioral responses of the animal (Lledo and Gheusi 2003).

16.6.2 Functional Diversity of Newly Generated Interneurons

Clear functional roles have yet to be defined for cells expressing TH, CalR, CalB, or ParV. Adult neurogenesis leads to the continuous renewal of CB+, CR+ and TH+ interneurons throughout life (Young *et al.* 2007; Merkle *et al.* 2007; Kohwi *et al.* 2007). These cell types differ not only in their neurochemistry, marker gene expression, and location in the olfactory bulb but also in their structural features and synaptic properties (Kosaka *et al.* 1998; Shipley and Ennis 1996; Kosaka and Kosaka 2005; Parrish-Aungst *et al.* 2007; Pressler and Strowbridge 2006). For example, it has been suggested that the different subtypes of periglomerular cells have specific functions (Kosaka *et al.* 1998).

Further studies will probably link these functional features with neurochemical markers, providing further evidence of periglomerular cell heterogeneity (Whitman and Greer 2007). However, it is noteworthy that the use of immunohistological markers to define cell types may be somewhat misleading, as these indicators may be transiently expressed at different levels during cell maturation, or in response to environmental stimuli. For example, glutamic acid decarboxylase (GAD), GABA, and CalR are present in neuroblasts from birth onward (Stewart *et al.* 2002) and are not affected by olfactory bulb odor deprivation (Stone *et al.* 1991). By contrast, TH immunoreactivity is not detectable before 10 days, peaks at 3 months (Winner *et al.* 2002; Baker *et al.* 1983), and is dramatically downregulated by odor deprivation (Philpot *et al.* 1997). Thus, the expression of neurochemical markers may, in some cases, be correlated with anatomical or physiological features of cell subtypes and may therefore be good markers of cell fate. In other cases, the expression of these markers may simply indicate the presence of transient neuronal features.

What are the functional implications of periglomerular cell integration for the first synapse in olfaction? We reported significant maturation in the postsynaptic features of olfactory nerve inputs to newly generated periglomerular cells (Grubb *et al.* 2008);

this indicates that immature connections between the nose and brain are different from their mature counterparts. Thus, in early postnatal development, despite the requirement for olfaction from birth for survival, which implies that the first synapse in olfaction is functional from birth, olfaction – and its first synapse in particular – needs not necessarily be fully mature from birth. Indeed, our data provide functional correlates for the considerable anatomical plasticity occurring postnatally in the primary olfactory projection (Zou *et al.* 2004). As in the granule cell layer of the olfactory bulb or the dentate gyrus of the hippocampus, newly generated periglomerular cells receive inputs functionally different from those of their resident neighbors, albeit transiently. Thus, for limited time periods, newly generated periglomerular cells in the adult brain can provide a functionally different connection for sensory information arriving from the periphery. However, as this different function is limited to the postsynaptic half of the synapse, presynaptic olfactory nerve terminals seem to be appropriate for the reliable signaling of olfactory information from the nose to the brain in the face of remarkable anatomical plasticity in terms of connections.

Granule cells constitute the largest interneuron subpopulation in the olfactory bulb, accounting for 90% of the total population of bulbar interneurons. In contrast to periglomerular cells, granule cells have no axon, and therefore make contacts exclusively through their dendrites and cell bodies (Shepherd *et al.* 2007). They are GABAergic interneurons mediating several types of inhibition over a wide range of spatial scales (Lledo and Lagier 2006) (see Box 2). Newly generated granule cells may be important for olfactory learning and memory, possibly through the formation of new combinations of projection neurons of increasing relevance during the life and experience of the subject (Lledo *et al.* 2005).

Adult olfactory bulb neurogenesis is affected by diverse neurological disorders and environmental stimuli, suggesting that newly generated neuronal cells may respond to these processes. The number of new granule cells in the olfactory bulb was shown to increase during olfactory learning (Alonso *et al.* 2006). Surprisingly, a simple exposure to olfactory cues did not affect the number of newborn granule cells; only odorants presented in a learning context could enhance the survival of a specific subset of granule cells (Alonso *et al.* 2006). These results imply that association learning may locally control the survival of newborn granule cells, thus potentially affecting the function of a given microcircuit while sparing the rest of the olfactory bulb network. Further studies are required to determine whether this heterogeneity in cell survival reflects interneuron diversity.

16.7 Concluding Remarks

We still do not know why new neurons are incorporated into some parts of the adult brain and not into others, but research into the regulation of adult neurogenesis, and an increasing number of studies investigating the potential function of the new cells, have provided two major pieces of information. First, adult neurogenesis is affected by relevant behavior, including behavior specifically related to olfactory bulb functions. Second, changes in the rate or extent of neurogenesis have in turn an effect on subsequent behavior. Thus, experience may change the entire structure of adult circuits, and changes in structure may modify or affect subsequent behavioral responses.

The correlation between neuronal activity and adult neurogenesis in the olfactory system suggests that new neurons are involved in general aspects of bulbar function, probably sustaining the ability of the adult network to adapt information processing to relevant ethological needs. Several studies have demonstrated that odor representation at the first central relay is dynamic and highly complex and may therefore require unique plasticity mechanisms. Neuronal production, migration, and the replacement of interneurons in the olfactory bulb are probably involved in this adaptive mechanism. However, the precise nature of the function required for adaptation remains unknown, which may be because we still know little about the chemical ecology of animals that rely heavily on olfaction. We still have very little idea of what constitutes a natural olfactory "scene" for a rodent. The bulbar interneurons are known to play a crucial role in transposing the spatial dimension of olfactory information into a temporal dimension. They also distribute this information throughout the entire bulbar circuit. However, we do not yet know whether the newcomers can accomplish both these tasks. The contribution of adult-born granule cells to the activity of preexisting networks may be different from that of established cells, at least during the early stages of their maturation.

Furthermore, adult neurogenesis does not simply introduce fresh cells into old circuits. Some selection also occurs as about half the new interneurons die within a month of maturation. It has been shown that the newly formed granule cells maintain synaptic contacts during this period of rapid cell death. The selective elimination of neurons may make it possible to sculpt the circuitry rapidly. As a result, most of the newly generated neurons never mature fully and may therefore never play a role similar to that of the preexisting neurons. An ability to mold the integration of new neurons and to eliminate other neurons continually, without depleting the neuronal population, may provide neuronal networks with an adaptive potential unequaled by synaptic plasticity alone and a substrate for learning.

16.7.1 Two Seemingly Contradictory Tasks: Ensuring Stability and Flexibility

The adult brain must guide the organism through an ever-changing world. In this context, it faces two conflicting challenges: maintaining behavioral responses and long-term traces, which requires preservation of the necessary synaptic organization of a given circuitry, and maintaining a capacity to adapt following brain injury, in learning conditions, or when faced with environmental changes. At first glance, neurogenesis in the adult brain would seem to bias the system in favor of flexibility. The introduction of a completely new cell into a functioning circuit or the removal of an existing neuron from the existing network is, after all, a major step.

However, computational models have suggested that adult neurogenesis, rather than increasing the flexibility of the network, may serve to stabilize an already plastic network confronted with a large volume of new information.

The discovery of adult neurogenesis has led to efforts to understand the relationship between the presence of new cells and learning in various model systems. However, conflicting results have been obtained. For instance, Gould *et al.* (1999) reported an increase in the survival rates of newly formed hippocampal neurons in rats subjected to spatial learning tests in the Morris water maze. By contrast, van Praag *et al.* (1999) observed no change in neurogenesis following water-maze training in mice. Other studies have confirmed the complex nature of adult neurogenesis, as both decreases and increases in neurogenesis have been reported with water-maze learning (Ambrogini *et al.* 2004; Ehninger and Kempermann 2006). These discrepancies may result from changes in the functions of newly formed neurons during their maturation. They may also reflect the temporal relationship between the labeling of newly formed neurons and exposure to the memory task (Döbrössy *et al.* 2003; Dupret *et al.* 2007). Many more new neurons than are actually needed may be produced, requiring the active and selective removal of those that have not established appropriate synaptic contacts. The learning-induced mechanisms involved are probably similar to those mediating brain development and maturation, shaping relevant and functional neural networks in the developing brain. There may be a balance between the addition and elimination of newly formed neurons during learning. According to this view, learning depends not only on the addition of new neurons and new synaptic connections, but also on regressive events culminating in the removal of neurons from the cellular network of the adult central nervous system.

16.7.2 Challenges for the Future

Improvements in our understanding of the role of adult neurogenesis in brain plasticity do not depend solely on the collection of larger amounts of higher-quality data concerning the dynamics, spatial extent, and longevity of experience-dependent structural changes in adult circuits. They also depend on (1) an extensive precise description of the integrative properties of individual adult-born neurons; (2) the design of better models taking into account the redundancies of neuronal network connections together with their sparse coding; and (3) a more complete description of the guidance and selection mechanisms acting on new neurons. Identification of the physical changes encoding particular elements of long-term information remains a central issue in neuroscience. Interdisciplinary approaches combining molecular, anatomical, physiological, and theoretical methods are likely to be required to address this and other questions relating to the physical substrate for long-term adaptation in the adult brain circuits.

Those like us who are trying to decipher the functional meaning of adult neurogenesis may find some clues by taking into account its extremely limited regional distribution. What is so special about the olfactory bulb and the hippocampus that these two adult regions are the only structures into which new neurons are continuously plugged? The answer to this question might be found when taking into account the specific task these networks have to fulfill. The olfactory bulb and hippocampal circuits are vital in the segmentation of olfactory and spatial information, respectively. Indeed, these circuits are both involved in decorrelating complex overlapping afferent inputs. It has also been suggested that both the olfactory bulb and hippocampus handle particularly large amounts of information (Chambers *et al.* 2004) or employ particular coding strategies to manage that information (Cecchi *et al.* 2001). However, for every similarity between the olfactory bulb and the hippocampus, large differences can be found between the two structures, not least in adult neurogenesis. More detailed investigations of the function of adult neurogenesis are required to determine whether this process plays a similar, specific role in the adult olfactory bulb and hippocampus.

Acknowledgments I apologize to those authors whose references, although relevant to this subject, have not been included in this review for reasons of space constraints. My laboratory is supported by the Fondation pour la Recherche Médicale (Équipe FRM), by the life insurance company "AG2R-La-Mondiale," and by Ecole des Neurosciences de Paris (ENP).

References

Ache BW and Young JM (2005) Olfaction: diverse species, conserved principles. Neuron 48:417–430.

Adrian ED (1942) Olfactory reactions in the brain of the hedgehog. J Physiol 100:459–473.

Alonso M, Viollet C, Gabellec MM et al. (2006) Olfactory discrimination learning increases the survival of adult-born neurons in the olfactory bulb. J Neurosci 26:10508–10513.

Altman J and Das GD (1965) Post-natal origin of microneurones in the rat brain. Nature 207:953–956.

Alvarez-Buylla A (1990) Mechanism of neurogenesis in adult avian brain. Experientia 46:948–955.

Alvarez-Buylla A and Garcia-Verdugo JM (2002) Neurogenesis in adult subventricular zone. J Neurosci 22:629–634.

Alvarez-Buylla A, Garcia-Verdugo JM and Tramontin AD (2001) A unified hypothesis on the lineage of neural stem cells. Nat Rev Neurosci 2:287–293.

Ambrogini P, Orsini L, Mancini C et al. (2004) Learning may reduce neurogenesis in adult rat dentate gyrus. Neurosci Lett 359:13–16.

Aungst JL, Heyward PM, Puche AC et al. (2003) Centre-surround inhibition among olfactory bulb glomeruli. Nature 426:623–629.

Baker H, Kawano T, Margolis FL et al. (1983) Transneuronal regulation of tyrosine hydroxylase expression in olfactory bulb of mouse and rat. J Neurosci 3:69–78.

Belluzzi O, Benedusi M, Ackman J et al. (2003) Electrophysiological differentiation of new neurons in the olfactory bulb. J Neurosci 23:10411–10418.

Beltz BS and Sandeman DC (2003) Regulation of life-long neurogenesis in the decapod brain. Arthropod Struct Dev 32:39–60.

Buck L, Axel R (1991) A novel multigene family may encode odorant receptors: a molecular basis for odor recognition. Cell 65:175–187.

Carlén M, Cassidy RM, Brismar H et al. (2002) Functional integration of adult-born neurons. Curr Biol 12:606–608.

Carleton A, Petreanu LT, Lansford R et al. (2003) Becoming a new neuron in the adult olfactory bulb. Nat Neurosci 6:507–518.
Cecchi GA, Petreanu LT, Alvarez-Buylla A et al. (2001) Unsupervised learning and adaptation in a model of adult neurogenesis. J Comput Neurosci 11:175–182.
Chambers RA, Potenza MN, Hoffman RE et al. (2004) Simulated apoptosis/neurogenesis regulates learning and memory capabilities of adaptive neural networks. Neuropsychopharmacology 29:747–758.
Changeux JP and Danchin A (1976) Selective stabilisation of developing synapses as a mechanism for the specification of neuronal networks. Nature 264:705–712.
Cleland TA and Linster C (2005) Computation in the olfactory system. Chem Senses 30:801–813.
Davenne M, Custody C, Charneau P and Lledo P-M (2005) In vivo imaging of migrating neurons in the mammalian forebrain. Chem Senses 30:115–116.
Davidson BL and Breakefield XO (2003) Viral vectors for gene delivery to the nervous system. Nat Rev Neurosci 4:353–364.
Döbrössy MDE, Aurousseau C, Le Moal M et al. (2003) Differential effects of learning on neurogenesis: Learning increases or decreases the number of newly born cells depending on their birth date. Mol Psychiatry 8:974–982.
Doetsch F (2003) A niche for adult neural stem cells. Curr Opin Genet Dev 13:543–550.
Doetsch F, Caille I, Lim DA et al. (1999) Subventricular zone astrocytes are neural stem cells in the adult mammalian brain. Cell 97:703–716.
Dupret D, Fabre A, Döbrössy MD et al. (2007) Spatial learning depends on both the addition and removal of new hippocampal neurons. PLoS Biol 5:e214.
Ehninger D and Kempermann G (2006) Paradoxical effects of learning the Morris water maze on adult hippocampal neurogenesis in mice may be explained by a combination of stress and physical activity. Genes Brain Behav 5:29–39.
Eisthen HL (1997) Evolution of vertebrate olfactory systems. Brain Behav Evol 50:222–233.
Enwere E, Shingo T, Gregg C et al. (2004) Aging results in reduced epidermal growth factor receptor signaling, diminished olfactory neurogenesis, and deficits in fine olfactory discrimination. J Neurosci 24:8354–8365.
Eriksson PS, Perfilieva E, Bjork-Eriksson T et al. (1998) Neurogenesis in the adult human hippocampus. Nat Med 4:1313–1317.
Falk N, Holmstrom M, Carlen R et al. (2002) Gene delivery to adult neural stem cells. Exp Cell Res 279:34–39.
Freeman WJ and Schneider W (1982) Changes in spatial patterns of rabbit olfactory EEG with conditioning to odors. Psychophysiology 19:44–56.
Frielingsdorf H, Schwarz K, Brundin P et al. (2004) No evidence for new dopaminergic neurons in the adult mammalian substantia nigra. Proc Natl Acad Sci USA 101:10177–10182.
Garcia AD, Doan NB, Imura T et al. (2004) GFAP-expressing progenitors are the principal source of constitutive neurogenesis in adult mouse forebrain. Nat Neurosci 7:1233–1241.
Garcia-Verdugo JM, Doetsch F. Wichterle H et al. (1998) Architecture and cell types of the adult subventricular zone: in search of the stem cells. J Neurobiol 36:234–248.
Gheusi G, Cremer H, McLean H et al. (2000) Importance of newly generated neurons in the adult OB for odor discrimination. Proc Natl Acad Sci USA 97:1823–1828.
Goldman SA and Nottebohm F (1983) Neuronal production, migration and differentiation in a vocal control nucleus of the adult female canary brain. Proc Natl Acad Sci USA 80:2390–2394.
Gould E, Reeves AJ, Graziano MS et al. (1999) Neurogenesis in the neocortex of the adult primates. Science 286:548–552.
Grubb MS, Nissant A, Murray K and Lledo PM (2008) Functional maturation of the first synapse in olfaction: development and adult neurogenesis. J Neurosci 28:2919–2932.
Hack MA, Saghatelyan A, de Chevigny et al. (2005) Neuronal fate determinants of adult olfactory bulb neurogenesis. Nat Neurosci 8:865–872.
Hensch TK (2005) Critical period plasticity in local cortical circuits. Nat Rev Neurosci 6:877–888.

Hildebrand JG and Shepherd GM (1997) Mechanisms of olfactory discrimination: converging evidence for common principles across phyla. Annu Rev Neurosci 20:595–631.

Hinds JW (1968) Autoradiographic study of histogenesis in the mouse olfactory bulb. II. Cell proliferation and migration. J Comp Neurol 134: 305–322.

Jakobsson J, Ericson C, Jansson M et al. (2003) Targeted transgene expression in rat brain using lentiviral vectors. J Neurosci Res 73:876–885.

Kempermann G (2006) Adult neurogenesis. Oxford University Press, New York.

Kohwi M, Osumi N, Rubenstein JL et al. (2005) Pax6 is required for making specific subpopulations of granule and periglomerular neurons in the olfactory bulb. J Neurosci 25:6997–7003.

Kohwi M, Petryniak MA, Long JE et al. (2007) A subpopulation of olfactory bulb GABAergic interneurons is derived from Emx1- and Dlx5/6-expressing progenitors. J Neurosci 27:6878–6891.

Koketsu D, Mikami A, Miyamoto Y et al. (2003) Nonrenewal of neurons in the cerebral neocortex of adult macaque monkeys. J Neurosci 23:937–942.

Kornack DR and Rakic P (1999) Continuation of neurogenesis in the hippocampus of the adult macaque monkey. Proc Natl Acad Sci USA 96:5768–5773.

Kornack DR and Rakic P (2001) Cell proliferation without neurogenesis in adult primate neocortex. Science 294:2127–2130.

Kosaka T and Kosaka K (2005) Structural organization of the glomerulus in the main olfactory bulb. Chem Senses Suppl 1: i107–i108.

Kosaka K et al. (1997) Chemically defined neuron groups and their subpopulations in the glomerular layer of the rat main olfactory bulb. II. Prominent differences in the intraglomerular dendritic arborization and their relationship to olfactory nerve terminals. Neuroscience 76:775–786.

Kosaka K et al. (1998) How simple is the organization of the olfactory glomerulus?: the heterogeneity of so-called periglomerular cells. Neurosci Res 30:101–110.

Laywell ED, Rakic P, Kukekov VG et al. (2000) Identification of a multipotent astrocytic stem cell in the immature and adult mouse brain. Proc Natl Acad Sci USA 97:13883–13888.

Lemasson M, Saghatelyan A, Olivo-Marin JC et al. (2005) Neonatal and adult neurogenesis provide two distinct populations of granule cells in the mouse OB. J Neurosci 25:6816–6825.

Lledo P-M and Gheusi G (2003) Olfactory processing in a changing brain. Neuroreport 14:1655–1663.

Lledo P-M and Lagier S (2006) Local interneurons transduce spatial coding into temporal patterning in the mammalian olfactory bulb. Semin Cell Dev Biol 17:443–453.

Lledo P-M and Saghatelyan A (2005) Integrating new neurons into the adult olfactory bulb: joining the network, life/death decisions, and the effects of sensory experience. Trends Neurosci 28:248–254.

Lledo P-M, Gheusi G and Vincent JD (2005) Information processing in the mammalian olfactory system. Physiol Rev 85:281–317.

Lois C and Alvarez-Buylla A (1994) Long-distance neuronal migration in the adult mammalian brain. Science 264:1145–1148.

Lois C, Hong EJ, Pease S et al. (2002) Germline transmission and tissue-specific expression of transgenes delivered by lentiviral vectors. Science 295:868–872.

Magavi SSP, Mitchell BD, Szentirmai O et al. (2005) Adult-born and preexisting olfactory granule neurons undergo distinct experience-dependent modifications of their olfactory responses in vivo. J Neurosci 25:10729–10739.

Malatesta P, Hack MA, Hartfuss E et al. (2003) Neuronal or glial progeny: regional differences in radial glia fate. Neuron 37:751–764.

Malnic B, Hirono J, Sato T et al. (1999) Combinatorial receptor codes for odors. Cell 96:713–723.

Mandairon N, Jourdan F and Didier A (2003) Deprivation of sensory inputs to the olfactory bulb up-regulates cell death and proliferation in the subventricular zone of adult mice. Neuroscience 119:507–516.

Marshall CAG, Novitch BG and Goldman JE (2005) Olig2 directs astrocyte and oligodendrocyte formation in postnatal subventricular zone cells. J Neurosci 25:7289–7298.

Mechawar N, Saghatelyan A, Grailhe R et al. (2004) Nicotinic receptors regulate the survival of newborn neurons in the adult OB. Proc Natl Acad Sci USA 101:9822–9826.
Merkle FT, Mirzadeh Z and Alvarez-Buylla A (2007) Mosaic organization of neural stem cells in the adult brain. Science 317:381–384.
Miwa N and Storm DR (2005) Odorant-induced activation of extracellular signal-regulated kinase/mitogen-activated protein kinase in the olfactory bulb promotes survival of newly formed granule cells. J Neurosci 25:5404–5412.
Mizrahi A and Katz LC (2003) Dendritic stability in the adult olfactory bulb. Nat Neurosci 6:1201–1207.
Mori K, Takahashi YK, Igarashi KM et al. (2006) Maps of odorant molecular features in the mammalian olfactory bulb. Physiol Rev 86:409–433.
Morshead CM, Reynolds BA, Craig CG et al. (1994) Neural stem cells in the adults mammalian forebrain: a relatively quiescent subpopulation of subependymal cells. Neuron 13:1071–1082.
Nissant A, Bardy C, Katagiri H et al. (2009) Adult neurogenesis promotes synaptic plasticity in the olfactory bulb. Nat Neurosci 12:728–730.
Oppenheim RW (1991) Cell death during development of the nervous system. Annu Rev Neurosci 14:453–501.
Parrish-Aungst S, Shipley MT, Erdelyi et al. (2007) Quantitative analysis of neuronal diversity in the mouse olfactory bulb. J Comp Neurol 501:825–836.
Peretto P, Merighi A, Fasolo A et al. (1997) Glial tubes in the rostral migratory stream of the adult rat. Brain Res Bull 42:9–21.
Peretto P, Merighi A, Fasolo A et al. (1999) The subependymal layer in rodents: a site of structural plasticity and cell migration in the adult mammalian brain. Brain Res Bull 49:221–243.
Petreanu L and Alvarez-Buylla A (2002) Maturation and death of adult-born OB granule neurons: Role of olfaction. J Neurosci 22:6106–6113.
Philpot BD, Lim JH and Brunjes PC (1997) Activity-dependent regulation of calcium-binding proteins in the developing rat olfactory bulb. J Comp Neurol 387:12–26.
Pressler RT and Strowbridge BW (2006) Blanes cells mediate persistent feedforward inhibition onto granule cells in the olfactory bulb. Neuron 49:889–904.
Rakic P (2002) Adult neurogenesis in mammals, an identity crisis. J Neurosci 22:614–618.
Rochefort C, Gheusi G, Vincent JD et al. (2002) Enriched odor exposure increases the number of newborn neurons in the adult OB and improves odor memory. J Neurosci 22:2679–2689.
Rogelius N, Ericson C and Lundberg C (2005) In vivo labeling of neuroblasts in the subventricular zone of rats. J Neurosci Methods 142:285–293.
Rubinson DA, Dillon CP, Kwiatkowski AV et al. (2003) A lentivirus-based system to functionally silence genes in primary mammalian cells, stem cells and transgenic mice by RNA interference. Nat Genet 33:401–406.
Saghatelyan A, Roux P, Migliore M et al. (2005) Activity-dependent adjustments of the inhibitory network in the adult OB following early postnatal deprivation. Neuron 46:103–116.
Schmidt-Hieber C, Jonas P and Bischofberger J (2004) Enhanced synaptic plasticity in newly generated granule cells of the adult hippocampus. Nature 429:184–187.
Schoppa NE and Urban NN (2003) Dendritic processing within olfactory bulb circuits. Trends Neurosci 26:501–506.
Scotto-Lomassese S, Strambi C, Strambi A et al. (2003) Suppression of adult neurogenesis impairs olfactory learning and memory in an adult insect. J Neurosci 23:9289–9296.
Seri B, Garcia-Verdugo JM, McEwen BS and Alvarez-Buylla A (2001) Astrocytes give rise to new neurons in the adult mammalian hippocampus. J Neurosci 21:7153–7160.
Shepherd GM (2006) Smell images and the flavour system in the human brain. Nature 444:316–321.
Shepherd GM, Chen WR and Greer CA (2004) Olfactory bulb. In: Shepherd GM (ed) The synaptic organization of the brain (5th ed.), edited by Oxford University Press, New York, pp. 165–216.
Shepherd GM, Chen WR, Willhite D et al. (2007) The olfactory granule cell: from classical enigma to central role in olfactory processing. Brain Res Rev 55:373–382.

Shingo T, Gregg C, Enwere E et al. (2003) Pregnancy-stimulated neurogenesis in the adult female forebrain mediated by prolactin. Science 299:117–120.

Shipley MT and Ennis M (1996) Functional organization of olfactory system. J Neurobiol 30:123–176.

Stewart RR, Hoge GJ, Zigova T et al. (2002) Neural progenitor cells of the neonatal rat anterior subventricular zone express functional GABA(A) receptors. J Neurobiol 50:305–322.

Stone DM, Grillo M, Margolis FL et al. (1991) Differential effect of functional olfactory bulb deafferentation on tyrosine hydroxylase and glutamic acid decarboxylase messenger RNA levels in rodent juxtaglomerular neurons. J Comp Neurol 311:223–233.

Tiscornia G, Singer O, Ikawa M et al. (2003) A general method for gene knockdown in mice by using lentiviral vectors expressing small interfering RNA. Proc Natl Acad Sci USA 100:1844–1848.

van Praag H, Kempermann G, Gage FH (1999) Running increases cell proliferation and neurogenesis in the adult mouse dentate gyrus. Nat Neurosci 2:266–270.

Wachowiak M and Shipley MT (2006) Coding and synaptic processing of sensory information in the glomerular layer of the olfactory bulb. Semin Cell Dev Biol 17:411–423.

Washbourne P and McAllister AK (2002) Techniques for gene transfer into neurons. Curr Opin Neurobiol 12:566–573.

Whitman MC and Greer CA (2007) Synaptic integration of adult-generated olfactory bulb granule cells: basal axodendritic centrifugal input precedes apical dendrodendritic local circuits. J Neurosci 27:9951–9961.

Willhite DC, Nguyen KT, Masurkar AV et al. (2006) Viral tracing identifies distributed columnar organization in the olfactory bulb. Proc Natl Acad Sci USA 103:12592–12597.

Wilson RI and Mainen ZF (2006) Early events in olfactory processing. Annu Rev Neurosci 29:163–201.

Winner B, Cooper-Kuhn CM, Aigner R et al. (2002) Long-term survival and cell death of newly generated neurons in the adult rat OB. Eur J Neurosci 16:1681–1689.

Yamaguchi M and Mori K (2005) Critical period for sensory experience-dependent survival of newly generated granule cells in the adult mouse OB. Proc Natl Acad Sci USA 102: 9697–9702.

Young JM, Friedman C, Williams EM et al. (2002) Different evolutionary processes shaped the mouse and human olfactory receptor gene families. Hum Mol Genet 11:535–546.

Young KM, Fogarty M, Kessaris N et al. (2007) Subventricular zone stem cells are heterogeneous with respect to their embryonic origins and neurogenic fates in the adult olfactory bulb. J Neurosci 27:8286–8296.

Zou DJ, Feinstein P, Rivers AL et al. (2004) Postnatal refinement of peripheral olfactory projections. Science 304:1976–1979.

Chapter 17
Control of Adult-Born Neuron Production by Converging GABA and Glutamate Signals

Jean-Claude Platel and Angélique Bordey

Abstract The production of adult-born neuron is an ongoing process accounting ~10,000 immature neurons migrating to the olfactory bulb every day. This high turnover rate necessitates profound control mechanisms converging onto neural progenitors and neuroblasts to achieve adequate adult-born neuron production. Here, we elaborate on a novel epigenetic control of adult neurogenesis via highly coordinated nonsynaptic cell–cell signaling. This communication engages the neurotransmitters GABA and glutamate whose extracellular concentrations depend on neuroblast number and high affinity uptake systems in neural stem cells. Neuroblasts release GABA providing a negative feedback control of stem cell proliferation and instructing them on the size of the neuroblast pool. Recent findings show an unexpected mosaic expression of glutamate receptors leading to calcium elevations in migrating neuroblasts and differential role on their development. Remarkably, stem cells act as lighthouses releasing glutamate onto neuroblast sailing by, thus providing migratory, survival, and proliferative cues. Finally, we propose that the timing of neurotransmitter release and their spatial diffusion will determine the convergent co-activation of neuroblasts and stem cells, and provide a steady-state level of neuron production. Upon external impact or injury this signaling may adjust to a new steady-state level, thus providing nonsynaptic scaling of neuroblast production.

J.-C. Platel and A. Bordey (✉)
Department of Neurosurgery, Yale University School of Medicine,
333 Cedar Street, FMB 422, New Haven, CT 06520-8082, USA
and
Department of Cellular and Molecular Physiology, Yale University School of Medicine,
333 Cedar Street, FMB 422, New Haven, CT 06520-8082, USA
e-mail: angelique.bordey@yale.edu

T. Seki et al. (eds.), *Neurogenesis in the Adult Brain I: Neurobiology*,
DOI 10.1007/978-4-431-53933-9_17, © Springer 2011

17.1 Introduction

The production of adult-born neurons persists in two brain regions, the subventricular zone (SVZ) and the dentate gyrus subgranular zone (SGZ) in the hippocampus. The SVZ contains the largest pool of dividing neural stem cells in the adult mammalian brain, including in humans (Sanai et al., 2004; Curtis et al., 2007). Divisions of stem cells generate intermediate progenitors (called transit amplifying cells), which in turn divide to give rise to neuroblasts (see Chap. 12). Neural stem cells express glial fibrillary acidic protein, a marker of astrocyte, and display other properties of astrocytes (Liu et al., 2006). The SVZ neural stem cells will be called stem cells or GFAP cells interchangeably throughout the text. It remains unknown whether every GFAP cell can behave as a stem cell. However, the properties of GFAP cells described below are found in all GFAP cells, suggesting that GFAP cells and stem cells are equivalent regarding our findings. Neuroblasts migrate along the rostral migratory stream (RMS) to the olfactory bulb where they differentiate into interneurons. Here, we discuss data obtained in the SVZ-to-olfactory bulb axis. Data discussing GABAergic signaling in the SGZ can be found in other reviews (Ge et al., 2007; Bordey, 2007).

Adult-born neuron production is an ongoing process. Estimates suggest that ~10,000 neurons migrate to the olfactory bulb every day (Lledo et al., 2006). This high turnover rate necessitates profound homeostatic control mechanisms. The production of neuroblasts and ultimately their number reaching the olfactory bulb is determined by the combined outcome of cell proliferation, migration, and survival, each of which is regulated by genetic and epigenetic mechanisms. For example, it is thought that each cell has an internal clock determining the number of divisions it can undergo or the length of time it can remain in cycle. Epigenetic factors, which include mitogenic or antiproliferative factors in the local environment, have been shown to control the length of the cell cycle or the number of cells cycling and the speed of neuroblast migration prior to their integration into the olfactory bulb circuitry [see (Pathania et al., 2010) for review].

Here, we will discuss the function of the local signaling molecules GABA and glutamate on cell proliferation, migration, and survival in the SVZ prior to the acquisition of synaptic inputs. While we start to understand how these signals operate at the single cell level (although much remains to be explored), we do not clearly understand how these signals are integrated at a population level to deliver the appropriate number of neurons. We propose that cell number controls the concentration and supply of local GABA and glutamate that determines the level of receptor activation thus providing a feedback control on cell production. We argue that convergent paracrine GABA and glutamate signaling is necessary for fine-tuning the level of neuron production upon external demand or injury, thus providing a nonsynaptic scaling of neurogenesis. Finally, we discuss that identification of various glutamate receptors in neural stem cells and neuroblasts provide novel pharmacological targets to increase the production of adult-born neurons and promote functional repair following injury or neurodegenerative diseases.

17.2 Expression of GABA and Glutamate Receptors in Neuroblasts

GABA's action is mediated by at least three types of GABA receptors (Chebib and Johnston, 1999): $GABA_B$ receptors that are coupled to Ca^{2+} or K^+ channels by means of G proteins and intracellular second messengers; and $GABA_A$ and $GABA_C$ receptors that are ligand-gated Cl^- channels. Neuroblasts express classical benzodiazepine-sensitive $GABA_A$ receptors (Stewart et al., 2002; Wang et al., 2003b) (Fig. 17.1). As shown in immature neurons, GABA is depolarizing in SVZ neuroblasts, presumably due to high level intracellular chloride in these cells (Wang et al., 2003b). Using calcium imaging, we recently reported that ~90% of SVZ cells respond to GABA with an increase in intracellular calcium (Platel et al., 2008). This result indicates that $GABA_A$ receptors are ubiquitous among neuroblasts presumably due to their critical function on neuroblast development (see section below). It remains to be determined whether SVZ neuroblasts express functional $GABA_B$ and $GABA_C$ receptors.

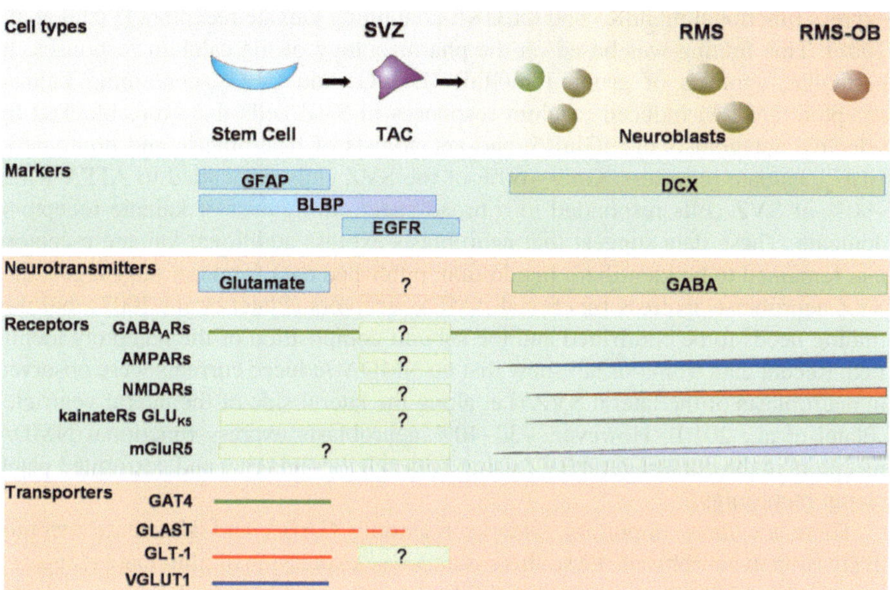

Fig. 17.1 Chart summarizing known GABA and glutamate signaling molecules along the SVZ cell lineage. GFAP stem cells contain glutamate while doublecortin (DCX)-positive neuroblasts synthesize and release GABA. Both GFAP cells and neuroblasts express functional $GABA_A$ receptors. Neuroblasts express a mosaic of glutamate receptors including ionotropic and metabotropic receptors. In addition, the level of receptor expression increases during their migration from the SVZ to the RMS of the olfactory bulb. GFAP cells express high affinity GABA and glutamate transporters not expressed in neuroblasts. The expression of other high affinity transporters remains to be explored. The presence of GABA and glutamate signaling molecules in transit amplifying cells (TAC), which express epidermal growth factor receptor (EGFR), remains unexplored

It has become evident that neuroblasts express an unexpected large repertoire of glutamate receptors much earlier than previously anticipated (Fig. 17.1). In neurons, glutamatergic signals are conveyed by different glutamate receptor subtypes, namely ionotropic NMDA and non-NMDA or AMPA/kainate receptors and metabotropic glutamate receptors. Eight genes encoding mGluRs have been classified into three groups: group I (mGluR1 and 5), group II (mGluR2 and 3), and group III (mGluR4, 6, 7, 8) (Coutinho and Knopfel, 2002). Group I mGluRs are coupled to phospholipase C and mobilize Ca^{2+} from IP3-regulated intracellular stores. Kainate receptors exist as multiple subtypes with a distinct biology and are composed of GLU_{K5-7}, GLU_{K1}, and GLU_{K2} subunits [IUPHAR nomenclature of the receptors that are also known as GluR5-7, KA1, and KA2, for review see (Bettler and Mulle, 1995; Lerma et al., 2001; Pinheiro and Mulle, 2006)]. GLU_{K5-7} either alone or in combination with GLU_{K1} or GLU_{K2} subunits form functional ion channels. Homomeric GLU_{K1} or GLU_{K2} receptors bind kainate with high affinity, but do not form functional ion channels. Recent findings from our group suggest the expression of functional AMPA/kainate receptors in neuroblasts in the SVZ and RMS (Platel et al., 2007). Using immunohistochemistry, calcium imaging, and more selective glutamate receptor ligands, we recently found that neuroblasts express functional mGluR5 and GLUK5-containing kainate receptors (Platel et al., 2008). This finding was based on the pharmacology of the calcium responses. In particular, agonists of group I mGluR (DHPG) and GLU_{K5}-containing kainate receptors (ATPA) induced calcium responses in SVZ cells that were blocked by selective antagonists of mGluR5, and antagonists of heteromeric and homomeric GLU_{K5} kainate receptors. Only ~60% of the SVZ cells responded to ATPA while ~90% of SVZ cells responded to a broad spectrum agonist of kainate receptors, domoate. These data suggest that neuroblasts express additional kainate receptors that remained to be identified. In addition, published data from our lab suggest that SVZ neuroblasts express functional AMPA receptors (Platel et al., 2007), but this finding needs to be confirmed and the subunit composition of the receptors identified. Recent data from our lab show that no NMDA-induced currents were observed in neuroblasts of the lateral SVZ (i.e. along the lateral side of the lateral ventricle) (Platel et al., 2010). However, ~30–40% neuroblasts express functional NMDA receptors in the dorso-lateral SVZ using both calcium imaging and perforated patch clamp recordings.

There are three important remarks regarding GABA and glutamate receptor expression in neuroblasts. First, there is a mosaic expression of glutamate receptors. About 40% of SVZ cells express $GABA_A$ receptors and both GLU_{K5} and mGluR5, 26% express either GLU_{K5} or mGluR5 with $GABA_A$ receptors while ~90% of neuroblasts express $GABA_A$ receptors either alone (24%) or with one these other receptors. These data were obtained by applying different agonists onto the same cells to obtain a relative receptor distribution. A subset of those neuroblasts expresses functional NMDA receptors. It remains to be determined whether neuroblasts expressing NMDA as well as AMPA receptors also co-express the other receptors. This different pattern of receptor expression further illustrates the diversity in neuroblast properties. In addition, neuroblasts have been shown to be predetermined to become certain types of olfactory bulb interneurons (Merkle et al., 2007). Second, neuroblasts progressively acquire more receptors during their tangential migration to the olfactory bulb. As a result in the RMS of the olfactory bulb ~90% neuroblasts express NMDA receptors prior to entering the synaptic circuit of the olfactory bulb. These findings suggest that neuroblasts become more differentiated towards a neuronal phenotype during their tangential migration and become primed to integrate into the olfactory bulb network. Third, the diversity of glutamate receptors in neuroblasts provides additional pharmacological targets for novel therapeutics aimed at controlling adult-born neuron production.

17.3 Non-Canonical GABAergic and Glutamatergic Signaling

Defining the organization of GABAergic or glutamatergic signaling requires the identification of source cells that release GABA or glutamate, target cells that express functional receptors, and uptake mechanisms to limit extracellular GABA and glutamate accumulation and receptor desensitization (Fig. 17.2).

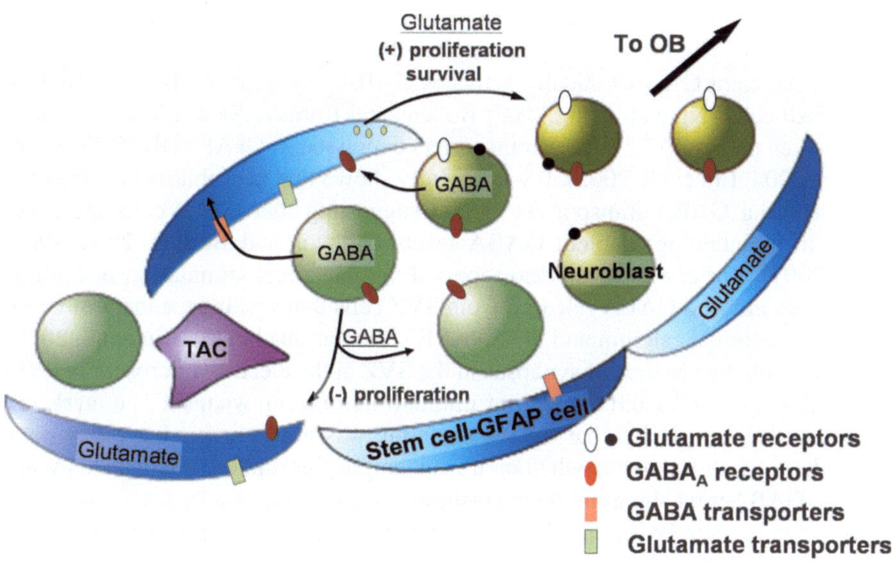

Fig. 17.2 Simplified diagram illustrating GABAergic and glutamatergic signaling in the SVZ and RMS. GABA released from neuroblasts activates GABA$_A$ receptors in both surrounding neuroblasts and GFAP cells leading to a reduction of their rate of proliferation. Glutamate released from GFAP cells (i.e. stem cells of the SVZ) activates glutamate receptors on migrating neuroblasts. Glutamate receptors include NMDA and AMPA/kainate receptors and mGluRs (not shown). Glutamate receptor activation promotes neuroblast survival as well as proliferation. Ambient glutamate is taken by high affinity glutamate transporters expressed in GFAP cells surrounding neuroblasts. The illustration also shows that there is a mosaic of receptors in the neuroblast population

Many components of GABAergic signaling have been identified in the SVZ (Fig. 17.1) and have been reviewed elsewhere (Bordey, 2006, 2007). SVZ neuroblasts synthesize and release GABA as shown by immunostaining for GABA (Stewart et al., 2002; Wang et al., 2003b) and by patch clamp recordings using $GABA_A$ receptors as sensors for GABA (Liu et al., 2005). To identify recorded cells as either neuroblasts or GFAP cells, we recorded these different cell types and reported striking differences in their biophysical properties and current profiles (Wang et al., 2003a; Liu et al., 2005). To specifically record astrocytes, we used transgenic mice expressing GFP under the human *GFAP* promoter (*hGFAP*-GFP) to visually identify GFAP cells prior to recording. Using patch clamp recordings, we can now determine the cell identity based on their passive properties and their ionic signature. Both neuroblasts and GFAP cells express functional $GABA_A$ receptors as shown using patch clamp and calcium imaging that induce cell depolarization (see above for neuroblasts) (Liu et al., 2005; Young et al., 2010). To image calcium dynamics in GFAP cells, we use transgenic mice where a Gq-coupled receptors (called Mas-related gene A1, MrgA1) fused to GFP selectively in GFAP-expressing cells using the inducible tet-off system (called MrgA1 mice). Mice expressing the tetracycline transactivator (tTA) under the *hGFAP* promoter were crossed to mice in which the MrgA1 receptor was transcribed off the tet (tetO) minimal promoter. In the absence of doxycycline, astrocytes express MrgA1:GFP. In these mice, GFP is expressed in the membrane allowing visualization of an outline of SVZ GFAP cells without masking the fluorescence from the calcium indicator dye. Ambient GABA tonically activates $GABA_A$ receptors on both neuroblasts and GFAP cells (Nguyen et al., 2003; Bolteus and Bordey, 2004; Liu et al., 2005), and is taken up by GAT4 GABA transporters expressed in GFAP cells (Bolteus and Bordey, 2004; Liu et al., 2005). It was recently shown that neuroblasts lack the classical neuronal GABA transporters GAT1, suggesting that GFAP cells are a key player in maintaining ambient GABA levels (Bolteus and Bordey, 2004; Platel et al., 2007). Liu et al. (2005) also reported that electrical stimulation or high K^+ application induced GABA released from SVZ cells that was independent of action potentials, external calcium and the SNARE vesicular machinery (Liu et al., 2005). Together with the absence of synapses in the SVZ at the electron microscopic levels (Doetsch et al., 1997) and the lack of immunostaining for synapsin 1, a marker of synapses, they concluded that GABA signaling was non-synaptic but rather paracrine. It remains unknown whether transit amplifying cells contain GABA and express GABAergic signaling mechanisms.

By comparison with GABAergic signaling, glutamatergic signaling had received very little attention until now, perhaps because of the ubiquitous expression of $GABA_A$ receptors in immature neurons as compared to the low expression level of glutamate receptors in the SVZ. In addition, ionotropic glutamate receptors suffered from the lack of selective agonists and antagonists until very recently. Nevertheless, several elements of glutamatergic signaling are now known to be present in the SVZ. Glutamate immunostaining is intense in GFAP cells while very faint in surrounding neuroblasts (Platel et al., 2007). Using immunohistochemistry and electron microscopy in collaboration with Dr. Mary Rubio, we also found that GFAP cells express vesicular transporter 1 (VGLUT1) in the SVZ, suggesting that GFAP cells may release glutamate via a vesicular mechanism (Platel et al., 2010). Using the MrgA1 mice (described above), it was clearly shown that glutamate release from GFAP cells is calcium dependent and released glutamate activates NMDA receptors in migrating neuroblasts (Platel et al., 2010). These data were shown using patch clamp recording of neuroblasts and applying a nonendogenous ligand of MrgA1 receptors. GFAP cells also express high affinity glutamate transporters GLAST and GLT-1 (Liu et al., 2006). The expression of neuronal glutamate transporters in neuroblasts has not been examined. As mentioned above, neuroblasts express different types of glutamate receptors. We recently reported that GFAP cells did not express functional AMPA receptors (Liu et al., 2006), but the expression of other types of glutamate receptors in GFAP cells remains to be thoroughly examined.

Collectively, GFAP cells in the SVZ, i.e. neural stem cells, exhibit a sophisticated release mechanism for glutamate as well as high affinity glutamate transporters. GFAP cells are thus in charge of glutamatergic signaling in the SVZ. By contrast neuroblasts synthesize and release GABA and are thus in charge of the GABA supply in the SVZ, but GFAP cells will maintain extracellular GABA levels through high affinity GABA transporters and thus its availability to SVZ cells.

17.4 GABA: A Stop Signal of Neuron Production But an Enhancer of Neuron Maturation

Tonic $GABA_A$ receptor activation reduces the number of proliferative neuroblasts (Nguyen et al., 2003) and stem cells (i.e. GFAP cells) (Liu et al., 2005). It also reduces the speed of neuroblast migration (Bolteus and Bordey, 2004) perhaps as a result of its enhancement of dendritic growth (Gascon et al., 2006). The intracellular mechanisms leading to these effects on cell development as well as the mode of Ca^{2+}-dependent GABA release are not clearly understood. It is in particular true for the antiproliferative effect of GABA on stem cell proliferation. A recent study reported that GABA blocks embryonic stem cell proliferation through S-phase checkpoint kinases and the histone variant H2AX independently of Ca^{2+} increases (Andang et al., 2008). This mechanism remains to be explored in SVZ stem cells.

17.5 Glutamate: A Mitogenic and Survival Signal for Neuroblast

It was recently reported that adult mice lacking mGluR5 or treated with mGluR5 antagonists showed a dramatic reduction in the number of proliferating cells in the SVZ (Di Giorgi-Gerevini et al., 2005). These findings suggest that mGluR5 are tonically activated by ambient glutamate presumably originating from SVZ and

RMS GFAP cells (see above). The function of GLU_{K5}-containing kainate receptors has never been explored on cell development prior synapse formation. Nevertheless, considering that neuroblasts express functional GLU_{K5} receptors, which leads to calcium increase, we explore its function on neuroblast migration. To do so, we used the whole mount preparation of the entire lateral ventricle from postnatal day 20–25 transgenic mice carrying green fluorescent protein under the doublecortin promoter. Using this new preparation, we confirmed that the $GABA_A$ receptor antagonist bicuculline increased the speed of neuroblast migration by an average of 27%. While the mGluR5 antagonist MPEP did not affect the speed of neuroblast migration, a GLU_{K5} antagonist increased the migration speed by an average of 38%. These data show that although both GLU_{K5} receptor and mGluR5 activation increases Ca^{2+} in neuroblasts, only GLU_{K5} receptors tonically reduce the speed of neuroblast migration along the lateral ventricle.

The function of NMDA receptors has only been recently explored on the development of neuroblasts in the SVZ and RMS. NMDA receptors in migrating neuroblasts are critical to transmit survival cue from surrounding astrocytes. Indeed, we found that a one-time intraperitoneal injection of a NMDA receptor antagonist results in a significant increase in the number of apoptotic cells in the RMS. In addition, single-cell genetic removal of functional NMDA receptors in neuroblasts resulted in apoptosis of neuroblasts along their migratory route prior to entering the olfactory bulb circuitry (Platel et al., 2010). To knockout NMDA receptors, we used transgenic mice expressing NR1 subunit floxed by LoxP sites. Expression of Cre recombinase in SVZ cells using neonatal electroporation resulted in NR1 excision and NMDA receptor knockout that was validated using patch clamp recordings and calcium imaging. Neuroblast apoptosis led to a 30–40% decrease in the number of neuroblasts that reach the olfactory bulb and up to 60% reduction in the number of synaptically integrated granule cells in 2 months-old mice.

Collectively, we propose that using distinct Ca^{2+}-dependent intracellular pathways each glutamate receptor type, GluR5 kainate receptor, mGluR5 and NMDA receptors differentially affects the development of neuroblasts. As in other developing systems (Schlett, 2006), we propose that glutamate acting at both ionotropic and metabotropic receptors positively affects the production of adult-born neurons.

17.6 Convergent Co-Activation of GABA and Glutamate onto Neuroblasts

GABA and glutamate receptor activation has been studied independently. However, activation of each receptor subtype may, in addition to have different outcome on cell development, modulates their respective function at the single cell level. Although there are no data available on this topic, we propose two hypothetical interactions based on the coincident release of GABA and glutamate in the SVZ and RMS. We indeed propose that GABA released from neuroblasts can activate both $GABA_A$ receptors on neuroblasts and GFAP cells. We have preliminary data suggesting that GABA's action on GFAP cells induces glutamate release resulting in the activation of glutamate receptors in neuroblasts. As a first example of interaction, activation of $GABA_A$ receptors may be necessary for removing the Mg^{2+} block of NMDA receptors thus allowing NMDA receptor activation in neuroblasts. It is thus conceivable that $GABA_A$ receptor activation in GFAP cells is necessary for neuroblasts to sense glutamate and receive a survival cue through NMDA receptor activation. As a second example, it is possible that $GABA_A$ receptor activation shunts the depolarizing effect of AMPA receptors in neuroblasts without affecting Ca^{2+} influx. GABA receptor activation may also interfere with the intracellular pathway activated following mGluR activation. For example, $GABA_A$ receptor activation in neuroblasts leads to Ca^{2+} increase, which may enhance the release of Ca^{2+} from intracellular stores following mGluR activation. It is indeed well known that Ca^{2+} acts as a co-agonist of IP3 receptors (Mak et al., 1998). Those are only a few examples of possible convergent interactions between GABA and glutamate on single cells through the expression of a mosaic of receptors. The functional significance of such interactions will need to be explored.

17.7 Integration of GABA and Glutamate Signals at the Population Level

At the population levels, the ambient GABA concentration provides information to stem cells about the size of the neuroblast pool (i.e. number of neuroblasts). As more neuroblasts are generated, more GABA is expected to be released into the extracellular space, resulting in increased $GABA_A$ receptor activation in stem cells. Thus, GABA carries information on the number of neuroblasts produced and provides a negative feedback control of stem cell proliferation and therefore neuroblast production. This negative feedback reconciles well with the constant migration of neuroblasts to the olfactory bulb, which would limit ambient GABA accumulation in the SVZ, and with increased proliferation of GFAP cells following elimination of neuroblasts (Doetsch et al., 1999). Assuming that stem cells release glutamate, glutamate has to diffuse variable distances to activate glutamate receptors on migrating neuroblasts. The glutamate concentration and receptor location may determine which receptors are activated resulting in selective effect on neuroblast development. It remains unclear how GABA and glutamate signals are integrated in neuroblasts. Both lead to Ca^{2+} increases but also shunt each other.

Based on our recent finding that stem cells (i.e. SVZ GFAP cells) do not migrate, we propose that stem cells act as lighthouses in the SVZ where neuroblast sail by. During their passage, neuroblasts receive signals from stem cells that are expected to positively affect their development. We propose that one of these signals is carried by glutamate. Interestingly, one stem cell may control or synchronize in a phase of development a group of neuroblasts while neuroblasts as a group can affect a single stem cell. In addition, GFAP cells, i.e. stem cells, are in charge of determining whether their progeny will reproduce and survive through glutamatergic signaling while neuroblasts as a group provide feedback information to stem cells through GABAergic signaling.

17.8 Upon External Impact or Injury Signaling Within the SVZ May Adjust to a New Steady-State Level, Thus Providing Nonsynaptic Scaling of Neuron Production

GABAergic and glutamatergic signaling provides a homeostatic control of adult born neuron production (Fig. 17.3). This signaling is complex and prone to regulations at different levels, e.g. receptor and transporter regulation, release regulation at the molecular levels or via interaction with Ca^{2+} dynamics. For example, upon injury it is now well-known that proliferation in the SVZ is increased, but this is not accompanied with a run-away increase in neuroblast production. This suggests that signaling within the SVZ reaches a new stated-state level of neuroblast production. We speculate that GABAergic and glutamatergic signaling provide a nonsynaptic scaling of neuron production controlled by external inputs. For example, cytokines and chemokines release from the injury site or activated microglia may induce the up-regulation of receptors in SVZ cells leading to a new steady-state of cell activation. Multiple feedback mechanisms are likely built-in to prevent overproduction of neuroblasts following injury, but lead to a new steady-state level of neuroblast production. Such a mechanism involving nonsynaptic communication among SVZ cell may provide a tightly controlled upregulation of cell production.

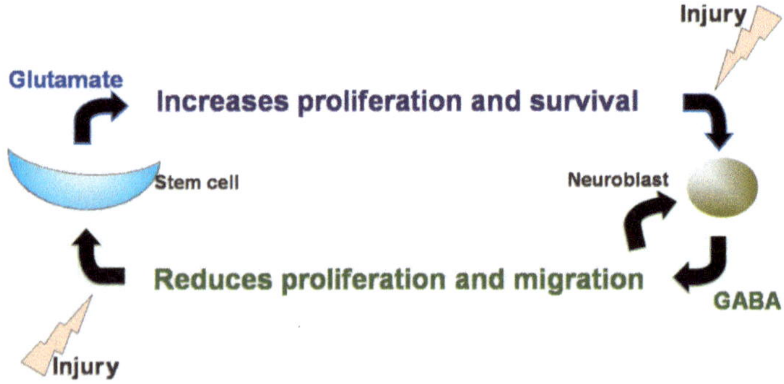

Fig. 17.3 GABAergic and glutamatergic nonsynaptic scaling of adult neurogenesis. This simple diagram illustrates that nonsynaptic GABAergic and glutamatergic signaling balance each other to maintain a stable level of neuroblast production and neurogenesis. Upon injury or insult to the brain, nonsynaptic signaling may be up-regulated to a new-steady-state leading to an enhanced production of adult-born neurons

17.9 Conclusion

We rarely get tumors in the SVZ suggesting very strong control mechanisms of neuroblast production. We propose that glutamate and GABA act as homeostatic signals to regulate neuroblast production. These signals are prone to regulation by other factors. For example, growth factors may up- or down-regulate the expression of transporter and/or receptors leading to new steady-state levels of ambient neurotransmitter levels and receptor activation. The production of neuroblasts may thus be scaled up or down following drug injection or injuries modulating GABA and glutamate signaling. Understanding the role of each of these receptors and transporters could offer the opportunity to pharmacologically control cell production in the SVZ under control conditions or following injury. Reciprocally, it is important to determine whether clinically approved drugs have negative effects on adult neurogenesis.

Acknowledgments This work was supported by grants from the National Institute of Health (NS048256 and DC007681, A.B.) and Yale Brown-Coxe fellowship (J-C.P.).

References

Andang M, Hjerling-Leffler J, Moliner A, Lundgren TK, Castelo-Branco G, Nanou E, Pozas E, Bryja V, Halliez S, Nishimaru H, Wilbertz J, Arenas E, Koltzenburg M, Charnay P, El Manira A, Ibanez CF, Ernfors P (2008) Histone H2AX-dependent GABA(A) receptor regulation of stem cell proliferation. Nature 451:460–464.

Bettler B, Mulle C (1995) Review: neurotransmitter receptors. II. AMPA and kainate receptors. Neuropharmacology 34:123–139.

Bolteus AJ, Bordey A (2004) GABA release and uptake regulate neuronal precursor migration in the postnatal subventricular zone. J Neurosci 24:7623–7631.

Bordey A (2006) Adult neurogenesis: basic concepts of signaling. Cell Cycle 5:722–728.

Bordey A (2007) Enigmatic GABAergic networks in adult neurogenic zones. Brain Res Brain Res Rev 53:124–134.

Chebib M, Johnston GA (1999) The 'ABC' of GABA receptors: a brief review. Clin Exp Pharmacol Physiol 26:937–940.

Coutinho V, Knopfel T (2002) Metabotropic glutamate receptors: electrical and chemical signaling properties. Neuroscientist 8:551–561.

Curtis MA, Kam M, Nannmark U, Anderson MF, Axell MZ, Wikkelso C, Holtas S, Roon-Mom WM, Bjork-Eriksson T, Nordborg C, Frisen J, Dragunow M, Faull RL, Eriksson PS (2007) Human neuroblasts migrate to the olfactory bulb via a lateral ventricular extension. Science 315:1243–1249.

Di Giorgi-Gerevini V, Melchiorri D, Battaglia G, Ricci-Vitiani L, Ciceroni C, Busceti CL, Biagioni F, Iacovelli L, Canudas AM, Parati E, De Maria R, Nicoletti F (2005) Endogenous activation of metabotropic glutamate receptors supports the proliferation and survival of neural progenitor cells. Cell Death Differ 12:1124–1133.

Doetsch F, Garcia-Verdugo JM, Alvarez-Buylla A (1997) Cellular composition and three-dimensional organization of the subventricular germinal zone in the adult mammalian brain. J Neurosci 17:5046–5061.

Doetsch F, Garcia-Verdugo JM, Alvarez-Buylla A (1999) Regeneration of a germinal layer in the adult mammalian brain. Proc Natl Acad Sci USA 96:11619–11624.

Gascon E, Dayer AG, Sauvain MO, Potter G, Jenny B, De Roo M, Zgraggen E, Demaurex N, Muller D, Kiss JZ (2006) GABA regulates dendritic growth by stabilizing lamellipodia in newly generated interneurons of the olfactory bulb. J Neurosci 26:12956–12966.

Ge S, Pradhan DA, Ming GL, Song H (2007) GABA sets the tempo for activity-dependent adult neurogenesis. Trends Neurosci 30:1–8.

Lerma J, Paternain AV, Rodriguez-Moreno A, Lopez-Garcia JC (2001) Molecular physiology of kainate receptors. Physiol Rev 81:971–998.

Liu X, Wang Q, Haydar TF, Bordey A (2005) Nonsynaptic GABA signaling in postnatal subventricular zone controls proliferation of GFAP-expressing progenitors. Nat Neurosci 8:1179–1187.

Liu X, Bolteus AJ, Balkin DM, Henschel O, Bordey A (2006) GFAP-expressing cells in the postnatal subventricular zone display a unique glial phenotype intermediate between radial glia and astrocytes. Glia 54:394–410.

Lledo PM, Alonso M, Grubb MS (2006) Adult neurogenesis and functional plasticity in neuronal circuits. Nat Rev Neurosci 7:179–193.

Mak DO, McBride S, Foskett JK (1998) Inositol 1,4,5-trisphosphate [correction of tris-phosphate] activation of inositol trisphosphate [correction of tris-phosphate] receptor Ca2+ channel by ligand tuning of Ca2+ inhibition. Proc Natl Acad Sci USA 95:15821–15825.

Merkle FT, Mirzadeh Z, Alvarez-Buylla A (2007) Mosaic organization of neural stem cells in the adult brain. Science 317:381–384.

Nguyen L, Malgrange B, Breuskin I, Bettendorff L, Moonen G, Belachew S, Rigo JM (2003) Autocrine/paracrine activation of the GABA(A) receptor inhibits the proliferation of neurogenic polysialylated neural cell adhesion molecule-positive (PSA-NCAM+) precursor cells from postnatal striatum. J Neurosci 23:3278–3294.

Pathania M, Yan LD, Bordey A (2010) A symphony of signals conduct early and late stages of adult neurogenesis. Neuropharmacology 58:865–876.

Pinheiro P, Mulle C (2006) Kainate receptors. Cell Tissue Res 326:457–482.

Platel JC, Lacar B, Bordey A (2007) GABA and glutamate signaling: homeostatic control of adult forebrain neurogenesis. J Mol Histol 38:602–610.

Platel J, Heintz T, Young S, Gordon V, Bordey A (2008) Tonic activation of GLUK5 kainate receptors decreases neuroblast migration in a whole mount preparation of the subventricular zone. J Physiol (Lond) 586:3783–3793.

Platel JC, Dave KA, Gordon V, Lacar B, Rubio ME, Bordey A (2010) NMDA receptors activated by subventricular zone astrocytic glutamate are critical for neuroblast survival prior to entering a synaptic network. Neuron 65:859–872.

Sanai N, Tramontin AD, Quinones-Hinojosa A, Barbaro NM, Gupta N, Kunwar S, Lawton MT, McDermott MW, Parsa AT, Manuel-Garcia VJ, Berger MS, Alvarez-Buylla A (2004) Unique astrocyte ribbon in adult human brain contains neural stem cells but lacks chain migration. Nature 427:740–744.

Schlett K (2006) Glutamate as a modulator of embryonic and adult neurogenesis. Curr Top Med Chem 6:949–960.

Stewart RR, Hoge GJ, Zigova T, Luskin MB (2002) Neural progenitor cells of the neonatal rat anterior subventricular zone express functional GABA(A) receptors. J Neurobiol 50:305–322.

Wang DD, Krueger DD, Bordey A (2003a) Biophysical properties and ionic signature of neuronal progenitors of the postnatal subventricular zone in situ. J Neurophysiol 90:2291–2302.

Wang DD, Krueger DD, Bordey A (2003b) GABA depolarizes neuronal progenitors of the postnatal subventricular zone via $GABA_A$ receptor activation. J Physiol (Lond) 550:785–800.

Young SZ, Platel JC, Nielsen JV, Jensen NA, Bordey A (2010) GABA(A) increases calcium in subventricular zone astrocyte-like cells through L- and T-type voltage-gated calcium channels. Front Cell Neurosci 4:8.

Index

T. Seki et al. (eds.), *Neurogenesis in the Adult Brain I: Neurobiology*,
DOI 10.1007/978-4-431-53933-9

O

P

MIX
Papier aus verantwortungsvollen Quellen
Paper from responsible sources
FSC® C105338

If you have any concerns about our products,
you can contact us on
ProductSafety@springernature.com

In case Publisher is established outside the EU,
the EU authorized representative is:
Springer Nature Customer Service Center GmbH
Europaplatz 3, 69115 Heidelberg, Germany

Printed by Libri Plureos GmbH
in Hamburg, Germany